Roads, Peoples,
Birds,
Mountaintops,
&
Billabongs

*To Marci
from Dean Fisher
with lots of luck
Mar. 25, 2020*

DEAN FISHER

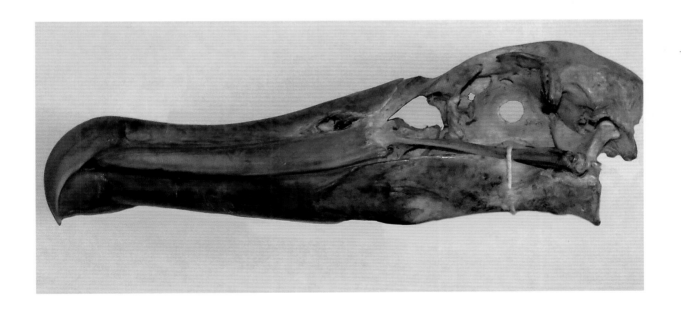

STEPHEN F. AUSTIN STATE UNIVERSITY PRESS

Printed in China
For more information Contact:

STEPHEN F. AUSTIN STATE UNIVERSITY PRESS
404 Aikman Drive, LAN 203
P.O. BOX 13007
NACOGDOCHES, TEXAS 75962
sfapress@sfasu.edu
sfasu.edu/sfapress
936-468-1078

ISBN: 978-1-62288-187-1

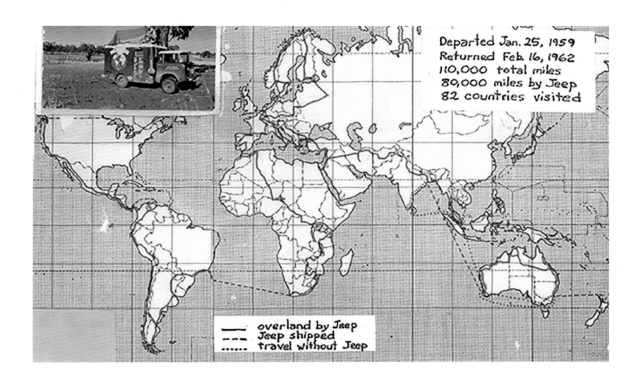

Departed Jan. 25, 1959
Returned Feb. 16, 1962
110,000 total miles
80,000 miles by Jeep
82 countries visited

—— overland by Jeep
- - - Jeep shipped
······ travel without Jeep

Foreword to Dean's book
(14 August 2017)

I first came to Stephen F. Austin State University in Nacogdoches, Texas as a freshman in the fall of 1969, and it was a lonely place for an avid young birder. To my good fortune, in the fall of 1970 a new professor arrived in the Biology Department, Dr. Charles Dean Fisher. I had heard that he was a "bird man", so I overcame my shyness and drifted by his office. We met and I knew right away that I had found a kindred spirit! Dr. Fisher was eager to get out in the field in a region new to him and invited me along to help him survey the overlooked avifauna of the East Texas Pineywoods. In the years to come I followed him through reptile-infested swamps and steaming mudflats, frozen grasslands in howling north winds, and dramatic spring and fall storms, as we roamed in search of the birds of the region. Always adventurous and intrepid, he let nothing hold him back. Finally, as an upperclassman, I was able to take his Ornithology course. In the lecture hall he was dynamic, but the best classes were always those when he digressed into the most remarkable tales of his travels around the world. I was captivated and encouraged to pursue my own dreams of seeing exotic birds in foreign places. Now, looking back over a friendship that has spanned decades, I reflect upon what an incredible mentor Dean was to me and many, many others. With his knowledge, enthusiasm and high energy he inspired generations of students for years to come, many of whom have gone on to serve as wildlife professionals in Texas and beyond.

Always amazing were the stories of his three-year expedition around the world, with only one companion and a vehicle that broke down endlessly. Dean spoke not only of the birds and natural history, but also the people and cultures encountered, not to mention the many challenges that had to be solved to continue their trip. This was all done long before international travel had become commonplace or bird guides were available for most of the places he visited. For many years those of us fascinated by his accounts urged him to write down these tales of adventure so they wouldn't be lost. Finally, with the publication of this book, he has done so, and now we can all share in his incredible journey, from a time that seems quite distant and more innocent. Dean, thank you!

David E. Wolf
Tour Leader, Victor Emanuel Nature Tours

To those who have navigated on unmapped roads of the world,
and to birdwatchers who have pursued birds in unfamiliar realms without a field guide

Acknowledgments

Our journey was conceived by Noble and would not have succeeded without his financial backing, mechanical skills, persuasive mannerisms, and never-give-up determination. Although we were an odd couple, we shared the thrill of adventure, the success of reaching the top of a mountain, and a glass of wine at the end of a day. We were comrades, and we defied the chances. A few of the photographs are his. My thanks go to Noble.

During our 3 years on the road we were provided gracious hospitality by people we met everywhere. Those with whom we spent the most time were (1) Dusty Rhodes & Clary Palmer-Wilson of an East African Game Safaris in Tanganyika, and (2) Andy & Ruth Andersen of the Kisangara Sisal Estate, also in Tanganyika. While traveling alone I was hosted for 2-1/2 weeks by Elliott McClure in Malaya, and for 1 week by Charlie & Mary Weaver at Riverdale Farm in Western Australia. All these occasions added greatly to our adventure, and it is a pleasure to acknowledge the generosity of our hosts. I apologize to all the others who befriended us for shorter periods and whose names are not mentioned here.

I examined study skins of birds in private or public collections whenever the opportunity was available, to identify birds from my field notes. I am particularly indebted to the following persons for allowing me access to their collections: (1) Dr. William H. Phelps in Caracas (in his own private collection), (2) Dr. Maria Koepcke in Lima (of the Museum Javier Prado), (3) Rafael Barros in Santiago (in his own private collection), (4) John G. Williams in Nairobi (of the Coryndon Museum), and (5) Dr. G. M. Storr in Perth (of the Western Australian Museum). To all these persons I am grateful. Mimi Hoppe Wolf skillfully sketched the Inca Wren, which is portrayed from Machu Picchu on Plate 14. I thank Rose Ann Rowlett for kindly offering suggestions in those chapters of my text which deal with the Neotropical Region. Being illiterate in modern computer technology, I could not have written my manuscript without almost daily technical support from Les Stewart, Troy Lagrotteria, Morgan Johnson and Emily Townsend who willingly solved my many problems. Will Godwin helped in designing the front cover, and Kimberley Verhines of the SFA Press enthusiastically organized the publication of this story.

Finally, I am particularly grateful to my wife for her encouragement and for her willingness to tolerate my long periods of writing, during which I gave up almost all other activities.

Prologue

Mt. Fuji is Japan's highest mountain, a symmetrical volcano rising all alone above the surrounding countryside to 12,389 feet above sea level on the island of Honshu. It is a sacred mountain with a Shinto temple and "torii" gate at the top, climbed by thousands of visitors every summer during a rather short hiking season (June through August) when the summit is free of snow. During the rest of the year, there is snow and ice on top of the mountain.

In September 1957, there was snow covering the top of the mountain to below 10,000 feet in elevation. I managed to obtain several days of leave from my aircraft carrier (the USS Kearsarge) and I traveled by train to a lovely hotel on Lake Yamanaka, one of 7 beautiful lakes surrounding the base of the mountain, at an elevation of 3,000 feet.

Here at the hotel, on the morning of October 31st, at 3:00 am, I awoke to my alarm clock, hurriedly got dressed, and went to the desk in the hotel lobby to telephone for a taxi. The young boy behind the desk, who spoke English quite well, asked where I wanted to go. I said "To the top of Mt. Fuji." He looked at me in horror and said I couldn't do that. When I inquired why not, he responded that it was not the right season, there were snow and ice on the top, the weather was unpredictable, and I would surely die if I attempted such a climb at this time of year! However, I had made up my mind and nothing the boy could say would persuade me otherwise, so eventually he relented and called a taxi. He told the driver (in Japanese of course since the driver spoke no English) to take me to the mountain.

After almost an hour's travel, we came to the end of our road, somewhere in a forest far below timberline. It was 5:00 am and very dark. I could see there was a hiking trail starting up the mountain at this point, so I paid the driver and bade him farewell. I was wearing a sweater, a lightweight jacket, and some simple hiking shoes. I had with me a little food and some water for the day's hike. It was a cold but clear morning, without much wind.

As I started up the mountain, I stumbled constantly over roots and rocks, because it was too dark to see the path clearly, and my initial progress was painstakingly slow. I decided I was probably somewhere between 5,000 and 6,000 feet in elevation, and I knew that tree line was at 7,000 feet, at which point I would know where I was in elevation on the mountain.

As it slowly became daylight I could begin to see much better, and my pace picked up considerably. By 7:00 o' clock, I had reached tree line. The weather, fortunately, was perfect. The top of the mountain didn't really look that far away (it never does), but it was obvious that there was a considerable amount of snow on the top. There was no one else on the mountain. I was all by myself. The climb was very pleasant and I told myself that the boy's fears were quite unfounded.

It was almost 10:00 o'clock by the time I reached the bottom edge of the snow, some 2,000 feet below the summit. At first, the snow was soft and posed no problem to my progress upward. The mountain slope, however, had become quite steep and my trail wound back and forth constantly. My rate of ascent slowed considerably at this higher altitude, as I continually had to stop to catch my breath. The day was still clear with unlimited visibility, and so far, I was enjoying my climb. Conditions on the ground, however, began to change. The snow became firmer and the mountain slope increased. To avoid slipping, I was forced to kick small steps in the snow, which was rapidly becoming ice. I looked around for anything that might help prevent me from slipping, but found only a very small stick, which was virtually worthless.

As I neared the top of the mountain, my trail became almost solid ice. It was very difficult to kick any kind of footstep at all. Then, when I was only several hundred meters from the summit, the mountain surface became solid ice. I could clearly see the torii gate not far away, at the entrance to the temple on top, but I could go no further. The ice was too solid. I was disappointed to be so very close to my goal, yet unable to get the last few steps. It was about 2:00 o'clock in the afternoon.

When I realized that I was not going to reach the very top, my thoughts for the first time all day turned toward my return trip down the mountain. I looked downward and a wave of fear swept through me as I realized exactly what my situation was, perched very precariously on a steep mountain slope which was covered entirely in ice, with 2,000 feet of snow and ice below me. One little slip, and there was nothing to stop a person from plummeting all the way down the mountain. I became very frightened and said to myself, "Dean, this is the dumbest thing you have ever done."

It took all of my effort to calm myself. I was certain that I would not be able to walk back down the steep mountain slope in a forward facing position without slipping. The only alternative was to proceed down the mountain backwards, facing upward and placing my feet in exactly the same steps that I had made on my climb upward. So, I began my descent extremely slowly and still very fearfully. It was more than an hour before I finally reached a level where the snow covering the trail became soft enough to allow me to increase my rate of descent, and to progress in a forward facing direction. However, it was already 4:00 o'clock by the time I reached the bottom of the snow, at an elevation of 10,000 feet. The sun was still shining, but I knew there were less than two hours of daylight left. It was obvious to me that I was going to be a long way from the bottom of the mountain by the time darkness came. I was not anxious to stumble down the mountain after dark or to sit huddled all night under a tree in the forest, in sub-freezing temperatures.

I knew that somewhere there was a tree line hotel at 7,000 feet, but I didn't really know where, from my present position, this hotel might be. Never-the-less, I decided to try and find it if I could before dark. I began running. Whenever I came to a fork in the trail, and there were many, I took the one I reasoned was most likely to take me to the hotel. When I arrived at the tree line, I chose to go left (rather than right) on a trail that was proceeding around the mountain. Luck was with me and shortly after dark, at 6:30 pm, I reached the hotel. I wasn't destined to die just yet.

There were no vehicles of any kind at the hotel, the lights were not on, and the big wooden front door was securely locked. The hotel was closed for the winter! My joy turned to despair. Then I noticed there was a dim light coming from a small upstairs window on the third (uppermost) floor of the hotel. I thought that surely someone must be inside. I found a log in the forest and began repeatedly pounding it on the big front door. I was about to give up when the door was opened by a small, smiling Japanese woman about 30 years old. She bowed courteously in the Japanese custom when she saw me standing there, and motioned for me to come inside. She took me upstairs to a little room where she lived with her two small children, sat me down at a table, and brought me a large bowl of steaming hot rice and vegetables. This was a welcome meal indeed, since I had eaten almost nothing all day. I was given a room for the night and slept between blankets on the floor, as was the Japanese tradition. The next morning the pleasant young woman prepared a hot breakfast for me, but adamantly refused to take any money for my night's lodging or meals. She must have wondered why I was on the mountain all by myself.

I started back down the mountain at about 8:00 am, the first day of November, and followed a narrow winding dirt road. It was another beautiful day and I was in no hurry, so I stopped frequently to take pictures or watch birds, squirrels, and other wildlife. Again, there were no other persons present on the mountain with me. The road descended at first through a coniferous forest of larches and firs, with scattered birches, then progressed downward through a dry pine forest before eventually joining a main road situated at about 3,000 feet elevation. It was mid-afternoon and I knew in which direction my hotel was located. So 30 minutes later when a somewhat dilapidated local bus came along the road headed in this direction, I flagged it to a stop, climbed aboard, and an hour later I arrived back at my hotel, in the late afternoon. The same boy was behind the reception desk and when he saw me walk through the front door he looked as if he were seeing a ghost, and it was apparent to me that he really did believe I would die on the mountain. This small adventure set the stage for the following story.

Chapter One – Getting Started

It was the spring of 1958, five months after I climbed Mt. Fuji. I was serving the last of my three years in the Navy, as an air control officer on the USS Kearsarge (CVA 33) in the Pacific. My primary function was tracking aircraft on the air search radar, which was located in the ship's Combat Information Center (CIC). Most of my time was spent directing the carrier's air defense fighter jets to a position where they could intercept any approaching aircraft that posed a threat to the carrier or task force. Our carrier was now headed from the Far East back to our temporary homeport in Alameda, California.

One of the ship's night fighter pilots was Noble Trenham, who flew F2H-3 "Banshees", a straight winged, twin-engine jet. Like me, Noble was going to complete his tour of active duty in the Navy in the forthcoming September. Noble and I were only casual acquaintances, although he knew my voice quite well from many hours of listening to my instructions while he was flying. (He told me, many years later, that I had once saved his life.)

One sunny afternoon Noble and I happened to encounter each other on the flight deck, during a period of leisure time for both of us. Noble walked up to me and said, in a direct and straightforward manner, "Would you like to fly around the world with me?" The question caught me completely off guard. He seemed serious and for a moment, I was too astonished to say anything. Noble went on to explain that it had been a lifelong dream of his to purchase his own small airplane and fly it around the world, that he needed another person to accompany him, and he would like for me to be that person. "I heard about your climb to the top of Mt. Fuji and that's the kind of spirit I want in the person who goes with me."

I was instantly excited at the thought of such an adventure and was very pleased by his invitation. However, I reminded him that I was not a trained aviator. He said "Leave the flying to me." I asked "What about the cost, I don't have very much money?" His response was "Leave the cost up to me." My questioning continued, when did he want to depart and how long did he plan to be gone? He said he wanted to leave as soon as possible after we both got out of the active Navy in September, and that the trip would probably take about one year. I told him to give me a little time to consider his invitation.

Noble was a graduate of the University of Southern California, where he studied economics, history, and political science. I graduated from the University of Michigan with a degree in forestry and wildlife management, and with a passionate hobby of birdwatching. ("You must be crazy," my father once said to me when I was 15 years old as I left our house in Decatur, Illinois, early on a cold mid-winter morning to go birdwatching, with the outside temperature below zero.) Our backgrounds and interests were thus quite different, as were our physical sizes. At six feet tall, Noble was six inches taller than I was, and he had a stronger body build from having played football in high school. We would both be 25 years old in a few months. Although we seemed ill suited for long-term travel together, we both possessed a very strong determination to succeed at those things we commenced. Neither of us were quitters. We also both enjoyed the challenge of trekking to the tops of tall mountains. Noble was a member of the Trojan Peak Club, and he liked telling me of his experiences and achievements.

A few days later, we talked again. I had decided that it would be better if the two of us were to travel by land and sea rather than by air. I proposed this idea to Noble, arguing if we drove around the world in a vehicle, instead of flying from airport to airport, that we would see more places and experience a greater part of the culture, history, and natural environment of the countries we visited. Noble thought about this proposed change and he agreed with me. We would travel by land and sea, rather than by air. Our excitement peaked and we began planning for the adventure of a lifetime. We were comrades!

Making preparations

In September, we were released from active duty in the Navy and immediately began making preparations for our journey. Noble was living with his parents in Pasadena, California, and I was living with my parents in nearby Long Beach. Both of us were very inexperienced in our planned undertaking. Noble took the lead in decision making. First and most important was the purchase of a suitable vehicle. We each had $3,500 saved for this purpose.

After much consideration and looking, we eventually chose a new FC-170 "forward control" model of a 4WD Willys built Jeep, a model that had been in production less than 2 years. It was an unusual looking cab-over-engine pickup truck with a big wrap around windshield. The small, 6 cylinder, 105hp engine was situated in the front cab between the driver's seat on the left and the passenger's seat on the right. Except for the large "JEEP" logo on the front of the vehicle, it would never have been recognized as a Jeep. The short wheelbase was only 10 feet in length, and the 4 relatively small tires were size 7"x16." The manual transmission had 3 speeds forward and 1 speed in reverse, with both a higher and a lower transfer case. To engage the 4WD it was necessary to stop the vehicle, get out, and manually turn the locking hubs on each of the two front wheels. We purchased only the cab and chassis of the pickup truck, without the pickup body on the back. The vehicle was designed and built for a maximum gross weight of 7,000 pounds. There were no power brakes or power steering.

We drew up our own plans for a camping van. Noble took the lead and contracted with two young men in a Glendale (Los Angeles) body shop to custom build it for us. We wanted it to be very sturdy, with the stipulation that the completed camping van should weigh no more than 5,500 pounds when the work was completed. This meant we could then add 1,000 pounds of our personal gear, equipment, supplies, and belongings to bring the total vehicle weight up to 6,500 pounds, which would be 500 pounds under the designed maximum gross weight. The cab and chassis weighed 3,000 pounds at the beginning of the van's construction, so that 2,500 pounds could be used in the building of the van. (The significance of these facts will become apparent after our departure.)

When our traveling home was finished the inside dimensions were approximately 10 ft. long, 6.5 ft. wide, and 6.5 ft. high (barely enough room for Noble to stand upright, or lie down on his bunk). There were two narrow bunk beds across the back of the van. The lower one (which I slept in) was only a short distance above the floor, and the upper one, which Noble slept in, was 2 feet above mine and folded up against the back of the van during the day, allowing a person to sit upright on the bottom bunk. There was a single door on the right side of the van that allowed entry from the outside of the van. It could be securely locked. There was a small-screened window capable of being opened on each side of the van, a narrow screened open and shut window above each bed, and a screened hatch in the roof that could be opened. There was a built-in sink and 12 gallon water tank for drinking water, a Coleman camp stove, a writing table, a safe for valuables, and an electric refrigerator (powered by a gasoline driven generator that was mounted outside the van at the top front, behind and just above the cab). There were also closets, cabinets, shelves, and drawers for storing clothes, food, dishes, eating utensils, books, typewriter, radio, stereophonic tape recorder, medicines, tools, spare parts, cameras, and film. Noble took pictures with a Bell & Howell 16mm movie camera, plus a Polaroid and other still cameras. I took all of my pictures (Kodachrome 64 color slides) with a small compact Canon 35mm camera weighing almost 2 lbs., equipped with a wide angle 35m lens (and a separate 300mm telephoto lens which I rarely used). It was entirely manually operated and had no light meter. I had purchased it a year earlier in Japan, for $300. One of our essential supplies was a cardboard box with 200 small bottles of 50 Halazone tablets each, for purifying our drinking water whenever we deemed it necessary. We tipped a whole bottle into the 12-gallon water tank.

To the underside of the van, at the back, we custom built a 40-gallon tank for gasoline. We also carried two spare five-gallon "jerry cans" of gasoline, thus allowing us a total carrying capacity of 50 gallons of gasoline. At an average of 10 mpg, we had a maximum driving distance of 500 miles without refueling. Under the van at the back was a narrow compartment extending from one side of the van to the other, in which we carried long handled tools and implements, such as shovels, a crowbar, and a small hydraulic jack for changing tires. This compartment had a door on either side, each of which could be locked. Two spare tires were mounted and locked on the back of the van, and there was a vertical ladder allowing us access to the flat top of the van (where I often slept in hot weather). On the front of the vehicle, mounted above the bumper, was a power winch and cable driven by the Jeep's engine.

When the body shop had completed our van, after four months of work, they presented us with an official weight slip from the California highway department, which stated that the total vehicle weight was slightly more than 5,000 lbs. We estimated that we then added about 1,500 lbs. of equipment, gear, and personal belongings, bringing the total weight to approximately 6,500 lbs. We trusted the builders, and since we were anxious to get underway, we did not verify the weight ourselves. (This proved to be a very critical error on our part because we discovered, much later after we had already begun our travel, that the total vehicle weight was actually 9,000 lbs!) Even before departure, we replaced the original springs on all four wheels with stronger, heavier duty springs.

We painted the van blue and then added the following features, painted on the outside:

(1) A large Mercator map of the world on the right side, on which we would paint in bright orange the route we navigated as we traveled around the world

(2) A 3 ft. globe of the world on the left side with the slogan "Seeking Understanding Across Six Continents" painted above it

(3) On the back of the van we painted a Roadrunner, a bird of the southwestern U.S. deserts, with the name "Roadrunner" painted under it. This became the nickname of our Jeep.

While the van was being constructed, we were busy obtaining important documents such as our passports, visas (all those necessary for Latin American countries), and a letter of credit from the Security First National Bank of Los Angeles, issued by the bank against the American Express Company for the sum of $20,000 (the value of a loan from the bank, in Noble's name). Our passports said that we were students. They were valid for four years, an initial two years plus an additional two years if renewed, and allowed us to visit any country or area in the world except China, North Korea, North Viet Nam, and Albania. My passport number was 1233336. (Persons copying this number frequently left out one of the 3's.) Before we finished our travels, it was necessary to have 72 additional pages added to the original number of 20 pages in our passport! From Brussels, Belgium, we were issued international driver's licenses and a "Carnet de Passage" for our vehicle, permitting us to take it into almost any country in the world, with the notable exception of Brazil. These documents were booklets with many pages, written in many languages but mostly in French and Flemish! It was also necessary for each of us to obtain permission from the U.S. Navy to travel outside the U.S.A. for an extended period, since we were both obligatory members of the inactive naval reserve. Another preliminary task was to obtain vaccinations and immunizations for as many exotic diseases as possible, and Chloroquine tablets to protect against malaria.

Finally, we sat down together and drew up a tentative itinerary, which we then had printed as a small pamphlet to distribute to friends and relatives just prior to our departure. This itinerary listed 75 countries on 6 continents. In the order of planned visit, these were North America, South America, Africa, Europe, Asia, and Australia. The planned duration of our adventure was 15 months!

Departing

On Sunday, January 25, 1959, Noble and I set out from Los Angeles to see the world, heading for the Mexican border at Nogales, Arizona, in our brand new, untested, Jeep camping van. We could never have envisioned the adventure on which we were embarking, like two tadpoles in a fishpond.

We drove all the way to Tucson that first day, a total of 500 miles. This was more than we would ever again travel in one day. Although the highway was straight and wide, with very little traffic, we never drove faster than 50 mph, which we had decided would be the maximum speed for our heavy camping van. That night we stayed in Tucson, at the home of an acquaintance of mine, Pat Jenks.

We spent the entire next day in Tucson, at a service station where many last minute maintenance, adjustment, and installation chores were performed. That evening, after putting in a new headlight, we headed south to the border crossing with Mexico at Nogales, where we arrived just after midnight. We were allowed entry only after agreeing to pay an extra fee of $3.00 for our permit because a border official had to be awakened! We then drove down the highway about 20 miles, pulled off to the side of the road, and slept for a few hours. We were on our way! It was decided that I would do most of the driving, while Noble wanted to handle all the transactions involving the van's maintenance and repairs, and would make all financial decisions and transactions. The major tasks for the success of our adventure were thus partitioned between us. We were both doing what we enjoyed. It was an exciting moment!

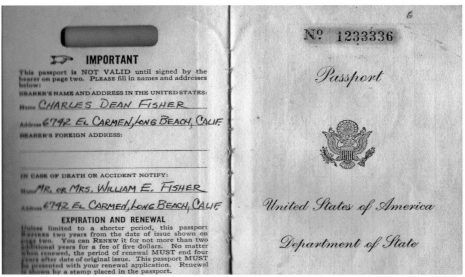

Plate 1: Noble and I departed in a new 4-wheel drive model "FC-170" Willys built Jeep with a custom built camping van, weighing a total of 9,000 pounds (2,000 lbs. more than the maximum designed weight). "My passport had the memorable number of 1233336, and when copying this number, border officials often left out one of the four 3s!"

Chapter Two – Mexico

As we started out in Mexico on Tuesday morning, Jan. 27, our highway in Sonora traversed a sparsely vegetated, arid plateau about 4,000 ft. in elevation, with large saguaro and organ pipe cactuses. The few dwellings were made of adobe with thatched roofs, dirt floors, no electricity, and no running water. People were cutting and burning the cacti to clear land for crops. The road was unpaved and dusty in places. Initially, there were frequent stops where federal police checked our travel permits. As we drove into the southern Sonora city of Navajoa during the 30-minute period before dark that evening, I counted 6 Great Horned Owls on the tops of telephone poles along the highway.

The next day (Jan. 28) I was particularly happy to observe a Purplish-backed Jay*, a local endemic corvid of the dry thorn forests. Two other species in this same thorn scrub ecosystem were the Golden-crowned Emerald* (a hummingbird) and a single flock of 60 White-fronted Parrots* which flew overhead. An asterisk denotes my first ever sighting of a species (a "lifer" in the vernacular of birdwatchers). All species seen throughout our seven months adventure in Central and South America are listed by country (or region) in Appendix A. (When we started out, I had a "life list" of 950 species of birds.)

The next morning (Jan. 29), we drove south into the state of Sinaloa, where people were growing sugarcane, tomatoes, peas, and wheat. Women carried water from the nearest river, stream, or well to their house in a large flask or jug on top of their head. Burros pulled carts with goods of all kinds, and it was necessary to drive with caution to avoid the many horses and cows in the middle of the highway. There were fields of crops along the rivers. The hillsides were covered in dry deciduous cactus-thorn forests with colorful pink flowers on leafless branches, since it was the end of the dry season.

Black-throated Magpie-Jays with conspicuously elongated tails flew back and forth across the highway. The most common and widespread birds were: Turkey and Black vultures, Crested Caracaras, American Kestrels, 3 kinds of doves (Mourning, White-winged, and Common Ground-Doves), White-fronted Parrots, Sinaloa Crows*, Common Ravens, and Great-tailed Grackles. My field guide for identifying Mexican birds (those not found in the U.S.) was Emmet Reid Blake's *Birds of Mexico*, published in 1953, mostly without any colored illustrations. (My only other "field guide" for anywhere in Central America was *Field Book of the Panama Canal Zone*, published in 1928 by Bertha B. Sturgis, again mostly devoid of colored illustrations). My optical equipment for viewing birds consisted of a pair of rather old 7x35 binoculars and a Bausch & Lomb telescope with a single 30-power non-zoom lens, which I very rarely used.

Late in the afternoon we crossed the Tropic of Cancer just north of the city of Mazatlán. An exciting moment occurred when a pair of Military Macaws* flew slowly above the roadside treetops, calling with loud, raucous vocalizations. We were in the tropics! At dusk two Rufous-bellied Chachalacas* flew across the highway, easily identified by their large size and long, fan-shaped tail. That night we camped along the roadside in an arid thorn forest not far south of the city of Mazatlán, in southern Sinaloa. We changed a low-pressured tire. Noble sat in the van and wrote in his lengthy journal, as he did regularly in the evenings (and occasionally at other times) for the whole of our travel. My own record keeping was devoted almost exclusively to the birds I saw, which were assiduously recorded in pencil throughout our entire adventure, in small (4.½" x 6.½") lined, spiral notebooks. In addition, however, I also kept a somewhat sporadic day-to-day diary for the initial six months of our three years of travel, as we drove from Mexico through southern Argentina.

The next morning (Fri., Jan. 30), we adjusted the accelerator cable, and then continued south, entering the state of Nayarit. The paved highway had many potholes, which I slowed to avoid whenever possible. Bridges were very narrow ("despacio -- Puente angosto"), with the approaches badly washed out. Trucks and buses were often stalled in the middle of the road, and burro-drawn carts were a common hazard. Women wore bright sweaters and dresses, carried firewood on their head, and washed their clothes in the river. Naked children played happily in front of their houses, which were made of adobe, sticks, and cacti. Roadside birds included a Social Flycatcher*, Thick-billed Kingbird*, and a half dozen Mangrove Swallows*

Our highway turned inland away from the coast and our route toward Mexico City climbed upward, back and forth, into the state of Jalisco, where we entered pine forests on the Mexican plateau at 5,000 ft. elevation. We arrived in Guadalajara in the early evening and passed by a colorful marketplace. We decided to pay for a restaurant dinner, which was a welcome change from eating bread and honey, and beans out of a can. That night we slept just outside the city.

The next morning (Sat., Jan. 31) we gulped down some fruit juice, bacon, and hard-boiled eggs, then continued on our way toward Mexico City. We passed by fields of sugarcane and corn along the western and southern shores of Lake Chapala at an elevation of 7,000 ft., and shortly thereafter, we entered the state of Michoacán. Navigating through the towns and villages was difficult because there were practically never any road signs or numbers. Very rarely, there would be an arrow painted on the side of a building with the name "Mexico" (referring to Mexico City) painted under it. Streets were often narrow and one-way, very poorly indicated, so it was easy to make mistakes. That evening we camped just off the highway east of the city of Zamora, in a lovely pine-oak forest at 7,000 feet. Some of the birds in the forest were Greater Pewee (Coues' Flycatcher), Tufted Flycatcher, Common Raven, Hermit Thrush, Brown-backed Solitaire, American Robin, Painted Redstart, and Hepatic Tanager. After we went to bed I could hear Northern Pygmy-Owls and Coyotes calling regularly throughout the long, cool, still night. It was exciting to listen to them. Life was great!

The next morning (Sun., Feb. 1), we stopped in the historic city of Morelia to take pictures of the scenic old Spanish aqueduct high above the city streets. Since Noble and I had decided we weren't going to shave during our travels, our beards were becoming noticeable. People passing us on the city streets looked at us and whispered "Fidelistas", in reference to Fidel Castro and his very popular revolutionaries fighting in Cuba (all of whom wore beards, which was not a custom among men anywhere else in Latin America at that time). Between Morelia and Mexico City we crossed over a mountain pass at 10,300 ft., necessitating the use of low gear most of the way. When we arrived in Mexico City that evening, we decided to celebrate the successful completion of our first week of travel by eating dinner at "Gitaneras" (gypsies), a very expensive restaurant with Flamenco dancers. We shared a bottle of Spanish Fundador brandy (a nostalgic drink for Noble from his navy days) with all those who passed by our table. By the end of the evening, when it came time to leave, the waiter brought us a bill equivalent to $30! That was the last time ever we did anything like that again.

We spent that night and the next two and a half days in Mexico City. One day we visited the Aztec ruins of Teotihuacan in the valley of Mexico, and climbed the 222 steps to the top of the Pyramid of the Sun. Virtually everyone was friendly. Our ability to speak Spanish gradually improved. We were interviewed by newspaper journalists, TV reporters, and photographers, all of whom wanted a story for the local media. Such experiences would become a regular occurrence as we traveled.

At noon on Wednesday, Feb.4, we left the many statues, more than 10 million people, and nightmarish traffic in Mexico City, and drove southeast toward the twin snow-capped volcanoes of Ixtaccihuatl and Popocatepetl, each over 17,000 ft. above sea level. Noble and I wanted to climb Popocateptl, the higher of the two at 17,880 ft. elevation. Our road wound back and forth up the mountain to near tree line at 12,500 ft. at the base of the cone-shaped peak, where there was a small stone refuge in which climbers could spend the night. On the way up, I got out at the 9,000 ft. level in a lovely forest of mixed pines, firs, and hardwoods. Then I walked and documented birds for the last 10 miles up to the refuge, arriving two hours after dark. Some of the birds which I saw on the way were 15 White-eared Hummingbirds, 3 Green Violetears*, 6 Russet Nightingale Thrushes*, 2 Olive Warblers, 7 Red Warblers*, (a gorgeous little bird of the forest mid-story and canopy, almost entirely red except for a white cheek patch), 1 Slate-throated Redstart*, 1 Golden-browed Warbler*, 1 Striped Sparrow*, and 6 Rufous-capped Brush-Finches*.

The next morning, Feb. 5, we arose at daybreak (7:30), ate breakfast, rented crampons (for hiking in the snow and ice), hired 2 guides (one for each of us, both with an ice pick), and started the long, steep, arduous climb to the snow-covered summit more than 5,000 ft. above us. Birds in the pines and grasslands as we began our ascent included 3 Strickland's Woodpeckers*, 2 Red Crossbills, and many Striped Sparrows. The vegetation disappeared at 14,000 ft., the trail became much steeper, and the substrate became bare volcanic ash and rock (scree). The going was tough. Soon the ground was covered with snow. A Common Raven soared past us at 17,000 ft. The trail zig-zagged back and forth, and we stopped frequently to catch our breath. Noble and his guide fell off the pace a little bit, but continued upward. My guide and I reached the top and waited for the other two. The view down into the smoldering bottom of the crater 2,000 ft. below the rim was spectacular. Popocatépetl is an active volcano and erupts regularly. The climb stretched Noble's strength and will power to the very limit, but he was determined to make it to the top, and he did. Afterwards, he remarked that never before had he been so exhausted.

We continued southeastward into the state of Oaxaca, where the countryside became more arid and the vegetation changed to tropical scrub. Oxcarts with wooden wheels posed a hazard on the highway. That night we camped along the roadside.

The next morning, Feb. 7, it was necessary to stop for 2 hours in the capital city of Oaxaca to replace a broken front brake drum. Fortunately, Jeep parts were available everywhere. As we descended toward the town of Tehuantepec on the Pacific coast, in the state of Oaxaca, the countryside became increasingly arid and more sparsely scrub-covered, with yellow-flowered Palo Verde trees brightening the landscape. There were many more oxcarts. Noble wanted to stop for a couple of hours to see the ancient archaeological ruins at Mitla, dating back almost 2,000 years. I chose to spend this time birdwatching in a small scenic roadside gorge. I was rewarded with a marvelous look at a Russet-crowned Motmot*. This colorful bird, about 12 inches in length and with elongated racket-shaped central tail feathers, perched obligingly low in an open scrubby tree, slowly swinging its tail back and forth sideways. This habit has given it the local name of "two o'clock, four o'clock bird." I gasped at the first sight of it in my binoculars. Motmots epitomized the neotropical avifauna. That night we camped just outside the town of Tehuantepec.

On Sunday morning, Feb. 8, we drove into the city of Tehuantepec for a day of leisure, two weeks after we had departed from Los Angeles. This hot, tropical location in southern Oaxaca was only a few miles inland from the Pacific Ocean. It would be a memorable day. We found a barbershop where we both had an hour-long haircut, for which we paid 2 pesos (about 18 cents) each. (We left our beards untrimmed.) Later, about noon, as we were wandering around the town sightseeing and taking a few pictures, we were approached by a young, simply dressed, medium statured boy of perhaps 16 or 17 years of age. He knew only a very few words of English, but he asked us, in Spanish, if we would like for him to be our guide in the town. He said he knew a large Catholic church where a wedding was being held, which we could watch if he took us there. He told us his name was Carlos Martinez and he seemed genuinely nice, so we agreed to go with him. He didn't ask for money but Noble and I agreed that we would give him a small sum at the end of the afternoon.

As dusk approached, Noble and I decided it was time for us to leave, so we gave Carlos a few pesos and thanked him for his services. To our great surprise he asked us (in Spanish), "Can I go with you?" He explained that he needed a ride to the city of Tapachula, in the state of Chiapas, where a friend of his lived and worked on a coffee plantation. This was more than a day's drive for us. After considerable discussion between us, Noble and I decided we would take him with us to Tapachula. There would be a place for him to sit in the cab, between the two of us on a small wooden seat with a cushion on top of the engine (which we had built to accommodate a third person if the occasion arose). First, we had to drive to his house so he could pick up a few items to take with him in a little duffel bag. That night

Plate 2: With one guide each, Noble and I climbed from 12,500 ft. elevation up to the 17,880 ft. summit of Popocatepetl, in Mexico, one of the world's highest active volcanos. After photographing the steaming crater we descended back down.

we all ate in a small, inexpensive restaurant in Tehuantepec, where Carlos paid for his own meal, and then we drove a short distance outside the town to a suitable campsite for the night. Carlos slept on the ground outside the van, on a mat.

The next morning (Mon., Feb. 9) Carlos got up early and voluntarily washed off our front windshield. In the early afternoon we arrived in Tuxtla Gutierrez, the capital city of the state of Chiapas, situated at 7,500 ft. in forested mountains. The city had a population of approximately 70,000. We took our Jeep to a service station to have the front springs adjusted and the pesky accelerator cable put back together, once again. We were told it would take several hours of work. It was now late afternoon. As we were driving past the big central square in the middle of the city, on our way out of town, Carlos suddenly declared, "I want to get out here." Noble and I were amazed, since this wasn't the destination he had given us. Never-the-less, I stopped alongside the square and he got out, carrying only his little duffel bag. We gave each other hugs and exchanged very small gifts. We also gave him a tiny sum of money, and wished him well. We then drove out of town, on the highway toward San Cristobal.

Forty minutes later, as the sun was going down, I pulled over to the side of the road and stopped to take a picture of the sunset, as I often did. While I was driving, I always kept my camera on a small middle upper shelf at the back of the cab, within easy reach. I reached for it but it wasn't there. I was puzzled and quickly looked inside the van, thinking I might have left it there, but I didn't find it. Suddenly the thought came to me - Carlos took my camera with him when he left! I was furious to think this could be true, but there was no other explanation. In my anger, I did not think rationally, so I said to Noble, "I'm going to drive back to the city and find Carlos to get my camera back." Really? What were the odds?

I drove back into the city after dark at about 7:00 pm, almost two hours after we had departed. I parked at the same location on the square where we had let Carlos out. I had no idea where he might have gone, or even if the name that he had given us was correct. Possibly, he knew a person in town with whom he could spend the night. In my minimal Spanish, I asked a policeman on the street "How many hotels are there here in town?" My question amused him, and he waved his hands in the air and said "Bastante" (a great many). That was no help. It was time for calm reasoning. I commented to Noble that perhaps Carlos just walked down the street and checked into the first cheap hotel he came to. That was the only option I could think of pursuing, so I said to Noble, "That's where we're going to look." At random, I chose a direction in which to begin. We walked several blocks, past a couple of moderately priced hotels, before coming to a much cheaper hotel. We walked into the small, vacant lobby and up to the somewhat dilapidated, wooden reception desk where a young boy was working. He asked me (in Spanish) "What can I do for you?" In my barely adequate Spanish, I said, "Did a person by the name of Carlos Martinez register here in the past two hours?" His immediate answer was "Sí" (YES)! Could this really be true?

I asked for his room number and was given a third floor number, so I walked up the deteriorating wooden stairs and knocked on his door. Noble remained in the lobby, sitting inconspicuously behind a newspaper to watch the front door in case Carlos should walk in. Knocking on the door produced no results, so I came back down the stairs to the front desk and asked if I could have a key to his room, explaining that I thought he might have mistakenly taken something of mine. The answer, of course, was no. Even when I said I would bring a policeman with me the answer was still no.

I joined Noble in the lobby corner and we both ordered a beer, and waited for Carlos to return. Finally, at 10:00 o'clock, Carlos walked in the front door and up to the reception desk, without seeing us sitting in the corner. We were prepared to chase him if he ran. The boy at the desk told Carlos there were two persons waiting to talk to him, and he nodded in our direction. Carlos turned and looked at us without showing any surprise, or fright, and walked calmly over to where we were sitting. His demeanor caught me off guard. I wasn't quite sure what to say to him. I was absolutely certain he had taken my camera. Our conversation was entirely in Spanish. I said to him that I was missing my camera and perhaps he had mistakenly placed it in his bag, among his personal items, when he left. I asked him if I could look in his bag, which was upstairs in his room. I was surprised by his quick consent, with no evident concern.

We walked up to his room together, he unlocked the door, and we walked inside. His duffel bag was lying on the floor in one of the corners. Feeling rather confident that if he had my camera it would be in his bag, I picked it up and looked inside. There weren't many items, but my camera was not among them in the bag. I was quite dismayed and perplexed. I looked around the room and asked myself where else he might have put it. The room was almost bare of furniture. I searched everywhere I could think of, including under the pillow and sheets, but without success. I stood there and carefully considered where I might not have looked the first time. A site came to my mind. I walked over to the small wooden bed and lifted the entire mattress and covers off the few wooden slats on which they were supported, without any bed springs. There, resting on top of a middle slat under the mattress was MY CAMERA! It had been very carefully hidden. I was jubilant. Carlos turned very white and stammered "¿Yo no tengo mucho dinero, está bien?" (I don't have much money, is it all right?) I was very angry with Carlos and I thought of punching him in the face, but refrained from doing so. I glared at him fiercely, but said nothing. I had my camera back, which was what really mattered. I walked downstairs to where Noble was waiting, told him I had retrieved my camera, and that we could depart now. Carlos came down the stairs, quickly glanced our way, then hurriedly walked past the boy at the desk, out the front door, and into the street. My only camera had survived the first of its nine lives!

That night we drove until 1:30 am, through San Cristobal to a campsite in a beautiful pine forest at 7,500 feet. Driving at night was quite hazardous. Not only were there the usual pedestrians, bicycles, dogs, cows, horses, carts, and stalled vehicles in the road, all without any lights, but trucks driving at night almost always turned off their lights as they approached oncoming vehicles! Whip-poor-wills sang from time to time during the early morning hours before daybreak.

The next morning (Tuesday, Feb. 10), we reached the Guatemala border about 1:45pm, at an elevation of approximately 5,000 feet. This was the only road from Mexico into Guatemala, and it had been open to public traffic for just one month, so we were obviously one of the very first tourists to travel overland in a vehicle from Mexico to Guatemala. It was part of the ambitious "Pan-American Highway" program connecting the United States with all the countries in Central and South America, by road. Formalities were very simple, even though Mexico and Guatemala squabbled over many matters.

While in Mexico for 15 days, I recorded a total trip list of 178 species of birds, 32 of which I had never seen anywhere previously (Appendix A). The last three of these were seen today, the Black-capped Swallow*, Yellow-backed Oriole*, and Rufous-collared Sparrow* (which was one of only 3 species, along with the Black Vulture and House Wren, which I documented in all 11 of the Neotropical countries we visited).

Chapter Three – Guatemala and Beyond

Our road in Guatemala was gravel, narrow, sometimes one lane, winding, hilly, extremely dusty, and very slow going. The dips and washes in the road caused the frame of our overloaded camping van to twist, bend, and creak. We were in mountainous countryside and there were canyons, streams, and waterfalls. Bridges were narrow and often incomplete, with only wooden planks on which to drive across a stream. Many areas of the road were still under construction. People along the roadside were curious and friendly. Men carried heavy racks of pottery on their back. We stopped in the late afternoon at a scenic roadside waterfall where the water looked so clean and clear that we boldly drank it without purification. We washed our clothes and then had a welcome, refreshing, but very cold shower under the waterfall. That night we camped about 10 miles SE of Huehuetenango.

The next morning (Wednesday, Feb. 11), it was necessary to change a tire before starting out, because of low tire pressure. This problem would plague us for much of our travel. Noble awoke with agonizing pains in his stomach and a very bad case of diarrhea, apparently from the water he drank at the waterfall. He took a dozen sulfasuxadine tablets and spent the rest of the day lying in bed in the back of the van while I drove, with frequent stops for him to jump out by the roadside. Our road was tortuous, winding back and forth in the mountains at elevations as high as 10,000 feet. It passed near the north side of Lake Atitlan, where I looked in vain through my telescope for the probably extinct, endemic, flightless Atitlan Grebe. The scenery was beautiful, with deep gorges, luxurious forests, and tall, cone-shaped volcanoes rising high above the distant skyline in many directions. I was excited to see lovely Pink-headed Warblers* in the high elevation pine forests. In the lower elevation, dry oak scrub forests, I drew sketches in my little field notebook of the head patterns of both Rusty Sparrows* and Prevost's (White-faced) Ground-Sparrows* (see illustrations), which were identified at a later time. That evening, about 8:00 pm, we arrived in Guatemala City, elevation 5,000 ft., and camped in a park.

Noble and I spent the next three days (Feb. 12, 13, & 14) in Guatemala City. There were many items of business to be taken care of. We visited the American Embassy, the post office, the Agricola Mercantile Bank, a laundry to have our sheets washed, and of course the Jeep agency for badly needed repairs. It was necessary to have additional leaves added to all of our springs, a new clutch, a tire repaired, and adjustments made to the emergency brake, idle speed, and accelerator cable (again). We were initially told the repairs would take 2 or 3 days - but they wound up taking much longer (as we eventually discovered would be true throughout all of our travel in Latin America). So we found various things to do while we waited. We visited a colorful Indian marketplace. We replenished some of our food supplies at a modern, expensive supermarket.

By great fortune, a very significant event was taking place in the city. An immense religious festival and celebration was being held for several days, presided over by Cardinal Spellman who had come to Guatemala for the occasion, a once in 50 years visit. Guatemalan Indians, dressed in their most colorful costumes, came from numerous different tribes all over the country (mostly the nearby mountains). Many of them came only by walking, often barefoot, while carrying personal belongings in a large blanket slung on the back. Women often carried a baby in a blanket, and wore brightly patterned shawls, tall bundled scarves, and long, wide-flaring skirts of many colors. Men also were often brightly decorated and wore large straw hats, of which there many shapes and varieties. There were long musical processions in the streets, with men playing horns, fifes, and drums. It was a great spectacle, and afforded Noble and me marvelous photo opportunities.

When it became obvious that repair of the Jeep was going to take much longer than we were initially told, we decided to fly for three days to the ancient Mayan ruins of Tikal, located in the low-lying jungle of the Petén at the base of the Yucatan peninsula in northern Guatemala. On Sunday, Feb. 15, a twin-engine DC-3 took us to a 1,500 foot dirt landing strip in the middle of the tropical forest at the headquarters of the on-going archaeological excavations and restorations. Our accommodations were surprisingly clean and comfortable, and there was a small information desk where I was able to purchase a newly published (Jan. 1959) *Check List of the birds of Tikal,* by Frank B. Smithe. This was my first opportunity to spend a whole day, and longer, looking for birds. It also allowed Noble and me time to photograph and inspect the magnificent limestone pyramids and temples, some of which towered more than 200 feet above the ground and were almost 2000 years old.

I awoke at early daybreak on my first morning (Mon., Feb. 16) to the loud, territorial, vocalizations of the Howler Monkeys. Soon thereafter, birds were singing or calling from all directions, fascinating sounds, which I couldn't yet name. Clad in a short-sleeved shirt, walking shorts, and hiking shoes, and carrying binoculars around my neck and a small brown canvas shoulder bag with my camera, Blake's Mexican bird guide, Smithe's checklist, and a little field notebook, I set out for a day in the tropical forest with great excitement and anticipation.

During the next 2½ days, I filled my field notebook with the names, descriptions, or sketches of the birds that I encountered. It was not easy. Although I had studied my Mexican field guide critically, it was possible for me to recognize only a fraction of the birds at first sight. It was necessary to write a short description of the others. Frustratingly, this took much time, consuming valuable minutes while other birds in the foraging flock went unobserved. There was no other solution if I were ever going to be able to apply names to the birds on which I focused my binoculars. I could usually guess the taxonomic family in which a bird was placed, and sometimes the genus, but identification at the species level required information that was mostly unavailable to me. To describe a bird in enough detail to allow future identification was not an easy matter. Many birds were too active or too hidden from view to permit any description at all. Very occasionally, I drew a simple sketch of the bird, or just its head, to accompany my written notes. It was a daunting and challenging task.

My most vivid bird memories from Tikal are of the far carrying, loud "hah, hah" vocalizations of the Laughing Falcon*, the gorgeous coloration and exotic racket-shaped tail feathers of the Blue-crowned Motmot*, the fantastic bill of the Keel-billed Toucan*, and the 3-6 ft. long nests of the Montezuma Oropendolos* which hung from the terminal ends of branches in the tallest treetops. Probably no birds characterize neotropical forests more than toucans, with their enormous, brightly colored bills. These birds are fruit eaters and the grotesque bill is not nearly as heavy as it appears, being mostly empty space with an outer covering and inner structural network of keratin, the same lightweight protein material of which feathers are comprised.

Toucans are zygodactyl (2 toes point forward and 2 toes point backwards) and they have the unusual ability to hold a fruit in one foot and convey it to their mouth while hanging onto a branch with their other foot (as do parrots).

Several examples from my field notes are:

(1) "parrot, medium, short tail, reddish under tail coverts, pinkish ring around the eye, bluish head, underparts greenish, wings green, forepart of crown cream"; while this description is not perfect, it allowed me to identify this bird, at a future time, as a White-crowned Parrot*.

(2) "very small, short black bill, yellow below, dark line through eye, white eye stripe, upperparts brownish (darker Gray on head)"; I later identified this bird as a Yellow-bellied Tyrannulet*.

(3) "tanager? rel. short tail, about 6in, throat bright yellow, rest of underparts yellow-green; head beautiful pale green, notes like a Tufted Titmouse; bill black, short, flat"; this bird was eventually identified as a Green Shrike-Vireo* (not in the family which I guessed).

During my three days at Tikal (Feb. 16, 17, & 18), I recorded nine species which were not on Smithe's Check List: Collared Forest-Falcon* (which I sketched), Lesser Swallow-tailed Swift*, Black-cheeked Woodpecker*, Yellow-bellied Sapsucker, Paltry Tyrannulet*, Ruddy-tailed Flycatcher*, Olive-backed Euphonia*, Yellow Warbler, and Green

Honeycreeper*. I observed jacamars and manakins for the first time. The most abundant bird seen in the forest was the Red-throated Ant-Tanager*, the most vocal diurnal bird was the Short-billed Pigeon*, and the most vocal nocturnal bird was the Pauraque. There were four kinds of trogons (two red-bellied species and two yellow-bellied species) and twelve species of wintering Nearctic wood warblers (including the enigmatic Yellow-breasted Chat). I searched in vain for an Orange-breasted Falcon, a species I was told nested on the tallest temples or pyramids. The total of 115 species which I recorded from Tikal are all noted in column two of Appendix A by having the "X" symbol superscripted with a small numerical "1." Sixty-six of these species were species I had never seen before (as denoted by an asterisk).

Noble and I flew back to Guatemala City on Feb. 18 and once again slept in the back of our camping van, which was still parked in the custody of the Jeep agency. Our repairs, after seven days in the garage, were finally completed. The head of the agency charged us only for parts, not for labor. We were ready to depart. Everyone we had spoken with about our intentions to drive by road all the way to the Panama Canal, told us that it was not possible and we would be unable to do this.

On Thursday morning, Feb. 19, we headed south and east from Guatemala City toward the border with El Salvador. Our road dropped steadily in altitude. There were low mountains and foothills with barrancas (ravines), and valleys with green fields. The "Pan-American Highway" was mostly gravel, hot, and dusty. We ran our electric generator to keep our food and film cool. We found a campsite at 7:00 pm, about 10 miles from the border, at 2,000 ft. elevation. I always looked for a site where it might be possible to see a few birds, either before dark or just after daybreak the next morning. That evening, I saw and heard both a Great Horned Owl and Ferruginous Pygmy-Owls*, and although Guatemalan Screech-Owls* called all night long I could never find one visually with my flashlight (which was frustratingly true of this nocturnal group of small owls throughout my entire travel). Pauraques and an unidentified caprimulgid (probably the Buff-collared Nightjar) were also heard but not seen.

I birded for an hour the following morning (Fri., Feb. 20) and found my first Stripe-headed Sparrows* and a Turquoise-browed Motmot*. The ten species of motmots (family Momotidae) are among the loveliest birds in the Neotropics, where they are slightly more diversified in Central America than in South America. They sit quietly in the shady lower or middle parts of the forest, searching for beetles, other large insects, or small frogs and lizards. Their nests are placed at the end of burrows, which are excavated by the birds in an earthen bank, or on the ground. Motmots were always a delight to encounter, and often afforded me a good view.

We left our campsite early and reached the El Salvador border at 8:00 am. For a very pleasant change, the highway was paved. Roadside houses were mostly adobe, with either tile or thatched roofs. Little boys played naked outside. Unlike Guatemala, we saw no Indians. Noble and I entered the modern capital city of San Salvador at noon, just as the two to three hour afternoon siesta period was beginning, and the streets were quiet and almost vacant. Therefore, we continued onward, through rolling hills, toward Honduras. Coffee "fincas" (plantations) and sugarcane fields dominated the countryside. Black Vultures, Turkey Vultures, and Great-tailed Grackles were the most numerous roadside birds. Ten miles from the border, at about 6:00 pm, we came to a nice wide stream and decided to stop for the night. We took the opportunity to have a bath and wash some clothes in the stream, hanging them on a makeshift rope clothesline, which we tied between our van, which was parked in the stream bed, and a shoreline tree.

An oxcart, which was being loaded with rocks from the river near our Jeep, awakened us early the next morning (Saturday, Feb. 21). Women were already washing clothes in the river. We were soon on the road again and reached the Honduras border at 8:15 am. We were able to complete all the necessary paperwork at the "inmagracion" and "aduana" offices within 45 minutes, and were then on our way once more, having spent only 24 hours in El Salvador.

There are fewer than 200 miles across the Pacific strip of Honduras to Nicaragua. The day was hot and dry, and the highway was gravel and very dusty. There were many detours. Bridges were mostly under construction or non-existent, making it necessary to drive through the stream beds, most of which were dry. Iguanas frequently walked lazily across the road, imperiling their life in spite of the slow vehicle speeds. There were hardly any road signs, so it

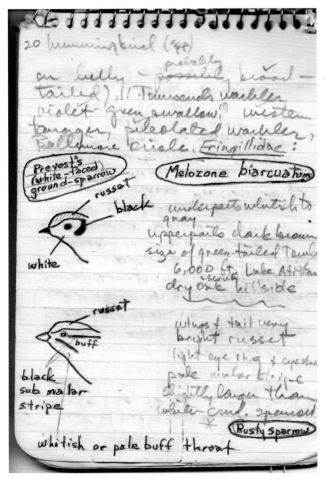

Plate 3: Our road traveled along the NE shore of Lake Atitlán in Guatemala at 5,000ft elevation. I drew sketches of two new birds for my list. Indian women wore elaborate, colorful dresses at the celebration for Cardinal Spellman in Guatemala City.

was necessary to stop frequently and ask directions. Once when I stopped and asked a local farmer how far it was to a town on my map, he answered "Dos dias" (two days). I knew this could not be true so I queried "En un carro?" (in a car). He shook his head negatively and responded "No, sobre un burro" (on a donkey)! He had apparently never traveled to the town in a car, only on a donkey. I also discovered not to ask if our road went to a particular town, by naming the town. The answer to this question was always "Sí" (yes), which often proved to be incorrect. So, I learned to re-phrase my question and ask "Where does this road go?" Usually I got no answer to this question at all, since a person did not like to admit he didn't know.

We reached the Nicaragua border by 2:00 pm (still Feb. 21), our third country of the day, having spent only 5 hours in Honduras! After clearing customs and immigration, we continued on our way to Managua, the capital city. Along our way were plantations of coffee and bananas, and fields of cotton and sugarcane. Oxcarts, pedestrians, cows, donkeys, pigs, horses, dogs, and chickens were constant hazards to driving, along with the usual stalled buses and trucks in the middle of the road. Roadside birds documented along the way during the short afternoon drive included 300 Black Vultures, 40 Turkey Vultures, 25 Great-tailed Grackles, and 15 White-throated Magpie-Jays. Late that night we camped along the Tipitapa River just outside Managua, near a popular hot springs.

During 2 hours of looking for birds the next morning (Sunday, Feb. 22) I was happy to see my first Squirrel Cuckoo*, a rather large, handsome, widespread neotropical cuckoo which slinked through the dense midstory vegetation. There were some Nicaraguan Grackles* (a local endemic species), a Peregrine Falcon, 20 Orange-chinned Parakeets* (a small green parakeet with a medium length, pointed tail, in the genus *Brotogeris*), 40 Ruddy-breasted Seedeaters*, and 500 wintering Scissor-tailed Flycatchers! There were also numerous other wintering species from the Nearctic, most notably Painted and Indigo buntings, Rose-breasted Grosbeak, Baltimore Oriole, and several kinds of warblers. I had a long, close look at a male Plain-capped Starthroat* with its iridescent red throat and distinctly striped face, and I drew a sketch of its head in my notebook. About mid-morning Noble and I took advantage of the warm water to bathe and wash our clothes in an overflow from the hot springs. Then we drove into the city and spent most of the day having the Jeep's engine tuned, sightseeing, and taking two young girls to a movie, at their request.

The next day (Monday, Feb. 23), we left Managua and progressed toward the border of Costa Rica. Some Black Vultures were eating a dead donkey at the side of the road. Our highway surface alternated every several hundred yards between gravel and pavement, with deep trenches where one surface met the other. Many vehicles were broken down and awaiting repair. We camped that night near Rivas, not far from the border with Costa Rica.

Early the next morning (Feb. 24), a bird near our camping van was singing with a distinctive series of trilled whistles dropping slightly in pitch and slowing at the end. I searched carefully for 5-10 minutes, but I could not locate the songster in the dry, wooded, roadside border. Therefore, I recorded in my notebook what the song sounded like, as best I could. It was many years later when I was listening to a recording of neotropical bird songs that I heard a song and immediately recognized it as my mystery bird. It was a Lesser Ground-Cuckoo*, and its song is one of my more pleasant memories. We left our campsite and within an hour, at 10:00 o' clock, arrived at the border of Costa Rica.

Chapter Four – Costa Rica and Panama

We followed the narrow, well-paved "Pan-American Highway" from the Nicaraguan border to San Jose, the capital city of Costa Rica, situated in the central highlands. We arrived on the morning of February 25, exactly one month after departing from Los Angeles. This was our staging point for the "impassable" section of road in southern Costa Rica, which connected the country to Panama. Everyone we had spoken with told us we would not be able to drive all the way to Panama because the road was still under construction and not yet open to public traffic. We were undaunted.

We talked with many persons in San Jose about our proposed travel. Among those were Mr. Foster (the head) and Mr. Ward (his construction chief) of the Foster-Williams construction company, who were under contract for most of the Pan-American Highway in southern Costa Rica. Both of these gentlemen were opposed to our travel plans, saying that tourists had posed numerous headaches by asking construction workers for aid in getting them out of all kinds of trouble. They advised us not even to try. We persisted, and they finally ended our long conversation with them by saying "We will help you if we can."

A Mr. Lincoln in the city was the assistant district engineer for the Bureau of Public Roads. He told us that two of the three large river crossings without a bridge were now fordable, but that the third crossing had four feet of water and would necessitate a "small float" to ferry us across. He said the road was not open to the public and would require permission, since the construction companies were working under contracts that assured them they would not be bothered by tourist travel. However, a construction worker he talked with on the radio was more enthusiastic and told him we could come ahead and they would provide us with a road grader if necessary.

Another person we spoke with was Mr. Keyes, the head of USIS (United States Information Service) in Costa Rica, whose office was in the U.S. Embassy. He was extremely friendly, cordial, and helpful, as was his beautiful Costa Rican secretary, Ana Cecilia, who was the recently crowned "coffee queen" of the country. Mr. Keyes had a letter of introduction typed for us, in Spanish, with a large embassy seal stamped on it. It was addressed "To whom it may concern" and said who we were and what we were doing. Mr. Keyes said the letter might help in difficult situations. He also told us how we could get past the guard at the entrance to the exclusive country club just outside of town, and make use of their facilities while we were in San Jose.

We took our van to the Jeep agency for two new rear shock absorbers and a washing, then drove to the country club to try our luck getting past the guard. There was no problem, as he was a very pleasant young man. So we took a shower, which was always welcome, ate a large dinner, and found a place to park our van for the night at the country club.

The next day (Feb. 26), we went by the American Embassy again, where Mr. Keyes introduced us to the American Ambassador, Mr. Wilhauer. A photographer and newspaper reporter recorded our adventure for publication in the local media. A moving picture of Ana Cecilia shaking hands with us was taken as we left the embassy to drive out of town. On our way out, many of the streets we encountered were one-way, usually very poorly indicated as such. We learned that when a policeman stood in the middle of an intersection facing us as we approached, we should stop, but if he were facing to our right or left we could continue on past him through the intersection.

Our destination for the day was the 11,000 ft. summit of a nearby volcanic peak, Volcán Irazú, from which we were told we would be able to witness a spectacular sunrise the next morning. The road up to the top was very steep and winding, and required us to drive in 4WD and 2nd gear, at a speed of 10 mph, for most of the way. We camped in a stunted "elfin" forest at the top of the mountain, on the rim of the volcanic crater. My supper consisted of two bananas and three peanut butter-jelly sandwiches. It was cold so I climbed inside my sleeping bag. All night long we heard the call of a caprimulgid, which I much later concluded was that of the Dusky Nightjar*.

The following morning (Feb. 27) was cold and foggy, without a sunrise. Noble cooked pancakes for breakfast, and then I birded for a couple of hours while he wrote in his trip journal. I documented the following species which I had not previously seen: Volcano Hummingbird*, Fork-tailed Flycatcher* (at an unusually high elevation), Blue-and-white Swallow*, Ochraceous Wren*, Sooty Robin*, Yellow-winged Vireo*, Flame-throated Warbler* (a particularly attractive little bird which reminded me of a Blackburnian Warbler), Flame-colored Tanager*, Sooty-capped Bush-Tanager*, Large-footed Finch*, Sooty-faced Finch*, and Volcano Junco*.

We descended the mountain very slowly, in 4WD low gear, and passed through the city of Cartago. We were on our way to attempt the "impassable road." First, it was necessary to travel along a high ridge of the Cordilleran Range of mountains, over the Cerro de la Muerte (summit of death) at 11,000 ft. elevation. There were moss- and epiphyte-covered forests with stunted trees ("elfin forests"), and occasionally hillside forests with towering trees more than 150 ft. tall. I vowed that one day I would return to these forests for bird observations. The brackets on both of our front shock absorbers broke, so we took the shock absorbers off and proceeded even more slowly.

In the city of San Isidro, we tried to have the shock absorbers repaired, but we were unsuccessful. So we filled up our 40 gallon gas tank, added oil and water, and continued toward the difficulties which were not far ahead, of which we had been forewarned many times. Temporarily we were on a new gravel road. Very soon we came to the first of numerous road barricades, all with the purpose of preventing any tourist vehicles like ourselves or others without a permit, from proceeding any further. When I stopped at the first of these barricades, a small statured, rather elderly Costa Rican guard approached me on the driver's side and said "Tiene usted permiso?" (Do you have permission?) I thought to myself that here was the time to present the letter (in Spanish) that Mr. Keyes had given us, with its big American embassy seal. I confidently told the guard "Sí" (yes) and showed him the letter. He studied it for a very long time, and then with a question in his voice he pointed to the signature and said "Firma?" querying whose signature it was, since it was obviously not the one he was looking for. I pondered how to respond and then said, in Spanish, that it was the signature of the "jefe grande" (big chief) in San Jose. The American seal looked very official. (I concluded the man probably didn't read at all, but merely was looking for a particular signature.) The guard was not convinced and he shook his head negatively. So I said to him, in my most authoritative manner (and poor Spanish) that he would be in big trouble if he did not let us through. After much thought and deliberation he finally nodded his head affirmatively, raised the wooden bar across the road, and motioned for us to continue. We had cleared our first hurdle.

An hour later down the road, in the afternoon, we came to another barricade, at the Río General where a highway bridge was under construction but far from completed. However, a temporary bridge across the river existed just upstream. Again, the guard was reluctant to let us pass. By very good fortune the highway bridge inspector, a Mr. Bennisch from Austria and a very important person, happened to come by just then. He had only recently arrived in Costa Rica. He was immediately sympathetic with our adventure and willingly gave us permission to continue. He then took us to the nearby office of the Beeche-Fait Ltd. Construction Company, which was in charge of building that particular section of the road. Here we were introduced to Señor Fait, half owner of the company. He was a likeable young Costa Rican who spoke good English, having studied for several years at LSU. He, too, was appreciative of what Noble and I were trying to do, and he invited us to have dinner with him at the construction camp located there at the bridge site. He also gave us a card that granted us permission to pass through all future barricades with no questions asked. How lucky could we be? That night we camped nearby along the river, where we bathed before going to bed.

The next morning (Saturday, Feb.28), I arose at dawn, 5:30 am, and was greeted by a myriad of tropical forest birds, which included 3 kinds of kingfishers (Green*, Amazon*, and Ringed*), many flycatchers, brilliantly colored tanagers (Blue-Gray, Cherrie's*, Bay-headed*, and Golden-hooded*), and the enigmatic Bananaquit* (a tiny nectar feeder of uncertain affinities). However, there was little time for taking notes on the birds. I gulped down a banana for breakfast and carried out my designated morning tasks of sweeping out the van and refueling the electric generator. We then said goodbye to the construction workers and continued navigating along the unfinished road, proceeding at a very slow speed toward the town of Buenos Aires. There were more than a half dozen streams of varying sizes and depths, which we forded, but none were greater than two feet deep. The road was hilly and very steep in places, with occasional deep potholes of mud. Noble requested that I take some pictures with his 16mm movie camera of him driving through some of the streams. Our travel was often watched by disbelieving construction workers. We were driving at speeds between 5 and 10 mph. The day became cloudy, humid, and hot. We were unable to purchase gasoline in Buenos Aires, as our road map indicated we could, but we had enough to continue onward.

Late in the afternoon we reached, for a second time, the Río General, which we had been more or less paralleling on its left bank as we followed it downstream in a generally southeast direction. Here in the foothills the river was known as the "Río General de Térraba," which we nicknamed the "terrible Térraba." The road ended at the river-bank. There was no bridge at all, only a very small, crude raft made of logs tied together, about 10 x 15 ft. in size. The fast flowing river was approximately 250-300 ft. across with a stony, rocky bottom and a maximum depth of just over 3 feet. The raft was designed to ferry people and small lightweight vehicles across the river. Our behemoth van did not qualify. Some native men and boys were standing by to assist. I looked at the raft and said to Noble that there was no way this raft could possibly float our Jeep across the river. His response was "We've got to try."

So we tried. With the aid of the power winch on the front of our Jeep we succeeded, with considerable effort, in getting our front wheels up over the nearest end of the raft, and on top of the first several logs. It was much more difficult to get the back wheels on also. After almost an hour of effort and with the help of our winch again, we drove entirely on board, at least on top of the raft. It immediately sank to the bottom of the river! So we unloaded some of our heaviest gear to a dugout canoe alongside the raft, to lessen the total vehicle weight, but the raft still remained stuck on the bottom. No amount of tugging or pushing or winching could budge the raft. So we backed off, with great difficulty, and pushed the raft farther out into the river, where the water was somewhat deeper.

With great effort, we then managed to get our van back onto the raft for a second time. The result was the same. The raft immediately sank again to the bottom of the river. We had tried for almost three hours to float the raft with our Jeep on top, but it was impossible. The evidence was clear. The raft was not going to ferry our van across the river. We had attempted to do this in every conceivable way, but our excessive weight was just too great. We were very disheartened. Our dream of driving all the way to Panama entirely by road seemed to have ended here on the bank of the Río Térraba. We gave our helpers a can of peanuts for their assistance. Darkness was approaching so we camped for the night there on the shore. We had a bath in the river, and then ate an evening meal of bread, marmalade, hard-boiled eggs, bananas, olives, and English walnuts. Our spirits were very low.

At daybreak the next morning (Sunday, Mar. 1), we made the decision not to give up quite just yet. Against all logic, we would attempt to ford the river! First, we needed to know exactly how deep the river was, so we both waded across, in shorts and tennis shoes. The water came up to just above my waist, or slightly more than 3 feet deep. Water depth was not the only difficulty. The current was very swift, and the river bottom had stones and rocks of all sizes, some almost a foot in diameter. We attempted to remove all the largest rocks from our proposed route across the river. We knew that the engine of our Jeep, situated between us in the cab, was quite high above ground level. The six spark plugs on top of the engine were just over three feet above ground level, or almost exactly the maximum depth of the river. Our thinking, in our inexperience and ignorance, was that if the spark plugs did not get wet we might make it all the way across the river.

We had been told by someone, somewhere, that if we were going to ford a river we should remove the radiator fan, to prevent water from being pushed by the fan backward over the engine. So we removed the fan. It was decided that I would drive and Noble would wade ahead of me with his movie camera in an attempt to get some moving pictures. I removed the cover from the top of the engine so I could watch the water level as it rose in the cab, and approached the spark plugs. I put the vehicle in 4WD, the lower transfer case, and 1st gear. This was the maximum power the Jeep possessed, at the slowest speed – barely a crawl. I was determined that once I started across the river I would keep the accelerator on the floorboard and not stop the vehicle for any reason at all, constantly maintaining forward progress.

Shortly after I started inching my way across the river two unforeseen problems occurred: (1) the fan BELT sprayed water into the cab and soon my windshield was so covered with water on the INSIDE that I could hardly see out, and (2) whenever my front wheels went over a large rock or boulder the swift current pushed me downstream (right to left), making it difficult to steer a straight course across the river toward the landing on the other side. Nevertheless, I DID NOT STOP OR REMOVE MY FOOT FROM THE FLOORBOARD, and the water level never quite rose above the spark plugs. My feet were under water and the van slipped over one rock after another, but the engine never coughed or quit, and the van maintained its very slow progress through the turbulent water. Almost ten minutes after I entered the river I drove out on the opposite bank. The Jeep made it! Noble and I embraced each other in great jubilation! The few workmen on the riverbank stared in disbelief. Maybe we would get all the way to Panama by road after all.

We had carefully transferred many low lying items inside the camping van to the top of Noble's bed or other high level places prior to setting out across the river, to prevent them from getting wet. However, a few items - clothing, bedding, dollar bills in our safe, etc. - did get wet. It was a sunny morning so we simply spread these out in the sun on the riverbank, where most of them dried out in a couple of hours. This gave me a little time for ferreting out birds, and I was happy to observe and take notes on the appropriately named Riverside Wren*, and the Rosy Thrush-Tanager* (though I could not initially guess in what family it belonged).

As we started out once again on our journey, we were faced with yet another problem. The new roadbed under construction was 50 ft. above the riverbank in preparation for the bridge to be built. The only way to get up to the new roadbed was a narrow, very steep, dirt ramp on the side of the embankment, built for small, lightweight 4WD vehicles. Our first attempt to make this steep ascent failed when ¾ of the way to the top our 105 hp engine coughed and died, having exhausted all its power. It was necessary to back all the way down to the bottom and try again. This time I got as much of a running start as I could, and with the Jeep in 4WD and our lowest gear ratio, we just barely made it over the top before the engine once again coughed and died. It no longer mattered as we were now on the newly graded level roadbed, high above the river below. Luck was still with us.

The next 25-30 miles to Palmar Norte was a spectacular, scenic road under construction, "Project 4" of the American owned Foster-Williams Construction Company. The road surface was mostly soft earth, which necessitated that we be in 4WD and low gear most of the time. The "highway" was cut out of the northern bank of the Río General. We sometimes had to stop and wait for Caterpillar bulldozers that were grading the road. Fortunately, the workers did not seem to object to our presence, and were usually quite willing to assist us if we needed it. There were no other private vehicles on the roadway. We were surrounded almost entirely by rainforest.

At the insignificant little village of Palmar Norte, it was necessary to cross the Río General once again, for the third (and last) time. No bridge existed, and the river was quite deep here, certainly not fordable. Tied up along the riverbank was an antiquated U.S. Navy World War II "LCM" (landing craft medium), which was used in the war to land heavy equipment on Pacific islands. It was just big enough and sturdy enough for our Jeep. The problem was that it was owned by the construction company and was not designated for private transportation. Fortunately, we were able to locate a Mr. Wallace, an American engineer, who after first denying us permission for its use, changed his mind and said it would be OK. I backed the Jeep onboard, and for the last time we crossed the Río General, to Palmar Sur, a somewhat larger town on the south side of the river.

We headed southeast through the jungle toward the city of Neily, about 40 or 50 miles away, not very far from the Panamanian border. This section of road was entirely under construction, and mostly newly graded earth. Fortunately, it had not rained for eight days. Our speed of travel was 5-10 mph, and we frequently encountered and waited for road building machines of all kinds. Private vehicles were non-existent. Luckily, nobody complained about our presence, in spite of what we had been told. The road grader drivers were good-natured and did not mind constantly pulling us through the softest spots. Our situation, of course, was greatly complicated because of our excess weight and small tires.

About mid-afternoon, we encountered an unusually large, deep, soft area of road surface, and our vehicle sank to its axles, at the base of a small hill. We had broken the clutch on our winch while attempting to get on the raft, so our winch was useless. We were very thoroughly bogged in the soft earth. However, a big D4 Caterpillar road grader came along shortly thereafter, and the friendly driver agreed to help us out. He hooked a very heavy chain to our front bumper. At this point, I made an error of judgment, in my ignorance of the matter, and thought I could help by adding power from our engine, so I turned it on. It required all the strength of the road grader to pull us out and up to the top of the hill, where the road surface was firmer. As we reached the top of the hill, I glanced at the temperature gauge for our Jeep, as I constantly did while I was driving, and I noticed it was far into the red on the gauge—well above the boiling point! So I instantly turned the engine off.

When Noble and I removed the engine cover to see what the problem might be, we immediately noted that one of the fan blades for our engine cooling system had gouged a circular cut in the core of our radiator, and our cooling system had lost almost all its water. Both the fan and radiator were beyond use. This situation occurred because the force applied by the road grader to pull us out, from the front bumper, had caused the frame of the jeep to bend upward just enough that the top of the radiator was sent backward into the rotating fan. Had our engine not been running, this disaster would not have happened. The road grader driver was very apologetic, but it was not his fault.

We asked the driver how we could get back to San Jose, to try and replace the fan and radiator. He pointed to the jungle on the southwest side of the road, to our right, and said that not very far from the road was a railroad track, paralleling our road. Banana trains came along the track regularly and could take a person either northwest to Palmar Sur or southeast to Neily. Each town had an airport with regular flights to and from San Jose. Our road map showed that we were broken down somewhere in the general vicinity of the village of Piedras Blancas, on the railroad track. It began to rain. So there we were, with a broken radiator on a muddy road in the middle of the Costa Rican jungle! Our situation was almost humorous.

Noble and I discussed our problem. It was decided that one of us would go back to San Jose for a new, or at least different, fan and radiator. The other would stay with the van, to avoid vandalism. The solution was easy. Noble enjoyed people-to-people interactions and handling financial and mechanical transactions; he wanted to be in charge of all such matters. On the other hand, my pleasure came from experiencing the natural world, and from the challenge of finding and identifying birds. Therefore, Noble would travel back to San Jose and I would stay with the Jeep. He departed shortly thereafter, wearing only rubber sandals and bathing trunks, and carrying a leather briefcase with his clothes and papers. What a sight! I went to bed early, after eating a peanut butter and jelly sandwich and some olives. It was very hot and humid, which made sleeping difficult, plus I still itched uncomfortably in many places from some unknown insect bite or plant toxin.

I awoke at daybreak the next morning (Monday, Mar. 2) to the sounds of the jungle – birds, monkeys, frogs, and insects. The tremulous, melancholy whistles of the Little Tinamou* were all around me. It was very exciting. After a hasty breakfast, and clad in rubber boots and a lightweight Gray jacket to avoid mosquitoes and other insects, I set out with binoculars, camera, notebook, and butterfly net. For precautionary measures, I carried a snakebite kit and insect repellent. I really didn't know quite what to expect. I was all alone in the rainforest; it was a wonderful feeling. Birds were all around me, vocalizing or flitting through the trees and understory, but they were almost impossible to observe long enough for me to write their descriptions. The intense heat and humidity kept me constantly soaked from perspiration. My clothes never dried out, day or night, and my itching never went away. There were peculiar looking 20-inch long lizards with large backward projecting casques on top of their heads, but I never saw any snakes, large or small.

I spent one whole day walking up and down a dry stream bed not far from the road, which afforded me a path through the rainforest and a forest edge where birds could be observed somewhat more easily. Now and then a Blue-crowned Manakin* would buzz by so fast his wings were blurred, and a White-tipped Sicklebill* with its long, greatly decurved bill would hover momentarily within arm's reach, and then dart off again. I gasped when a White Hawk* suddenly glided low over a forest opening just above the canopy. An Ochre-bellied Flycatcher's* nest was hanging over a stream, and a banana frond had a hummingbird's nest attached to the underside of it. I tracked down a Rufous Piha* (in the cotinga family) which was calling loudly. Striped Woodhaunters* foraged on tree branches and in palm fronds. Fiery-billed Aracaris* fed on treetop fruits, and a White-throated Shrike-Tanager* appeared briefly in the canopy. Tiny lizards leaped from leaf to leaf, and small Grayish squirrels chattered from tree trunks. Tropical rainforests were exciting, with the greatest avian diversity of all terrestrial ecosystems.

Gorgeous, huge, glistening, pale blue Morpho butterflies bounced back and forth across sunlit openings in the forest, and along streamside edges. Now and again a Coati meandered along on the ground searching for almost any organism to eat, then ran noisily up the nearest tree trunk when it saw me. A bright red, locally endemic Baird's Trogon*, with a pale blue eye ring, was an exciting discovery in the forest midstory. The uncommon Long-tailed Woodcreeper* was described in my field notes, along with the somewhat more numerous Black-striped Woodcreeper*. Late one afternoon, two noisy Scarlet Macaws* flew very high over the forest, silhouetted against the sky. During the middle of the hot days, when bird activity declined in the forest, I kept active by catching butterflies with my net, and preserving them in paper envelopes. Afternoon rain showers were frequent. One of the most common birds in the forest was the Black-cheeked Ant-Tanager*, a bird endemic to the Golfo Dulce region of southwestern Costa Rica, with one of the most limited geographical ranges of any Central American avian species (as I eventually discovered). My brief field notes described the bird as: "tanager: sides of face blackish, dark brown above, Grayish brown below, rose pinkish throat."

About 4:30 pm on Mar. 4 (Wednesday), my third day in the forest, Noble suddenly appeared along the road, carrying a radiator under one arm and two fans under the other. He had just descended from a train that had stopped to let him off a half mile away in the forest. His mission had been a pleasant success, and he had delighted in talking with everyone he met. Inexperienced as we were, the old radiator was removed and the new one installed in a couple of hours, just before dark, with Noble doing most of the hard work. We washed our sweat, dirt, and grease off in a nearby stream. Supper was a can of tuna, bread with olive-pimento spread, walnuts, and toffee candy for dessert. Because of the oppressive heat and humidity, I slept naked on top of my bunk, as I had for the past two nights. Not since Tikal had I seen so many birds in such a short span of time. (These are all included in column three of Appendix A).

Daybreak the next morning (Thursday, Mar. 5) was cloudless. I walked a short distance into the forest for one last memory. Looking through my binoculars, I followed a Fiery-billed Aracari as it flew into the top of a nearby tree. I gasped at what I saw. There, basking in the sunlight on a large horizontal branch of the tree, about 100 ft. above the ground, was one of the most amazing creatures I had ever seen. It looked like a prehistoric reptile from the Mesozoic Era, about 7-8 feet long. It had a long row of narrow sharp-pointed spines sticking up dorsally from head to tail. Its head was grotesque in appearance, with a prominent dewlap hanging under its chin. I was in disbelief. Surely such animals had long ago become extinct. I called for Noble to come and look, and bring our cameras and long lenses. He, too, was quite astounded. Unfortunately, it proved impossible to obtain a satisfactory photograph, so I took time to elaborately sketch a line drawing of the creature in my little field bird notebook. Only much later did I learn that it was a Green Iguana (*Iguana iguana*), and was actually quite a common resident reptile throughout much of Central America.

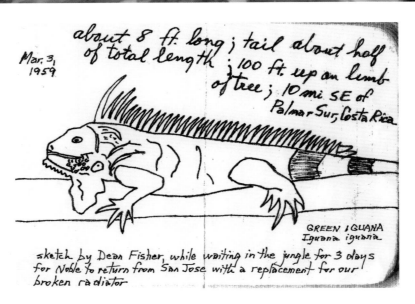

Mar. 3, 1959

about 8 ft. long; tail about half of total length; 100 ft. up an limb of tree; 10 mi SE of Palmar Sur, Costa Rica

GREEN IGUANA
Iguana iguana

sketch by Dean Fisher, while waiting in the jungle for 3 days for Noble to return from San Jose with a replacement for our broken radiator

Plate 4: We surveyed and then forded the wide, 3 feet deep Río Térraba in Costa Rica. Shortly thereafter we became stuck on a muddy, newly graded road in the Costa Rican jungle, with a broken radiator. Here I located and sketched a Green Iguana resting 100 ft. up in a very tall tree.

Noble and I set out on the road once again and soon discovered that our new radiator, which was not a Jeep product, was less efficient in cooling our engine than the original radiator (an extremely frustrating situation that caused us continual problems for a very long time). Our road was more of the same soft, muddy, freshly graded earth, mile after mile. It was very slow going and tortuous travel for us, and we frequently needed assistance from construction workers and road graders. Having learned from our serious mishap, we now had all tow chains fastened to our front axle rather than the front bumper. A good natured bulldozer driver once traveled along with us for several miles, to assist us whenever it was necessary, which was often! Our rate of travel was the pace of a turtle. Without assistance from the road graders, we would not have succeeded.

We eventually reached the village of Neily in the early afternoon. Here we encountered an American road engineer who told us that the new roadbed under construction from there to the Panama border, a distance of about 20 miles, was much worse than anything we had yet been over. This information was, to say the least, most disconcerting. However, the engineer said there was an alternate way to get to Panama, a "bypass" route leading up away from the coast into the mountains, to the Costa Rican village of Agua Buena, and from there to the Panamanian city of Volcán, in the mountains. A well-established road would then bring us back down to the Panamanian city of David on the Pacific coast, and from there the Pan-American Highway was complete to Panama City. He said we would have to ask directions to the border of Panama when we were in Agua Buena, but he did not mention the city of San Vito, in Costa Rica, as being on our route. Our map did not show this "bypass" road at all.

Noble and I discussed the situation and decided that any route at all would have to be better than 20 more miles of soft mud. Little could we imagine the adventure that lay ahead. Thus we turned north from Neily, up into the mountains toward Agua Buena and away from the very muddy, incomplete "Pan-American Highway" along the Pacific coast of southern Costa Rica.

The ascending mountain road was narrow, rocky, and steep, but it was dry and firm! As we approached the village of Agua Buena, about 4,000 ft. elevation, three Swallow-tailed Kites*, with very long, deeply forked tails, glided low across our road. What stunning birds! We stopped in Agua Buena and asked directions to the Panama border, as we had been advised to do. Of course there were no road signs of any kind. People were amiable and curious, as always. They looked at us and our Jeep in astonishment. Children were sometimes frightened and ran away. It was obvious that very few tourists, if any, had ever passed this way. One man confidently gave us some directions to "Panama", so that was the road we followed out of the village, as we had been instructed to do.

Less than an hour after leaving Agua Buena, we arrived at the Panama frontier, at 4:00 pm. Here there were two National Guard stations, one on each side of the border, facing each other. The two border personnel on the Panama side, at the gate, were unfamiliar with "tourist" vehicles and did not seem to know exactly what to do. A captain was roused from his bed inside the station. He came outside, casually surveyed our vehicle, stamped our passports, and motioned for us to continue. We had made it to Panama somewhere in the mountains, on an unmapped "bypass road" which we hoped was taking us to Volcán.

The road in Panama deteriorated almost immediately. It became not much more than a track for oxcarts. Almost all the tracks in the road were those of horses and cows and carts, only very rarely of a vehicle. Surely this was not the route the engineer had mentioned to us. Maybe we had been given the wrong road in Agua Buena. Nevertheless, we decided to continue rather than turn around and go back. There were deep ruts, and our sinuous track constantly wound back and forth up or down steep hills. It was very narrow in places and sometimes the roadside scrub brushed against both sides of our van simultaneously. We were in low gear and 4WD the entire time, proceeding at 5 mph. Our top-heavy jeep swayed precariously from side to side on the uneven ground. All four of our shock absorbers had long since been broken, but we no longer were worrying about this situation. Bridges over small streams were constructed of crude rotting logs, and on several occasions we chose to ford the stream rather than trust the bridge. Our only consolation was that so far the dirt road surface was relatively firm and dry.

This situation soon changed when we experienced a small shower, at which time our road surface became temporarily wet and exceptionally slippery. I was very worried we would slide off the track and down a steep hillside. It required several attempts for me to get to the top of some of the hills, backing down and then starting up again. We were not having any fun! There were no other vehicles on the road, or even any oxcarts, only occasional pedestrians or people on horseback, all of whom stared at us in wonderment. Children hid behind their mother's skirt. The countryside was devoted predominantly to cow pastures, with occasional patches of bananas or natural scrub. Houses were wooden, usually raised somewhat above the moist ground, with thatched roofs. By nightfall, we were still quite some distance from where we thought Volcán might be. We certainly did not want to drive after dark so we camped for the night, somewhere in the highlands at 4,500 ft. elevation. We hoped adamantly that it wouldn't rain again.

The next morning (Friday, Mar. 6) a Scarlet-thighed Dacnis*, Silver-throated Tanager*, and Yellow-throated Brush-Finch* were attractive little birds foraging around our van. Several men walked by carrying pails for water, to be gathered from a nearby stream. Two young girls were already there, washing clothes even before the sun was up. They giggled as I walked past them. I greeted people with "Buenos días", and they would return my greeting, sometimes with just "Buenos", or simply with "Días."

We finally reached Volcán in mid-morning, at an elevation of 4,000 feet. We breathed a huge sigh of relief. The most difficult section of our "bypass road" was behind us. We never found out whether there was another route or not between Agua Buena and Volcán, longer but more passable, perhaps through San Vito. The local resident in Agua Buena who confidently pointed out to us the "road to Panama" had no doubt traveled it on horseback, or in an oxcart, but probably never in a motorized vehicle.

We turned west from Volcán and began descending from the mountains toward David, on the Pacific coast of western Panama. Initially the road was steep and a little muddy, but it gradually improved as we progressed. By noon, we were back on the Pan-American Highway, just outside David. Happily, there was no more new construction. We filled our gas tank and calculated that our average miles-per-gallon since the last fill-up, 36 gallons ago, was 6 miles per gallon!

I pointed the van eastward toward Panama City, for the very last leg of our "impassable road" between Costa Rica and Panama. Our highway now, as we began, was gravel and badly corrugated in places, but the surface was firm and became smoother as we progressed. For the first time in two weeks we made very good mileage, and by nightfall we were ten miles east of Santiago. We camped by a river and finally, after six days, we took the jeep out of 4WD.

Roadside birds the next morning (Saturday, Mar. 7) included a Yellow-headed Caracara* and flocks of Brown-throated Parakeets*. Our road changed to an excellent paved highway 70 miles from Panama City. That afternoon we drove across the Miraflores locks of the Panama Canal and into Panama City; 42 days and 5,500 miles after departing from Los Angeles. We had conquered the "impassable" Pan-American Highway entirely by road, all the way from the U.S.A. to the Panama Canal, never putting our Jeep on either a railroad car or a coastal ferry. According to Mr. Tomas Guardia, the Chief of Panamanian Public Roads with whom we spoke personally the next day, we were, to his knowledge, the very first travelers ever to accomplish this feat! That night Noble and I celebrated with a bottle of wine.

South America was our next destination. Although eastern Panama and Colombia are connected by a narrow stretch of land, the Darien Gap, no navigable road existed through this mostly undeveloped, sparsely populated region in 1959, from one country to the other. (To this day, it still doesn't!) Therefore, to travel from Panama to South America it was necessary to go a short distance by coastal transportation (boat, barge, or ship) either in the Pacific or the Atlantic (Caribbean Sea). So Noble immediately began searching for the most economical way for us to get to South America. He talked with steamship agencies, ferryboat companies, and even tugboat and oil barge offices. We drove back and forth between Panama City and Balboa (on the Pacific side of the canal), and Colón and Cristóbal (on the Atlantic side). Everything was too expensive. Finally Noble succeeded, and he booked us and our Jeep from Balboa to La Guaira, Venezuela, on the "Reina del Mar", a British luxury passenger liner of the Pacific Steam Navigation Company, departing on the morning of Mar. 18, and arriving on the morning of Mar. 21. It was en route from the west coast of South America to London, with brief stops in Balboa, Cartagena (Colombia), the island of Curaçao, La Guaira, and beyond. The total cost for us and our van was only $320!

Our departure date gave us quite a few days to spend in Panama, both for business and pleasure. So we had some of our Jeep problems repaired and went shopping, sight-seeing, and nightclubbing (once or twice). We ascertained that the three most frequent responses to our many questions were "No sé" (I don't know), "No hay" (there isn't, or aren't, any), and "Mañana" (tomorrow). Once again, we found the local people to be both curious and friendly. We were treated to dinner, on a couple of nights, by a very personable man who managed the Braniff Airways Office in Colón.

Since Noble didn't need my assistance while he was busy talking with shipping and boating companies, I decided to spend three days (Mar. 9, 10, & 11) at a biological reserve and scientific research station on Barro Colorado Island, run by American institutions. This small island was situated in Gatún Lake, a man-made body of water in the Panama Canal that was created when it was built. I was thus afforded another opportunity to observe birds.

As previously, I wrote descriptions of the birds I encountered (whenever they stopped moving for a few seconds). The book by Sturgis was very little help. A White-whiskered Puffbird* sat motionless in the lower forest midstory for a very long time, allowing me ample time to take extensive and detailed notes. I described almost every feature of the bird, not knowing which character might be the one to separate it from its nearest relative. As I later discovered my considerable effort had been unnecessary, and less detail would have been sufficient for identification as there was no other Panamanian puffbird similar in appearance. Contrastingly, I took very few notes on a "pigeon" which I saw, also perched low in the forest, because it possessed very few features which I thought could be definitive. My brief description was "pigeon, head and chest pale pinkish Gray; underparts white; wings deep maroon-chestnut; legs red; tail very short, maroon-chestnut." When I finally identified this bird, years later, I was pleasantly surprised to conclude that it was a Violaceous Quail-Dove*, which was considered to be rare and local in Panama.

Black Vultures and Orange-chinned Parakeets were the two most numerous species, largely because they could easily be seen as they flew above the forest canopy. Within the forest, 3 species of "antbirds" (family Thamnophilidae) were frequently sighted: the Western Slaty-Antshrike*, Checker-throated Antwren*, and White-flanked Antwren*. These birds foraged in pairs or small groups in the densely foliaged forest understory or midstory. Larger forest birds included the Rufous Motmot*, Chestnut-mandibled Toucan*, and Crimson-crested Woodpecker* (a large species in the genus *Campephilus*). Hummingbirds were the most diverse family with 7 recorded species, including the striking White-necked Jacobin* with its blue head, green back, white collar, white belly, and almost completely white tail. Blue-Gray Tanagers were abundant at the forest edge and in the more open areas outside the forest. Column 3 in Appendix A denotes with an X^2 symbol all of the 80 species I eventually identified on Barro Colorado Island.

On Tuesday, March 17, our overweight camping van was hoisted on board the 600-ft. long "Reina del Mar" by a very sturdy dockside crane in Balboa, and placed in a hold for the 3-day voyage to Venezuela. We departed early the next morning (Wed., Mar. 18). It required almost the entire day, 8 hours, to travel the 50-mile length of the Panama Canal. Surprisingly, the canal heads in a generally northwesterly direction as one travels from the Pacific Ocean to the Atlantic Ocean (Caribbean Sea). The ship passed through three locks - the Miraflores, Miguel, and Gatún - rising in the first two and dropping in the third, and it followed the canal through Gatún Lake. A Black Hawk-Eagle* soared leisurely above the lake, giving me just enough moments to sketch it in my little lined notebook, for future identification (see illustration). The South American continent lay ahead!

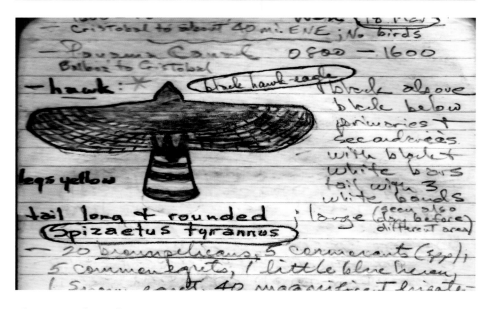

Plate 5: We drove from Costa Rica to Panama on an umapped "bypass road" through dense jungle at 4,000 ft. elevation. While we were navigating through the Panama Canal on the "Reina del Mar" I sketched a soaring raptor, which I later identified as a "Black Hawk-Eagle."

Chapter Five – Venezuela

Noble and I sailed from Panama to Venezuela on the "Reina del Mar" for 3 days and nights (March 18, 19, & 20), stopping briefly on the way in Cartagena, Colombia, and in Willemstad on the Dutch island of Curaçao. As shipboard passengers we enjoyed the pleasant change of three square meals a day, including wine, fresh bread, and butter. Seabird watching on deck was not very productive. The winds were always high and the water was continuously covered in whitecaps, making seabirds almost impossible to distinguish against such a background. I observed only a single species of procellariiform, the Audubon's Shearwater, just two separate individuals. However, I enjoyed watching the numerous Magnificent Frigatebirds, one of my favorite species with their long, narrow, pointed wings and their effortless gliding. Of several jaegers that flew past the ship, I concluded that at least one of them was a Pomarine Jaeger, although none of them had their central tail feathers elongated. My reference for identification of seabirds was *Birds of the Ocean* by W. B. Alexander, published in 1954.

On Saturday morning, March 21, we docked at La Guaira, Venezuela. The nation's capital city of Caracas, with a population of about one million people, was situated eleven miles inland at an elevation of 3,000 ft. in the coastal mountains. Owing to the country's oil industry it was a very wealthy city, as reflected by its great many expensive high-rise buildings. The highway connecting La Guaira to Caracas was a very modern, six lane "autopista" which had cost 60 million U.S. dollars to build! Noble and I spent several days in this capital city. On our first day there, I set out early in the morning by bus for the Caracas Country Club and its park-like surroundings, excited at my first opportunity for pursuing birds in South America. Identifying birds was going to be a challenging undertaking as I had no books or references of any kind for the birds of Venezuela (or for that matter anywhere else in South America, north of Chile).

Birds were everywhere in the flowering trees; singing, foraging, and chasing one another. Many of them were species I had never seen before. I began describing them in my notebook as fast as I could – those that would sit still long enough! It was a daunting task. In several hours at the country club the new birds I recorded (and subsequently identified), in the sequence in which they were recorded in my notebook, were: Saffron Finch* (one of the most common species), Stripe-backed Wren*, Carib Grackle*, Oriole Blackbird*, Barred Antshrike (my first male), Pale-breasted Thrush*, Lesser Seed-Finch*, Yellow-headed Parrot*, Yellow Oriole*, Scaled Piculet* (see notes below), Common Thornbird* (identified in part by their conspicuous long, hanging stick nests placed at the ends of low branches), Shiny Cowbird*, Green-rumped Parrotlet*, Lazuline Sabrewing*, Sparkling Violetear*, and Black-and-white Becard* (a common species). My field notes for the Scaled Piculet read as follows: "Furnariidae?: very small (about 3 ½ in), exceedingly short tail; upperparts Grayish-brown, back spotted, wings edged with buff, top of head blackish finely spotted with white, underparts dirty white scalloped with black (at least on breast); tail black middle feathers white." I drew a sketch of its short, pointed bill. When I finally identified this sighting, I was surprised how hopelessly wrong I was about the family, which I had guessed at for this enigmatic little "woodpecker." In addition to the above new species, I recorded two dozen birds with which I was already familiar, such as the Ruddy Ground-Dove (very common), Collared Trogon, Great Kiskadee, Green Jay (always a pleasure to see), Red-legged Honeyeater, Bananaquit (ubiquitous), Grayish Saltator, Blue-black Grassquit, and wintering Yellow Warblers.

Late in the morning, I left the country club and walked several miles to a rather dry, wooded hillside at the outskirts of the city. Here I followed a small stream upwards for a mile or more, once again taking notes on new birds for my list. These were Cocoa Thrush*, Plain-brown Woodcreeper*, Buff-fronted Foliage-gleaner*, Sooty-capped Hermit*, Variegated Flycatcher*, Rufous-vented Chachalaca*, Strong-billed Woodcreeper*, Pectoral Sparrow*, and Glittering-throated Emerald*. A Squirrel Cuckoo, which came slinking through the midstory, was great fun to watch.

Many colorful butterflies brightened the sunny spots along the stream but, alas, my net was in the van. Finally, I walked all the way back to the hotel, arriving about 4:00 pm, with very tired, sore, aching feet since I had to wear my leather soled street shoes all day as my regular field shoes were locked up in our Jeep waiting to be cleared by Customs.

Noble had enjoyed his day by taking long walks on many of the downtown avenues, talking to, and exchanging viewpoints with, almost everyone he met; taxi drivers, flower vendors, shopkeepers, and businessmen in coffee shops. His two main topics, as always, focused on the politics and economics of the country and provided him with new knowledge and insight, and inevitably with controversy and criticism of American policy. He inquired at the Willys agency if they had a new radiator to fit our Jeep. Unfortunately, they didn't.

One afternoon I visited the downtown office of the Creole Petroleum Corporation, the largest oil company in Venezuela. Its president was an American businessman, Dr. William H. Phelps, who had lived in Venezuela for many years and was the country's foremost ornithologist, well known and respected internationally. At the corporation office I was given a free copy of a small paperback publication, *Aves Venezolanas, Cien de las Más Conocidas* (Birds of Venezuela, 100 of the best known), published in 1954 and written entirely in Spanish by Kathleen D. Phelps, the daughter-in-law of William H. Phelps. There were rather simple colored illustrations, by the author, of 92 species (17 of which bred in the United States). These were my only paintings of any birds in South America north of Chile! I was invited to visit the very large Phelps estate on the outskirts of the city, where I could observe birds and examine avian specimens in Dr. Phelps' extensive private collection.

I took a taxi to the estate and met Dr. Phelps, who was 84 years old at that time. I was welcomed as an esteemed visitor. He presented me with two of his major publications on Venezuelan birds: *Lista de las Aves de Venezuela y su Distribución, Passeriformes*, published in 1950 with 427 pages, and *Lista de las Aves de Venezuela y su Distribución, No Passeriformes*, published in 1958 with 317 pages. Both were co-authored by his son, William H. Phelps, Jr., whom I also met. These two paperback publications were entirely in Spanish and without any illustrations. The 1950 volume was autographed by Dr. Phelps on the front cover of the copy he gave to me, and signed by him. It is one of my most treasured bird publications. I spent several hours looking through many of the trays of bird specimens in the Phelps collection, attempting to identify some of the Neotropical birds I had recorded in my field notebooks, from my day in Caracas and from Panama. My success was not very great, and of course the only names on the labels were Latin names of the genus and species.

I returned to the Hotel Rio Azul late in the afternoon via bus (with some difficulty) and joined Noble who was there with our Jeep, having finally recovered it from Customs. The American Embassy had arranged for us to be interviewed that evening on Caracas TV channel 2, for 10 minutes as part of a weekly "de pueblo a pueblo" (from town to town) program. We spoke mostly in English, which was then translated into Spanish by a translator, and occasionally in our broken Spanish.

The next morning, Mar. 24, there were a few errands Noble wanted to take care of in the city, including purchasing some new tools to replace those which we discovered had been stolen from the cab of our van somewhere during its transit from Panama to Venezuela. So I visited the Phelps estate again, for another two hours of birding. New species which I recorded were the Bare-eyed Thrush*, Silver-beaked Tanager*, Orange-billed Nightingale-Thrush*, White-lined Tanager*, Ochre-breasted Brush-Finch*, Fulvous-headed Tanager*, and Rufous-browed Peppershrike*. I spent quite some time tracking down this latter species as it sang continuously while hidden in the tree tops, a sweet melodious tanager-like song. My written description, after I finally located the singing bird, read as follows: "cotinga? About 5 ½"; bill broad at base, tail medium short, head Grayish, patch through eye chestnut, underparts whitish with throat yellow, upperparts olive-green." I was able to identify this bird from the colored illustration in Kathleen Phelps' booklet, and it came as somewhat of a surprise to me to discover it was a member of the Vireo family. I said goodbye to all of the Phelps family and thanked them for their kindness and hospitality.

Prior to leaving Caracas, just before noon, we stopped by the American Embassy one last time, so that we could be interviewed, once again, by a newspaper reporter and photographer. We drove out of the city just as the two-hour midday siesta was beginning. At last, we were on the road in South America! It was almost exactly two months since our departure from Los Angeles. Our excitement level rose. We were headed for the southern tip of the continent, at Punta Arenas, Chile, on the Straits of Magellan.

The autopista heading west from Caracas climbed upward over a mountain pass at 5,000 ft. in the "Cordillera de la Costa Central", then down to the city of Valencia. It had four lanes with new pavement, enabling us to travel at 50 mph, which was a luxury for us. At Valencia, we had a choice of two routes to the Colombian border, one on either side of the "Cordillera de Mérida" range of mountains (which were an easternmost extension of the Andes). We chose, initially, to take the southern (left hand) route, through San Carlos. Both routes, often considered spurs of the Pan-American Highway, would eventually arrive at San Cristóbal in western Venezuela, a short distance from the Colombian border.

The countryside was dry, brown, and mostly barren. It was predominantly cattle ranches. Naked children played in front of crude, wooden, one-room houses. People watched us go by in a somewhat disinterested manner, though our noisy electric generator mounted on the top front of our camping van attracted their attention. Roadside birds included White-tailed Kites and Crested Caracaras. We made good mileage and passed through San Carlos by dusk.

Ten miles beyond San Carlos, we entered an area of open, second growth, dry woodland where we decided to camp for the night, about 100 yards off the highway. It was so hot that I decided, for the first time in all our travel, to sleep outside on top of the flat roofed camping van. My sleeping bag was employed as a mattress and because of numerous mosquitoes, I rigged a mosquito net over me, using poles I cut from the woody vegetation, and my telescope tripod, to support the net. This sleeping arrangement worked very well and I would subsequently utilize it frequently on hot nights. Noble preferred to sleep inside, no matter how hot it was.

As the full moon came up, I heard some loud, harsh sounds coming from the open woodlands not far away. I pursued these, carrying my binoculars with me, and soon encountered a group of four Double-striped Thick-knees* running rapidly on the ground in front of me, through the open woodlands in the moonlight. What a thrilling experience! I was exuberant. Here was a new family of birds for me, pan-tropical in their worldwide, local distribution. I recorded these in my notebook as follows: "about size and shape of a lapwing; white patches in the primaries (?); feet extended slightly beyond tail [in flight]; wings rounded; aroused from a dry barren area; notes loud and harsh." I returned to the van and climbed back to my bed on top, where I happily went to sleep listening to the constant calling of Pauraques. It was truly the adventure of a lifetime.

A great variety of birds were singing the next morning at daybreak (Mar. 25) but there was no time to record any notes, as we needed an early start for our day's travel. I drank a can of tomato juice and ate an orange, which Noble graciously peeled for me, and then we were on our way. Our good paved highway alternately passed between cattle country and tropical deciduous woodlands. Long, pendulant nests of caciques and oropendolas hung from the outer canopy of tall roadside trees, and bulky stick nests of an unknown furnariid species were conspicuous in the lower branches of small trees at the woodland edge.

Throughout Venezuela, and elsewhere in South America, there were frequent vehicle inspection posts along the highway, by both local and federal policemen. Often we would be motioned onward without being asked to stop, but quite regularly we were also required to stop for interrogation and inspection of documents and vehicle. We had to show our passport and tourist card, and the inside of our van would be searched for prohibited items. One person looked through my binoculars, back to front! The attitude of the officials varied from friendly to brusque and antagonistic. Noble and I were both shirtless because of the very hot morning and at one stop the officials were very irritated by this fact and instructed us to put our shirts on, saying that not wearing a shirt was prohibited by law. Of course we complied but were amused by the fact that small children ran around completely naked.

As we drove through Acarigua, we stopped for 15 minutes to repair a flat tire and buy a large bunch of bananas. We continued westward toward the city of Barinas, along a paved highway traversing an area of hot, tropical lowlands with cattle ranches and patches of deciduous forest. En route, I saw my first King Vulture*, a magnificent large black-and-white soaring bird, which is arguably the most attractive of the New World vultures. While we were stopped for lunch along the roadside, I spotted a bird perched conspicuously not far away on a bare branch in the top of a tree. After studying it carefully in my binoculars I was quite puzzled as to what it might be. I wrote in my notes: "about size and shape of a martin but head rather crested & tail rather short and square (perhaps very slightly notched), bill slender & pale (reddish?); all very dark (black glossed with purple?) with posterior underparts lighter (brownish?); sat upright in top of a tree." While this description was not completely accurate in all details, it allowed me to identify the species, quite some time later, as a Swallow-wing*, a most unusual member of the puffbird family, Bucconidae.

At the city of Barinas, we decided to take a more scenic route to Colombia, which would take us up into the Andes through the city of Mérida, then down the other side to the northern spur of the Pan-American Highway, along the southern edge of the Lake Maracaibo basin. So we turned right onto a good gravel road and began climbing upward all the way to a pass at 12,000 feet, and then down slightly into the town of San Rafael. The scenery at times was spectacular, with very steep mountainsides, sparsely vegetated valleys, rivers, and long, ribbon-like waterfalls. Grassy hillsides were dotted with small tin-roofed adobe houses. Occasionally we passed through clouds where our road was barely visible. Indians trudged along the road wearing colorful red-and-black ponchos for warmth. We offered a ride to an old woman carrying a very heavy sack of corn on her back. She was with her young granddaughter, who was frightened by our bearded appearances and refused some crackers that we offered her.

As we approached Mérida, the countryside became wetter, with roadside forests and streams. There were stone fences and the adobe houses had red-tiled roofs. Birds were scarce, but Great Thrushes* and Chestnut-capped Brush-Finches* were two roadside species I had not seen previously. We passed through Mérida at dusk, stopping only long enough to fill up our gas tank at a Creole station. The man in a vehicle behind us came up to us and said he had seen us on TV in Caracas! We continued down the mountain after dark for several hours, finally picking a campsite along the Mucuchíes River at an elevation of 1,300 feet.

The next morning (Thursday, Mar. 26), we decided to have a bath and wash our clothes prior to getting underway, in a clear running side stream coming into the river. After I finished hanging up my wet clothes there was some time available to me for birding prior to our departure. I was rewarded with my first look at a very approachable Snail (Everglade) Kite*, which surprisingly I had not encountered previously anywhere in Central America (or in Florida). Leaving our campsite, we drove the short distance to the city of El Vigía, where we joined the paved northern spur of the Pan-American Highway, and then followed it westward along the southern edge of the Lake Maracaibo basin. Our plan for the day was to reach the frontier of Colombia before dark.

Patches of moist tropical forest bordered our highway. I persuaded Noble to let me pull off the road and stop for an hour to look for birds in a particularly dense patch of forest, while he caught up on his journal notes. Brightly colored Crimson-backed Tanagers*, red with a shiny silver lower mandible, were very common and conspicuous. Finding birds in the "jungle" was a difficult task, although Black-chested Jays* were relatively easy to observe. Other species present included Blue-crowned Parakeets*, Citron-throated Toucans*, Black-mandibled Toucans (often separated from the chestnut-mandibled populations), a pair of White-necked Puffbirds*, and a lovely Blue-necked Tanager*, one of the many vividly colored species in the endemic neotropical genus *Tangara* (which became one of my most favorite genera as we journeyed through Central and South America).

From the Maracaibo basin, we climbed back up into the Andes on a gravel road toward the city of San Cristóbal. As was characteristic of mountain roads almost throughout Latin America, there were many small altars and shrines along the roadside, with flowers and burning candles. Small crosses beside the road designated where a person had lost his or her life, usually from a vehicle accident. The road was narrow and winding with hairpin curves and very few stretches where it was safe to meet or pass another vehicle. We were in low gear most of the time, going up or going down. Unidentified swifts and swallows twisted, turned, glided, and dived high and low over the mountainsides, in and out of the clouds.

Plate 6: In Venezuela we journeyed SW from Caracas to Cúcuta (Colombia), traveling within a NE extension of the Andes. We washed clothes at a river and I caught colorful butterflies at one of our campsites.

We reached San Cristóbal at an elevation of 3,000 ft. about 5:00 pm, with the late afternoon sun not far above the horizon. It was less than an hour on the paved Pan-American Highway to the Venezuelan/Colombian border post of San Antonio.

When we arrived at the border, it was after working hours and the personnel had all gone home. However, the border was not barricaded and traffic could proceed. There was no one in the small Venezuelan immigration office, and loudly sounding our horn did not cause anyone to appear. We wanted to continue on to the Colombian city of Cúcuta, about 20 miles inside the country, so without carefully considering the situation we decided to just proceed onward along the highway, into Colombia.

Shortly thereafter, we came to a Colombian check post where we were motioned to stop. Here the border official was not familiar with travelers from the USA and we were told that it would be necessary to pay the equivalent of 1,000 U.S. dollars to bring our vehicle into the country. We protested and presented our carnet de passage, the function of which was to prevent such entrance costs. Another official was summoned who, fortunately, was more knowledgeable, and after examining all our documents, he confirmed that we did not have to pay such a fee since we were "turistas" (tourists). However, it was necessary for us to complete entrance requirements at the customs office in Cúcuta, 20 miles away, from the "Capitán del Puerto" (captain of the port) upon our arrival there. A border guard from the check post rode with us to show us the way, for which we had to pay him 10 pesos. Cúcuta was at an elevation of only 1,000 ft. above sea level.

We arrived long after dark, about 9:00 pm, and not surprisingly the customs office was closed for the night. Since we did not yet have official permission to continue on our journey it was necessary for our vehicle to be impounded for the night, inside a locked, fenced enclosure at the customs office. We were given permission to sleep inside our van, and were told that the Captain would arrive at his office at 7:30 the following morning.

Not wanting to wait in a long line, we arose early the next morning and arrived at the customs office by 7:00 o'clock. It was Good Friday, Mar. 27. We were too late, and a long line had already formed. We felt that our business, as foreign tourists, might be given some priority, but that didn't happen. The Capitan did not arrive on time, and finally showed up at 9:00 o'clock! Apparently, this was quite normal. (His military rank was actually only that of a lieutenant.) When we were finally admitted to his office he critically scrutinized our passports and discovered, correctly, that we had no "salida" (exit) stamp from Venezuela. We told him that there were no persons at the border post of San Antonio to stamp our passports. He explained to us, quite emphatically, that he could not permit us to enter Colombia until we had officially checked out of Venezuela. Our protests were to no avail. It would be necessary for us to return to the border. (Of course we should have foreseen this when we boldly proceeded past the checkout post late yesterday afternoon in our impatience to continue a bit farther in our travel.)

So there we were, driving from Colombia back to the border of Venezuela to ask for an EXIT stamp from Venezuela. I became very amused at the thought and laughed aloud as I pondered just how we would explain our situation to the Venezuelan authorities. We walked into the border post and I said authoritatively that we needed a stamp in our passports so that we could LEAVE Venezuela. Of course, the officials thought I had made a mistake with my Spanish and said that what we needed was a stamp to *enter* the country. So I carefully explained what had happened, and showed them our entry stamps from La Guairá. Much to our pleasant surprise they, too, found some amusement in the situation and gladly consented to give us each an exit stamp. A final problem occurred when they could not find their appropriate stamp, so they improvised with 2 other stamps which together gave the word "salida" with the date under it, and the passports were then signed. We now officially left Venezuela and drove back to Cúcuta, once again.

Chapter Six – Colombia

We returned to the customs house in Cúcuta, where it was necessary to stand again in a long line until we could see the "Capitan." He quickly stamped all our documents and issued us the necessary permits for travel in Colombia. Shortly after noon, we departed the city on the paved Pan-American Highway and began climbing south and west in the Eastern Andes toward the city of Bucaramanga. The highway constantly wound back and forth, ascending then descending the dry, steeply sloped mountains between 3,000 and 11,000 ft. elevation, in and out of the clouds. We drove at an average speed of 20 mph, mostly in 2nd gear. The grassy mountainsides were generally brown, but there were scenic waterfalls and picturesque valleys interspersed with occasional patches of green forest. Birds were scarce at the higher elevations, except for Great Thrushes, which were very common along the roadside wherever there was some woody vegetation. Indians walking along the road wore black hats, (both men and women) and brightly colored blanket-like "ruanas" (capes) draped over their shoulders and upper body for warmth. In the towns and villages, people were dressed in their finest black suits and dresses, since it was Good Friday, and large crowds were gathered around churches and in the principal squares. We purchased a bunch of green bananas at one of the markets. Along the highway there were the usual "reténs" (police check posts), where we were occasionally asked to stop and show our documents.

We descended to 3,000 ft. and entered Bucaramanga an hour after dark, just before 8:00 pm. It was a fairly large city with a population of about 250,000 people. Our plan was to let Noble look around town for an hour while I took a nap in the front seat, being tired after driving in the mountains all afternoon. However, Noble no sooner left the van than a crowd of curious people gathered around to look at the Jeep and stare through the front window at me. Children climbed everywhere over the truck, while most of the adults just peered and said nothing, although I tried to converse with them a little in Spanish. Finally, I drove off, thinking I could find a somewhat more secluded spot until it was time to pick up Noble. Yet, when I parked at another location, it was not long before a crowd gathered around the van once again. Since I was not getting any rest I decided to drive around the city streets slowly, looking at the sights.

Bucaramanga was an old city with many ancient buildings and narrow, poorly lit streets, many of them one-way, inconspicuously indicated as such. As I turned down a narrow street, a policeman standing in the shadows blew his whistle loudly at me. I stopped, wondering what I might have done wrong, and I climbed down from the cab as he approached me. No other persons or vehicles were anywhere around. The policeman was obviously very annoyed at me and he pointed to a small, dirty, white arrow high above the street painted on the corner of a building, indicating that the street was one-way and I was going the wrong way. He acted as if this were a very serious offense, and he pulled a narrow citation booklet out of his pocket and began writing me a ticket. He did not look at the English on the sides of our van or our California license plate, or ask to see my driver's license. In a friendly, non-confrontational manner, I apologized, in my limited Spanish, and explained that I was an ignorant tourist and promised never to do it again. He was not sympathetic and continued writing the ticket. Suddenly a thought occurred to me, from out of nowhere, and I said to him "Pero yo vengo de Cuba" (but I come from Cuba). He could see my beard, of course. He stopped writing, looked directly at me, and queried "Cubano?" (Cuban?). In a very confident manner I responded "Sí, soy Cubano" (yes, I am Cuban). Without another word he put the ticket booklet back in his pocket and with a wide sweeping movement of his arm he motioned for me to go on down the street, in the wrong direction, saying "Siga no más" (an idiom for "go right ahead"). It was obvious that if a person were from Cuba he could drive the wrong way on a one-way street in Bucaramanga! Fidel Castro, and Cubans in general, were very popular throughout Latin America in 1959. I still smile broadly whenever I think of this incident. Soon thereafter, I picked up Noble and we drove a short distance outside the city to a mountain stream where we camped for the night. I was tired from the day's drive and went straight to bed.

The next morning (Sat., Mar. 28), we bathed and washed our clothes in the stream. Birds in the vicinity included a small group of diminutive Spectacled Parrotlets* and several Black-faced Tanagers* (*Schistochlamys melanopis*), a handsome, unmistakable-looking somberly hued Grayish tanager with the entire front of its head black. After a can of tomato juice, we were underway again, with a destination of Bogotá. Our Pan-American Highway was no longer paved, but gravel, and it climbed upward from 3,000 feet elevation through green steep-sided valleys with sugarcane fields and narrow high waterfalls to a dry grassy plain ("puna") at 10,000 feet. It then curved back downward all the way to an arid valley with thorn-cactus vegetation at 2,500 feet, where two Bicolored Wrens* flew across the road. This species is in the genus *Campylorhynchus*, containing the largest members of the New World wren family. Horses and burros carried heavy burdens of firewood, water jugs, or rocks. Children hid behind their parents as we drove past them. Black hats, worn by both men and women, were still the custom, as were adobe houses.

Our travel was slow and it was not until after dark, at 8:00 pm, that we reached Tunja, a city with many narrow cobblestone streets and a population of approximately 50,000 people, situated at an elevation of 9,000 ft. in the Andes northeast of Bogotá. We were unprepared for the events which were about to follow. Simply wishing to purchase a loaf of bread on our way through the city, we stopped along one of the main streets, somewhere outside the city center, at a small store stocked with groceries, beer, and other items. We walked into the store together, bought a loaf of bread, and headed back toward our van, parked in front of the shop. In the corner of this small shop was a little wooden table around which four men were sitting drinking beer. There were many empty bottles on the table, indicating the men had probably been there for quite some time. The men were all small in stature, of native ancestry, and very modestly dressed, each covered with a heavy "ruana" (cape) to keep out the cold night air. They looked at us with interest and one of them called out "Amigos, vengan aquí" (friends, come here), motioning for us to join them. Noble and I hesitated and then decided we would join them to see what they wanted. They pulled up two chairs for us and ordered each of us a beer.

It was not easy trying to engage them in conversation, because there was not only a language barrier, but also they were all quite intoxicated. Noble and I finished our beers then repaid their hospitality by ordering another beer for each of them. They asked us where we were going to spend the night, saying that there were some good hotels in town. We responded that we had our "hotel" with us and would drive a short distance from the city to a suitable place where we could camp for the night. This answer seemed to irritate at least one of them, who told us that it would be dangerous to do so because of "bandidos." When we had finished our beers, Noble and I got up, thanked them for their friendship, said goodbye, and walked outside to our Jeep, which was parked with the passenger side next to the sidewalk. Noble said he would drive, for a change. He got in first and then I opened my door, climbed up to my seat, and pulled on the door to shut it. To my surprise the door would not shut. I looked to see what the problem might be and I saw that the door was being held open by one of the men from inside with whom we had been drinking. He said to me "Venga conmigo" (come with me). I queried "¿Por que?" (Why?). He only repeated "Venga conmigo." I replied "No, Señor." I knew he was drunk and I saw no reason why I should go anywhere with him. At this point, he pulled a very tattered, frayed little card out of his pocket and held it up in front of my face. The only words I could see clearly, at the top of the card, said "Intelligencia de Colombia" (Colombian Intelligence). I thought to myself that surely this person could not be any kind of federal agent. So I said to Noble that I was going to shove him away from the cab and shut my door, and he should immediately drive off, leaving him behind. Noble agreed to this, so I gave the little man a very vigorous push, which caused him to fall backwards and land on his bottom at the side of the street. Noble quickly drove away.

We thought the incident was finished. What we didn't know, however, was how to get out of town. As usual, there were no road signs providing highway directions. We arrived at the city square and stopped to ask directions. We were told it was necessary to go back almost the same way we had just come. So we returned, and ten minutes later we were back to a location not far from the store where we had purchased the bread, heading along a principal two lane street which was taking us out of town, toward Bogotá. To our great surprise, and horror, here came the four men from the shop, walking together down the middle of the street directly toward us. I'm sure they were as surprised as we were. There was no way to avoid them. Our initial plan was to slow down, but not stop, thinking they would move out of our way. But, as we got close to them, the man whom I had shoved down at the side of the street reached under his ruana, pulled out a pistol, and pointed it directly at us. We were perfect targets sitting in the cab behind our big wrap-around windshield. So, we stopped.

The angry little "federal agent" came up to my door, opened it, put the point of his revolver against the side of my head, cocked it and said, once again, "Venga conmigo." Obviously, this was a time for rational reasoning, not for innate stubbornness. However, against all logic, I once again repeated "No, Señor." Noble instantly screamed out, in both English and Spanish at the same time, "Don't shoot, I'll go with you." He jumped out of the van and ran around to my side, where our adversary still had his pistol against the side of my head. By extremely good fortune an American geologist by the name of Sam Pratt came driving along the street in his Volkswagen at that very moment. Sam lived in Bogotá, where he worked for Texaco. He saw our situation, stopped, and asked if he could be of assistance. He spoke excellent Spanish.

The pistol was slowly lowered from my temple. Sam talked to the man and ascertained that he was, in fact, an authentic federal agent, and that he wanted us to accompany him to the police headquarters, so our passports and tourist cards could be inspected. Sam offered to go with us and help with the translation. The intelligence agent agreed that only one of us, (Noble or me), needed to go with him, and Sam could come along. Noble volunteered to go. I was to remain behind and guard the vehicle against a not too friendly crowd of onlookers who had gathered around.

When Noble and the others arrived at the police station, on foot, the chief was not there. He was soon located drinking beer at a nearby restaurant, from where he returned to his office. Sam's knowledge and language skills helped with the interrogation. After inspecting our documents and finding them in order the chief told Noble he could leave, but not before he severely chastised him for the manner in which I shoved and defied his agent. That the agent was quite drunk did not seem to matter. The government officials were all very indignant, and without Sam's help we might both have spent the night in jail.

Noble walked back to the jeep with Sam, who warned us to be careful of driving at night and camping along the roadside, because of the many "bandidos." Sam invited us to have a shower at his apartment the next day after we arrived in Bogotá. That night Noble and I drove about 10 miles out of town and found a secluded campsite well off the highway, on a small hidden side road (which in fact may not have been the best choice to avoid bandidos). Supper consisted of a peanut butter and jelly sandwich, after which we turned into bed, at 11:30pm. The events of the evening will always be a vivid memory for me, providing evidence of my irrational stubbornness (and a genetic makeup contrary to evolutionary survival).

The next day was Easter Sunday, Mar. 29, and Rufous-collared Sparrows were singing from all directions in the high mountains. Everyone was on their way to church, mostly walking (often barefoot), but also by horseback, bicycle, or standing in the back of open bed pickups. The square in Villapinzón was an open marketplace crowded with vendors selling hot soup and a great variety of fruits (mostly oranges and bananas), green vegetables, corn, and goods such as firewood, blankets, hides, and wool. While Noble took pictures, I spent 30 minutes putting our accelerator cable back together, once again, because we had not yet found how to prevent it from regularly coming apart. We purchased gasoline (very low octane, the only available) at a cost equivalent to U.S. 13 cents a gallon, and then drove out of town.

Ten miles from Bogotá, I convinced Noble to stop for two hours while I looked for birds at the forest edge along the highway, at an elevation of 8,000 feet. The single most striking species was the Scarlet-bellied Mountain-Tanager* with its gorgeous red, blue, and black coloration. Happily, this was a very common bird, which often foraged in small groups, and brightened the forest, and my soul! Other species which were first encounters for me included more than a dozen White-throated Tyrannulets* the Superciliared Hemispingus*, Golden-fronted Redstart* ("Whitestart"), Black-crested Warbler*, Purple-backed Thornbill* (a hummingbird with a short, tiny bill and the unusual scientific name of *Ramphomicron microhynchum*), Pale-naped Brush-Finch*, and Thick-billed Euphonia*. These birds, like my other first time sightings, were identified from my field descriptions when I eventually had access to specimens, lists, guides, or colored illustrations. In 1959, I had almost no such information. Even when I did, names were usually only scientific names. Birdwatching was difficult and challenging, but always exciting. Noble didn't understand my passion for this hobby and he accepted it a little reluctantly at times, when he thought I was stopping too often to look at a roadside bird. From my ornithological discussions with him, he was most fascinated with the term "zygodactyl", and he put this term in his memory and delighted in reminding me, at odd times, that this feature was a characteristic of toucans, woodpeckers, and a few other birds, such as the Roadrunner! It was the extent of his ornithological knowledge except for occasional remarks about the Yellow-bellied Sapsucker, which was the only bird name he could ever recall!

It was late afternoon when we arrived in Bogotá, the capital city of Colombia, situated at 8,600 feet in the eastern Andes with a population of approximately 2 million. Like many other Latin American cities, it was a combination of modern and ancient architecture, with both wide, newly paved boulevards and narrow old cobblestone streets. Traffic was horrendous. The great many taxis were unbelievably cheap - - 5 cents to get in and 50 cents for a 30 minute ride. There were an inordinate number of funeral parlors! All the shops and stores in the city which were selling a particular item or category of goods tended to be congregated together on a single street or two, virtually side by side. Because of our beards we were still widely considered by onlookers to be Cubans (or perhaps Europeans), but only very rarely were we perceived as Americans, except when we were next to our van. All attempts to replace our inadequate radiator, or repair the broken one, were once again in vain, although as usual we were assured by the Willys agent they could do that for us.

Almost immediately after our arrival in the city we went to the apartment of Sam Pratt, at his invitation from last night in Tunja. Here we had a shower and washed a few clothes, although we were hampered in our activities by a water shortage in Bogotá due to a severe drought in the area. Afterwards, we all ate in a small restaurant together. Because it was Easter, the traditional Sunday evening fireworks were unusually extravagant. It was midnight by the time Noble and I arrived at a large city park, the "Parque Nacional", which we had chosen as a place to spend the night. It was necessary to convince two policemen that we would not make any noise. They told us to lock the front cab because of thieves.

I got up at daybreak the next morning (Mon., Mar. 30) for a half hour of birdwatching before Noble arose and we commenced our workday. There were several kinds of hummingbirds feeding on the many flowers in the park, but their constant motion made it difficult to describe them. Viewing them so that the light would reflect their iridescent colors was a difficult task. Consequently, I was only able to adequately describe one species, the stunning and very distinctive Green-tailed Trainbearer*, a small green hummingbird with a relatively short bill and a greatly lengthened tail, which was twice as long as its body. If only all hummingbirds were so easy!

We were still plagued by constant mechanical problems with our vehicle, and it was necessary, once again to find the large "Agencia de Willys" (Jeep Agency) in the city for some essential repairs and maintenance. The most critical item was the radiator, which constantly overheated on the steep Andean roads. As usual, we were assured this problem could be solved in one or two days. However, after waiting five days with no success we gave up and on Friday, Apr. 3, we departed (still with our inadequate cooling system) for the distant destination of Cali in the upper Cauca valley. We were wished "Buen viaje" (good travel) as we left the Jeep agency, and were warned yet another time to beware of bandits along the highway.

We descended the western slope of the Eastern Andes all the way down to the city of Girardot in the upper Magdalena River valley, at an elevation of only 1,000 feet. We were in second gear most of the distance, and our engine was already overheating badly. In Girardot, Noble succeeded in making a telephone call to the head office of Willys in Toledo, Ohio, where he talked to a Mr. T. S. Day, who was in charge of engineering specifications. Noble explained the difficulties we were having with our radiator, and the reasons. Mr. Day promised to try and have some help for us in Quito, Ecuador.

As we left Girardot, our road began winding westward up the eastern slope of the Central Andes. I saw my first

Gray Seedeater* (*Sporophila intermedia*). Our radiator overheated and boiled over several times and it was necessary to stop frequently to let it cool off. We added over a dozen canteens of water from mountain streams to the engine. A heavy rain shower helped cool things off. We passed through the city of Ibagué at 4,000 feet. Against all the warnings we had been given, we continued to drive the mountain roads for some time after dark. Approaching trucks would turn their headlights entirely off, leaving the truck outlined in small colored lights like a Christmas tree. We were still climbing upward and it was necessary to travel in 1st or 2nd gear almost the entire time. Eventually we camped for the night along the Coello River at an elevation of 6,000 feet, not far from the city of Cajamarca. As precautionary measures against marauders, we moved all valuables from the cab into the van with us, which we always kept very securely locked at night. Our money and small valuable items were locked in a sturdy little safe inside the van with us. As an additional safety measure, we covered our windows with shades, and finally we each slept with a large kitchen knife at our bedsides. However, the night was uneventful, except for additional heavy rain showers.

Our campsite was surrounded by sugarcane fields, banana plantations, and patches of wet subtropical forest. I got up at daybreak and had a couple of hours to search for birds while Noble busied himself with personal tasks. Hummingbirds and tanagers were particularly numerous. My sightings included four species of hummingbirds (Green Hermit*, Long-tailed Sylph*, Andean Emerald*, & Blossomcrown*), four species of tanagers (Blue-necked, Scrub*, Golden*, & Flame-rumped), Glossy-black Thrush*, and Rusty Flowerpiercer* -- a small nectar-feeding specialist with a slightly upturned bill and a slender hook on the tip of its upper mandible, which is an adaptation for feeding on the corolla of flowers. My field notes for the Blossomcrown read as follows:

"Hummingbird: rather small; bill dark and rather short, straight(?); dingy Grayish below; tail medium, somewhat rounded, & outer feathers with prominent white tips; forehead bright chestnut; upperparts?; prominent white spot behind eye." Except for the slight misplacement of the chestnut on the crown, this description fits that of a Blossomcrown extremely well. I discovered, years afterward, that this species was a rare and local endemic in Colombia.

After a bath in the cold river, and washing a few socks and underwear, we departed our campsite and commenced the day's drive to Cali. A sweet little girl along the roadside, about 7 or 8 years old, was selling cherimoyas and other fruits. We stopped and bought some from her, and gave her a peso. Our radiator overheated constantly, requiring us to make frequent stops while we let it cool off. There were intermittent rain showers as we passed in and out of the clouds. High elevation roadside birds included the following new species for me: Brown-bellied Swallow*, Rufous-breasted Chat-Tyrant*, and Lacrimose Mountain-Tanager*.

We were now on the western side of the Central Andean range, and we began a long descent all the way down to the beautiful, green, luxurious upper Cauca valley at 3,000 ft. elevation. The Cauca and Magdalena rivers parallel each other inside Colombia, separated by the central Andean range of mountains. Both flow northward, merging together just before flowing into the Atlantic Ocean (the Caribbean Sea). The most conspicuous features of the fertile Cauca valley were cattle ranches and sugarcane fields. At long last, we were able to travel on a good paved highway in normal gear, at speeds of 40 or 50 mph. Pedestrians were mostly barefoot. The most numerous roadside birds, in order of abundance, were Black Vultures, Cattle Egrets, White-collared Swifts, Turkey Vultures, Southern Rough-winged Swallows, Yellow-headed Caracaras, Smooth-billed Anis, and Great Kiskadees. Other species identified were the Tropical Kingbird, Ruddy-breasted Seedeater, Blue-black Grassquit, Blue-Gray Tanager, Flame-rumped Tanager, Great Egret, and either Common or Lesser Nighthawk (a single bird flying overhead in the daytime, which I could not identify to species).

At 7:30pm, we arrived in Cali, situated in the upper Cauca valley at an elevation just above 3,000 feet. It was a modern, clean city with wide streets and the reputation for the prettiest girls in all of Colombia. Not long after our arrival, we were approached by a young newspaper reporter who wanted a story for the local newspaper, which we obligingly gave to him. That night we camped a short distance outside the city. The next morning, Sun., Apr. 5, Noble drove me to a nearby non-residential area where I could spend several hours searching for birds, while he returned to the city to have our radiators switched around (again). We decided to persevere one more time with the replacement

Plate 7: In the NE Andes of Colombia we drove from Cúcuta to Bucaramanga, where I convinced a policeman that I came from Cuba! The next evening in Tunja was our harrowing experience with the intoxicated federal agent.

Plate 8: Our route from Bogotá to Cali took us SW down to the Magdalena River Valley, up over the Central Andes, then back down to the Cauca River Valley.

radiator which Noble had purchased in Costa Rica (on his train ride and airline hop between the jungle and San Jose).

My area for finding birds was one of fields, ponds, and scattered trees. Here I recorded my first Rufescent Tiger-Heron*, an immature bird which I studied and described very carefully before separating it from the other two tiger-herons. Another exciting bird was the Dwarf Cuckoo*, which I discovered later was generally considered to be an uncommon and local species in Colombia. My notes for this bird said: "cuckoo: rather small (about 7-8); upperparts pale bluish-Gray, wings and lower back tipped with brown; tail dark brownish-black narrowly tipped with white; bib ruddy chestnut; rest of underparts white, flanks tinged with buffy; bill black; eye deep red." Other birds seen and described (in part) were an estimated 80 Wattled Jacanas*, one Striated Heron*, a single Blackish Rail*, 2 Southern Lapwings* (in the genus *Vanellus*, which became one of my favorite genera), 1 Black-necked Stilt, 6 Solitary Sandpipers, 200 Ruddy Ground-Doves, 30 Eared Doves, 8 House Wrens, 3 Vermilion Flycatchers, a dozen Pied Water-Tyrants*, 1 Rusty-margined Flycatcher, 3 Yellow-bellied Elaenias, 8 Grassland Yellow-Finches*, 1 Giant Cowbird*, a single Grassland Sparrow* (*Ammodramus humeralis*, which I very carefully observed and described to distinguish it from the Grasshopper Sparrow which I also found and described at this same site), and an estimated 250 Yellow-hooded Blackbirds* (at the southwestern limit of their range here). I also observed the only Yellow-billed Cuckoo I saw during our neotropical travels, 6 Northern Waterthrushes, and a Swainson's Thrush.

Noble returned and picked me up at 11:00 o'clock and we began a two-day journey southward to the border of Ecuador. Before very long we began climbing up into the foothills of the Andes, at the very upper end of the Cauca valley. Our sinuous gravel road constantly went up and down, generally between the elevations of 1,500 ft. (in the Río Sambingo valley) and 5,000 feet. Our average speed and gasoline mileage were both very low. Roadside scales confirmed that our camping van weighed a total of approximately 9,000 pounds. Much of the countryside was devoted to cattle ranches, but there were also coffee, sugarcane, and banana plantations. Rural houses, as usual, were adobe with tile roofs, or mud and stick with thatched roofs. Some houses were painted white, and often had commercial advertisements of all kinds, particularly for tonics, which were said to cure just about everything -- "aches, pains, and miseries." People along the roadside were friendly and often waved at us. Many carried machetes for cutting firewood, sugarcane, or bananas. On one occasion we stopped to watch a very aggressive roadside cockfight, where intense onlookers were betting on the outcome, and paying much more attention to the fight than to us.

Birds along our route were such widespread species as Great Egret, Black and Turkey vultures, Yellow-headed Caracara, American Kestrel, Roadside Hawk, Southern Lapwing, White-tipped Dove, White-collared Swift, Smooth-billed Ani, Vermilion Flycatcher, Great Kiskadee, Tropical Kingbird, Southern Rough-winged and Blue-and-white swallows, Tropical Mockingbird, Ruddy-breasted Seedeater, Lesser Goldfinch, and Rufous-collared Sparrow. There were also Spectacled Parrotlets (a diminutive, green, short-tailed parrot in the genus *Forpus*). That night we camped beyond Popayán, near the town of El Bordo at an elevation of just 2,500 feet. We were at the extreme upper end of the Cauca valley.

The next morning (Mon., Apr. 6), I birded for an hour and a half in the surrounding dry scrub and trees before leaving our campsite. Here I recorded 25 Scarlet-fronted Parakeets* (genus *Aratinga*) which flew past, Pale-breasted and Slaty* spinetails (genus *Synallaxis*), 3 species of wintering Nearctic passerines (Yellow Warbler, Blackburnian Warbler, and Red-eyed Vireo), 2 Mouse-colored Tyrannulets*, and 2 Crested Oropendolas*. A long, deep, hollow vocalization went unidentified.

Noble and I continued toward the border of Ecuador, traveling at various elevations in the Andes but gradually gaining altitude on our way to the city of Pasto. As we drove through the small town of La Unión at 5,000 ft. elevation, we were stopped by a chain across the road at one of the numerous police check posts. Here our Jeep was sprayed with two different fluids which the officials told us were to help control the spread of both yellow fever and hoof & mouth disease (though I wasn't sure just how).

In the town of Pasto, at an elevation of over 10,000 ft. the telephone poles were constructed of giant bamboo, cut from neighboring forests. Indians of both genders and all ages wore black, wide-brimmed hats of felt or straw.

We stopped in town to ask a federal policeman who was dressed in an old U.S. Marine uniform where the post office was located, since Noble wanted to mail a postcard. The policeman not only got in our van and rode with us to show the way, but he then insisted on riding with us around the town while he pointed out some of the sights of the city. After showing us the road out of town, he saluted us both twice and said "Buen viaje, pues." It was a local custom to add the "pues" on the end of this cordial goodbye, wishing us well. Such friendly encounters were among the many pleasant memories of our travel.

Birdlife at the higher elevations beyond Pasto, between 6,000 and 11,000 feet, included the following: Rufous-collared Sparrow, Brown-bellied Swallow, Glossy-black Thrush, Hepatic Tanager, and an Acorn Woodpecker (a species typical of California; which caught me completely by surprise). Late in the afternoon I suddenly felt sick at my stomach while driving, so I stopped at the roadside, got out and immediately vomited, apparently because the only food I had eaten all day were 20 little green bananas which we had purchased from a small girl along the roadside. I was fine again immediately thereafter. (As it turned out, this would be the only occasion during our entire three-year adventure that I was ever sick at my stomach.)

We were high in the Andes, not far from the border town of Ipiales as darkness approached. We crossed a steep-sided gorge over the Río Guambuyaco, on a well-built suspension bridge. Our narrow gravel road was under construction and there were rocks and gullies in the middle of the roadway, necessitating caution to avoid them. A steep cliff abutted the right side of the road. As I was circumnavigating to the right around a large rock in the middle of the road, our right wheels entered a deep gully, causing the Jeep to tilt severely to the right, and our heavy steel van veered into a low overhanging rock protruding from the cliff, causing a very noticeable dent on the top back right corner of the van. It was an identifying feature for the rest or our adventure, but otherwise did no damage that required repair. Not long after this incident, we found a campsite for the night, at an elevation of 10,500 feet, about 20 miles from Ipiales.

The following morning (Tuesday, April 7), there was an hour available to me to look for birds prior to our departure. The high altitude air was very invigorating and it was a great way to start the day. The pleasant songs of Rufous-collared Sparrows and Great Thrushes carried through the still morning air. A Carunculated Caracara* came gliding and flapping above the open "páramo" (high elevation wet grasslands). This species is the northernmost of three mostly allopatric, similarly appearing caracaras in the genus *Phalcoboenus* (with a geographic range extending almost the entire length of the high Andes from north to south). To my astonishment, a Short-eared Owl flew past, low over the grasslands. (This owl is found throughout the world, on all continents except Australia.) Three species of small passerines in the woody shrubs at the edge of the páramo were the Tufted Tit-Tyrant* (with a long slender forward-curling tuft of feathers on top of its head), the Black Flowerpiercer, and the Cinereous Conebill* (a small, warbler-like bird of uncertain affinities, now placed with the tanagers). I was happy also to record three species of hummingbirds: the Sparkling Violetear (which I had seen previously), the Black-tailed Trainbearer* (the higher elevation species of the two trainbearers, which I described in my notes as having a tail "like a Fork-tailed Flycatcher"), and the tiny, all green, coppery-tailed Tyrian Metaltail*.

As Noble and I were leaving our campsite an Indian man came walking along the road and wished us "Buen viaje, pues", the friendly farewell which always gave me pleasure. Our smooth, gravel road took us south across the páramo between 10,000 and 11,000 ft., and by mid-morning, we reached the town of Ipiales on the border with Ecuador. To our pleasant surprise, the border officials were cordial, knowledgeable, efficient, and swift. Our documents were stamped within a few minutes, and a brief inspection of our van elicited the complimentary comments of "Qué bonito", "Qué bueno", and "Qué lindo." It was another happy experience. Noble offered the "jefe" (senior official) one of our postcards, whereupon all the other border officials wanted one. Luckily, we had plenty. For a final time in Colombia, we were wished "Buen viaje, pues."

Chapter Seven—Ecuador

We arrived at 11:00 am on the morning of Apr. 7 at the entry port into northern Ecuador from Colombia, at the city of Tulcán, in the Andes Mountains at an elevation of 11,000 feet, on the Pan-American Highway. Our passports and visas were examined and stamped by the immigration authorities, and our vehicle was inspected and approved by the customs office (aduana), where we were given a stamp in our carnet de passage booklet. As usual, no one asked to see a driver's license.

Much of the Pan-American Highway from the Colombian border to Quito was hand built of cobblestone. The roadway meandered across the páramo of the Ecuadorian altiplano at elevations up to 12,000 feet, and sometimes descended into dry cactus covered valleys as low as 5,000 feet. There was very little traffic. The higher altitudes were characterized by two prominent, tall herbaceous plants: (1) a terrestrial bromeliad called "Puya", with a whorl of bayonet like leaves at the base of a tall slender flowering stalk, and (2) "Espletia", with a narrow basal trunk and a bushy looking top of flowers and leaves, in the botanical family Asteraceae. Both of these plants could be as tall as 7 or 8 feet. Conspicuous birds of this habitat which I had not encountered previously were the Bar-winged Cinclodes* (a furnariid), Plumbeous Sierra-Finch* (in the genus *Phrygilus*), and "Variable" Hawk* (a common, widespread soaring *Buteo* which like other members of the genus came in a variety of plumage patterns). Although Clements in his 2007 Checklist recognized two separate species in this complex (the Red-backed and Puna hawks), I will refer to all populations and color variations as a single species, the Variable Hawk. Their overlapping patterns made it impossible for me to separate them in my field notes. Rufous-collared Sparrows were everywhere, as they were throughout much of the neotropics, but they were particularly conspicuous here in the high Andes where their pleasant singing made them a favorite bird of the rural inhabitants.

Colorfully dressed Indians, usually barefoot, walked along the highway. Many women carried a baby on their back, wrapped in a blanket. Men wore ponchos and wide-brimmed black or brown hats, and often were heavily burdened with a basket of corn or a bundle of firewood on their backs. Everyone was friendly, and they all wanted us to give them a ride. Burros were often used for transporting people and heavy goods of all kinds. Rural houses were customarily made of adobe or sticks, with thatched roofs, as they were throughout the northern Andes.

Slowly we descended to 5,000 feet in an arid, dusty river valley with cactus and mesquite. After crossing a river, we began an equally slow ascent back up again, to the city of Ibarra at 7,300 feet. This was an area of montane scrub, where I stopped briefly once to record my first Black-billed Shrike-Tyrant*, a large, mostly brownish-Gray flycatcher with very conspicuous white sides to the tail in flight. We continued upward to a grassy plateau (puna) at 9,000 feet, which we then followed southward some 30 or 40 miles to the equator, where we arrived shortly after dark, about 8:00 pm. Here, at 9,000 feet above sea level, was a large, spherical concrete monument marking this separation of the northern and southern hemispheres. We wanted to photograph this monument the next morning so we chose a nearby site to camp for the night, a narrow, eucalyptus lined drive leading to a distant private estate. Our first day of travel in Ecuador came to a pleasant end. It had taken us 97 days to get from California to the equator. "Ecuador" is the Spanish word for the equator, which transects the northern part of the South American continent and extends westward in the Pacific Ocean through the Galapagos Islands 600 miles off the coast of Ecuador. These were the islands which Charles Darwin visited in 1835 as a naturalist on the HMS "Beagle", where he arrived at his thinking about biological evolution and natural selection.

As was my custom, I got up at daybreak on Apr. 8 and spent an hour looking for birds before Noble got up. I recorded notes on the following species which were new for my trip list: Rufous-naped Brush-Finch*, Sapphire-vented Puffleg* (a hummingbird), Band-tailed* and Plain-colored* seedeaters, Golden-bellied Grosbeak* (a cardinalid), and Hooded Siskin*. Other birds present were Cinereous Conebill, White-throated Tyrannulet, Great Thrush, Black-tailed Trainbearer, Brown-bellied Swallow, Tufted Tit-Tyrant, and the ubiquitous Rufous-collared Sparrow.

We returned to the large, spherical, concrete equatorial monument in the middle of the highway and spent 45 minutes photographing the two of us, our Jeep, and some passing Indians in front of the monument. Then we were on our way south again toward Quito, not far away. As we crossed a muddy river, we stopped for a much-needed bath. The countryside was mostly ranchland, with cattle, horses, and sheep. Adobe fences separated properties. Tall, non-native eucalyptus trees (originally from Australia) bordered the highway and formed windbreaks around ranch houses. People walking along the road looked at us with curiosity and occasionally held up an arm for a ride. Indian men wore their hair in a long braid that hung down the middle of their back. I recorded my first Blue-and-yellow Tanager* (a member of the genus *Thraupis*) as it flew across the highway at 7,000 ft., distinctively plumaged in bold yellow, olive, and blue colors. This species inhabited the edge or outside of forests and scrub, at moderate to high elevations.

Quito, the capital city of Ecuador, was situated on an elevated plateau at 9,000 ft. between high Andean peaks to the west and east. It was a city of about ¼ million mostly friendly people, although policemen sometimes blew their whistles at us for no apparent reason. (We just waved in an innocent fashion and kept going.) Side streets were so steep that sometimes we had to stop, get out, lock our front wheels in 4-wheel-drive, and ascend the hill in our lower transfer case. We went to a movie our first evening and then found a place to spend the night in an open lot at the edge of the city, between a radio station and a school.

On Thursday, April 9, we kept busy with our usual chores – bank, American Embassy, post office, supermarket, media interviews, and further attempts to solve our radiator problem. A telegram waiting for Noble from Mr. T. S. Day at the Jeep factory in Toledo, Ohio, said that a new radiator similar to our "hot climate" original could perhaps be air freighted to us in Quito, or in Lima where we could pick it up on our arrival there. Unfortunately, no radiator arrived for us during our stay in Quito.

While Noble was exchanging some money at a principal bank in Quito he met one of the more prominent bank officers, a Señor Bermeo who also happened to be the current president of the "Club Nuevo Horizontes" (New Horizons Club), which was an outdoor activities environmentally oriented organization.

He invited Noble and me to join a group of about ten members who were going to travel to the base of Mt. Cotopaxi over the forthcoming weekend because some of the group wanted to climb to the top of the mountain, one of the world's highest active volcanos at an elevation of almost 19,500 feet. Others in the group, including his wife and daughter, would simply hike, take pictures, or go duck hunting at Lake Limpiopungo. The group would camp at 13,000 ft. at the base of the mountain and the 6,500 ft. climb to the top and back down would occur on Sunday. Of course, Noble immediately said yes, and became very excited at the thought of planting a Trojan Peak Club flag at the top of Cotopaxi. He was told that all the necessary equipment (including climbing ropes and crampons) would be provided for us, and I would be loaned a shotgun if I also wanted to shoot some ducks. The plan was for Noble and me to drive to the lake tomorrow afternoon, on Friday, and arrive there a day ahead of the others, who would join us early on Saturday afternoon. The climb up and down the mountain (for those who wished to do so) would be on Sunday. Noble was given directions to a hacienda ("El Porvenir") near Lake Limpiopungo and was told to hire a local Indian boy there as a guide to ride with us and show us the way across the open páramo to our camping spot on the lake.

Noble and I got underway the next day for Lake Limpiopungo, at mid-afternoon on Friday, April 10, as planned. We drove 30 kilometers south of Quito on the Pan-American Highway to the city of Machachi, where we turned left off the highway and began proceeding upward on a narrow, undulating, dirt and cobblestone road. About an hour later, we arrived at the "El Porvenir" hacienda, following the directions we had been given. At the hacienda, we continued to follow our instructions and, for a very small fee, we hired a young Indian boy to ride with us and show us the rest of the route to Lake Limpiopungo. There was no longer a road to follow, only a few tire tracks leading across the páramo, 8-10 miles or so to the lake. Our guide was a pleasant young man in his mid-twenties, married with two small children. In addition to his native Indian tongue, he spoke a little Spanish, but of course no English at all. We offered to pay him 10 sucres (about 60 cents, more than a full day's wage). He would ride with us to the campsite, sitting in our middle front seat on top of the engine, and then walk back to the hacienda.

At first, our track across the páramo was relatively dry and firm, but as we slowly progressed it gradually became damper, with patches of spongy, tundra-like bog. As always in such situations, I was in 4-wheel-drive and our lower transfer case. Our track was gradually gaining in elevation and the base of Cotopaxi was not far away. Our unmarked track wound back and forth, avoiding the wettest areas. However, I became quite apprehensive about the capability of our heavy van with its small tires to travel over such terrain. The boy kept motioning us confidently onward, to go straight ahead, but of course, he was completely unfamiliar with a vehicle such as ours. I should have turned around, but my determination to reach our destination overshadowed common sense. Then the inevitable occurred. In the late afternoon glare of the sun, I failed to detect a small, boggy, tundra-like patch of terrain directly ahead of us. Although our front wheels got across the wet, soft depression, the back wheels did not, and the rear of our Jeep sank deeply into the mud and came to a halt. Our most powerful gear did not even begin to move us forward, and there was nothing at all to which we could attach our winch. Not only were we stuck, but the back end of our heavy vehicle began to slowly sink deeper into the bottomless muck. We were at an elevation of almost 13,000 feet. I said to myself that this was the end of our adventure, and our Jeep would eventually just sink out of sight! I could blame no one but myself.

Noble and I did the only thing we could think of. We got out our shovels and pick from the tool compartment, got down on our hands and knees, and began to dig the mud out from around the back right wheel, which was the more deeply bogged. The van had acquired a very noticeable backward leaning tilt to the right. The more mud we removed the deeper the wheel sank and the greater the tilt became. The hole we were making began to fill with water, and before long, the water was over the axle! Our efforts were only making the situation worse! The sun had gone down and the air temperature dropped rapidly. We were working in the dark and eventually we became cold, wet, and exhausted because of the high altitude. There was nothing more we could do that night. Our Indian guide had long ago set off on foot for the hacienda, but at our request, he promised to return the next morning (for another 10 sucres). Our spirits were very low and we went straight to bed with only a can of juice for supper. It was a cold night but we were warm in our sleeping bags.

Upon awakening the next morning, I looked up from my bed with great amazement at the towels on our drying rack. They appeared to be hanging at a considerable angle, rather than vertically. I thought to myself that's impossible, maybe I am suffering from overwork last night. The realization then came to me that the towels were hanging straight downward as they should, but that the rack was tilted at a 20 degree angle from the horizontal. I rushed outside to look at our Jeep. To my horror, I observed that the right rear wheel had almost completely disappeared from view. A wave of despair passed over me and I said to myself that this would indeed be the end of our adventure. The front wheels were still on firm ground, but the back wheels, particularly the right one, were submerged in mud and water, as were the rear bumper of the van, the differential, and most of the rear axle. Our plight seemed hopeless, and the situation was slowly getting worse by the minute. Maybe our mountaineering friends could conjure some sort of a miracle when they arrived later in the day.

A happier moment came when an Andean Condor* soared not far above my head, providing me with one of my most memorable moments. During the next several hours, I saw more than half a dozen of these awe-inspiring scavengers, including both adults and sub-adults, singly or in pairs, soaring or gliding above the páramo and the mountainsides, searching for food items. Occasionally one would land briefly on the ground. It took 5-6 years for young birds to reach breeding maturity and attain adult plumage, with a white collar and a wide white area in the wings. These majestic, almost prehistoric-looking birds were the world's largest raptor, weighing 30-35 pounds and with the largest wing area of a flying bird. Their 10½-foot wing length was very slightly shorter than that of a Wandering Albatross (which had narrower and more pointed wings). Andean Condors mated for life, and lived up to 60 years or more in the wild. Unlike most large raptors, males were larger than females. The species was uncommonly distributed throughout the entire length of the Andes, all the way to the southern tip of South America (where they were found down to sea level, as they were also along the coast of Peru). The Andean Condor was worshiped and revered in the culture of the Incas, and was the national bird of Argentina. (Its slightly smaller Nearctic relative, the California Condor, had a much more restricted geographical range, and is now almost extinct in the wild.)

Another exciting species for me was the Chimborazo Hillstar*, a hummingbird endemic to very high mountains in Ecuador, up to 16,500 feet (5,000m), one of only two hummingbirds (along with the Bearded Helmet-crest) occurring at such a high elevation. It possessed a gorgeous, entirely purplish head, snow white underparts (with a central vertical black streak), and mostly-white tail. It was one of six virtually allopatric hillstars in the genus *Oreotrochilus*, which inhabited very high elevations along the whole chain of the Andes. I was also happy to observe Andean Lapwings*, a conspicuous, noisy plover in the genus *Vanellus* which lived at high elevations.

Additional birds which I recorded for the first time were: the Cinereous Harrier* (a close relative of the Northern Harrier), Andean Ibis* (a very handsome, dark Gray ibis with a distinctive, bright, buff-colored head, neck and chest), Speckled Teal*, Andean Gull*, Paramo Pipit* (in the worldwide genus *Anthus*), Andean Tit-Spinetail* (a furnariid), Streak-backed Canastero* (another furnariid), and Plain-capped Ground-Tyrant * (an open grassland, ground-dwelling, rather lark-like "flycatcher", one of numerous similar looking species in the genus *Muscisaxicola*). I walked to the shoreline of Lake Limpiopungo, a disappointing, large, shallow lake with a very muddy shore, which was almost lacking in vegetation and birds. The few ducks were a long way off.

About 6:00 pm, Señor Bermeo finally arrived with his wife, daughter, and group of about eight mountaineers. They were all traveling in an old 4-ton, open-bed, dual-wheeled, Chevrolet truck with wooden seats constructed for use in carrying passengers. Everyone immediately pitched in to try and help us out of our mud hole. More digging only resulted in the right rear wheel sinking lower and more water filling the hole, which then had to be bailed out. Obviously, this was not the solution. A chain was hooked from the back of the Chevrolet truck to the front of our Jeep, and an attempt was made to pull the van forward out of the hole, using power from both vehicles. When this failed, an attempt was made to pull it out backward. This effort, too, was unsuccessful. The Jeep did not budge in either direction. When our winch was fastened to the Chevrolet truck and our engine was started, the only vehicle that moved was the truck! Since it was now well after dark, tents were pitched nearby for the Quito group and the rescue attempt was terminated until tomorrow morning (Sunday). Noble and I ate a cold supper and climbed into our sleeping bags for another night at a tilt.

Sunday morning, (Apr. 12), was gray and cold. The decision was made that we should try and jack the right rear wheel up out of the hole to a height where the van was level with the ground, and no longer on a tilt. This would necessitate jacking up the right rear end of the Jeep. All we had was our little 9-inch hydraulic jack we used for changing a tire. We found a short, wide sturdy board to place under the jack to prevent it from sinking into the mud, and positioned the jack in the hole we had dug, under the right side of the rear bumper. After

Plate 9: The Pan-Am. Highway crossed the equator just north of Quito, at an elevation of 9,000 feet. Fifty miles to the south was the snow-covered volcanic peak of Cotopaxi, rising to 19,500 ft. above sea level.

Plate 10: On our way to the base of Cotopaxi we saw our first Llama, at 13,000 feet. We became very seriously bogged on the spongy, tundra-like terrain which our Indian guide had directed us across.

we had raised the back of the van about 6 inches, it was necessary to find some rocks to put in the hole. The Chevrolet truck was dispatched to a small, nearby mountain river about 20 minutes away, where there were some medium sized flat rocks, long flat boards, and some wooden poles, just what we needed to put under the Jeep as it was slowly raised, bit by bit. The right wheel was prevented from sinking back into the hole again by placing a great many rocks underneath it, filling the hole.

Finally, the Jeep was on a level surface. Rocks were placed on the ground in front of the two back wheels for them to drive on as the Jeep was slowly winched forward for a distance of about 75-feet, to bring it out of the boggy patch and back to relatively firm terrain. First, we had to construct a very sturdy three legged tripod structure to which the cable of our winch could be fastened. We made this out of two poles gathered from the river and an iron bar from our jeep tools. The poles and bar were hammered several feet into the ground and were bound together with our tow chain, hopefully giving our crude, 3-foot-tall, handmade, tripod structure the great strength which would be needed for its use as a winch attachment. It required five hours of hard work to complete this task. We were proud of our accomplishment, but the test was yet to come. We pulled out the free end of our winch cable and fastened it to the man-made tripod. I then turned the jeep engine on, to power the winch. Noble put the winch in gear. It worked! Slowly, inch by inch, our jeep moved forward. Within ten minutes, all four wheels were once again on firm ground. Cheers went up by everyone, and Noble and I broke out several bottles of wine to be shared by all. Our journey had not yet come to an end.

It was now midday. An ascent of Cotopaxi was no longer feasible, not only because of the time spent getting us out of the boggy tundra, but also because the alpinistas discovered they had forgotten their climbing ropes! Without these, the climb was not possible. (Noble confessed to me that he probably did not have the necessary stamina to achieve the top anyway.) It began to rain. Therefore, we all hugged, said goodbye to each other, and departed on our separate ways. We had survived another calamity in our adventure, and we had only barely begun. Quitting never entered our minds.

I navigated cautiously down the narrow, muddy, slippery road back to Machachi. En route, our right front wheel slipped in and out of a deep rut in the steep downhill road, putting sidewise stress on the wheel and creating a noticeable, persistent wobble in the wheel. On one occasion during our descent, we had to stop for 15 minutes to help a local farmer put on a wooden wheel which had come off his oxcart, which had broken down in the middle of the road and was blocking our progress. He was hauling very smelly manure for fertilizer! At Machachi, elevation 10,000 feet, we reached the cobblestone Pan-American Highway again and turned south, toward the city of Latacunga some 40 miles away. At dusk, (7:00 pm), we stopped for the night on a small side road, at an elevation of 11,000 feet. I was tired and beginning to come down with a slight cold from all my exertion and late nights of the past two days at a very high elevation, so I took a couple of antihistamine tablets with a glass of juice and went straight to bed.

It was a cold, windy, and mostly cloudy morning the next day, on Monday, April 13. Our Pan-American Highway went up and down between 9,500 and 11,500 feet as we progressed toward Latacunga. The lower elevations were drier, with irrigated fields. Livestock included cattle, horses, sheep, and our first encounter with llamas. The most common and conspicuous roadside birds were the Cinereous Harrier, Great Thrush, Brown-bellied and Blue-and-white swallows, Rufous-collared Sparrow, Carunculated Caracara, Bar-winged Cinclodes, and Plumbeous Sierra-Finch. A new species for my trip list was the Black-winged Ground-Dove*.

An attempt to have the wobble removed from our wheel at a small mechanic shop in Latacunga did not succeed, so we proceeded onward, hoping it would not worsen before our arrival in Guayaquil. We had decided that because of the tortuous Andean roads and our radiator problem it would be better for us to travel in the coastal lowlands as much as possible, rather than in the Andes. Our single road map for Ecuador was entirely unsatisfactory, since the Ecuadorian government prohibited the sale of any Esso Oil Company road maps of Ecuador, objecting to the position of Ecuador's international boundaries as shown on the Esso maps. Therefore, we had only a crude hand drawn map showing a road connecting Latacunga in the mountains with Quevedo in the Pacific lowlands. Never-the-less, Noble and I decided to follow this route down from the Andes. We turned westward off the Pan-American Highway in Latacunga and immediately climbed to a pass at 13,000 feet before starting a long descent to the Pacific lowlands.

The road downward was very narrow and steep. For almost two hours, we wound continuously in and out of the clouds, in low gear virtually the entire time. We passed between patches of mature forest which looked like enticing habitat for birds, but there was no time to stop, nor were there any spots wide enough along the shoulder for us to do so. I would have to return to this location for birding sometime in the future. As we reached lower elevations, there were many clearings with banana plantations. The population became mostly European, with few Indians, and people paid less attention to us as we drove past. Children along the roadside held out their hands for money. Houses were mostly wooden structures built on stilts, with thatched roofs. That evening we camped near a small stream next to a very large banana plantation at an elevation of only 1,000 feet, about 30 miles prior to (east of) the town of Quevedo.

The next morning, Noble wanted to do some reading and writing, and to wash some clothes in the stream. That gave me an opportunity for recording birds, and butterfly collecting. The habitat around our campsite was a mosaic of small patches of forest, streamside edge with trees, and many acres of bananas. It was hot, humid, and overcast. For the first few hours, birds were quite abundant and active, but recording their appearance was not easy. The bright overcast sky caused birds in the treetops to appear as dark silhouettes, and birds in the forest understory always seemed to be in motion or behind the vegetation. I soon was dripping with perspiration. Observation was frustrating, but I persevered, adding many descriptions to my notebook, which I hoped that one day I would be able to identify.

Two of my descriptions are as follows. (1) "flycatcher: very small (about 4"); rather chubby; tail short; bill short & heavy; head grayish, crown and loral area darker; large whitish spot in front & above each eye; narrow yellow median stripe; back & wings olive (wings darker); lower back & underparts yellow; base of tail rufous, blackish terminally; throat grayish; breast olive." With such a detailed description of this very colorful little flycatcher, it was easily identified, eventually, as an Ornate Flycatcher*, one of the most attractive, friendly, and delightful little tyrannids. It immediately became one of my favorites. (2) "wren: upperparts, except head, bright rufous-chestnut; head black with large white spot below and behind eye & rather indistinct white superciliary; underparts whitish, flanks rufous-chestnut with black bars; medium small." It was subsequently identified as a Bay Wren* (*Thryothorus nigricappilus*). Some of the other birds, in their notebook order, were the Pacific Antwren*, Rufous-tailed Hummingbird, Black (Chestnut)-mandibled Toucan (very common), Chestnut-backed Antbird, White-whiskered Hermit*, Chestnut-fronted Macaw*, Orange-bellied Euphonia*, Long-wattled Umbrellabird* (2 females or sub-adults seen well, perched conspicuously together in a treetop), Bat Falcon* (surprisingly my first record of this widespread neotropical species), Tawny-breasted Flycatcher*, and Tawny-crested Tanager*. Many unidentified *Amazona* parrots were constantly flying back and forth overhead, silhouetted against the sky, calling noisily to each other. Both members of a mated pair characteristically flew next to one another, even within a large flock. Like all parrots, pairs remained mated for life.

By mid-morning, I was glad to stop and have a refreshing cool bath in the stream, and catch a few of the many brightly colored tropical butterflies. These were labeled as to date and location and preserved in envelopes, which were placed in empty "Ryvita" cracker tins. Just after 10:00 o'clock Noble and I climbed into the "Roadrunner's" cab and got underway again, for Guayaquil. En route, we passed through many miles of banana plantations. It was plain to see why Ecuador was the world's number one exporter of bananas. There were also extensive rice paddies where jacanas and egrets foraged, and occasional fields of corn. Goats and pigs were a constant hazard in the middle of the road. A ping in our engine was a new source of worry. The weather remained hot and humid. Roadside birds included a Cocoi Heron* (closely related to the Great Blue Heron),

Croaking Ground-Doves*, Baird's Flycatchers* (somewhat resembling a kiskadee), Scrub Blackbirds*, Long-tailed Mockingbirds*, a Peruvian (or Red-breasted) Meadowlark*, Yellow-rumped Caciques, and Pale-legged Horneros*, which were a mud-nest building furnariid.

By early evening we arrived in Guayaquil, Ecuador's largest city with a racially mixed population of about ¾ million people. It was a sprawling, dirty, smelly, generally unpleasant city where gangs of small boys constantly followed us everywhere we went asking for money and shouting "Fidel Castro" at us. We couldn't find a movie we wanted to watch so we walked inside a small sidewalk cafe where we drank a beer and ate a hamburger, an ice cream cone, and some popcorn for Noble. Afterwards we had to settle for a muddy, vacant lot to park for the night, with hordes of mosquitoes. Luckily, our van had mosquito proof screening on all the windows, which we needed open because of the repressive heat. The site had virtually nothing to recommend it except the fact that there were relatively few people in the immediate vicinity.

The next day there were two major items of business on our agenda: (1) to find a service agency for our Jeep, and (2) to book passage for us and our van on a coastal ferry to Puerto Bolívar, a short distance south on the Ecuadorian coast near the Peruvian border. We had decided to bypass the mountains as much as possible until we could obtain another radiator. The coastal ferry would allow us to do this, plus the fact we had been told that the Pan-American Highway in the Andes was not yet open to traffic between Ecuador and Peru. By good fortune, we met a young boy on the street who informed us that the Chrysler-Plymouth agency had the best mechanics in town, and he rode with us to show us the location. The manager at the agency spoke very good English, having spent some time in the United States. He confirmed there was no radiator available to us in Guayaquil, so we explained that our other major problems were the shimmy in the right front wheel and the engine ping. It required 2 days for these problems to be corrected but we were allowed to sleep inside our van at the repair garage, as we had often done throughout our travels. The engine ping was found to be caused by a tiny hole in piston #6, probably resulting from the very low 60-70 octane gasoline which we had to purchase throughout the Andes (where no higher octane was available).

The extra day in Guayaquil afforded me an opportunity for some time in the field, albeit the local habitat was not my favorite. I was awakened early at 6:30 am. by some noisy workmen, so I hurriedly got dressed, downed a can of juice, grabbed my binoculars, put my notebook and some "6-12" mosquito repellent in my field bag, and set off on foot in the relatively cool morning air for the nearby waterfront. Here I encountered my first Chestnut-collared Swallows*, which were foraging abundantly with lesser numbers of Gray-breasted Martins and migrant Barn Swallows from North America. Also present was an adult plumaged Franklin's Gull, a winter visitor and migrant from the Nearctic region.

From the waterfront I made my way, by walking, hitch-hiking, and bus, to a relatively unpopulated area on the northwest outskirts of the city where there were some trees and shrubs. Here I spent almost six hours, mostly in the middle of the hot day, before returning to the city in the late afternoon. Avian activity became very slow by noon. I didn't have my net with me to catch some of the many vividly colored butterflies. One of my most exciting observations was an excellent view of a Savannah Hawk*, my first definite record of this very handsome, approachable, rufous-hued raptor. I also saw a Harris's Hawk, which somehow seemed out of place here along the Pacific coast of Ecuador. Two pint sized, mostly green parrots with short tails were the Gray-cheeked Parakeet* (Brotogeris pyrrhopterus) and the Pacific Parrotlet* (Forpus coelestis). I flushed a Pauraque from the middle of a dry, grassy foot path and when I looked down, to my amazement there was a single egg lying on the bare ground, in what must have been the bird's "nest." My first Short-tailed Swifts*, with their rather bat-like shape, became an appealing new favorite. The red-and-black plumaged Crimson Finch*, and the yellow-and-black plumaged Yellow-tailed Oriole* were the two most colorful species I encountered. Other birds new for my list for were the Fasciated Wren* (a member of the genus Campylorhynchus), Amazilia Hummingbird*, One-colored Becard*, Chestnut-throated Seedeater*, Sooty-crowned Flycatcher*, Scarlet-backed Woodpecker*, and Ecuadorian (formerly Ruddy) Ground-Dove*. At the city dump there were almost a thousand Black Vultures picking through the garbage!

The "Don Antonio" was under the command of a short, fat, surly, cigar-smoking captain. Because of the weight of our vehicle, it took considerable effort to load it on board, and it was placed so close to a bulkhead, on the right side of the Jeep, that we could not open the door to get inside our van. There were only three other vehicles on the boat – two little Jeeps and a small International truck driven by an unshaven but personable American with his Ecuadorian wife. Barefoot dockside workers carried 150-200 pound bags on their backs, walking up a narrow wooden plank from the pier onto the boat. These were filled with all kinds of cargo: salt, rice, wheat, sugar, bananas, and bottled beverages. There were more than 200 mostly dirty, smelly passengers on board, taking up every inch of space. Many of them strung up hammocks in every conceivable niche, where they could sleep for the overnight 8-hour passage. Since Noble and I could not get to our beds in the van, it was necessary for us to try sleeping sitting up in the front seats of our cab.

Surprisingly, the ferry left right on time, at 8:00 pm. The total cost of our trip, for the two of us and our Jeep, was only 600 sucres, or about $35. It soon became apparent that I could not sleep sitting in the cab, no matter how much I tried to position myself comfortably. I got cramps in my legs and back. So, I took the two wire-meshed front seat cushions that Noble and I usually sat on, and climbed up to the flat roof top of our camping van and lay down horizontally on the cushions. This worked satisfactorily until 30 minutes after departure, when it began to rain quite heavily. So, I grabbed the two cushions and slithered under the Jeep, in the very small space between the Jeep's underside and the deck. Sure enough, it wasn't long before the deck, too became very wet from the rain. As a last resort, I climbed into the back seat of the vacant little Jeep next to ours, and here I was able to curl up horizontally and actually fall asleep for a couple of hours. Noble managed somehow inside the cab. At 4:00 am, right on time, there was a very loud, long blast from the ship's whistle as we approached the lights of Puerto Bolívar.

Because the tide was out, the sturdy concrete pier at the shoreline was too high above the ferry for us to tie up. So, we had to utilize an adjacent, lightly constructed floating pier. This lightweight pier was not strong enough to support our heavy van, so while the other three vehicles and all the passengers and cargo were unloaded, we sat and waited 4-hours for the tide to come in sufficiently for the ferry to tie up at the concrete pier. I amused myself by watching the dozen or so immature Laughing and/or Franklin's gulls foraging in the harbor, plus I took notes on my first Gray-hooded Gulls*. When the time finally arrived for our Jeep to be unloaded, some very heavy wooden planks were placed between the ferry and the concrete pier, to support the weight of our van. Even then, I drove off with difficulty, while Noble took moving pictures and laborers looked on with great interest.

Puerto Bolívar was the checkout point for Ecuador, even though the Peruvian border was still almost 50 miles away. The customs and immigration offices opened at 8:00 am. An hour was required before all the necessary formalities were completed and we were allowed to depart. We drove away on a bare, unmodified dirt road, which was badly rutted and would have been impossible for us if it had been muddy. Fortunately, it was quite dry. We hoped it wouldn't rain. As we neared the Peruvian frontier the countryside became more arid and the roadside vegetation changed to dry, cactus-thorn scrub where there were some Saffron Finches and my first Long-tailed Mockingbirds*. We were in a vastly different ecosystem as we exited southwestern Ecuador rom that of where we had entered the country in the north, on the high páramo. In our 10 days of travel, we had glimpsed only a few of the many contrasts within this fascinating nation, and we had survived a near catastrophe. Sixty-two species of birds were added to my trip list (mostly identified from my field notes at a later time). I would have to return.

Chapter Eight—Peru

At 1:00 pm on Fri., Apr. 17, we arrived at the insignificant little town of Huaquillas on the hot, dry, dusty frontier near the Pacific Ocean between Ecuador and far northwestern Peru, at the latitude of 3.5 degrees south. This was the entry post into Peru. It required 45 minutes to pass through the "inmigración" and "aduana" offices, where our passports and carnet de passage were stamped. Tumbes was only 15 minutes up the road, where once again we were required to have our documents inspected. Here we filled up with gasoline and water, and then proceeded onward. To our great surprise, amazement, and pleasure the Pan-American Highway here was new asphalt, smooth and without potholes!

We drove several hours southwestward, paralleling the nearby Pacific coastline. The scenery was generally featureless and the countryside was mostly barren with little vegetation, but there was a pleasant cool breeze from the ocean and at times we could see endless white beaches and the very blue Pacific. Roadside birds included Magnificent Frigatebird, Brown Pelican, Turkey and Black vultures, Ringed Kingfisher, Red-breasted Meadowlark, Tropical Mockingbird, Scrub Blackbird, Vermilion Flycatcher, Crested Caracara, Groove-billed Ani, Saffron Finch, Pacific Parrotlet, and Pale-legged Hornero. Just before sunset, about 7:00 o'clock, we stopped for a welcome bath and short swim in the cold waters of the Pacific's Humboldt Current. Shortly thereafter, we chose a small sandy hilltop for a campsite, about 50 miles southwest of Tumbes. It had been a long, eventful day and we both slept very soundly, with the ocean breeze blowing through the open windows of our van.

I arose at daybreak the next morning, Saturday, April 18, and spent a delightful hour searching for birds in the desert while Noble slept a little longer. It was a beautiful, cool, sunny morning and a variety of birds were active in the scattered vegetation along a nearby dry streambed. Desert environments always provided me with a particular thrill and a feeling of remoteness, of being alone with nature. It was particularly important to get an early start before the mid-day heat appeared and activity diminished. Several species of endemic birds lived in this arid northwestern corner of Peru. The five species that I encountered were the Coastal Miner* (a very pale, sandy colored, beach loving, lark-like furnariid in the genus *Geositta*), Cinereous Finch* (a large Gray finch with yellow legs and bill, in the monotypic genus *Piezorhina*), Superciliated Wren* (a plain, light colored *Thryothorus* wren), Tumbes Sparrow* (one of the 14 species in the New World genus *Aimophila*, only 2 of which lived in S. America; I sketched its distinctly marked head), and Necklaced Spinetail* (a pale, slender, wren-like furnariid in the genus *Synallaxis*, with a band of black streaks across its chest). Desert birds are characteristically paler in color than their relatives in other environments, since matching the low color intensity of their arid surroundings is critical to avoid predation in a region where the cover of vegetation is sparse. It was one of my most pleasant mornings. (These species were all identified one week later when I compared my field notes with specimens in the natural history museum "Javier Prado" in Lima.)

As Noble and I proceeded south on the newly surfaced Pan-American Highway we were allowed the very rare luxury of travelling at 50 mph. However, our progress was initially slowed by numerous police highway check posts where we were required to stop. Our vehicle was inspected and forms were filled out which provided authorities with our vehicle license plate number, names, nationality, passport, visa numbers, marital status, and age. For our occupation, we said "student." All this paperwork amused me somewhat and I always wondered what anybody ever did with it. Stops at control posts became less frequent as we drove farther south of the Ecuadorian border. With so many stamps in our passports, it was time consuming for the inspecting officials to find our Peruvian visas, so to assist them we attached paperclips to the correct pages. Most of the officials were friendly and curious, and they usually wished us "Buen viaje" (have a good trip) or "Feliz viaje" (happy travel) as we drove away.

The entire Pacific coast of Peru is a narrow strip of very dry desert, with less than an inch of rainfall annually. This desert resulted from the fact that Amazonian rain-bearing winds coming from eastern Peru lose their moisture as they rise and go up over the very high Andes, and moist oceanic winds coming from the west lose their moisture as they cross the very cold northward flowing Humboldt Current, which originates in the Antarctic and parallels the entire coast of Peru not far offshore. Therefore, the continental coastal environment is one of very sparse rainfall. The vegetation and sand dunes were somewhat reminiscent of the Imperial Valley of southern California or the Sonoran Desert of southern Arizona. Tall columnar cacti and mesquite were the most conspicuous features of the landscape. The region was irrigated only where water was available from streams and rivers coming down out of the Andes and across the desert. In these stream valleys, cotton was cultivated and such livestock as cattle, pigs, goats, horses, and donkeys were raised. Domestic animals often wandered across the unfenced highway, where locally they posed a considerable hazard to traffic. The few houses were constructed of adobe with tile or mud roofs, and had dirt floors, no glass windows, and electricity only from gasoline generators. Water came from wells or a stream. A local farmer whom we encountered along the road complained to us about the current lack of rainfall!

As we drove south along the coastline, toward the town of Talara, I was surprised to observe an Osprey, although this worldwide species was a not uncommon migrant in the area (with no known breeding). Another raptor with a worldwide range, which I saw along this route, was a Peregrine Falcon, a regular visitor from the Nearctic during the northern winter and a rare breeding resident in the Andes (and perhaps also locally here along the coast). Two Variable Hawks were seen, lazily soaring overhead. Birds which were foraging along the beach or just offshore were 300 Peruvian and/or Brown Pelicans (which I did not attempt to separate), 7 Peruvian Boobies*, 60 Magnificent Frigatebirds, 2 Killdeers, 5 Gray Gulls*, 2 Kelp Gulls*, 500 Franklin's Gulls (a non-breeding visitor from their nesting region in the upper Great Plains of N. America), and 1 Elegant Tern (another boreal migrant).

At the coastal town of Talara, our Pan-American Highway turned inland to the city of Piura, and from there it climbed upward slightly in elevation for 15 or 20 miles and entered an arid temperate habitat with woody, chaparral-like scrub in the foothills of the western Andes, through which we traveled to the town of Olmos. We stopped to bathe in a freshwater stream. Along this route I recorded a Streak-headed Woodcreeper (which I had first seen in Costa Rica) and two species I had not previously seen, the Snowy-throated Kingbird* and the Pacific Dove* (formerly considered a race of the White-winged Dove). Other birds were a Variable Hawk and a Green Kingfisher. Croaking Ground-Doves and Peruvian Meadowlarks were particularly common and Long-tailed Mockingbirds were abundant, with an estimated count of 300 individuals in 3 or 4 hours of travel. From Olmos we continued to meander briefly through the foothills and then we descended back to the coast again, to the city of Chiclayo. That night we camped just south of the city, along the shore.

The next day (Sun., Apr. 19), we drove along the coast for 6-hours from Chiclayo to Chimbote, arriving in the afternoon about 2:00 o'clock. The highway continued to be new, and there was little change in the coastal scenery or the birdlife. An American Kestrel and an Andean Swift* (*Aeronautes andecolus*) were the only two additional species which I observed. Noble and I spent two hours in Chimbote, drinking a beer at a nice little restaurant and photographing the operation of a very large anchovy factory, which was processing these little fishes for fertilizer. Small fishing boats were unloading their catches at a wharf, which attracted almost a thousand Franklin's Gulls and a few Gray, Kelp, and Gray-headed gulls, plus more than a hundred Peruvian Boobies, a dozen Peruvian/Brown pelicans, and a few Guanay Cormorants*.

Noble and I left Chimbote and drove another 2 hours to a campsite near the town of Barranca, about 150 miles from Lima. Along the way, we passed an Osprey on the top of a roadside telephone pole. At our campsite on the beach, we chased sand crabs by moonlight and amused ourselves by shining the beams of our flashlights on flying nightjars, in an area of sparse vegetation. The reflection from their eyes produced bright orange spots which darted and zigzagged in the night sky, creating trails of light in unusual and weird patterns as they flew past us. It was almost surreal. These nightjars had a white band in their primaries and white at the corners of their tail. They were not nighthawks and I eventually identified them as Band-winged Nightjars*. It was a refreshingly cool evening and I climbed inside my sheet "sleeping sack" instead of lying on the top of it, for a pleasant change.

The next morning (Mon., Apr. 20), I counted the marine birds, which were foraging along the beach at our campsite: 30 Peruvian/Brown pelicans, 25 Peruvian Boobies, 125 Red-legged Cormorants*, 450 Gray Gulls, 1 Belcher's Gull*, 4 Kelp Gulls, 5,000 Franklin's Gulls (migrating northward just offshore, toward their distant breeding grounds in the central Great Plains of N. America), and 120 Black Skimmers. The cormorants were perhaps the most handsome and striking of all cormorants in their overall Gray appearance, conspicuous white neck stripe, orange bill, and bright red legs and cere. They frequented rocky shorelines with cliffs.

We departed our campsite at 8:00 o'clock for the 5-hour drive to Lima. En route, a spectacular adult Andean Condor came soaring low over the highway. This magnificent bird resided in small numbers along the Pacific coast of S. Am. (as I have previously mentioned), where it scavenged on the carcasses of marine mammals. Another sighting was that of a single Tricolored Heron which was foraging at the edge of a marine estuary. This species occupies the warmer parts of the western hemisphere, in both North and South America. Additionally, approximately 3,000 Gray Gulls were tallied along the shore between Barranca and Lima.

We arrived in Lima at 1:00 pm, almost three months and 10,000 miles after our departure from California. The 12°S latitude was approximately halfway between Los Angeles and the southern tip of South America. It was the beginning of the 2-hour siesta period and almost all businesses were shut, except restaurants. Therefore, we chose a place to sit and enjoy a leisurely, hot roast beef lunch and a cup of coffee, for a total cost of just 14 soles (50 cents). Afterwards, we picked up our mail at the American Embassy and then filled up our van's 40-gallon tank with gasoline, at the most welcome low price of 12 cents per gallon. Lima was a very clean, modern city, the third largest in South America (with a population of more than 2 million), and the cultural center of the continent. Many Americans and Europeans were living here. Like everywhere else we visited in Latin America, the girls were said to be the prettiest in the world!

Once again, it was necessary for us to locate the local Jeep agency in the city, for various maintenance and repairs, most significantly to try yet again to solve our radiator problem. The Jeep agency here in Lima was ultra-new and modern, under the name of "Ferreyco." We arrived there at 6:00 pm, right at closing time, but were directed to the office of Mr. Neves, one of the senior managers. We explained our radiator problem to him, and he told us that he had not yet received any correspondence, or radiator, from Mr. T.S. Day at the head office of Jeep in Toledo, Ohio, but that tomorrow he would send our broken radiator out to a radiator specialty shop for repair. He agreed to let us leave our camping van at the Ferreyco agency, and sleep in it at night while it was being repaired (as we had done on many prior occasions during our travel). We were in a guarded, locked compound where the guard at night wore a scarf around his neck for warmth because the night breeze coming in from the ocean felt quite cool to him even though the outside air temperature was 70 degrees.

The following morning (Tue., Apr. 21, our 2nd day in Lima), we were introduced to "Tex", a big American who owned a small nearby radiator shop and certainly looked like a Texan, with a cowboy hat and boots. He had been in Peru for 10 years and had once studied at Texas A&M University. After critically examining our original (once repaired) radiator, which we had broken in Costa Rica, he declared that the problem was not enough cooling fins per inch, and he could correct this for us. We told him we might be receiving a new radiator from the U.S. before we left, and he said "No problema"; if it should arrive, he would simply keep our old radiator and not charge us for his work. Tex was a pleasant, cooperative person with whom to work, and he was sympathetic with our adventure. The mechanics at Ferreyco began working on our Jeep, the major necessity being to add an additional leaf to each of the two front springs, since they were still not strong enough to adequately support our overweight camping van. We also needed two new tires, an engine tuning, and another attempt to eliminate the shimmy in our right front wheel (since the repair work done in Guayaquil proved faulty). In addition, it was necessary to send our Briggs & Stratton electric generator to an outside shop for routine maintenance and servicing. We were told that all these tasks would require 2 or 3 days, but of course, we knew from previous experiences that this period of time would be an underestimate, even though the manager emphasized his estimate by emphatically declaring "claro." So we sent our laundry out for washing and busied ourselves with writing notes and letters home.

To our great astonishment, we encountered two young French boys who also were having some work done on their vehicle at the Ferreyco garage. They were travelling in a 1952, 2WD Jeep station wagon which they had purchased in Connecticut several months earlier. They had painted a large French flag on the back of their station wagon. Jean (John) Roumanteau and Raymond Abecassis were recent graduates of a business school in Paris, and were spending the graduation money given to them by their parents to buy a vehicle in the U.S. and travel in it from Canada south through Central and South America. Like us, they were constantly troubled with all kinds of mechanical problems. We had dinner with them that evening and we all agreed that the four of us should travel together in the Andes for a couple of weeks or so, meeting in Cuzco in about 6 or 7 days. Both of them were multilingual and spoke Spanish quite well (along with French and English). They introduced us to three girlfriends of theirs and we all ate dinner together.

At 8:45 the next morning (Apr. 22, our 3rd day in Lima), we were surprised by an unexpected knock on our van door, and when I opened it there stood a middle aged man who identified himself as Dr. Aguilar, an entomologist with a Ph.D. from San Marcos University in Lima (founded in 1551). Very coincidentally he, too (like Tex), had studied for a while at Texas A&M University. He worked for the government on insect control in cotton crops. It was not clear who had informed Dr. Aguilar of our presence at the Jeep garage (maybe Tex since there seemed to be a small group of acquaintances in town who had all attended "A&M" at some time in the past). Dr. Aguilar offered to assist us in any way he could, so I told him of my interest in birds and my desire to visit the bird guano islands just off the coast, and of my wish to visit the Museum of Natural History. He said he could help in both matters.

Later that afternoon, Dr, Aguilar (whose first name I never learned), took me in his Jeep to the office of the government's "Compañía Administradora del Guano." We could not find any officials of importance there, so we gave up and he then took me to the home of Drs. Hans and Maria Koepcke, a middle aged German couple who, for many years, had worked as research scientists at the Museum Javier Prado, he as an ecologist and she as a world-known ornithologist. Both were friendly and they invited me to visit the museum tomorrow morning, where I could examine bird specimens. They suggested that Dr. Aguilar and I should talk to Mr. Basca of the Dept. of Agriculture for obtaining permission to visit the Guano Islands. So, we drove to his house and spoke to him. He promised to do his best, on such short notice, and would call me at the museum the next morning. Finally, Dr. Aguilar drove us to the "Domino Café", where we met Noble (as previously arranged), and all three of us ate dinner together, with Dr. Aguilar being our guest. Afterwards, at 11:00 pm, Noble and I were driven by Dr. Aguilar back to the Jeep agency for another night.

Streetcars cost about 5 cents a ride so I took one to the museum at 10:00 o'clock the next morning (Apr. 23, day 4), as scheduled. Here I once again met Maria Koepcke, who very graciously spent 2 hours helping me to identify the birds which I had recorded in my field notebook from Peru (during our drive along the NW coast from the Ecuadorian border to Lima). We were very successful. She also presented me with a gift, a copy of her 120 page paperback publication *Memorias del Museo de Historia Natural 'Javier Prado'*, No.3, Parte 1, 1954, which dealt with the ecological distribution of birds in central Peru, from the Pacific shoreline to the high Andes. It was written entirely in Spanish and contained no color illustrations, only a few black-and-white line drawings of birds, depicting some of them in their appropriate habitat. Never-the-less, this was my first and only reference to any birds in Peru, and I spent quite a bit of time studying it prior to departing Lima. I have preserved this little booklet in my library as a treasured memory. I thanked her for all her assistance when I left that morning, and said I would try to return at least one more time.

On Saturday morning, Apr. 25 (day 6), I returned to the museum for one more hour, to continue working on my notebook identifications. While I was at the museum, I received a telephone call from a Mr. Gamarra, who was second in command at the "Compañia Administradora del Guano." He said he would accompany me to visit some of the Guano islands on Monday (two days away), and that I should meet him at the southeast corner of the Plaza de San Martín at 8:30 in the morning. That was very good news indeed, as Noble had decided that he also wanted to see the Islands.

That same morning we were told by Ferreyco that our Jeep and radiator were all repaired and ready for a test drive. To test drive the Jeep I decided to try and find a lagoon about 20 miles south of Lima, which Maria Koepcke had mentioned to me as being a good site for observing birds, in the Lurin River Valley. Soon after we left the city we discovered, to our great dismay, that although the performance of the radiator was somewhat improved, it was still not operating efficiently. We were not satisfied. Tex had not succeeded. I couldn't find the lagoon, so we decided to go to the beach for a swim and I could look for birds there. In a couple of hours, I counted (in groups of a thousand birds each) more than 300,000 Guanay Cormorants flying in a continuous stream of birds paralleling the shore, in a northerly direction toward the islands where they roosted and nested. These birds were the primary producers of the guano that sustained the lucrative guano fertilizer industry. In addition to the cormorants, I also counted 5,000 Gray Gulls and 40 Inca Terns*, a very handsome, distinctive tern which was an inhabitant of the rocky Pacific coasts of Peru and Chile. It was sooty Gray all over with a black crown, bright red feet and bill, and a conspicuous long, white line of feathers on its face below the eye, which curled backward, down, around, and behind the auriculars. There was no other similar tern in the world, and it was easily the most outstanding tern I had ever seen.

That evening we visited a nearby bawdy nightclub where there were drinks of all kinds and topless dancing girls with bikini bottoms. Afterwards, we drove back to the beach and found a campsite for the night. The next day (Sun., Apr. 27, day 7 in Lima) was a day of leisure for us since our boat trip to the Guano Islands was not until tomorrow. I took the Jeep to the fields and pastures in the nearby Lurin River valley and spent most of the day recording the names or writing descriptions of almost 50 species of birds. These included 10 new species for my trip list: Yellowish Pipit*, Plumbeous Rail*, Masked Yellowthroat*, Hooded Siskin*, Parrot-billed Seedeater*, Drab Seedeater*, Short-tailed Field-Tyrant*, Collared Warbling-Finch*, White-crested Elaenia*, and Peruvian Martin* (*Progne murphyi*, which was not identified until *Birds of Peru* by Schulenburg, *et. al.*, was published in 2007). We drove back to the Ferreyco agency after dark.

The next morning (Mon., Apr. 27, day 8), was our much-anticipated visit to the Guano Islands just off the coast from Callao. Our alarm clock woke us up at 7:00 o'clock. Breakfast consisted of an orange, a pear, and bread with grape jelly. We arrived at the SE corner of the Plaza San Martin as scheduled, at 8:30 am. With us we carried still and movie cameras, telescope, binoculars, notebooks, and a 300mm lens for the more distant pictures with my little Canon camera. Mr. Gamarra was unable to meet us, but he sent his brand new 1959 Ford with a driver to take us to the pier in Callao. Our chauffeur was a dark-skinned man of African descent which was a rare sight in Peru. It was only a 30-minute drive.

At the pier in Callao we were transferred to a small diesel powered boat with a driver who wore a white shirt and tie, and spoke only Spanish. He took us about three miles to the largest of the nearby Guano Islands, which had a rugged, steep, rocky shoreline. Here we were transferred to yet another boat, with a different guide. For more than an hour we patrolled around the shore of this island, and several others nearby, observing the activities of both the birds and the guano workers, and taking pictures from the boat. The water was sometimes rough and there were many rocks and breakers, making photography difficult. I was excited to see two Humboldt Penguins* standing on shore. This penguin is one of only two penguin species to nest inside the tropics (the other being the Galapagos Penguin), both of which take advantage of the cold water in the Humboldt Current.

The Humboldt Current is very rich in minerals and nutrients which have been brought upward to the surface from the bottom of the "Antarctic Ocean." These minerals are then transported northward in the surface waters of this current all the way into the tropics, in the eastern Pacific. Because of the nutrients, an abundance and variety of oceanic marine life are produced, particularly plankton and small fishes. It is this super abundant food source which attracts tens of thousands of marine birds (some say millions) to forage along the Peruvian coast, and to nest on the small, barren rocky islands (and coastal mainland). Here on these islands the nesting birds produce hundreds of tons annually of "guano" (digestive excrement). Because there is no rainfall the guano is not washed away or dissolved, but simply builds up on the surface of the ground, until it can comprise a layer over 100 feet thick. The guano is extremely rich in nitrates and phosphates. Once this fact was discovered by the Peruvian Indians centuries ago, they began to harvest it for use as agricultural fertilizer. The industry eventually became one of the most important sources of revenue for the Peruvian government, which owns the islands and since the early 1900's, has closely regulated the harvest and sale of the guano.

The vast majority of the nesting seabirds, perhaps 90%, were Guanay Cormorants. Peruvian Boobies were the second most abundant, followed by the many fewer Peruvian Pelicans, Red-legged Cormorants, Kelp, Gray, and Belcher's gulls, and Inca Terns. On the guano islands, these birds produced a magnificent spectacle, all crowded together side by side, covering almost every square foot of space on an island surface. In addition to these marine, nonpasserine birds, I observed a single Peruvian Seaside Cinclodes* along the rocky shoreline of one of the islands. This furnariid species is regarded as the most marine-foraging of all passerine birds (along with its "twin", the Chilean Seaside Cinclodes, with which it was formerly lumped under the name "Surf Cinclodes"). These birds are not unlike a Ruddy Turnstone in their behavior and habitat.

Our boat landed on one of the larger islands, not without some difficulty because of the rocky shore and large breakers. Noble and I spent two hours walking around the island, taking many photographs of the birds and workers. The guano was harvested entirely by Andean Indian workers who were employed by the government for periods of only 3 months at a time, to limit their exposure to the harmful guano dust. They lived in gunnysack tents on the islands and were mostly barefoot, but they wore long sleeved shirts, long lightweight cotton pants, and either a wide-brimmed hat or a scarf tied around their head. Some also wore a scarf over their face to avoid inhaling the fine dust. The guano was harvested by brushing or scraping it from the ground, using picks, shovels, and brooms. It was then shoveled into upright gunnysacks standing on the ground, in closely spaced lines. The sacks were eventually hand-carried to the shore and onto small lighters, which then transported them to an offshore freighter, for further transport to processing plants.

As guests of the government, Noble and I were treated to a noonday meal on one of the islands. The featured item on the menu was "criadillas" (bull testicles)! There were also fried eggs, potatoes, steak, coffee, and bananas for dessert. In mid-afternoon, we boarded our small boat for the return trip to Callao. We arrived back at the pier sunburned and covered in very fine guano dust. It was a day to be remembered.

Tuesday, April 28, was our 9th day in Lima. Two astounding events occurred. First, it rained! For a city with an average of less than 2" of rainfall annually, this was great cause for celebration. Some of the city streets even flooded since there were no drainage ditches to carry off rainwater. The second event was even more remarkable. A brand new radiator and fan for our FC-170 Jeep arrived for us from Mr. T.S. Day in Toledo, Ohio! We could not believe our good fortune. After two months of struggling with inefficient engine cooling, our problem

Plate 11: Nothern and central Peru have a long, sandy coastline. Guano islands provide a substantial income. The Inca tern, to my eye, is the most attractive of all the 44 species of Terns in the world.

was finally solved! The 59 day saga of our broken radiator finally came to an end. It had been a long, arduous, 8 weeks, which had tested our resolve to the limit. Noble had taken the lead in our many efforts to solve the problem. It had afforded him the opportunity to meet and communicate with other entrepreneurs like himself.

Along with the radiator and fan, Mr. Day had sent a letter (which we did not see) to the Ferreyco agency. The only expense the agency asked us to pay was the cost of the airfreight for the radiator. Everything else was free of charge. Juan, of the radiator shop, true to his promise did not ask us to pay him anything for his efforts. We gave him our old radiator. We celebrated by offering everyone a glass of wine. By the time the radiator was out of customs and installed in our Jeep, late that afternoon, there was not enough time left in the day to depart Lima, so we spent one more night at the agency, sleeping in our Jeep. It was the first night in a very long while that we could sleep without any major vehicle worries on our minds.

We arose at dawn the next morning (Wednesday, April 29), and critically perused our very primitive road map for Peru. It showed a land route of unpaved roads from Lima to Cuzco, via the Andean towns of La Oroya, Huancayo, Acostamba, Huanta, Ayacucho, and Abancay. Also shown was a railroad track going up the steep western slope of the Andes directly from Lima to Huancayo, traversing a mountain pass at 15,686 ft. above sea level. It was the world's highest standard gauge railway. Noble very much wanted to travel this train route and add the adventure to his list of accomplishments. I could drive the Jeep to Huancayo and he could take the train. We both agreed to this idea, so he booked himself a one-way first class ticket (at a cost of under $4) for the 9-hour train ride tomorrow, which would arrive in Huancayo just before evening. If I left immediately in the Jeep today I would have two full days for travel before meeting him at 6:00 pm, at the Hotel de Turistas in Huancayo.

I departed Lima shortly after this decision was made, at 8:00 am, for the 200-mile drive by myself. Our only map showed the highway to be paved from Lima all the way to La Oroya. I soon discovered, however, that this was far from true. The pavement ended after the rather short 15 or 20-mile distance to Choisica. From there it was mostly narrow and in very poor condition. An American priest I met in the town warned me to drive very cautiously, as the road beyond was extremely hazardous. It was gravely, wet, slippery, full of potholes, ruts, and boulders, with numerous landslides, washouts, and detours because of recent heavy rains. I drove almost entirely in 4WD, in lower gears between 5 and 15 mph. Once I stopped for 10 minutes to change a flat tire (which both Noble and I had learned to do very quickly and efficiently, with relatively little effort using our small hydraulic jack). On another occasion, at a detour, I went the wrong way and shortly thereafter had to turn around and go back. The road was conly going up, at times quite steeply and with sharp switchbacks. The good news was that the radiator never once overheated, even the slightest!

The road followed the Rímac River almost all the way to a pass at 15,885 feet elevation (very slightly above the elevation of the railway pass). The countryside changed from coastal desert to arid subtropical woodlands, to temperate scrubby woodlands, and finally to high elevation, dry "puna" grasslands above 11,000 feet. Almost everywhere the vegetation was sparse and limited, and the mountainsides were mostly brown and quite barren. There were a few cacti and some red or yellow flowering shrubs, but very few trees. The few people who inhabited these steep mountain slopes were mostly Indians, dressed in bright, warm dresses, shawls, and capes, though often barefoot. Any pedestrians along the roadside almost always wanted me to stop and give them a ride. They stared at my vehicle as I drove past. A few small roadside mines of copper, lead, or zinc were in operation. I reached the pass at 2:00 pm, after an exhausting 6-hour drive of 75 miles. It was very cold and wet, with intermittent rain, sleet or snow. Nearby mountain peaks were all snow-covered. My head was throbbing and my heart was pounding. Uncharacteristically, I had a headache. Obviously, I had ascended from sea level to almost 16,000 feet too rapidly.

On my way up to the pass, I stopped very few times to investigate roadside birds. My first stop was at 5,000 ft. elevation where there were some scattered trees. Here I was happy to see and write a description of the local, endemic Black-necked Woodpecker* (*Colaptes atricollis*, in the same genus as flickers). My notes for this species were: "woodpecker: top of head and nape red; cheeks pale grayish; throat black; upperparts more or less olive (bronzy?); underparts barred with dark (and tinged yellowish)." At this stop I also recorded a Purple-collared Woodstar* (a hummingbird with a slightly decurved bill and rather long, somewhat forked tail), 3 Golden-bellied Grosbeaks, and of course a Rufous-collared Sparrow. A second brief stop was at 8,000 ft., where I saw my first Greenish Yellow-Finch*, along with 10 Ash-breasted Sierra-Finches and a single Band-tailed Seedeater. An Andean Swift was swooping overhead at 10,000 feet, Black Siskins* were at 13,000 feet, and a White-winged Cinclodes* was documented at 13,500 feet. At an elevation of over 14,000 feet, I took notes on a group of a dozen Cinereous Ground-Tyrants* (see comments below) and Peruvian (formerly Gray-hooded) Sierra-Finches*, two species characteristic of high elevation puna grasslands. In addition, there was a particularly exciting White-winged Diuca-Finch* at 14,500 feet. This handsome species is said to breed at higher elevations, up to 18,000 feet, than any other South American passerine, and to roost in the crevasses of glaciers! It was a lovely subdued bluish-Gray color with a large white throat patch and a conspicuous white wing patch, reminding me somewhat of N. Am. rosy-finches in its overall demeanor and ecological niche. I added it to my list of favorite birds.

I had considerable difficulty trying to figure out which species of ground-tyrant I saw (in the genus *Muscisaxicola*, family Tyrannidae). My field notes read: "about 5 inches; lark-like; pale bluish-Gray above; whitish below; black lores; primaries blackish; tail medium short (flicked open and shut), squarish, black with narrow white sides; bill slender, black; legs black; terrestrial; 14,000 ft." I counted a total of 14 individuals, all at the same stop. After much study, I eventually decided they were Cinereous Ground-Tyrants* (*M. cinereous*), an austral migrant to Peru from the south, from the Andes of Bolivia, Chile, and Argentina.

A potentially life-threatening incident occurred suddenly about 4:00 pm when I was driving at 25 mph toward La Oroya, at an elevation of 12,500 feet about 5 miles outside the town. Quite suddenly, without any warning, my brakes failed completely. The brake pedal went all the way to the floor board with no resistance, and there was no stopping power to the brakes at all. Because of the van's tremendous weight, the emergency hand brake had been virtually useless since the day we departed California. Very luckily there was no incline to the road and I was able to coast slowly to a stop, using low gear and the soft earth on the shoulder to gradually slow my speed. I got out of the Jeep and put it in 4WD, then proceeded onward toward La Oroya very slowly, in my lowest gear. Our heavy vehicle had great momentum and was always difficult to safely bring to a stop.

Travelling at a snail's pace without any brakes I entered the town of La Oroya, late in the afternoon. I looked for any sign along the street that read "servicio" or "taller de mecánico." Surprisingly, there were none at all. I stopped to ask a passing pedestrian and he said "mas alla" (farther ahead) and motioned up the street in the direction I was going. I continued, and not long afterward I came to the outskirts of the town, on my way out, without having encountered any repair facilities. At that point I happened to notice that I was just outside an entrance to a large copper mine belonging to the "Cerro de Pasco" company. On the spur of the moment I drove up to the main gate and explained my problem to the guard. He directed me to the main office, where I found several Americans and a Canadian with whom I could converse. I was taken to the Chief Engineer, who told me it was against company policy to service private vehicles, but because I was "on a trip" he would agree to do so, and he took me to the company garage where vehicles were repaired. I arrived there 15 minutes after official closing time but the young Peruvian supervisor, who had studied in the United States and spoke English, agreed to work on my Jeep if I paid his workers overtime compensation. Of course, I said yes. But when he saw my camping van and learned of the adventure on which I was embarked he immediately said there would be no charge at all for any of the repair!

The problem was that one of the small copper brake fluid lines had broken, and the jeep had lost all of its brake fluid. The line had broken because road vibration had caused the left rear shock absorber support bracket to lose both of its bolts and come loose. The broken bracket then banged against the brake fluid line causing it to become separated from the brake drum, and all the brake fluid ran out. By 7:15 the repairs to the brake line had been completed, but it took another 45 minutes for a mechanic to refill the brake fluid reservoir. I thanked all the people who had provided me with assistance, and I presented a bottle of Pisco to the two mechanics who had performed the labor. As I left the garage I was told that the road from La Oroya to Huancayo would be better than the one I had just traveled, from Lima to La Oroya. I fervently hoped this would be true.

I drove 5 miles out of town before pulling off to the side of the road for the night, at 9:00 o'clock and an elevation of 12,000 feet. My first full day of driving all by myself had been quite an adventure. The last 30 minutes provided even more excitement, as I experienced the insane Latin American driving custom where oncoming trucks at night would turn off their headlights entirely as they approached one another, leaving both vehicles in almost complete darkness (except for colored lights distributed all over their trucks like a Christmas tree). If I didn't reciprocate and also turn off my headlights, which I didn't, the oncoming driver would shout at me and shake his fist as we passed. I tried to keep my composure and told myself it would not be an adventure if everything were done the same way as in the United States. On the more positive side was the fact that our new radiator never once registered a temperature above the normal mark, even though it had been driven almost nonstop upward on the steep western slope of the Andes for over 6 hours, from sea level to 16,000 feet. Three cheers and a hallelujah!

I got up at daybreak the next morning, 6:00 am on April 30. The road to Huancayo was quite good gravel, with even some pavement toward the end. It fluctuated slightly in elevation between 11,000 and 12,500 feet on the altiplano (high plain) as it followed along the Mantaro River valley. There were fields of corn and wheat, yellow flowering shrubs, rock fences, and eucalyptus trees for windbreaks. The rural population was almost entirely Quechuan Indian, and these indigenous people compromised more than half the total population of Peru, which had more Indians than any other South American nation. Both men and colorfully dressed women worked in the fields. Women spun yarn onto a spool everywhere they went, often while carrying a baby on their back. Donkeys laden with wheat, corn, grasses, or reeds were herded along the road. People stared at my Jeep curiously.

There was no hurry for me to get to Huancayo, so I took the opportunity to stop regularly for roadside birding. My list of 32 species for the day's 100-mile journey included 14 lifers for me (see Appendix A). Not surprisingly, Rufous-collared Sparrows were the most numerous birds, followed in abundance by Ash-breasted Sierra Finches, Hooded Siskins, Black-winged Ground-Doves, Blue-&-white Swallows, Chiguanco Thrushes*, and Bare-faced Ground-Doves*. Less common species included the Cinereous Harrier, American Kestrel, Andean Lapwing, Andean parakeet*, Andean Flicker*, Bare-winged and White-winged cinclodes, Plain-breasted Earthcreeper*, (a furnariid in the genus *Upucerthia*, reminiscent of N. American thrashers), d'orbigny's Chat-Tyrant*, Andean Negrito*, Taczanowski's Ground-Tyrant* (I struggled greatly before I arrived at this identification), Andean Swallow*, Golden-billed Saltator*, Grassland Yellow-Finch*, Rusty-bellied Brush-Finch*, and mourning Sierra-Finch* (which, with its prominent black bib, brought back memories of the wintering Harris's Sparrows I saw in my early boyhood days in eastern Kansas).

The Giant Hummingbird was the world's largest hummingbird, an inhabitant of high-elevation, open, shrubby areas in the Andes, up to 13,500 feet above sea level. Although not colorful in its bronzy-green and rufous-brown plumage and pale whitish rump, it was nevertheless an exciting bird to see. Its wing-beats were much slower than in other hummingbirds, and it often sat on flowers or branches to feed, rather than hovering. There are 340-350 species of hummingbirds in the world, all of them confined to North and South America, mostly in the tropics and predominantly at lower elevations. A graph which plots the number of breeding hummingbird species on the vertical axis versus degrees of latitude on the horizontal axis, between 60° N latitude and 55° S latitude, would result in a remarkably symmetrical bell-shaped curve, with the highest part of the curve centered almost exactly at 0° (on the equator) and the two extreme latitudes having only one species each (the Rufous Hummingbird in southern Alaska and the Green-backed Firecrown in Tierra del Fuego)!

I met Noble at the hotel on schedule, at 5:15pm. We searched for the highest-priced gasoline station, hoping it would have higher octane gasoline, but costs were just about the same everywhere (15 cents a gallon) and the octane was never very good (70-80, occasionally less). Sometimes it was pumped by hand into our gas tank from a 50-gallon drum sitting on the ground behind a private house. After filling up, we left town in the dark, about 9:00 o'clock, on a dirt road that we hoped would take us to Ayacucho. As always, there were no road signs of any kind anywhere. We stopped twice to ask pedestrians for directions. One person said yes the road was going to Ayacucho, and the other person was too drunk to respond at all. We drove less than an hour before pulling off to the side of the road for the night, about 12,000 ft. on the altiplano.

The following summary of our travel during the next five days may help in understanding the sinuous, undulating, tortuous, 600-mile route we navigated between Huancayo and Cuzco – Day 1 (Friday, May 1): from just outside Huancayo (at 11,000 ft.) to a pass at 13,000 ft. to Acostambo (11,000 ft.) to the Mantaro River, where we drove within a steep sided gorge all the way down to 7,500 ft., then back up to just beyond Huanta (at 9,000 ft.); Day 2 (Saturday, May 2): from just beyond Huanta (9,000 ft.) to Ayacucho (9,000 ft.), then up to a pass at 14,000 ft., then down all the way to the Rio Pampas at 7,000 feet; Day 3 (Sunday, May 3): from Rio Pampas (7,000 ft.) to Chincheros, up to a pass at 13,900 ft., then down to 9,500 ft., then back up to 13,700 feet at the km 100 marker before Abancay; Day 4 (Monday, May 4): from 100 km before Abancay (13,700 ft.) to a pass at 13,900 ft., then down all the way to the Río Pachachaca at 5,500 ft., then up to Abancay (7,800 ft) and on up to a pass at 13,000 ft., from where we descended back down all the way to the Rio Apurimac at 6,500 ft.; Day 5 (Tuesday, May 5): from the Río Apurimac (6,500 ft.) to Cuzco (11,000 feet). If reading this makes you a little dizzy, you can imagine how it was for me to drive our camping van over this very primitive road!

The morning of May 1, (Friday), was overcast and chilly. An Indian man came walking along the road by our campsite carrying a wooden plow over his shoulder and told us, to our joy, that the gravel road we were on would take us to Ayacucho. After getting underway we soon ascended to 13,000 feet, and then began a gradual descent to Acostambo, at 11,000 feet. Indians were herding sheep, cattle, and horses. Women sat along the roadside weaving with spools of yarn. They looked at us with curiosity as we drove by. In Acostambo, Noble and I stopped and held up traffic while we took pictures (with both our movie and still cameras) of our first encounter with llamas being herded down the middle of the street. It was a scene with which we would become familiar in the following weeks of our travel through the high Andes. The road beyond Acostambo was narrow and winding as it descended to the upper region of the scenic, steep Mantaro River gorge at 10,000 feet. When we entered the gorge, the road became one-way for 75 miles, all the way to Huanta. Traffic proceeded in only one direction on any given day, and in the opposite direction on the following day. Fortunately, on the day we arrived the road was open in the direction we wanted to travel. We headed down the gorge, a truly magnificent, scenic, and picturesque natural wonder, comparable to any in the world. At times our road was hewed right out of the vertical rock wall of the canyon. The thorny shrubs and cacti in the bottom of the canyon reminded me of Arizona. After several hours of tortuous and hazardous driving we reached the lower end of the gorge, which widened and was joined by another river, at 7,500 feet elevation. We stopped by a waterfall, under which I bravely took a very brief shower in the frigid water, much to the amazement of two small children and an old woman who watched me from nearby. We also took the opportunity to wash a few clothes.

Our road was still one-way as we left the river and began ascending toward the town of Huanta, where we arrived at 5:00 pm, at an elevation of 9,000 feet. At this point, we passed through the exit control for the one-way road, and it became two-way again. Soon after leaving the town we pulled off onto a side road and stopped for the night, in an arid area still dominated by cactus and thorn scrub. We had covered only 125 miles all day, after 10 hours of long, tedious, and strenuous driving. I was quite exhausted. It was a warm evening, 65 degrees, and we ate a watermelon before retiring. Ayacucho was not far ahead. My bird notes for the day included 2 Cinereous Harriers, 12 American Kestrels, 8 White-tipped Doves, 7 Scarlet-fronted Parakeets (first seen in Colombia), 1 Striated Earthcreeper*, 1 White-capped Dipper* (one of five species of dippers in the world, two in S. America, one in N. America, and two in Eurasia), 4 Black-backed Grosbeaks*, and 4 Yellow-rumped Siskins* (generally considered to be a rare bird).

The next morning (Saturday, May 2), there were a couple dozen Eared Doves and 10 Croaking Ground-Doves foraging on the ground around our campsite. In the shrubs and small trees were Chiguanco Thrushes and Black-backed Grosbeaks. We hastily ate some cereal and an orange each, and started along the road toward Ayacucho. Our route continued to wind through arid cactus covered mountains. As usual we had to avoid the many domestic animals which meandered freely across the road - - pigs, chickens, goats, ducks, sheep, cows, and donkeys. People along the road were mostly barefooted, and dressed in bright reds, blues, purples, and many other colors. We arrived in Ayacucho at 10:30 am, at an elevation of 9,000 feet. Women at a fountain were drinking water out of their hats. We stopped to buy gas and to purchase more bread, and then continued onward toward Abancay. Our unchartered road became an adventure which tested our perseverance and my driving skills. We were continuously in 4WD and practically never out of low gear, travelling between 5 and 15 mph as we constantly wound back and forth on sharp hairpin turns and switchbacks on steep mountain slopes. From the top of a ridge at 14,000 ft. elevation we would wind down to the bottom of a river valley at 5,500-7,000 ft., cross a bridge and then start back up again all the way to the top of the next high ridge. There were frequent road signs that read either "sinuoso" (winding) or "angosto" (narrow). It was beautiful, scenic, and exciting, but it was necessary for me to navigate the narrow, gravel, uneven surfaced roads with extreme caution.

Noble and I would sometimes engage in little games of competition. On one occasion as our road commenced a very long steep descent with many switchbacks which were visible for several miles below us, in barren countryside, the thought suddenly came to me that I could probably walk rapidly straight down the mountainside faster than Noble could drive the distance, with all the switchbacks. So we stopped and picked out a roadside landmark far below us, and I set out at a fast pace straight down the mountain, while Noble carefully drove down the mountain. Sure enough, 20 minutes later, I reached our predetermined destination point five minutes before Noble arrived, after traversing more than 4 miles. Not to be outdone, Noble said he could do the same thing, and soon thereafter he did (not without some wobbly legs and heavy breathing on his arrival). He then decided to refresh himself with a quick bath in a mountain stream where we measured the water temperature to be 40 degrees! (I did not join him.) Our adventure was fun and we enjoyed each other's companionship most of the time, as different as we were.

The most conspicuous and widespread birds were furnariids, tyrannids, and emberizids. Giant Hummingbirds were occasionally seen, and American Kestrels were prominent along the roadside, as were White-tipped Doves, Andean Flickers, and Black-billed Shrike-Tyrants. I had scattered observations of Mountain Caracaras* (the middle population of the three allopatric species of *Phalcobaenus*), Ornate Tinamous* (at 14,000 ft. elevation), and an unexpected Aplomado Falcon* at an elevation of 13,500 ft. (quite a handsome raptor, with a local distribution all the way from the extreme southwestern U.S. to the Straits of Magellan, but my first ever sighting). I stopped at some highland lakes and set up my telescope to observe Crested Ducks*, Speckled Teal, and a Puna Snipe*. Noble took a picture of me at one of these stops, as I was looking through my telescope. Common Miners* and Bar-winged Cinclodes were abundant in the puna grasslands, as were many unidentified "canasteros" (small, slender, brownish, wren-like furnariids with long graduated tails, in the genus *Asthenes*) which constantly flitted from one low shrub to the next. A cute little pale rusty-colored fox with a black tip to its tail trotted across the open grasslands not far from the road, paying little attention to our camping van as we drove past it on the high altiplano. That night we camped near the Río Pampas Bridge at 7,000 ft. elevation, between Ayacucho and Abancay.

The next morning (Sunday, May 3) was a beautiful, bright clear day. Noble and I both awoke with itchy welts from an unknown insect (or spider) in the night. I recorded the following birds in the vicinity of our campsite:

16 Andean Swifts, 1 Andean Tinamou*, 6 White-bellied Hummingbirds*, 1 White-bellied Woodstar*, (a male with a beautiful iridescent purple gorget), 12 Scarlet-fronted Parakeets, 1 Black Phoebe, 2 Brown-capped Vireos, and 9 Black-backed Grosbeaks. After departing our campsite, we once again started up a series of steep, winding switchbacks. There were fragrant, yellow flowering shrubs along a streamside. Indians were herding donkeys along the road with a switch. They didn't speak any Spanish, only Quechua. Pedestrians seemed frightened by our vehicle and moved well off the roadside as we approached them. Roadside houses almost always had a small cross on the rooftop. Catholicism was widespread everywhere throughout Latin America, among persons of all ethnic groups and incomes.

At mid-morning, we entered the small village of Chincheros, about 9,000 ft., where a crowd of people were attending a colorful, bustling, weekly Sunday market. This event was one of the most distinctive features of Quechuan life in the Andes, in almost all towns and villages. People came from all over the countryside, usually by walking (often barefoot), but also on horses and burros, or in the back of a truck if one should come along. The marketplace was situated on the outskirts of the town, in a large open area adjoined by shops of all kinds, where people mingled to sell every imaginable article or good. The varied items included livestock (particularly sheep and llamas), animal hides, blankets, hats, textiles, clothing of all varieties (shawls, dresses, shirts, etc.), rope, salt, and a large variety of vegetables, fruits, and meats, among which were beef, pork, lamb, potatoes, yuca (or "manioc", a native starchy tuber), corn, barley, wheat, peppers, and oranges. Two traditional foods endemic to the region were "quinoa" (a hardy, very healthy small grain) and guinea pigs (a small native rodent eaten as a delicacy since ancient Inca times). Also available for sale were all kinds of drug store or dime store articles (combs, pins, scissors, pens, pencils, medications, razors, etc.). Women traditionally sat on the ground and spread out their goods for sale all around them on a blanket, sitting there all day long as they spun yarn, weaved, or nursed a baby. The Sunday marketplace was a truly marvelous spectacle.

As we left the market and reached an elevation of 10,000 feet, a Black-chested Buzzard-Eagle* came gliding overhead, which I described in my notes as being light colored below with a black throat and flight feathers, reminding me somewhat of a Swainson's Hawk. It was a large, distinctive raptor with a very short tail. Further up the mountain, at 11,000 ft. on the altiplano, were Mountain Caracaras (5 adults and 2 juveniles), Variable Hawks, Black-billed Shrike-Tyrants, Andean Ibises, Andean Flickers, and Spot-winged Pigeons*. This last species was seen flying in the distance (a group of four) at 12,500 ft. elevation. I recorded them in my field notes as White-winged Doves, but wrote that they looked larger, with dark wings and tail. It was many years later that I discovered the White-winged Dove (now called Pacific Dove) did not occur here; the species in my field notes was almost certainly a Spot-winged Pigeon. Some of the other birds in my recorded observations for the day were Cinereous Harrier, Aplomado Falcon (again), Puna Ibis* (an all-dark *Plegadis* ibis), Andean Lapwing, Andean Gull, Speckled Teal, and Ornate Tinamou (a conspicuous and widely distributed inhabitant of the open puna grassland).

Our road descended into a wide valley at 9,500 ft., where there were fields of corn and grain, with cattle, goats, pigs, and horses. Eucalyptus trees bordered the roadside. Our road then climbed back up to a pass at 13,900 feet elevation, where the ground was white with hailstones from a recent shower. Llamas, and sometimes alpacas, dominated the high puna grasslands, where they were herded by mostly barefoot men or colorfully dressed women, or children, all wearing knit or brimmed felt hats. Llamas and alpacas are closely related domestic animals which were derived from ancestral populations of wild Guanacos and Vicuñas, respectively (both species being native New World members of the disjunctly distributed camel family). Both animals were bred for their wool, and the larger llama was also often used as a beast of burden, being capable of carrying loads up to 100 pounds in weight on its back. The clipped wool was dyed (using natural dyes) and then spun into yarn by women with hand held spools. This traditional activity by women was carried out all day long, wherever they went and no matter what else they might be doing. Women spun wool while they walked along the road, sat on the ground in the marketplace, or stood talking with friends, often while they were carrying a baby on their back. It was one of the most visible features of life in the Andes. Afterwards, the yarn was used for weaving shawls, skirts, ponchos, caps, blankets, handbags, mats, decorative wall hangings, and many other items. The articles were not only practical, but also a source of income.

As the sun was setting we camped at a scenic overlook, 13,700 feet in elevation where there was a 100 km roadside marker, indicating the distance still to go until we reached Abancay. In almost 8 hours of driving on this date (Sunday, May 3), we had traveled only 100 miles, for an average speed of just 12 miles per hour. Our adventure was not progressing very rapidly!

We arose at 5:30 am on the next morning (Mon., May 4), hoping for a brilliant sunrise, but the sky was too overcast. Our outside weather thermometer was exactly on freezing, 32° F. Noble braved a very quick bath in a nearby stream, with a White-capped Dipper and a water temperature of only 42 degrees. I walked ahead on the road with my binoculars, trying to write some notes on the furtive grassland furnariids - - without any success. As we started out toward Abancay, the very high snow-covered mountains ahead of us (to the northeast) were particularly spectacular, and we stopped at 13,000 ft. for a picture. Here I encountered a small group of Bright-rumped Yellow-Finches* (one of a dozen species in the genus *Sicalis*). There were also some Brown-bellied Swallows, House Wrens, and Rufous-collared Sparrows. Our road was beginning a long descent all the way down to the Río Pachachaca at 5,500 feet. Some of the birds along this route were Variable Hawks, Mountain Caracaras, Bar-winged Cinclodes, Giant Hummingbirds, Rufous-naped Ground-Tyrants* (at last, a ground-tyrant that was easy to recognize, with its conspicuous bright rufous hind crown), Ash-breasted Sierra-Finches, Plain-breasted Earthcreepers, White-browed Chat-Tyrants*, and White-collared Swifts (circling above the river).

The vegetation at the river was arid, cactus-thorn scrub. After crossing the river on a bridge, we began a long, slow, winding climb through dry countryside upward to Abancay, at 7,800 ft. elevation. We stopped in the city only long enough to purchase gasoline (for a price of 15 cents per gallon) at the one service station in town. As we proceeded east five miles outside of Abancay, we encountered an oncoming vehicle with 3 young people (2 boys and a girl) from South Africa, driving a diesel powered 4WD Landrover custom built camping van. They were on a 2-year adventure in South America, North America, and Europe. Of course, we all stopped for a conversation and to exchange information. To our great surprise, they had talked with our two French friends (Jean and Raymond) in Cuzco, who had told them to look for us along the road! We exchanged post cards, wished each other *bon voyage*, and continued on our independent journeys. Such infrequent meetings with other young world travelers were one of the highlights of our adventure.

It was midafternoon and we were disheartened by the very slow progress we were making, so we decided to make no more stops until we camped for the night. Thus we drove all the way up to a pass at 13,000 feet, then began a long, tiring, switchback descent down to the swift flowing Apurímac River at 6,500 feet, arriving an hour after dark. Here we camped for the night, after covering only 125 miles during the entire day's travel. The river was flowing from southeast (our right) to northwest, where it would eventually curve sharply eastward through a mountain gorge and into the Ucayali River in the tropical rainforest of eastern Peru, then flow northward and join with the Marañon River to form the Amazon River, upriver from Iquitos in Loreto Province, Peru. It had recently been determined that if the Apurímac River was followed all the way upstream to its beginning glacial rivulets in the high Andes of southern Peru, that this was the SOURCE OF THE AMAZON RIVER, over 4,000 miles from its mouth!

The next morning (May 5), I took a bath in the river and washed out a few clothes prior to departing our campsite. We climbed into the cab of our vehicle, I sat down in my customary seat behind the steering wheel, and we began the last 75 miles of our five-day journey between Huancayo and Cuzco. The day was cloudless, and the snow-covered Vilcabamba range of mountains was clearly visible to the north of us, with highest elevations above 20,000 feet. We

Plate 12: Our narrow gravel Andean road wound back and forth, up and down between 5,000 and 14,000 ft. elevations. Llamas were common along the roadside. Colorful Sunday markets were traditional in all the towns.

ascended once again to the altiplano, between 11,000 and 12,000 feet in elevation. A small pond had 5 Speckled Teal, 100 Andean Gulls, 15 Andean Lapwings, and 2 Black-crowned Night-Herons (a species of worldwide distribution). In the valleys were fields of corn and grain, and ubiquitous yellow-flowering shrubs. Houses, as usual, were constructed either of adobe or of cane, with tile or thatched roofs, and were bordered by stone fences and eucalyptus trees. The rural population was almost entirely Quechua Indians, descendants of the Incas.

We drove into the small village of Izcuchaca at mid-morning. A colorful, week-long, annual religious celebration known as the "Fiesta de la Cruz" (festival of the cross) was underway, with three men in stocking-like masks performing the "dance of the bear." Such festivities were of very regular occurrence in the lives of the Quechua people. Music was provided by traditional drums, reed flutes, and a mandolin-like stringed instrument known as a "charango." As always, virtually everyone was intoxicated and in a state of euphoria from drinking alcohol and chewing coca leaves. These two narcotics were part of the culture, dating back to the early Incas, and were incorporated into the religion and considered divine. They were indulged in by both men and women, throughout the day or night, and were used in almost all religious rites, ceremonies, and festivals. Drinking, feasting, and dancing were part of the everyday life of the Quechua people, and drinking to drunkenness and stupor was socially approved and encouraged.

"Chicha" was the cheapest, most traditional and popular alcoholic beverage. It was beer-like, pale yellowish in color with a slightly sour taste, and was most commonly fermented from a specific kind of maize known as "jora", but other grains, potatoes, or even peanuts could be used. Coca leaves contain cocaine (an alkaloid) and come from a tea-like native shrub. The dried leaves, when chewed, act as a stimulant and produce a mild euphoria. They are said to suppress hunger, thirst, pain, and fatigue. Their use dates back for more than a thousand years, and the dried leaves were constantly chewed, being bought and sold in large quantities in the marketplaces. Prolonged use produced very foul-smelling breath and eventually stained one's teeth green in color. Whenever Noble and I stopped at a roadside celebration we were always offered chicha. However, we both found the taste to be bitter and unpleasant, so we would sip only a very small amount so as not to offend anyone. People were intoxicated much of the time and we had to be very vigilant when driving to avoid pedestrians who stumbled onto the road in front of us, or were even sleeping in the middle.

The traditional dress for a Quechuan woman was a colorful, long, very wide-flaring multi-layered skirt, which reached to the ground. Women wore several kinds of hats, the most popular being a black bowler type hat, but another style was a flattened, platter-shaped felt hat (typically worn by the poorest women). Sometimes their hair was braided. Women carried a baby (or goods) on their back in the traditional manner of tying a blanket across their chest and draping it over their shoulders and back. They also transported many kinds of items in a large basket balanced on top of their head, as they had been doing for centuries. Except for festivals and special occasions, men were not as brightly dressed as the women, and much of the time, they wore western style work clothes. However, traditional knee length, hand woven, baggy "bayeta" pants were very popular. Their brimmed hats were mostly rather ordinary looking, but for special occasions they would wear a colorful, pointed knit cap with earflaps, known as a "chullo", which was ordained with beads and tassels. For ceremonies, men wore a brightly colored, elaborate, decorative vest called a "chaleco." Cape-like "ponchos" were worn by both men and boys, for everyday use and for such special occasions as festivals, weddings, and village meetings. These were often hand woven and of intricate pattern and design. Men also carried a small woven pouch in which they transported their coca leaves. Although both men and women were often barefoot, they also wore traditional sandals made of animal hide, or (in more recent times) even from old automobile tires!

We arrived in Cuzco in the early afternoon, situated on the altiplano of southern Peru at an elevation of 11,000 feet. Cuzco was a historic city, the ancient capital city of the Inca Empire. It was the cultural and economic center of the Peruvian Andes, a designated World Heritage Site, and a popular tourist destination. The heart of the city was the Plaza de Armas, surrounded by government buildings, flanked by two magnificent cathedrals, and entered under two ancient stone arches that were constructed during the Inca Empire. The current population of the city was about 80,000 people, more than half of them indigenous or of mixed descent. To our surprise, we encountered Jean and Raymond, the two young French boys we had met in Lima. They had experienced engine problems -- a broken piston -- on their way to Puno and had become detained in Cuzco for repairs, at a service garage for Caterpillar tractors. We needed a broken accelerator cable mended, so all four of us spent the night together at this garage. It was best for everyone since there was a citywide strike scheduled for tomorrow and we were told that all unattended vehicles in the streets would have rocks thrown at them.

Both Noble and I wanted to visit Machu Picchu, so we booked a 2nd class ticket on the train for the next morning (Wed., May 6), which departed Cuzco at 7:30 am. The train left on time and followed the spectacular canyon of the Urubamba River steeply downstream for over 50 miles, while descending more than 4,000 feet to our destination at the tiny village of Aguas Calientes, on the fast flowing Urubamba River at 6,500 ft. elevation, at the foot of the peaks on which Machu Picchu is situated. The train was old and slow, and there were many stops where vendors crowded outside our open window to sell us all kinds of souvenirs and food items, including complete meals of rice and fried bananas. We had saved a little money by purchasing a second class (rather than a first class) ticket, and the train was overcrowded with unwashed Indians. At 1:00 pm, five and a half hours after departure, we arrived at Aguas Calientes, the gateway to Machu Picchu.

The ruins of Machu Picchu were reached from the railway station by taxis or buses which shuttled tourists on a very narrow road with many switchbacks, all the way up the very steep side of the river gorge to the ruins at 8,000 ft. above sea level (1,500 ft. vertically above the river). The road was an engineering marvel and not for the faint-hearted, at least not in the year 1959. At the top of the road a magnificent Andean Condor came gliding low along the mountainside, looking for a corpse to scavenge. It was a bird revered by the ancient Incas, who built these ruins 500 years ago.

Machu Picchu is justifiably considered one of the seven "New Wonders of the World." It is a remarkably well preserved stone city or estate built by the Incas in the 1400's, the purpose of which is still being debated by archaeologists. It is a masterpiece in civil engineering, stone masonry, and urban planning. Of all the amazing features, perhaps the most astounding is the way in which 200 pound blocks of stone were flawlessly placed together without mortar and with not the slightest space between stones. On the 80,000-acre site were plazas, gardens, underground tombs, temples, cathedrals, residences, parks, and irrigation channels. The complex was self-sustaining, where about 1,000-2,000 people could live. Curiously, the Incas did not have a written language.

On the north side of the complex is a tall pointed peak with very steep sides, Huayna Picchu, rising 800 ft. above the temple grounds. An extremely narrow (1-2 foot wide) stepped rocky trail built by the early Incas led perilously up one side of the vertical peak to the top, with a fall of almost 2,000 feet down to the Urubamba River far below, if a person lost their balance. In 1959, there were no guard railings or cables to grab if one slipped. Noble and I, being young and foolish, decided we would take the risk, so we spent an hour walking very carefully along this trail all the way to the top, without losing our balance and falling over the edge! However, it was much more difficult to keep one's balance on the steeply descending return walk, facing downhill, and we proceeded even more slowly and cautiously, using a stick to help maintain our balance. (Today, the number of people ascending the peak is carefully regulated by the authorities and there are numerous safety features to prevent a person from falling over the edge.)

Of course I always carried my binoculars with me to observe birds wherever I went in the ruins. One of the small birds which I saw well on a half dozen different occasions, and described carefully in my notebook, was a wren which I thought probably belonged in the genus *Thryothorus*. My field notes read as follows: "wren: somewhat larger than a house wren; tail medium short; bright rufous above; top of head Gray; white below, becoming Grayer posteriorly & flanks tinged with rufous; white superciliary; black line through eye; breast streaked with black; black malar streak; cheeks indistinctly variegated." Over the years, I tried many times to identify this bird, whenever a new bird book or publication covering this region was published. Yet I never found an illustration or description of a *Thryothorus* (or any other) wren from Peru with a prominently streaked breast. I was very puzzled.

In 1970, Rodolphe Meyer de Schauensee published *A Guide to the Birds of South America*. However, there was no description of any wren that matched my field notes. In 1982, John Dunning, in collaboration with Robert Ridgely, published *South American Land Birds, a photographic aid to identification*. This book had a photograph or a written description of all recognized South American land birds. There was still no mention or description of a wren that matched my field notes from Machu Picchu. Finally, in 1989, a comprehensive, well-illustrated book of South American oscine birds was published which allowed me to identify my unknown wren as an Inca Wren* (*Thryothorus eisenmanni*). The book in which I found the bird was *The birds of South America, the Oscine Passerines*, by Robert Ridgely and Guy Tudor. This book gave a written description of the Inca Wren but there was no illustration. They wrote that this wren was first collected by scientists in 1974, and first named in 1985 (by Ted Parker and John O'Neill). They said that it had been "seen at Machu Picchu as early as 1965," but gave no names or references. This unnamed sighting was six years after I saw and wrote a description of the bird there. Thus, it may be that I was the first person to observe and write a description of this species, 15 years before it was first collected and 26 years before it was named! Of course, I was unaware of these facts until more than 30 years after my encounter with the Inca Wren. At that time, I decided it really didn't matter to anyone but me. It was just another challenge, to find it, to write a description of it, and eventually to identify it. My father's comment, "You must be crazy," was verified yet again.

During our two days at Machu Picchu (May 6 & 7), my time for birdwatching was quite limited, mostly around the ruins at 8,000 ft. elevation. However, I spent a couple of hours observing along the Urubamba River at 6,500 ft. elevation on the afternoon of May 7, before we caught the train back to Cuzco. This was the eastern side of the Andes in the upper subtropical life zone, with tall trees in the moist cloud forest. Particularly note-worthy of this zone were tanagers, including a variety of colorful species in the genus *Tangara* (which I have mentioned as one of my favorite genera). Therefore, I was particularly delighted to observe two different species in this genus that I had not previously seen. These were the Beryl-spangled Tanager* and the Saffron-crowned Tanager*. I described the exquisite Beryl-spangled Tanager as: "tanager: small (about 4-1/2"); green above (appearing blue in the shade); back streaked with black; wings blackish; yellowish-green crown and nape; loral area black; pale green below, paler posteriorly and heavily mottled with black anteriorly." The Saffron-crowned Tanager was described as: "tanager: small (about 4-1/2"); top of head very bright orange-buff; face narrowly black; throat & nape black; upperparts green; back with heavy black streaks; wings mostly blackish; breast pale green fading to orangish posteriorly." (Note that my estimated sizes are a little too small, as they often were in my field notes.) My observations of these two colorful little tanagers in the misty cloud forest along the rushing Urubamba River are among my fondest memories. It was such encounters that kept me motivated in my nev-er-ending pursuit of birds.

Noble and I returned on the evening train to Cuzco, this time travelling 1st class. The next morning (May 8), we picked up Raymond and Jean in Cuzco, and all four of us set off together in the cab of our Jeep to drive down the steep, forested eastern slope of the Andes, all the way to the edge of the Amazon rainforest, outside the village of Pilcopata, on the Pilcopata River. The French boys had obtained the name of a plantation owner there, a Mr. Palomino, whom they wanted to visit and who would arrange for us to visit the Shintuya Catholic mission further downstream on the Alto Madre de Dios River. I occupied the driver's seat, Jean sat on top of the engine in the middle "seat", Noble sat in the passenger seat, and Raymond sat straddled between Noble's legs, all of us packed together like sardines in a can! (The French had to leave their station wagon behind at the garage in Cuzco while their repair work was being completed.) It would be a journey of 125 miles, which we hoped to complete in a single day, descending from 11,000 ft. to about 2,000 ft. elevation.

Our road initially took us south out of Cuzco through an inter-montane valley, then turned eastward and

Plate 13: Cuzco, at 11,000 ft. elevation in the central Peruvian Andes, was once the capital of the Inca Empire. It was now a tourist destination. The poorest Quechuan women went barefoot and begged for money.

Plate 14: Machu Picchu rises 1,500 ft. above the Río Urubamba, a masterpiece of architecture built in the 15th century. In my bird field notes I recorded what may have been the first written description of an Inca Wren. Nearby, a lone shepherd boy played a flute to himself.

began climbing slowly upward a short distance to a pass at 12,000 ft., on the other side of which was the town of Paucartambo, in a river valley at 9,500 ft. A Blue-and-yellow Tanager flew across the road. I stopped on the altiplano to take a picture of an Indian boy standing all alone just above the edge of the road, playing a reed flute to himself, for his own joy, in the solitude of his empty environment. He was barefoot, wearing knee-length home -made pants, a long-sleeved shirt, a red poncho, and a brimmed hat. He seemed to be at peace with the world. I envied him. It became an all-time favorite photo of mine.

We stopped again for all of us to get out when we came to several thatch-roofed houses, where a small number of people were gathered for a religious celebration. Everyone was dressed in their most festive, ornate, traditional clothing, predominantly red or black in color. As was the custom, women wore a long, ground-length, multi-layered dress with a shawl, and a very wide-brimmed, woolen, black, hand-made, saucer-like hat. Men wore traditional "chalecos" (vests), knee-length "bayeta" pants, ponchos, and "chullos" (knit, often pointed, caps with earflaps and tassels, including one on top). Most wore sandals. Women were all sitting together on the ground in a circle drinking chicha, while the men were dancing and playing drums, flutes, and "charangos." It was an extremely traditional festivity, as had been performed for centuries, and everyone was intoxicated from the chicha or pisco, and most were chewing coca. All four of us got out of our Jeep and photographed the event, with no objection from anyone, and we were offered to join them in the drinking, which we politely refused. No one was able to explain to us what event was being celebrated. Such ceremonies were almost weekly among the indigenous people of the Peruvian Andes. It must have compensated somewhat for their poverty and hardships, and lack of any goals in life.

The town of Paucartambo was a historic village, very popular with Peruvian tourists who came here for major annual religious celebrations. I stopped briefly to take a photograph of a very small barefoot girl, perhaps 5 or 6 years old, who was standing alone on a narrow cobblestone street, carrying an infant sibling wrapped in a shawl on her back, bending forward under the weight. She seemed much too small and frail for such a task. Her mother was nowhere to be seen, but presumably only inside a nearby shop. Children had to assist in family responsibilities at a very early age. After leaving Paucartambo, our route climbed upward again to yet another pass, at 12,000 ft. Then it began a very steep, treacherous 35-mile descent on what had to be one of the worst roads anywhere in the world, all the way down to the little town of Pilcopata, where it ended along the river at the edge of the hot, steamy Amazon rainforest at 2,000 ft. elevation. As we started downward, the road passed through puna grassland to treeline at the top edge of the "elfin" temperate forest at 10,000 ft. elevation, then continued downward through the moist subtropical cloud forest with its epiphytes and tree ferns. It continued to descend through the wet upper tropical montane forest with bamboo, to finally arrive at the edge of the vast Amazon rainforest, which stretched almost 2,000 miles to the Atlantic Ocean. It was a spectacular transition of life zones in a very short distance, like nowhere else on the continent.

The scenery on our descent was sometimes spectacular, with many waterfalls cascading from the very steep forested sides of the road. The viewing pleasure was offset, however, by the nature of the very wet, slippery, muddy road with numerous rocks and potholes, which was often so narrow that only one car could pass, and oncoming traffic (if any) had to stop and back down to give right of way to descending traffic. It was impossible to see around turns. Driving required all of my concentration and effort. Attesting to the great hazard of the road were numerous skeletons of wrecked buses, trucks, and cars lying in the forest down the mountainside below the roadway. There was no protective barrier at the edge of the road, which simply suddenly dropped off, hidden by vegetation. Numerous little white crosses lined the roadside where travelers had died. Only the most adventuresome tourist would attempt this journey. There was no roadside space to stop for birding. In 1959, this was definitely not a tourist road. I would have to return one day.

Pilcopata was a small, dirty little town at the end of the road on the left bank of the Pilcopata River, with almost nothing to recommend it. There was no bridge across the river, but there must have been a raft available somewhere to ferry small vehicles and supplies to the few plantations on the other side. A cable with chairlifts stretched across the narrow, fast -flowing river, carrying people back and forth. Not many persons had reason to continue, but long dugout "lanchas" with an outboard motor were available to transport people downriver. The river was flowing in a northerly direction here, from our right to left. From Pilcopata, it was not far down the river to where the small Pinyipini River came in from the left bank, after which the river was known as the Alto Madre de Dios River for a short distance. Before very long, it was joined by the larger Manu River, again coming in from the left, and the two then became the Madre de Dios River. Much farther downstream, the river flowed into the wide Madeira River, which was a major tributary of the middle Amazon River (on its right bank). Thus, one could travel by boat all the way from Pilcopata to the Atlantic Ocean! We located a small clearing on the outskirts of the village, where we camped for the night. All four of us slept inside the van, where it became quite warm and steamy in our very small living space. The two French boys slept next to each other on the floor, side by side in the narrow aisle.

Nowhere on the surface of earth are there more species of plants, animals, and other living things than in the Amazon rainforest. Biologists are constantly discovering new varieties and it is likely that in some groups, such as fungi, flowering plants, and insects, the number of unknown species greatly exceeds the number so far described. The vertical layers of the forest, from the forest floor to the canopy, each supports its own unique ecosystem. Evolution has favored every imaginable life style and predator-prey relationship. Among invertebrate animals, there is a great diversity of insects, spiders, millipedes, and other arthropods. Beetles overwhelmingly outnumber all the other groups, but ants, butterflies, termites (with conspicuous terrestrial and arboreal houses), bees, and wasps occupy almost every niche. Within vertebrates, birds are the largest and best known group, containing such familiar and colorful species as toucans, macaws, hummingbirds, trogons, tanagers, and cotingas. Among the many mammals are a great diversity of bats, which feed on insects, fruit, nectar, fish, and even mammalian or avian blood (vampire bats). Spider monkeys swing through the trees by their arms, legs, and prehensile tail, while the forest reverberates with the loud vocalizations of howler monkeys. Capuchin monkeys feed on the nests of arboreal termites. Marmosets and tamarins peer curiously through the foliage at an intruder. Brocket deer, peccaries, capybaras, agoutis, coatis, porcupines, anteaters, armadillos, opossums, and swarms of army ants all roam the forest floor. Squirrels scamper through the trees, while slow moving sloths climb at the rate of only inches per minute. Along the rivers are alligators, turtles, jaguars, and tapirs. Two species of dolphins swim in the Amazon River and its larger tributaries. Snakes come in all sizes and food habits, from 25-foot-long boas and fatally poisonous bushmasters and fer-de-lances to slender green treesnakes, some of which are five feet in length and no more than ¼ inch in diameter. Iguanid, teiid, and gekkonid lizards of many designs and shapes inhabit a great variety of niches from the forest floor to the canopy. There are bromeliad inhabiting tree top forms and those that run across the surface of streams. The most notable amphibians are small, brilliantly colored treefrogs in the family Dendrobatidae, with highly poisonous skin secretions, which are used by the indigenous forest Indians to tip their arrows. Freshwater fish diversity in the ponds, streams, and rivers of the Amazon rainforest is among the greatest in the world. In pondering the immense variety of life in the rainforest and the fact that so many of all the animals are green, to match the color of their environment, I am reminded of an examination question posed to an undergraduate class of ecology students many years ago by Marston Bates at the University of Michigan: "Why are there no green mammals?"

The botanical diversity in a rainforest is equally diverse. Bromeliads and other epiphytes, vines, creepers, and lianas (including strangler figs), and wide-buttressed trees are common and conspicuous features of a rainforest. Orchids and heliconias are among the most widespread of the non-woody flowering plants. Palms and ceiba trees grow along the edge of swampy areas, and there are many kinds of fig trees (often pollinated by bats or birds). The forest canopy averages 150 feet in height, but single trees sometimes extend above the canopy to tower 200 feet above the ground. Contrary to popular belief, a mature rainforest is not an impenetrable

Plate 15: Frome Cuzco we began a long travel & descent to Pilcopata, on the western edge of the Amazon rainforest. Near the town of Paucartambo we encountered a typical, frequent roadside celebration where everyone (both men and women) was intoxicated from drinking chichi or chewing coca leaves.

"jungle", but in fact has an open understory with little ground cover, and it is not usually difficult to traverse on foot, except along the tangled edge of the forest which receives more sunlight. Fungi are an important component of the forest floor.

On Saturday morning, May 9 (our first day in the rainforest), Raymond got up and went across the river on a cable chair to locate Mr. Palomino's "finca" (plantation). He took with him a bottle of wine and some of Noble's freshly popped popcorn as gifts. He returned 45 minutes later bringing us some breakfast of fresh papaya juice and fried eggs and rice. We left our vehicle where it was and all four of us then cabled across the river, two persons at a time, and walked to the hacienda of Mr. Palomino, where we were warmly greeted. In front of the large hacienda house there were coffee beans and cacao spread out on a spacious cobblestone pad, drying in the sun. The plantation also produced many fruits such as bananas and papayas, as well as manioc (known also as cassava or yuca), chickens, ducks, and pigs.

We left Mr. Palomino's finca in mid-morning and went with him downriver in his dugout canoe, powered with a 50-hp outboard motor, to visit a friend of his, Herr Gerlach. He had come here 25 years ago and, with the help of many Indian laborers, converted 1,000 hectares of jungle into a ranch for growing coffee, cacao, sugarcane, and tea. His hand-hewn house had no glass windows and very little electricity. He had his own still for producing rum! While Noble enjoyed talking entrepreneurship with Herr Gerlach on his veranda, I took advantage of the sunny afternoon for my first-ever birdwatching in the Amazon rainforest. It was both exciting and frustrating. While I was studying one of the birds in a foraging flock, and then writing a description of it, the other members of the flock went unobserved. I was able to record only a very small fraction of the birds around me. Observing birds in an unknown realm without a bird book was truly an adventure. Some of the species I eventually identified were Buff-tailed Sicklebill* (a remarkable hummingbird with a sickle-shaped bill used for probing the corollas of Heliconia flowers), White-banded Swallow* (foraging low over the rocky river), Blue-and-yellow Macaw* (flying over the forest in pairs or small groups), Roadside Hawk (perched at the forest edge), Swallow-wing ("flycatching" from exposed treetop branches), Red-billed (Cuvier's) Toucan* (common, noisy, and conspicuous in tree tops, with loud yelping), and Fasciated Tiger-Heron* (hiding along the riverbank). As the afternoon waned, large flocks of Crested Oropendolas flew over the river, returning to an unseen roost.

I also amused myself part of the afternoon by chasing and catching some of the many brightly colored butterflies. These flitted in and out of the forest openings, and along the edges of streams, stopping on flowers, patches of moist sand, or animal dung on the forest trails. As dusk approached, we returned upriver to Mr. Palomino's. Noble and I went back across the river to sleep in our Jeep, but the French boys preferred to stay and sleep on the floor of the more spacious hacienda. It had been a very exciting and memorable day.

The next morning (May 10), the four of us "tourists" joined a small group of Peruvian agronomy scientists, and we all traveled together downriver in a "lancha" equipped with a 35hp outboard motor. Our river became the Alto Madre de Dios River. Before very long, the scientists came to their study site and left our boat, carrying their equipment and camping gear. The rest of us continued downstream for another two hours to the Shintuya Catholic mission for Indians, which was located on the right bank of the river, where we spent the rest of the day and night. Riverbank birds included the very handsome, intermediate sized Capped Heron*, immaculate appearing with its delicate Gray plumage, buff-colored neck, black crown, and striking blue bill and loral area, it was truly one of the most attractive ardeids in the world. Other riverside birds were extremely noisy Black Caracaras*, Anhingas, Green, Ringed, and Amazon kingfishers, Bat Falcon (in a distant treetop), and a Wood Stork, which flew across the river. We spent most of the next day (May 11) at the mission, before departing in the late afternoon.

Amazonian Indians lived an extremely primitive lifestyle, in small segregated "tribes", often isolated from one another. Those at the Shintuya mission came from the surrounding area of rainforest and were known as the Matsiguenkas (Machiguengas), a tribal group of no more than 1,000-2,000 people. They lived naked in the jungle, in elevated cane houses with thatched roofs, subsisting almost entirely on what they found in the forest. They had no written language and lived almost exclusively by hunting, fishing, and gathering such items as fruits, nuts, and leaves from the forest. Peccaries, tapirs, capybaras, monkeys, guans, and catfish were some of their more important food items, which they obtained by hunting or fishing. Bows and arrows were the most frequently used weapon for hunting. The arrows were very long (generally about six feet in length) with a bamboo shaft and brightly colored macaw feathers attached to the back for guidance. Different kinds of tips were used for different kinds of game -- mammals, birds, turtles, or fish. These were frequently tipped with poison obtained either from tree sap or the small "poison arrow" frogs. People suffered severely from intestinal parasites and eye diseases, which often left them blind in one eye.

The Catholic mission had been established for about 20 years, and was unusual in having a priest (padre) assigned to it. It had a school (where Christian doctrine was taught) and a dispensary for medications, but very little electricity, although there was a wireless for communication with other missions. A few basic food items were cultivated at the mission, such as bananas and manioc. Indians could come and go from the mission as they pleased, but those staying at the mission were required to wear some sort of covering over their genitals, though women did not need to wear tops and children were allowed to remain naked. We spent just one night at the mission and then left the following afternoon. Before I left, I bargained with a teenage Indian boy for his bow and some of his arrows, in exchange for my pocketknife. He readily agreed and today these souvenirs are among my most cherished items, prominently displayed in my house, some of the arrows as curtain rods!

My two days in the forest at the mission (May 10 & 11) allowed me a tiny glimpse of the Amazon rainforest. At times, the forest would be teeming with foraging flocks of birds at all levels in the trees, and on other occasions (which I found difficult to explain), there would be almost complete absence of sound or movement. Birdwatching was an extremely daunting task. Characteristic of neotropical forests in the tropical zone almost everywhere, there were a large variety of small to medium-sized "antbirds" (family Thamnophilidae) foraging in the understory and the forest edges, where they hid from my view. Their presence was revealed only by their persistent calling and chattering. The ones I found were Dot-winged Antwrens, Warbling*, Scale-backed*, White-lined*, and Spot-winged* antbirds, and Great, Barred, and Chestnut-backed* antshrikes. Surprising, however, was the fact that no foliage-gleaners (family Furnariidae) were seen, and only one woodcreeper (the distinctive Red-billed Scythebill* with its greatly decurved, sickle-shaped bill).

Vermilion Flycatchers and River Tyrannulets* were along the river, as was a Collared Plover* (surprisingly my first sighting of this species). At the edge of forest clearings I enjoyed watching the distinctive little Long-tailed Tyrant*, all black in color except for a white crown, with its two central tail feathers greatly elongated. Large groups of Giant Cowbirds foraged in the middle level and outside of the forest. Yellow-rumped Caciques and Russet-backed Oropendolas* were conspicuous at the edge of the forest, and a single Amazonian (Olive) Oropendola* flew into the top of a tall forest tree. A Gray-necked Wood-Rail* ran across the forest floor, a Bluish-fronted Jacamar* sat quietly in the understory, and Pale-vented Pigeons* called from the canopy. My adrenalin level was raised on two separate occasions: first, when a ghost-like White Hawk soared low over the forest; and, second, when a streamlined Swallow-tailed Kite came gliding low across the river. What awesome raptors! A single male Amazonian Umbrellabird*, entirely black in color and the largest cotingid, gave me a thrill as it flew over the river. The forest was brightened by such colorful species as the lovely, deep blue Plum-throated Cotinga*, the brilliant vermilion-and-black Masked Crimson Tanager*, and the absolutely gorgeous multi-colored Paradise Tanager* (another of the 49 species in the genus *Tangara*). A single pair of either Scarlet or Red-and-green macaws flying high over the canopy could not be distinguished to species because they were silhouetted against the sky. Other parrots flying above the forest were groups of Chestnut-fronted Macaws (first seen in Ecuador), Dusky-headed Parakeets* (an *Aratinga*), and Mealy Parrots. A Crimson-crested Woodpecker

Plate 16: We crossed the Alto Madre de Dios River at the tiny village of Pilcopata, on a highline, and went down the river in a "lancha" to the Catholic "Shintuyu" Mission for the primitive Matsiguenka Indians, at the western edge of the vast Amazon rainforest.

foraged low on a large tree trunk, and handsome Yellow-tufted Woodpeckers* (in the genus *Melanerpes*) were conspicuous in forest clearthe forest midstory. Groups of Red-throated Caracaras* along the river edge were even noisier than were the Black Caracaras, with their loud parrot-like screeches and screams. Chestnut-collared Swifts circled above forest openings, and Red-billed (Cuvier's) Toucans once again perched conspicuously on high bare branches in the early morning and late afternoon, calling one's attention to them by their loud yelping, which was one of the most characteristic sounds of the rainforest. An exciting moment for me, years later, was to identify a Black-backed Tody-Flycatcher* (*Poecilotriccus pulchellus*) from my field notes written at the mission. It was carefully described as: "flycatcher: small (about 4"); black above; yellow below; throat white; short white line behind eye; lesser coverts yellow; primaries widely edged yellowish." I learned later that it was a local species with a very limited geographic range, endemic to southeastern Peru. My one night at the mission allowed me the opportunity to experience nightfall along the river, in the Amazon rainforest. It was a memorable experience, which I described in my notes as "the changing of the guard" when, as dusk approached, the diurnal aerial insectivores foraging above the river (swallows and swifts) were gradually replaced by the nocturnal aerial insectivores (caprimulgids and bats). I found it a time for reflection and philosophical questioning of man's place in the universe.

We departed the mission on the afternoon of May 11. On our way back upstream, we stopped to pick up the group of agronomists, who had finished the work at their campsite. When we were about five miles from Pilcopata, late in the afternoon, we suddenly encountered a very severe torrential rainstorm, characteristic of rainforests everywhere. The motorman did not want to proceed in the rain. Fortunately, there was a very primitive open thatched hut on stilts on the riverbank, which belonged to him. So he pulled ashore there, and we all went inside the hut, sheltered from the rain. Darkness came, and the motorman did not want to continue to Pilcopata after dark, saying we could all sleep there for the night. This displeased one of the scientists, a Peruvian of Chinese ancestry by the name of Mr. Chang, who said he did not want to do this and he would prefer to walk through the jungle the remaining five miles, on a badly overgrown trail paralleling the river not far from the bank. He asked for volunteers to go with him. After talking it over between us, Noble and I decided we would go with him.

Mr. Chang had only one flashlight, a machete for each of us, and a pistol, which he said would protect us from jaguars, snakes, or robbers! It was very dark in the rainforest, particularly for the two of us following Mr. Chang with his one flashlight. We were in a procession of three persons walking in single file. (If my memory is correct, I was the last person in line.) To get to the overgrown trail it was first necessary to hack through some very thick undergrowth at the river's edge, but with the use of our machetes, we eventually succeeded. The forest was full of strange, unfamiliar sounds, which Noble and I tried not to contemplate. The rain had stopped. Our nocturnal progress in the rainforest was uneventful, and two and a half hours later, we arrived safely at the village. Here we headed for the only restaurant in town, and a beer or two. Much to our surprise, we found Mr. Palomino there; he had come into town from across the river. Noble and I joined him for a beer, and then, shortly afterward, we retired to our Jeep for the night, having survived yet another adventure. During our three days in the rainforest, I had documented and eventually identified a total of 81 bird species, an exciting and yet very small percentage of all the species that lived there. These are listed in Appendix A, column 7, superscript 3.

Jean and Raymond arrived the next morning (Tuesday, May 12), and we all piled into our Jeep for the return trip back up the incredibly bad road to Cuzco. Once again, I was behind the wheel. Although it was necessary for me to concentrate carefully on my driving, I never-the-less was able to observe three species of roadside birds en route, between 6,000 and 6,500 ft. elevation, which I had never seen before, at the edge of the forest. I stopped very briefly to record their descriptions. These were: (1) Highland Motmot*: an incredible bird (like all motmots) with its gorgeous pastel green coloration, wide opal blue eyebrow, black mask, and elongated central spatula-like tail feathers; it was the highland representative of the very closely related and quite similar appearing Blue-crowned Motmot of lower elevations; (2) Cinnamon Flycatcher*: a handsome, bright rufous-buff colored little flycatcher with two wide wing bars of the same color, somewhat reminiscent of *Empidonax* flycatchers in its size, shape, and behavior; and (3) Collared Inca*: a relatively large, striking, dark green hummingbird with a prominent wide white chest patch and an almost entirely white tail.

Just as we entered Cuzco, the accelerator cable broke (for the umpteenth time). However, by now, Noble was an expert at putting it back together, and he soon had it fixed. We dropped off the French boys at the Caterpillar garage, and later we stayed there with them for the night. The following morning (Wednesday, May 13), Noble and I departed for Puno, with a planned 2 days of travel. The French boys would drive there to meet us tomorrow evening, after their repairs were finished in Cuzco.

As we departed for Puno, Noble decided that he wanted to begin the day's drive. Our road across the altiplano was mostly flat and dusty with occasional shallow, dust filled depressions, at an elevation of 12,000 ft. Noble was travelling at a speed a little faster than I normally did, and as he turned the wheel slightly to the right to avoid a small depression in the road, our overweight Jeep van skidded in the dust; because of our high center of gravity, the Jeep tilted very steeply to the left and almost overturned. The incident caused the hangar bracket for the left front spring to break off from the frame, and the main leaf of the spring snapped in two. We were forced to limp along very slowly all the rest of the way to Puno, with the broken bracket clanging and banging against the frame. It was welded back on, and the spring repaired in Puno, for a cost of $14.

Our first night out of Cuzco, we camped along the roadside at 12,500 ft. elevation, at a location 25-30 miles before the town of Ayaviri. To my surprise, a Great Horned Owl called outside our window several times during the night. The next day (May 14), we drove all the way to Puno, passing through Ayaviri and the rather large town of Juliaca. The grass-covered puna was quitescenic, with views of distant snow-covered mountains and high volcanic peaks. Herds of alpacas grazed along the roadside, tended mostly by young Indian children. I stopped several times to take pictures. An Indian man and his wife were standing in the grasslands near the roadside, working to readjust a load on their packhorse, which appeared to be carrying tents, tent poles, and household belongings, from one location to another across the plains. They were bundled up in warm clothes, wearing brimmed hats and sandals, with crude packs on their backs. My picture taking did not cause them to stop what they were doing or pay me any attention. Further along the road, I stopped again to take a picture of a tiny little girl standing near the road, watching over some alpacas. She was wearing not just one bowler-style hat on top of her head but two, one on top of the other!

I was able to record some of the more common or conspicuous birds of the altiplano, most of them only as descriptions in my notebook. My list for the day was as follows: 6 Chilean Flamingos* (see comments below), 4 Cinereous Harriers, 120 Common Moorhens, 95 Puna Ibises, 1 Andean (Slate-colored) Coot*, 110 Puna Teal*, 30 Yellow-billed Pintails*, 3 Greater Yellowlegs, 290 Andean Gulls, 1 Andean (Ruddy) Duck*, 27 Andean Lapwings, 50 White-backed Stilts* (considered a race of the Black-necked Stilt at the time), 1 Andean Negrito, 2 Paramo Pipits, 4 Turkey Vultures, 12 American Kestrels, 1 Aplomado Falcon, 80 Andean Flickers, 15 Common Miners, 2 Variable Hawks, 10 Bar-winged Cinclodes, 2 White-tufted Grebes*, 5 Black Siskins, 15 Puna Yellow-Finches*, 7 Black-crowned Night-Herons, and 1 Plumbeous Rail. I was delighted to see my first wild flamingos. Six species of flamingos are currently recognized around the world, three of them living sympatrically in the southern Andes. Flamingos are an aberration of nature, a biological anomaly, seemingly illogical in design, marvelous to watch and among my favorite birds. They filter their food from the bottom of shallow pools and lakes with a sideways motion of their head, while holding their beak underwater in an upside-down position. I smile because no one can mention a flamingo without referring to it as a "pink" flamingo, as if there were any other color, which there isn't! We spent that night in Puno, at the garage where our broken spring was to be welded the next day.

The next morning (May 15), we walked the two blocks to the hotel where we had arranged to meet Jean and

Plate 17: Flocks of Alpacas frequented the high alpine lakes on the paramo. Women spun wool all day long, wherever they went. This roadside celebration was taking place in the town of Juliaca, near the shore of Lake Titicaca.

Raymond. Since our Jeep was not going to be available tomorrow for travel, the four of us decided we should visit the Orurra (Uro) Indians, who lived in *totora* reed houses on man-made floating reed islands in Lake Titicaca, as their pre-Inca ancestors had done for many centuries. The Peruvian government sanctioned and regulated tourism to the islands, supplying motorboats for transportation. However, it was much cheaper to walk down to the waterfront and bargain with a private boat owner to take you out in his small wooden boat equipped with a sail for propulsion. So, this is what we did. Along the waterfront were five new birds for my list: Short-winged (Titicaca) Grebes* (endemic to this small area), Puna Plovers*, Mountain Parakeets* (genus *Psilopsiagon*), Correndera Pipits* (eventually identified with considerable effort), and Yellow-winged Blackbirds*. I was surprised to also see 5 Black-crowned Night-Herons.

We found the owner of a crude 30 ft. long, wooden, fishing sailboat: a cantankerous, sullen, coca-chewing Indian, whom we nicknamed "Tonto", who spoke very little Spanish. After much bargaining, he finally agreed to take the four of us out on the lake tomorrow for a price equivalent to about 70 cents each. It would turn out to be a 14-hour round trip boat ride on the lake. We were told to arrive at the pier at 11:00 o'clock that evening for a departure at midnight, in order to take advantage of wind direction at that time. We were required to obtain written permission in Puno to visit the islands from both the "policía" and "capitán del Puerto" in Puno, who gave us a stamped official paper ("licencia") which listed our names, the name of the boat, and the name of the boat's owner ("Tonto"). That evening, we ate dinner at a small restaurant where we all ordered "churrasco," a delicious meal of steak, eggs, rice, onions, and fried potatoes -- all for the price of 20 cents each.

Lake Titicaca is situated in the Andes at 12,500 ft. above sea level, on the Peruvian/Bolivian border. It is the highest navigable lake in the word, 125 miles long and 50 miles wide, with a surface area of 3,260 sq. miles, a maximum depth of 920 feet, and an average depth of 350 feet. The Orurra (Oru) Indians began living in the lake several thousand years ago, for protection and safety, constructing small floating islets from the native totora reeds (which are a member of the sedge family of plants) growing in the lake. They built their houses and boats from these reeds, and their way of life has changed very little over the centuries. Here on these islands, they secured most of their own food and brought up their families. They subsisted on potatoes, other vegetables, chickens, ducks, goats, pigs, and fish, which they raised, hunted, or caught. The floating reed mats were supported by their own underwater root system, although fresh reeds had to be continually added to the top layer of the mat because the bottom layer constantly rotted away. In 1959, the Uro Indians acquired considerable income from tourists who visited the islands to photograph their lifestyle, with permission from the government. Approximately 40 floating islands were inhabited, varying in size from 100-300 feet across, with a total population of several hundred people. These numbers have changed very little over time.

We met Tonto on schedule at 11:00 pm, at the pier where his boat was moored. We arrived with sleeping bags, all our warmest clothes, cameras, and food for the day (bread, honey, beans, soda, and of course Noble's popcorn, primus stove, and cooking pan which he took with him everywhere he went). It was a very cold night and it would be almost six hours before the sun came up. Tonto's breath was very foul smelling, because of the coca leaves he chewed. He brought two other persons with him for the day's trip, both Indians, a young boy who was quiet and shy but bright looking and smiled readily, and a helper by the name of Antonio, who assisted Tonto with the boat. All three were barefoot. Noble made popcorn for everyone before the boat departed, on time at midnight. As soon as the boat got underway, we all crawled into our sleeping bags, huddled together on a blanket in the bottom of the boat side by side as closely as possible to keep warm. In spite of the cold, we each managed to get a little sleep.

We awoke six hours later just before sunrise, as our little boat was gliding up to a small island of totora reeds with several houses on it. The world around us was like a dream. To the east across the lake was a brilliant orange sunrise, and to the west was a vast expanse of totora reed marsh within which were many tiny floating man made islets of reeds, each with reed houses and boats. The air was still and quiet, except for the crowing of some roosters. Tonto motioned for us to be quiet, then left the boat and disappeared into one of the houses. The four of us got up and took pictures of the sunrise, from the boat. Tonto returned to the boat 15 minutes later and said we must leave the island! This, of course, was not the purpose of our trip or what we had paid for, or wanted. After a short discussion between the four of us, we decided to stay, and in spite of Tonto's protests we climbed out of the boat and onto the island. We did not know what kind of reception we would receive, as we had been warned that the Orurras were unpredictable and not always friendly toward strangers. Jean and Raymond each had a small, hidden .22 caliber pistol with them, and they got these ready for use if necessary.

We were quite surprised, therefore, when a powerfully built Uro Indian man suddenly appeared from inside one of the houses with a big grin on his face and greeted us, while chewing coca, with a very friendly "Buenos dias." He was in charge of the island, and a friend of Tonto's. Since we didn't know his name, we called him "Jefe" (chief). Like Tonto, his teeth were stained green and he had very foul smelling breath. We all shook hands and explained to him, in Spanish, that all we wanted to do was to take a few pictures of the island, from outside the houses. (All the other family members living on the island had remained out of sight, inside the houses.) He considered our request and then said it would cost us a total of 20 soles (about 70 cents) for all of us together. In our usual response, we argued that we were poor students without much money, not rich tourists, and had already paid Tonto to bring us to the island. So, he came down to 10 soles, and we agreed.

We waited patiently for the sun to appear from behind the clouds so we could take some pictures. The jefe sat outside with us, and talked to Tonto while they both drank chicha and chewed coca. It was about 8:00 o'clock in the morning. We four visitors walked around the small island looking for things to photograph. There were several small houses, several reed boats, rock fireplaces, clay pots, vertical stacks of dried totora reeds tied together at the top, a few chickens, a pig, a dog, a cat, and not very much else. The reed "balsas", about the size of a canoe, were quite unusual in their entirely reed construction. Long poles were used for pushing them through the reeds, and crude wooden paddles were available if the water got too deep. Both men and women usually kneeled or sat cross-legged on top of the boat to propel it through the water, usually only one person at a time.

Two women, one of whom was old and wrinkled, suddenly appeared from inside one of the houses. As soon as they saw us, they became very angry and began waving their arms and screaming at us, in an Indian dialect. They each picked up a long boat pole, and wielding these over their shoulders they ran toward us with the obvious intent of hitting us over the head if they could. I had to run very fast to stay ahead of the older woman (while Noble attempted to photograph the event with his movie camera). Fortunately for us, the jefe quickly came to our rescue and chased the women back inside. However, the younger one reappeared shortly thereafter and sat down outside the entrance to her house and began nursing an infant. When I approached her to take a picture, she immediately grabbed the baby by one arm and dragged it back inside the house with her.

We had been on the island for more than an hour and had pretty much exhausted all the photographic opportunities when a young boy came out of one of the houses and threatened to hit us with sticks. At that point, the jefe stood up and announced it was time for us to leave. By then, he was quite drunk. He said we must give him another 10 soles, contending that we had taken pictures not only of his half of the island but also the other half, which belonged to his parents. We argued, falsely, that we took pictures only of his part of the island because the women chased us away from the other houses with their "palos." We won the dispute, thanked him, shook his hand, said goodbye, and climbed into our sailboat.

The jefe then went back inside his house, accompanied by Tonto and Antonio, and they all commenced drinking chicha once again. The young boy from Puno stayed in the boat with us. After 45 minutes, the four of us "tourists" in the boat became impatient, and we sent the boy back inside the house to tell the others that if they did not come out, we would leave without them and sail back to the wharf by ourselves. When there was no response to this threat, we took the long poles that were in the bottom of the boat and began pushing the boat out into the lake, away from the island. This immediately brought Tonto and Antonio out from inside the house on the island, and they agreed to leave with us if we would come back to pick them up. So we did. Then they spent five minutes embracing and saying goodbye to the jefe, and they all kissed each other warmly on the cheeks and the palms of their hands, as was a traditional custom. Tonto insisted we do the same thing, embrace and kiss the headman. In the spirit of friendship, we all managed to carry out this ritual, in spite of the very unpleaant odor of his breath. Tonto further insisted that we all sign a piece of paper, a "certificado," to give to the jefe, which read that we were well received by him on the island, stating his name, and that we were permitted to take pictures. Raymond wrote up such a paper (in Spanish), and we all signed it, using fictitious names. Everyone was happy. In fact, Tonto was so intoxicated that he fell into the lake while attempting to climb into the boat.

Our two boatmen pushed the sailboat away from the island and headed it toward Puno, which was far in the distance and not in view. There was very little wind, so they had to push it with poles, and when the water became too deep they propelled it with paddles. At our request, they navigated close to some other floating islands with houses, where we took more pictures as we passed, particularly of women sitting on the top of their reed "balsas", maneuvering and propelling them with long poles. The balsas were used for gathering reeds, visiting neighbors, or fishing. Suddenly, to our great surprise, the "jefe" from the island we had just visited, came paddling up to us from behind, in his own little balsa, completely out of breath from the exertion of overtaking us (and the fact he was still intoxicated). Immediately after he caught up, three other Orurra men from neighboring islands, each in his own balsa, also caught up with us. We were very puzzled, and apprehensive about their intentions. The two Frenchmen readied their pistols. One of the three new men asked (in Spanish) to see our "licencia" (from the Capitán del Puerto). Tonto showed this to them, somewhat reluctantly. We were then advised, probably correctly, that the government of Peru prohibited tourists from passing within a certain distance of the Indian islands without a permit for each island, and it was demanded that we pay all of them for passing close by. We argued that we had no more money with us, having given it all to Tonto and the jefe (which was almost true). So, once again, we were told to write a "certificado" to give to each of the four islanders, saying that we took pictures of their island. Each certificado included their name and the names of their parents (which they wrote for us)! Raymond, for a second time, undertook the task of writing up these four documents, in Spanish, then we four "tourists" each signed them, again with fictitious names. No payment was made to anyone. The local islanders were satisfied that their dignity and culture had been preserved, and they left contentedly in their balsas.

The afternoon wind came up, so our sails were raised and we began a more rapid return to the waterfront. Noble and I took a quick swim in the very cold lake water, just to say that we did! The Frenchmen amused themselves by shooting their pistols at cormorants and coots, with no success at hitting any. I was delighted to see a group of 30 flamingos flying across the lake in the distance, appearing as giant flying crosses with their very long slender necks extending far in front of their wings, and their long slender legs stretched out equally far behind them. We reached the pier in Puno at 2:00 pm, 14 hours after first setting sail the previous night. It had been another adventure about which Noble and I would write in our journals.

Birds which I recorded while on our boat trip were: 15 Many-colored Rush-Tyrants* (see comments below), 9 Wren-like Rushbirds* (a furnariid), 5 Plumbeous Rails, 75 Common Moorhens, 50 White-tufted Grebes, 30 Chilean Flamingos (see comments above), 25 Puna Teal, 10 Andean Ducks, 50 Andean Gulls, 2 Mountain Caracaras, 20 Neotropic Cormorants, and 200 Andean Coots. The Many-colored Rush-Tyrants, as their name implies, were decorative little flycatchers of the reed beds, one of the most highly patterned of all tyrannids. My description said "flycatcher: about 4 to 4-1/2 inches; upperparts blackish; back olive green; white superciliary; long narrow white line longitudinally alongside of back (across wings); tail medium with rounded corners and prominent white sides; rump & under tail coverts bright chestnut tawny; underparts yellow; long black mark on side of breast; cheeks bluish?" While not all of this description was completely accurate, this distinctive little flycatcher of the reed beds would be much easier to identify, one day, than most of its relatives.

Our Jeep was ready for us at the garage in Puno, but since it was late in the afternoon we decided not to leave until tomorrow morning. The total cost of all our repairs was $14. Raymond and Jean could not get permission to

Plate 18: We went in a little sail boat to visit the Orurra Indians who lived on small floating islands of totora reeds in Lake Titicaca, as they had done for centuries. Our coca chewing boatman became very intoxicated from drinking chichi with the island's headman.

take their station wagon into Bolivia, so they decided to ride to La Paz with Noble and me. Afterwards, they would return by bus for their Jeep, which they could then drive by road from southern coastal Peru directly into Chile, bypassing Bolivia. Our tentative plan was to rendezvous with them again in Chile. (However, as it turned out, this never occurred.) All four of us ate dinner together in Puno that evening, at the same restaurant as the night before, where the owner was quite drunk and kept us amused with his antics, though he successfully brought us the food we had ordered. We turned in early with our alarm clocks set for 5:30 am.

The French boys left their station wagon in Puno at a Catholic mission, and then crowded into the cab of our Jeep in the same positions as before, and we all departed Puno at 6:30 am on a bright, sunny Sunday morning (May 17), heading for the border with Bolivia. It would take us three days to reach La Paz. We started out in a southeastward direction, following the nearby shore of Lake Titicaca. It was market day and once again all the towns and villages, large or small, had local marketplaces full of activity and color. The most spectacular was in the town of Ilave, where the market was the largest and most splendid of any we had seen, with several hundred indigenous people gathered together, standing or sitting on the ground, selling or buying an infinite variety of goods. Many of the women were hand spinning yarn, and men, as usual, were drinking chicha and chewing coca. Some were already quite drunk. The atmosphere of the market was like no other, and we left with many pictures.

Our route paralleled the shore of Lake Titicaca, fluctuating slightly between 12,000 and 13,000 ft. elevation. The countryside was picturesque, with fields of barley and snow-capped mountains, and herds of llamas, alpacas, and sheep. Women working in the fields were often barefoot in spite of the rather cold, harsh climate, and almost always they were dressed in traditional bright colors. They wore hats of several different sizes and shapes, and frequently they had a baby on their back wrapped in a blanket or shawl, which they carried with them as they worked, bent over in the fields cutting barley by hand with a sickle. Thatch-roofed houses were encircled by stone fences.

On the way to the small town of Juli, on the shore of Lake Titicaca, one of the nuts on the recently repaired bracket which supported the left front spring, came off, and the four of us were unable to reposition the bracket correctly. Shortly thereafter, the bracket broke again, and the spring once more banged against the bottom of the Jeep. Obviously, the garage in Puno had not solved our problem satisfactorily, as was so often the case during our adventure. It was going to be necessary to have the bracket welded a second time, tomorrow in the city of Yunguyo.

We arrived in Yunguyo just after dark, but the welding shop was not open on Sunday, so we would have to wait until tomorrow to have our spring repaired. The border with Bolivia was just 10 km beyond the town, and it was necessary for us to check out of Peru with the customs and immigration offices here in Yunguyo. We located these, which were still open, and the border authorities there stamped our documents for us. For dinner that night, we ate at a small restaurant in town, ordering soup, fried potatoes, and the Peruvian equivalent of spaghetti and meatballs, for 20 cents a person. Since there was very little room in our van, the French boys asked for, and received, permission to sleep inside the small local police station! We loaned them some sleeping bags.

It required five hours the next day (Monday, May 18) for a welder to repair our broken spring. While we were waiting at the repair shop, an official from the immigration office came rushing up to us to say that the authorities last night had forgotten to stamp our passports with a "salida" (exit) stamp, and he then proceeded to do this for us! Finally, our Jeep was declared "listo" (ready) by the mecánicos at the garage, and the two French boys bargained the asking price of 150 soles down to 85 soles, or the equivalent of about $3.00 for the total cost of the five-hour repair!

In the 10km (6-mile) distance from Yunguyo to the border, there were four different Peruvian police check posts, where our passports and vehicle documents were inspected and re-inspected. Peru, with its great diversity of people, scenery, birds, and adventures, would become my favorite country in Latin America. Of the 257 birds I recorded from Peru, 147 were species new to my trip list, all but nine of them lifers (Appendix A). This tally was only a tiny fraction of the country's 1,700-1,800 known species, but the satisfaction that came from eventually identifying many of the birds described in my field notebook made the task worthwhile. Birdwatching was a challenge.

Chapter Nine—Bolivia and Chile

Noble, Jean, Raymond, and I entered Copacabana, Bolivia, in the late afternoon on Monday, May 18. Here, in this popular tourist town on the south shore of Lake Titicaca, we officially checked into the country, at the immigration and customs offices. The formalities were professional and without delays, requiring no more than 30 minutes. We then proceeded out of town, on a relatively good gravel road, heading for the ferry landing across the Strait of Tiquina, on the south shore of Lake Titicaca, about one hour away. The scenery was spectacular, with our road climbing up to 13,000 ft. elevation for a short distance. There were many miles of golden barley fields against a background of distant snow-covered mountains, and occasional views of the blue water in Lake Titicaca. Yellow-flowering shrubs brightened the puna grasslands, above which Variable Hawks and Mountain Caracaras glided and soared.

Soon after we left Copacabana, we overtook a long funeral procession in the middle of our road, on their way to a cemetery not far ahead of us. Men at the front of the procession were carrying a casket over their shoulders and women walked behind, mostly barefoot and wearing black, bowler style hats. Almost everyone was drinking pisco, and many of them, both men and women, were quite intoxicated. When we stopped to take a few pictures, eleven of the men climbed on top of our van, where they rode with us the rest of the way to the cemetery. As he left, an older man kissed my hand, which was a polite, traditional custom.

The Strait of Tiquina connected the much larger northern part of Lake Titicaca ("Lago Grande") to the much smaller southern part ("Lago Pequeño"). The strait was 125 miles long and a little less than ½ mile across at its narrowest width. A very small, wooden ferry boat equipped with a sail, ferried vehicles back and forth across the strait, mostly one vehicle at a time. We arrived at the ferry landing at 6:30pm, which unfortunately, was 30 minutes after the last ferry of the day had left. Therefore, we had to camp at the very small town of Tiquina which was situated there. Jean and Raymond again obtained permission to sleep at the police station.

The next morning (May 19), we caught the first ferry across the strait, at 7:00 o'clock. There was barely enough room for our Jeep to fit on the small boat. After crossing the strait, our road headed eastward along the north shore of Lago Pequeño, to the town of Huarina, where it turned SE toward La Paz. Foraging in the roadside shrubbery were a Black-throated Flowerpiercer*, a Giant Hummingbird, several Mountain Parakeets, a pair of d'Orbigny's Chat-Tyrants, and a dozen or so other species.

Bolivia was the fifth, largest country in S. America, and one of only two (along with Paraguay) without a seacoast. The western portion of the country was situated in the Andes Mountains, where volcanic peaks were sometimes more than 20,000 ft. above sea level, but eastern provinces dropped all the way down to tropical wet forests, dry deciduous forests, and tropical savannas. Vast salt flats existed on the high altiplano at 13,000 ft. elevation. More than half of the total population of Bolivia was Indian (including 1½ million Aymaras) and a majority of all the others were "mestizos" (persons of both Indian and white ancestry). The country was one of the poorest in S. America, with the primary source of income being from the production of tin. Inflation was astronomically high and ordinary currency transactions were almost impossible to conduct. One US dollar was equivalent to 12,500 bolivianos! There were no coins in circulation and paper currency came in bundles of ten, 50 boliviano notes (totaling 500 bolivianos per bundle). Each bundle was worth so little that nobody ever bothered to count the actual number of notes in a bundle. The situation could not possibly have been worse.

In the early afternoon, we drove down into the mountain basin which harbors La Paz, the highest capital city in the world at an elevation of 12,500 feet, with a population in the vicinity of ½ million people in 1959. We entered the city under a metal arch with 2 stone bell towers supporting a large welcome sign "Bienvenido a La Paz." The outskirts of the city were narrow cobblestone streets with small, adobe, tile-roofed houses. In the city, there were many neon signs in French. Americans were particularly disliked at this time because two months earlier, just after the Bolivian government had issued an unpopular law, "Time" magazine in the U.S. wrote in one of their articles, jokingly, that what the neighboring countries should do is divide up Bolivia and each take a piece. The Bolivian government published this article to distract attention away from its unpopular laws, and Bolivian citizens became so angry at the U.S. that they broke all the windows in the U.S. embassy with stones and it had to be evacuated. Now, 2½ months later, their anger had subsided only a very little and there were still anti-American pamphlets plastered to the sides of downtown buildings, and spray-painted anti-American graffiti on walls. Fortunately for Noble and me, nobody recognized us as Americans. (We were still widely considered to be Cubans.)

Noble and I drove Jean and Raymond to their friend's house, a Mrs. Pinto, who graciously allowed all of us to have a shower. Noble and I then said goodbye to Jean and Raymond, who promised to meet us again in Santiago. I asked Jean if I could borrow his pistol until then, so I might shoot a rabbit, goose, or duck for Noble and me to eat as an evening meal. He agreed, and Noble and I departed. (Our planned rendezvous in Santiago never materialized and I carried Jean's pistol with me, well hidden in my belongings, for the next 1½ years until Noble and I finally returned it to his parents at their home in France, as we toured that country.)

Noble and I departed La Paz on Friday, May 22, and drove south on the high, open, grass-covered Bolivian altiplano to the city of Oruro, at 13,000 ft. elevation. En route, we passed many herds of llamas and alpacas, but there were very few birds. The next day we visited the nearby Huanini tin mine, one of the largest mines in Bolivia, which, like all the others, was owned by the government. Here we were given a special 3-hour guided tour of the mine, both above and below the ground, through tunnels and down shafts. We were given a miner's hat and rubberized orange waterproof garb to wear, and were photographed underground next to a life-sized mannequin of a devil sitting on a chair drinking pisco, also in a rubberized orange suit. Indian women laborers outside the mine wore traditional long flaring skirts, shawls, and tall hats. Workers lived in crude, squat, stone huts with flat, tin roofs. Our guide constantly provided us non stop (in English) with facts of the mine's operation, profits, and government control, all of which Noble listened to intently and enthusiastically, frequently asking questions and somehow storing all the information in his head. Nothing gave him greater pleasure.

While we were walking outside between mine sites, I observed a group of six birds running across the open grassland. I was not carrying my binoculars with me, but I recorded the following notes: "shorebird: about the size and shape of a killdeer; upperparts patterned brown; wings with wide whitish stripe in flight, prominent but not sharply outlined; underparts buffish; tail tawny with black band terminally; bill not observed; flew across road and alighted in open, rather barren dry grassy area (several hundred yards from a shallow lake shore)." I identified this sighting that evening as a Tawny-throated Dotterel* from my two volume set on Chilean birds (by Goodall *et* al, 1951 & 1957), which I had brought with me from California). This handsome species was placed in the monotypic genus *Oreophilus*. For me, the bird seemed to be a free spirit on the open plains, with its slim, upright stance, distinct pattern, and lovely subdued buff coloration. It became an all-time favorite. I could not have been happier. The day at the tin mine was a success for both Noble and me, emphasizing our contrasting interests in the world around us.

The next day (Sunday, May 24, Noble's 26th birthday) we departed Oruro, with the hope of finding a shortcut route to Chile. We had been told in Oruro that our flat road south to Challapata was flooded and impassable because of recent rains. Sure enough, as we skirted the edge of Lake Uru Uru at 13,000 ft. elevation, not far south of Oruro, the lake waters had risen to slightly over our roadway, which had become flooded by the lake. It was impossible to see exactly where the sides of our road were and it was necessary for me to drive very cautiously for 5 or 6 miles through 4-8 inches of water, where I was fearful of running into a hidden roadside ditch. After I finally navigated through this flooded section, without mishap, our travel was still very slow for quite a long distance because the muddy road was exceptionally slippery. We were travelling on the mostly featureless, level, grassy altiplano, and I enjoyed watching a small fox as it patrolled the grasslands looking for an unsuspecting rodent or tinamou. Because of our slow travel (and a few brief stops), I was able to record the following birds: 600 Chilean Flamingos, 2 Mountain Caracaras, 4 Crested Ducks, 35 Yellow-billed Pintails, 40 Puna Teal, 12 Cinnamon Teal (my only sighting in Central or South America), 10 Andean Lapwings, 30 Puna Plovers, 50 Andean Avocets*, 5 Greater Yellowlegs, 2 Andean Gulls, 10 Blue-and-white Swallows, 7 Black-crowned Night-Herons, 3 Puna Yellow-Finches, 60 Andean Negritos, 1 Andean Flicker, 5 Golden-spotted Ground-Doves*, and 40 Common Miners.

After travelling for 5 hours, we arrived in Challapata at the southeast end of Lake Poopo, the largest of the shallow, alkaline lakes in the region. Here, we purchased gasoline from behind the house of a local vendor, hand pumping it from 50 gallon drums into a five gallon pail. We then filtered it through a cloth rag as we poured into our gas tank. I queried a truck driver who was unloading more drums where the nearest road was that would take us to Chile. He took out a pencil and a blank page of paper and sketched a map showing me the location of a "ruta de contrabando" (contraband route) that was very little traveled, veering off to the right from our main road just 15 or 20 miles south of Challapata. He said it would take us to Chile. The map we had with us did not show this route at all. I thanked him and we departed in search of the "ruta de contrabando." A short distance out of town, I chose a campsite for the night in a wide open area, off to the side of the road, still at 13,000 feet. Fearing sub-freezing temperatures that night, we drained all the water out of our radiator into a new red pail we had been given in Oruro, and carried it inside the van with us for the night.

The next morning, the outside air temperature, as registered by our thermometer, was 17°F (fifteen degrees below freezing). I prepared a cooked breakfast for the two of us, bacon and eggs, which Noble much appreciated. We were soon underway, on Monday, May 25, searching for the ruta de contrabando. There were no road signs or markers of any kind, only flat open space in all directions, with the distant snow-covered Andes clearly visible on the far western horizon. Our main road was proceeding in a generally southward direction on the Bolivian altiplano, leading eventually to the city of Uyuni almost 200 miles away, from where a principal road went to Antofagasta on the coast of Chile. We wanted to reach Chile further north on the coast, at Iquique.

We passed through the small village of Huari and soon afterward there was an unsigned vehicle track angling off our main road to the right, across a flat, open, dry barren area with virtually no topographic relief or vegetation as far as one could see. This was our ruta de contrabando, which was hardly more than two tire tracks leading west into the unknown. This was no adventure for the timid. Our only encouragement was derived from the fact that the ground was solid, at least temporarily. To our surprise it was not long before we came to the tiny settlement of Quillacas with a population of a few hundred people or so, a short distance from the southern shore of Lago Poopo. It was mid-morning and there was no reason for us to stop, so we continued westward. There were many forks in our track, and it was not always apparent which way we should proceed. Fortunately, we always guessed correctly. The sandy countryside gradually acquired more topographic relief, and low hills and ridges appeared, with scattered clumps of dry grasses and occasional flat patches of salt, across which I navigated very cautiously.

Four hours and 75 miles later, in the middle of the afternoon, we arrived in the rather sizeable town of Salinas de Mendoza, having encountered no other vehicles on our route. It was certainly situated in the middle of nowhere, far removed from civilization, with the apparent function of salt production. We were on the northern edge of the Salar de Uyuni, a vast salt flat extending north-south for 75 miles and east-west for 50 miles, at an elevation of 13,000 ft. above sea level. Here, to our amazement, was a vehicle inspection station where our truck was examined very carefully, inside and out, for almost an hour before we were permitted to continue our journey. However, no one wanted to look at either our passports or our carnet de passage. It appeared as if our vehicle track was only partially a "ruta de contrabando."

As we left the town behind us, we entered a vast, white, dream-like world of salt, stretching south to the far distant horizon as far as the eye could see. This was the Salar de Uyuni. The crusted surface was unbroken, except for the two tire tracks which we were following westward across the northern fringe of the salt lake. Once again, there were forks in the "road" and we had to decide which way to go. (Perhaps they all later converged with each other.) At times, there were patches of water lying on top of the salt. I was extremely fearful and apprehensive that the surface would not support the tremendous weight of our truck, on its narrow wheels, and that we might just sink out of sight - - forever! There was no way to turn back even if we wanted. My heart beat increased to twice its normal rate. I said to myself that we had definitely made the wrong decision this time and I drove a little faster, almost 50 mph. The fact that there were a very few vehicle tracks on our salty route was of no consolation. The last thing I was going to do was take my foot off the accelerator, not until we had safely reached the other side, which seemed much farther away than it really was. So it was with tremendous relief that I finally navigated off the surface of the Salar de Uyuni, without having sunk to the bottom, out of sight forever. I calmed myself with a big gulp of wine. Our adventure, once again, had survived and not yet come to an end.

Noble and I reached the town of Llica at 6:00 pm, still at 13,000 ft. and not very far from the border of Chile. Once again our van was given a very critical search for over an hour, by 3 customs officials at an inspection station. Although the truck driver who drew the map for us referred to this route as the "ruta de contrabando", our camping van was never-the-less very thoroughly inspected at two different towns not far from the Chilean border, as we were leaving the country. (Perhaps goods coming into the country were not as carefully regulated.) Maybe we looked particularly suspicious since they had never previously seen a vehicle such as ours. However, the only item of ours which they took away was a spray gun we used for killing insect pests. As he confiscated the spray gun, the inspector remarked (in Spanish) "My wife will like this." Again, we were not asked to show either our passports or our carnet de passage. Border crossings in Latin America were always unpredictable, and an adventure in themselves. Patience and ingenuity were required (and monetary bribes were almost always welcome if they were offered, but neither Noble nor I ever did this). We chose a campsite that night only a short distance outside Llica, in a hilly area with scattered cedar-like woody shrubs.

The next morning, May 26, we passed through the very small settlement of Bella Vista, which was situated at 13,500 ft. on the western edge of the Bolivian altiplano, at the base of the high Andes which separate Chile and Bolivia. This was our last populated site in Bolivia, but there were still no offices of the government here for checking out of the country, so we started our ascent up to the pass at 15,750 feet. The countryside was barren and rocky, devoid of shrubs or trees, reminding me of pictures from the surface of the moon! Near the top of the ridge was a very small, square stone hut situated close to the road, which appeared to be abandoned. Out of curiosity I decided to stop and look inside. The door was not locked but I thought I should knock first before entering. When I received no response, I opened the door and walked in. Noble followed. It was a one room structure with a crude wooden table, a single wooden chair, and some wooden shelves with food, eating utensils, and sundry supplies. A tall stack of blankets on the floor in one of the corners was available for sleeping, and there was a wood burning stove to provide heat. To my astonishment, a man was sitting in the wooden chair, sound asleep, leaning over the edge of the table. A burning candle on the table was providing a small amount of light. Various papers, pencils, and office items were on the table, along with an enormously thick book.

In a very loud voice I said "Buenos dias." The man immediately raised his head and looked at us with great amazement. In Spanish, he said "What can I do for you?" Since I presumed he was Bolivian, I responded (in my nearly forgotten high school Spanish) that we would like to "Leave your country." He asked "What do you need?" I pondered this question for a moment and then answered "We need a stamp in our passports and in the travel document for our car." To my surprise and disbelief, he nodded affirmatively, opened a drawer in his table and pulled out an ink pad and several kinds of stamps. Noble and I immediately retrieved our documents from the van, and I handed him my passport. He opened it to the first page, inside the front cover, and stared at it without saying anything. Of course all the information was written in English. After several minutes had gone by, I finally said to him (in Spanish) "Do you read English?" His absolutely classic answer (in Spanish) was "I read it, but I don't understand it"! I never cease to smile when I think back to this moment.

He stamped our documents for us, and then decided he should look inside our van. Of course, we allowed him to do so. He stood inside and looked all around him, in obvious bewilderment. His eye fell upon my small portable typewriter, which he picked up and carried with him back inside the hut. When I followed him and protested, he said "prohibido" (prohibited). He opened up the enormous book on his desk, with many hundreds of pages, and quickly turned to a page which depicted typewriters of countless models and makes, all of which were prohibited from leaving the country. To my complete disbelief there was my portable typewriter, a Hermes model 300! I thought this can't really be true. However, it was true, and I was about to lose my typewriter. A thought instantly came to my mind and I rushed outside to the van and returned with a small flask of partially consumed bourbon whiskey (which I had purchased in California before our departure and had not yet finished drinking). I sat the flask on the table in front of him and said "Señor, the whiskey is yours; the typewriter is mine." I then picked up my typewriter and carried it back to the van with me. The man did not protest, but merely nodded his head affirmatively. This amusing event would be another story for my book.

Chile is the most narrowly shaped of any country in the world. It is 2,500 miles in length (north to south) and only 75-220 miles wide (and even less in the fiord country of the far south). It is bordered on the east by the Andes (with peaks over 20,000 ft.) and on the west by its long, often rocky, Pacific shoreline. It extends on the South American continent from 18°S latitude (in the tropics) all the way south to 54°S latitude at the tip of the continent on the Straits of Magellan (and even slightly farther south on the island of Tierra del Fuego). The Atacama Desert (said to be the driest desert in the world) runs as a narrow ribbon the entire length of the coastal northern portion of the country. Roads in Chile, overall, were the poorest in all of Latin America. It was a driving nightmare for me, and caused us numerous minor to major vehicle breakdowns. Noble's expertise and patience as an auto mechanic were tested to their limit.

We entered the country unofficially, somewhere before the pass at 15,750 ft., as we crossed the Andes from Bolivia. There were no road barriers, entry stations, or posts of any kind in Chile where our documents could be stamped, and no flags or road information whatsoever to indicate we had entered the country. (We learned after our arrival in Santiago that the government was not aware of this road at all.) Indeed, it may well have been "la ruta de contrabando." From the pass, we began a long descent of over 15,000 ft. all the way down the western slope of the Andes to the city of Iquique on the Pacific Ocean. The countryside was very dry everywhere. As we started down, at 13,000 ft., I stopped to look at a group of grayish, medium-small, partridge like birds which flew across the road and landed nearby on the open mountain slope. I described the males as handsome in appearance, with black markings on their pearly Gray head and breast, outlining a white throat. The birds remained motionless for a long time, providing me with the opportunity to write a very detailed description. I discovered later that they were Gray-breasted Seedsnipes*, relatives of "shorebirds" in the family Thinocoridae, an endemic S. American family. Other birds in the vicinity were Bright-rumped Yellow-Finch, Cordilleran Canastero*, Ash-breasted Sierra-Finch, Black-hooded Sierra-Finch*, Mountain Parakeet, 80 Golden-spotted Ground-Doves, Common Miner, and White-winged Cinclodes. Further down the mountain, between 10,000 and 8,000 feet, were an Aplomado Falcon, Grayish Miner*, and a female Oasis Hummingbird* (distinguished in part by a long, slightly curved bill and a pale tawny rump coloration).

Our dry gravel road gradually widened, but the surface deteriorated very badly, becoming horribly corrugated ("calamino" in Spanish - - like a washboard), and almost impossible to drive at any speed above 5-10

Plate 19: Sheep pastures and barley fields covered the Bolivian shore of southern Lake Titicaca, at 12,000 ft. elevation. We ferried across a narrow inlet, & several days later drove from Bolivia to Chile on an unmapped "ruta de contrabando" across the Salar de Uyuni salt flat at 13,000 ft. elevation.

mph. The corrugations caused a constant vibration and put an enormous stress on our overweight van's support system and joints. By travelling at 25 mph or faster the vibration was decreased a little, as we skimmed the tops of the corrugations, and I decided to try this for a while. It was a critical mistake. Our front axle housing cracked at the ball joint, causing the front wheels to toe outward from the vertical, and making it difficult to steer. We tried to hobble down the road to the town of Pachica, where there might be some sort of repair facility. Sadly, we didn't make it that far. The crack was widening all the time. We could go no further, and I pulled over to the roadside at 8,000 ft. elevation and parked our crippled van for the night.

We were still about 10 miles from Pachica, and were very dejected.

The next morning (Wed, May 27), we discussed our situation and decided that we would both walk together into the town for help, leaving our van unattended by the roadside for the first time ever. We transferred a few articles from the front cab to the back van, and we put all our small valuables, documents, and other items in our small, very sturdy, securely locked safe, which was well hidden inside the van. We then securely locked all windows and doors, front and back, before setting out on foot for Pachica. The two of us each carried a small backpack with water, personal items, clothes, and money - - enough for several days if necessary. As always, my binoculars and camera were with me. It was a 10 mile, two hour, 2,000 foot descent into Pachica, at 6,000 ft. elevation. We took shortcuts across some of the hairpin bends in the road, and arrived in the town before mid-morning.

Pachica was a very small, backward little town isolated from the rest of the world on the dry, sparsely vegetated, western slope of the Andes in the far north of Chile. There was virtually no through traffic, and very few resident vehicles of any kind. Not surprisingly, there was no welder anywhere in the village, nor for that matter was there any electricity either, except as supplied by privately owned generators. The one café in town had a dirt floor, mud walls, many stray dogs, and lots of flies. It was obvious that Noble and I would have to proceed farther down the mountain for assistance, probably all the way to Iquique. In the meantime, we would have to wait in Pachica for whatever local transportation might come along to give us a ride. Noble found a bench outside on which to sit and write in his journal, while I meandered around the village outskirts with my binoculars to see what birds I might encounter.

In Chile, for the first time in any of my travel in South America, I possessed a two-volume set of books describing the birds in Spanish. There were a few color plates and illustrations scattered throughout the pages, depicting some of the species, their habitats, and their eggs. These books were *Las Aves de Chile* by Goodall, Johnson, and Philippi, volumes 1 and 2, published in 1951 and 1957. They provided Latin and English names (as well as Spanish) for all the species. I had transported these two publications with me in our van, all the way from California. Even with these books, I continued to write descriptions for most of the birds on my first encounter with them.

After two hours of observing around Pachica, where there were only a few scattered trees and shrubs, I tallied just 16 species: Eared Dove, Rufous-collared Sparrow, Grayish Miner, Pacific Dove, Plain-mantled Tit-Spinetail*, Bare-faced Ground-Dove, Black-throated Flowerpiercer (confusing, immature-plumaged birds), Yellow-billed Tit-Tyrant*, Cinereous Conebill, White-crested Elaenia, American Kestrel, House Wren, Andean Swift, Chiguanco Thrush, Turkey Vulture, and Groove-billed Ani (heard only). Of interest to me was the fact that here in this isolated little town at 6,000 ft. elevation in the middle of nowhere, with only sparse dry vegetation and very few birds, were 2 of the 3 species I saw everywhere throughout my travels in Central and South America, the House Wren and the Rufous-collared Sparrow. (Of the three species, only the Black Vulture was missing, although Turkey Vultures were present).

At noon, a big yellow Ford truck came by with stake sides and wooden benches in the back on which passengers could sit. It was greatly overcrowded, with a Canadian priest, 15-20 Indians (mostly standing), and several sheep. Never-the-less, people graciously made room for Noble and me to sit on one of the benches. The truck was coming from a festival in a nearby valley and traveling all the way to Iquique, which was great for us. The road was still very rough, corrugated, and dusty all the way down the mountain to the principal north-south highway, where the town of Huara was situated. Forty miles to the north was the coastal city of Arica, not far from the Peruvian border, which was often said to be the driest city on earth, receiving an average annual rainfall of only 0.02 inches!

Our truck headed south from Huara on the main north-south "highway", which paralleled the coast for almost 20 miles to the town of Humberstone, where we turned west off the highway, and drove another 15 miles through undulating coastal sand dunes to the city of Iquique, situated on a scenic rocky seashore of the Pacific Ocean. Iquique was a very popular marine sport fishing site for sportsmen from all over the world, with a population of 40,000 people. The world record for the largest marlin ever caught comes from Iquique (pronounced "eekeekay"). We arrived there on May 27 in the late afternoon.

We descended from the back of the Ford truck, thanked the driver, and looked for anyone who might be able to direct us to the nearest automotive repair facility. People stared at us curiously because of our beards. It was mid-afternoon and siesta time. It didn't take us long to find a pedestrian who directed us to the Ford agency in the city. Here we were introduced to a very nice young man, Gus Medrano, who was the son of the agency's owner and had once lived in southern California for three years, and therefore spoke perfect English. He and Noble shared a fondness for the Pasadena area, and they immediately became friends. Gus told us there was no electric welder at the Ford Agency, but he knew a person, Señor Lizama, who could supply us with a welder tomorrow, who would go back to our truck with us to weld our cracked axle housing. We were optimistic that our problem would be solved rather quickly, without too much expense. (By now, we should have known better.) That evening Gus took us to the "Gran Hotel España", where we checked in for the night.

The next morning (Thursday, May 27), we began searching for a welder to take us back up the mountain to our broken-down Jeep, 75 miles away. It was a national holiday in Iquique, and there was a military parade with a marching band (resembling that of the Salvation Army). Horse-drawn carriages served as taxis. Nobody was working. Therefore, it was not until late afternoon that, with Gus's help, Noble found a welder who agreed to go in a taxi with him to the site of our broken Jeep to weld the axle housing, for a total cost of just 15 dollars. Not expecting any difficulties, Noble told me it was not necessary that I go with him, and that he could drive back to Iquique by himself. As always, he wanted to supervise the work and to handle the financial transactions. He was still optimistic.

The welder was slow, not very competent, and it was long after dark, 9:30 pm, when he finally finished, and departed with the taxi driver to return to Iquique. Confident that our mechanical problem was solved, Noble climbed into the cab, started the engine, and began proceeding slowly down the mountain. Almost immediately, he became aware that the brakes were not functioning. He stopped at the side of the road, got out, and quickly determined that the welder had burned a hole in the rubber brake fluid hose, and almost all our brake fluid had leaked out! Fortunately, such a hose was included among our many spare parts. By working on his back under the truck, with light from a Coleman lantern, Noble amazingly succeeded in replacing the broken hose, a task with which he had no previous experience! However, the operation of the brakes was still considerably impaired, because he had not been able to bleed all the air out of the system. Never-the-less, he started back down the mountain again, driving very slowly and cautiously on the rough, corrugated road, and he once again passed through the village of Pachica. He continued his slow descent downward, but 13 miles before he reached the main highway at Huara disaster struck once again. THE NEW WELD BROKE! Noble could go no further. With the last of his energy, he maneuvered the Jeep once again to the side of the road, jacked up the front end, and finally collapsed into bed at 1:00 am. For Noble it was one of the worst days of our adventure so far. (Tomorrow was yet to come.)

Plate 20: We crossed from Bolivia into Chile on a gravel, barren, horribly corrugated road at an elevation of 15,750 ft., with no Chilean officials or border post. Shortly thereafter our front axle housing cracked, and we broke down entirely.

The next morning (Fri., May 29) Noble pondered his problem. The easiest solution was for him to leave the Jeep by the roadside again and catch a ride back to Iquique for assistance. This would require both more money and more time, which he considered very precious commodities, so he made the decision to remove the entire front axle assemblage by himself and bring it with him to Iquique for repair. This would be another completely new experience for him and would test his mechanical skills to their limit. The daytime temperature climbed rapidly and Noble became covered in sweat and dust, but he persevered. Three hours later, he accomplished his task! Noble removed, all by himself, the 200 lb. axle (with the differential and hubs), the steering arms, shocks, U-bolts, and brake hoses. It was a monumental feat. He had passed his apprenticeship as an auto mechanic. Three cheers for him! It was almost noon and the air temperature in the sun was now over 120°F. Noble had succeeded where almost anyone else would have failed. It was such determination and effort that kept us on the road, and within our very limited budget.

At 4:00 pm, the first vehicle of the day came down the road, a pickup truck. The driver took Noble, with the broken axle and parts, only as far as Huara, on the main north-south gravel highway. Here he tried without success to find a bus or any other vehicle to take him to Iquique, so he had to spend the night in Huara, sleeping in a filthy, little hotel, which cost him 20 cents. The next morning (Saturday, May 30), Noble caught a bus for 50 cents all the way to Iquique, which dropped him and his heavy vehicle parts at Gus Medrano's Ford agency in town. Gus took him to a machine shop where he was told a sleeve could be made to fit inside the cracked housing, and a new axle could be built from the old one. The repair was estimated to take one or two days. Noble said OK.

This finally gave us a little leisure time together, on Saturday, to look around the town. I took pictures of the horse-drawn carriages, which were conveying people to and from downtown stores and their homes. We enjoyed some ice cream from a little sidewalk shop, and then watched with amusement as young people strolled slowly around the central plaza, boys in one direction and girls in the other, flirting with one another as they passed, but rarely stopping to talk. Not surprisingly, Noble was very tired and wanted a bath and early night at the hotel. Imagine his disappointment when we discovered there was no water for a bath or shower (or even the toilet) because the hotel had run out of water! Noble collapsed on his bed and immediately fell asleep, dirty clothes and all, then announced the next morning (Sunday) that he had never slept more soundly.

During my time in Iquique, I walked down to the seashore on several occasions, to stroll along the sandy coves and to take pictures of the scenic rocky shoreline, to photograph and document birds, collect seashells, and take pictures of starfish on the rocks. A few fishermen waded along the shore, catching small fish in a butterfly-like net at the end of a long pole. Sometimes large waves crashed over the rocks and spilled out over the sandy coves, but at other moments, the ocean was quiet and serene. Highlights of my birdwatching were: 6,000 Guanay Cormorants, 85 Red-legged Cormorants, 24 Humboldt Penguins, 1,200 Kelp Gulls, 3,600 Gray Gulls, 40 Belcher's Gulls, 50 Inca Terns, 15 Peruvian Diving-Petrels* (family Pelecanoididae, see comments below), 7 Blackish Oystercatchers*, 1 Chilean Seaside Cinclodes*, and 35 Dark-faced Ground-Tyrants* (a winter visitor from southern Chile, feeding in a garbage dump). I particularly enjoyed watching and taking pictures of the Inca Terns and Red-footed Cormorants. Both were striking birds with their dark plumage, red legs, and red bill, providing me with photographic memories as they dived for fish or sunned themselves on the black lava cliffs and rocky ledges. The Peruvian Diving-Petrel is one of four species of diving-petrels in the world, occurring only in the colder waters of the southern hemisphere, where they were ecological counterparts of the northern hemisphere auklets and murrelets, being small, dumpy seabirds which flew low over the water on rapid fluttering wing beats.

The axle repair was completed by the end of the afternoon on Sunday, May 31, and we picked it up at the machine shop. After much bargaining, Noble agreed to pay a total of $45 for all the work. We wanted to take the axle back to our Jeep that very evening, but the buses to Huara were already booked out and there were no seats for us. Thus it was necessary for us to check into the Gran Hotel España for one more night. We transported the heavy axle from the bus station to the hotel in a donkey cart, with a very drunk driver. Noble finally got the hot shower he so badly wanted.

Monday morning (June 1), we boarded a bus for Huara, with our axle in the luggage compartment. However, the bus broke down in the town of Humberstone, about halfway to Huara. Like all the other passengers, we had to either wait for another bus or find alternative transportation. Luckily, we succeeded with the latter option. We located a small, dilapidated, 1926 Studebaker pickup truck with wooden sides. The driver agreed to take our axle and us all the way to our Jeep, 30 miles away, for a cost of only 2,000 pesos ($2). It was an hour's drive over dusty, bumpy roads. The driver had a helper with him. We arrived about noon and I took several photographs of Noble and the other two men unloading our axle from the ancient little truck. We paid the driver and the two of them left. With my help, Noble reinstalled the axle, which was a much simpler task than removing it because there were two of us and Noble had more knowledge of what needed to be done and how to do it. We completed the task in about two hours, and confidently set off down the road once again. We hoped fervently that nothing more would break! I drove with extreme care and caution and by 6:30 pm we reached Huara - - for our last time.

After a quick meal at a local restaurant, we filled up with gas at a service station and commenced our 1,000 mile journey south to Santiago. Our dusty, gravelly, washboard road paralleled the coastline, which was never far away to our right. Two hours later, we stopped for the night and camped along the highway, on a sandy rise between two salt flats in the Atacama Desert, at an elevation of 1,000 feet above the nearby Pacific Ocean. There were no birds.

The next morning (Tuesday, June 2), our "highway" south in the desert was still as bad as ever, with deep, evenly spaced ruts which vibrated our teeth and our van at every speed over 5 mph. A road sign said it all - - "Camino destruido" (road destroyed). The sign recommended a maximum speed of 40 kph (25 mph), but this was much too fast for us. The road surface never improved, and continued to be "destruido" for mile after mile. I assumed that the government had no funds to hire road graders. Driving demanded all my attention and concentration, and was not any fun. It was not possible to enjoy the desert landscape, or watch for any bird to fly across the road.

Several hours down the highway, in the middle of absolute desert, was a police control post, where all vehicles were required to stop for inspection. I did so with a certain amount of trepidation, knowing that neither Noble nor I had an entry stamp in our passport, and there was no stamp in our carnet de passage. We had entered the country on a "ruta de contrabando", which would be very difficult to explain to a government official. I feared the worst, but got my passport ready to show to the policeman as he approached my open, driver's side cab window. This was our first inspection stop since we entered Chile eight days ago. The policeman looked directly at me and said in a very firm, authoritative voice, "Licencia de manejar." I recognized the word "licencia" (license) but could not recall the meaning of the word "manejar." So I said "Que?" (What?). He repeated "Licencia de manejar" in a more rapid, somewhat irritated voice. Suddenly, out of the depths of my memory from a high school Spanish class ten years ago came one of the meanings of "manejar" (to drive a car). The policeman wanted to see my driver's license!

This was the first time in four months of travel in Latin America that I had been asked for my driver's license, and it caught me completely by surprise. However, both Noble and I had an international driver's license, issued to us from Brussels before we departed California. It was actually a little booklet with 37 pages in it, each page in a different language. Having originated in Belgium, the only languages on the front cover, inside or out, were French and Flemish! There was no English and no photograph of me. I had not examined it very critically at all. However, I handed the policeman my license, whereupon he spent more than five minutes examining every page in the booklet, one of which was in Spanish. With a puzzled look on his face, he looked at me and said with a querying tone "Numero"? He spoke so rapidly that I had to ask him again what it was he said. He repeated "Numero?" (number). Even when I understood what word he was saying, it was not immediately clear what he wanted. Then it came to me. He wanted to know the number of my license, so he could record it on the clipboard he was holding in his hand, with a pencil ready to write.

I had not examined my license for the number, and I asked him to hand me the booklet so I could look at it, which he did. To my utter astonishment, I could not locate an identifying number anywhere, inside or outside of the booklet. In fact, there was no such number. I found this incredible, but thought to myself there was no way I could explain this to the policeman, who obviously did not want to give us permission to continue our journey until he had recorded my driver's license number. A thought suddenly came to me. I said to him "Momentito" (just a moment), and I got out of the cab with my license and went back to the inside of our van, out of his sight. Here I took my own pen, and in ink, I wrote the number "70" at the top of one of the pages near the front of the booklet, not too conspicuously. (Why I chose this unlikely two digit number I have no idea.) Then I returned to my driver's seat, handed him my license, pointed to the number "70" I had just written, and said to him "Aqui es el numero" (here is the number). He got a big grin on his face, recorded the number "70" on his clipboard sheet, and with a big wave of his hand and a cordial "Buen viaje" (have a good trip) he motioned for us to continue down the road. I smiled to myself and took a deep breath of relief that he had not asked for my passport, which would have been much more difficult to explain. Life was an adventure!

Later in the day, as we approached the coastal city of Antofagasta, Noble and I crossed the Tropic of Capricorn at 23½° S latitude. We had been inside the tropics for 125 days (just over 4 months). Why 23½°? Because that is the angle at which the earth's axis of rotation is tilted to its plane of rotation around the sun. This angle defines the climatic regions of the earth (tropical, temperate, and arctic/antarctic), and the seasons.

In Antofagasta we took a scenic, sightseeing drive along the rocky, rugged seacoast. There were four or five spectacular natural rock bridges, of varying sizes, just offshore. Waves splashed high off the rocks and spewed across the small sandy coves. Both Noble and I took quite a few photographs. Among the birds which I recorded were a Humboldt Penguin, a half dozen Wilson's Stormne-Petrels (following a small fishing boat just offshore), several Blackish Oystercatchers, a single Ruddy Turnstone, six Snowy Egrets, a few Whimbrels, nine Gray Gulls, a small group of Dark-faced Ground-Tyrants, and four Chilean Seaside Cinclodes. We drove out of town a short distance that evening to a campsite along the roadside, in the desert hills away from the coast, at an elevation of 1,300 feet.

The Atacama Desert is one of the most desolate places on earth, and our highway was one of the worst, still very badly corrugated. There was hardly any sign of human life, only a few widely scattered, abandoned, adobe houses. We climbed up to 8,000 ft., and then back down again. Our van vibrated relentlessly. Then once again, it succumbed, and the front differential casing cracked! For yet another time, our front wheels toed outward, at an angle of 5-10° from the vertical. Our plight seemed almost hopeless, and our spirits were at an all-time low. Nevertheless, we were determined to succeed with our adventure, and we were not going to quit. The two of us surveyed the situation, and evaluated our options. We were in the middle of the desert with no help in sight, or likely. We needed another weld. The decision we made was to continue driving, as slowly as possible, to try to make it to the coastal city of Chañaral, 30-40 miles ahead of us. Here, there were large copper mines where we might find assistance. I drove at a snail's pace. The off-vertical angle of the wheels to the ground did not increase. As darkness approached we stopped along the roadside for the night.

The next morning (June 5) we successfully hobbled into Chañaral. Here we located a sympathetic mine manager, Mr. Woodroffe, who agreed to work on our vehicle in his machine shop. He was Chilean but of English descent with a British passport, who considered himself a subject of the queen, and he spoke excellent English. Chile was the world's second largest producer of copper, behind the United States. Our Jeep would not be ready for departure until late the next day, so Noble and I accepted an offer to sleep in some bunk beds in

Plate 21: The nearest town for repairs was Iquique, 75 miles away on the rocky, scenic, desert coast of N. Chile. Seven days later after numerous delays and faulty repairs, we were ready to continue our adventure.

one of the empty worker's quarters. In the meantime, Noble went on a tour of the mine with Mr. Woodroffe, who was happy to discuss with him Noble's favorite topics of entrepreneurship, government, and investment opportunities in Chile.

I walked down to the seashore to photograph and watch seabirds, and ponder our predicament. I carried with me Noble's movie camera to take some footage of the seabirds as they came and went from the rocky cliffs. Noble complained, afterwards, that I used up too much of his precious film! I also succeeded in getting color slides of the cormorants, pelicans, boobies, and Inca Terns. A new bird for my trip list was the South American Tern*, which was much like a Forster's Tern but with a bright red bill. Also present were an Oasis Hummingbird, Chilean Seaside Cinclodes, and Dark-faced Ground-Tyrants.

The work on our differential was completed by the end of the next day, with a beautiful new arc weld. We hoped it would last! We slept one more night in the worker's quarters. That evening, at Mr. Woodroffe's home, we were offered some pisco sour, and Noble made popcorn for everyone. We left early the next morning (Sunday, June 7) and followed the coastline southward for some 70 miles before turning inland again. Our highway surface had not improved. It was a constant worry what part of our heavy camping van would break next. We climbed upward through arid, sandy hills to the city of Copiapó, situated in the green Copiapó River Valley at 1,500 feet. This was our first really green vegetation since entering Chile, and here at last were a few non-marine birds. I recorded my first Chilean Swallows* (distinguished by their conspicuous white rump) and Common Diuca-Finches* (a more southern, lower elevation relative of the White-winged Diuca-Finch).

We continued our travel, heading south toward the city of Vallenar, approximately 100 miles away. Once again we were on a rough road, in arid desert hills which were almost devoid of vegetation. Our elevation fluctuated between 1,500 and 2,000 feet. Sure enough, the constant road vibrations re-created a minor mechanical problem when the left front spring hangar bracket broke loose, again! (Do you remember the Peruvian Andes?) We found a place to camp for the night just outside Vallenar, at 1,500 feet elevation in the Huasco River Valley.

The next morning (Monday, June 8), we had to stop at a police check post as we entered the city, where the police officer was a little unfriendly. An English speaking, retired farmer of British descent, Mr. Wodehouse, and his wife, came driving by just then in their 1930 Ford Model A. Upon seeing our van and the police officer, they stopped to ask if we needed any assistance. We were having a little difficulty explaining our situation to the inspecting officer, who initially had asked to see our passports. Mr. Wodehouse spoke rather gruffly to the policeman, in Spanish, and we were immediately motioned on our way. The good natured ex-farmers then invited us to their very modest house, which was surrounded by eucalyptus trees, just outside the town on a small, rather rundown piece of land. They had no electricity, only old kerosene lanterns with worn out wicks. We informed Mr. Wodehouse that we needed to have some welding done on our Jeep in town, and he obligingly took us to a welding shop, where we explained our problem to the owner and then left the Jeep with him.

When I mentioned my interest in birds to Mr. Wodehouse, he immediately said that I should meet a good friend and neighbor of his, who was also of British descent and owned 800 acres of good farmland in the river valley. So I was introduced to William Mille at his home. By a remarkable coincidence and stroke of good fortune, Mille was an avid ornithologist and egg collector! He showed Noble and me some of the eggs in his extensive private collection, which were very carefully preserved and catalogued in his own custom made trays, all of which were stored in a single handsome, tall, case which he had built by himself and was justifiably proud of. Furthermore, he pointed out to me that he had accompanied A. W. Johnson (one of the authors of *Las Aves de Chile*) on an ornithological expedition just two years earlier, in January 1957, to study flamingos on the Chilean altiplano and collect their eggs. In the front of volume one, Mille is acknowledged as a contributor to the two books, and elsewhere in the volume he was named and pictured with Johnson (and others) on the flamingo expedition. Some of his eggs are shown on color plates in volume two. (I was puzzled, initially, why he was referred to as "G. Mille" in these publications, until I remembered that "Guillermo" is the Spanish equivalent of "William.") What were the odds that I should make his acquaintance by happenstance?

With his permission, I spent three hours on his extensive farm that day, recording the birds that I encountered, while our Jeep was being repaired in the town. Almost a dozen species were new for my trip list, including the Black-faced Ibis* (a very handsome buff and black colored ibis with a black neck wattle, sometimes considered conspecific with the Andean Ibis), Green-backed Firecrown* (the most southerly distributed hummingbird in the world), Chimango Caracara* (the southern counterpart of Yellow-headed Caracara), Fire-eyed Diucon* (a dapper Gray, phoebe-sized flycatcher with a prominent white throat and brilliant red iris), Picui Ground-Dove* (a small dove with a conspicuous white band in the wing in flight), and Dark-bellied Cinclodes* (a species which Mille told me did not occur this far north in Chile and I must be wrong, but I persisted with my identification and eventually I read in volume two of *The Birds of South America* by Ridgely and Tudor, 1994, that the range of this species extended as far north as Antofagasta, and I felt exonerated.) Once again, I was surprised to see four Harris's Hawks, a species which I could not become accustomed to seeing in South America.

We retrieved our Jeep the next morning (June 9). The welder had simply re-welded the old weld, which made us quite skeptical that it would last very long. We said goodbye to William Mille and the Wodehouses, and thanked them for their hospitality and assistance in my pursuit of birds. We then continued south through arid, rocky, cactus covered hillsides toward the twin cities of La Serena and Coquimbo, which were approximately 100 miles away on the central Chilean coast. Thirty miles north of La Serena, I convinced Noble to let me stop for a few minutes to see what birds I could find in a rocky area of desert where, at last, there was a little green vegetation. The site proved a good choice, for I discovered a Moustached Turca* (*Pteroptochos megapodius*), a species endemic to central Chile and my first ever encounter with a member of the family Rhinocryptidae. It was an exciting moment for me. Since I had studied this bird in my *Las Aves de Chile*, it was not necessary for me to write a long description in my field notebook of this distinct ground dwelling, rather large "tapaculo" with oversized feet. It was a very handsome bird of rocks and boulders, with a boldly barred belly and conspicuous white moustaches outlined by its tawny chest. Its moderate length tail was often carried cocked wren-like vertically over its back. I watched the bird for several minutes, obtaining great views as it ran rapidly over the open ground and rocks, between clumps of cacti. It was one of my most thrilling birding experiences, and a lasting memory.

As we were nearing La Serena, a pair of White-tailed Kites flew above the arid countryside, one of my few encounters with this species in South America. After we passed through La Serena, it was only a few miles to Coquimbo. Sure enough, the new weld of our hangar bracket broke as we were entering the city. It was late afternoon, but we found a blacksmith shop which was owned and operated by a very large, powerfully built man and his equally robust wife. They said they could repair the break for us, so we left our Jeep with them and walked down the street to a small restaurant where we ordered a bite to eat. Two hours later when we returned to the blacksmith shop Noble discovered that the spring had been replaced backwards! They promised to correct the mistake "mañana" (tomorrow). Unhappily, it was necessary for us to drive out of town to find a campsite for the night.

The next morning (Wed., June 10), the owner of the shop said he had a plan to fix the broken hangar bracket in a manner which would prevent it from ever breaking again, by drilling a new hole in the centerline of the spring. Noble pondered this idea and decided to give it a try. (We didn't have much to lose since the bracket had already broken almost a half dozen times since the altiplano of Peru!) The work, however, would require all day to complete. While the repair was being carried out, I hitchhiked to the outskirts of the city for a day of recording birds. At 30°S latitude we were in a latitudinal belt where the temperature, rainfall, and physical features of the vegetation were all very similar to each other around the world, north and south of the equator, on the western side of continents. The distinctive climate of this belt (with hot dry summers and cool wet winters) was the prevalent climate along the northern and southern shores of the Mediterranean Sea, and it was thus given the name "Mediterranean Climate." Here in Chile, the dense, scrubby, woody vegetation characteristic of this climate was known as "matorral." (In southern California it was called "chaparral.")

This was the habitat in which I meandered for my day of birdwatching at Coquimbo, along the rocky and sandy Pacific coast. There were arid, scrub-covered hillsides, scattered trees, cacti, and small ponds. By the end of the day I had tallied 44 species of birds, 8 of which I had not previously seen. These were the White-winged Coot* (one of six species of coots in Chile), Austral Negrito*, Rufous-tailed Plantcutter* (one of three species of peculiar finch-like birds of uncertain affinities found only in southern South America, and most recently placed with cotingas in the family Cotingidae), Scale-throated Earthcreeper*, Crag Chilia* (a handsome furnariid of rocks and boulders which was remarkably like a Canyon Wren in coloration, appearance, habitat, and behavior, though it was somewhat larger), Dusky-tailed Canastero*, Sharp-billed Canastero*, and Gray-hooded Sierra-Finch*. The Crag Chilia and both species of Canasteros were all endemic species of arid, rocky, or shrubby areas in central Chile. I became particularly fond of the aptly named Crag Chilias, as I watched them hopping in and out of rock crevices and perching momentarily on the tops of small boulders, displaying their wide snow-white throat and flashing the rufous in their wings and tail. The most abundant birds of the day were Picui Ground-Doves, Green-backed Firecrowns, Chilean Swallows, Chilean Mockingbirds, Mourning and Band-tailed sierra-finches, Common Diuca-Finches, Rufous-collared Sparrows, and Yellow-winged Blackbirds. Two additional species of particular note were a pair of Burrowing Owls and three Moustached Turcas, my second and last sighting of this special species.

As we drove out of Coquimbo the next day (Thursday, June 11), we had been travelling for 17 days and more than 1,000 miles in Chile and no authority had yet looked at our passports. By now, we were quite amused by this fact and we decided to try and reach Santiago without having our documents officially inspected. Of course, we were a little worried what would happen when it was discovered that we had no entry stamps. We drove the rest of the day all the way south to the border of Aconcagua province, where we camped for the night. Raptors which I saw en route were 1 Andean Condor, 20 Turkey Vultures, 3 Harris's Hawks, 15 Chimango Caracaras, and 12 American Kestrels.

We entered the large, coastal city of Valparaíso on Friday, June 12, having driven 13,319 miles from Los Angeles according to our Jeep's odometer. Along the harbor were a half dozen Brown-hooded Gulls*, my first sighting of this rather small gull of southern South America, an inhabitant of lakes and bays (and not to be confused with the Brown-headed Gull of southern Asia). We proceeded to the nearby city of Viña del Mar, where there was a Chilean naval air station. Noble was excited to stop by and say hello to two of the pilots with whom he had trained at Pensacola, Florida, in 1955-56 (only a few years earlier). He was warmly received and introduced to some of their buddies. We were invited to stay awhile as their guests at the BOQ (bachelor officer's quarters), and everyone could party just like old times. So we did! It was great fun, with lots of good Chilean wine, fish, shrimp, story-telling, and recalling of old memories. From the BOQ we telephoned the two French boys (Jean and Raymond) in Santiago. They were just about to leave for Argentina and Jean said to me not to worry about his pistol, that maybe we could all catch up in Buenos Aires.

Very early on the following morning (Sunday, June 14), I got up and drove our camping van the short distance to the fishing boat wharfs in Valparaíso, from where the shrimp boats left each day to catch fish and shrimp ("langostinos") in the cold, off shore Humboldt Current, at the edge of the continental shelf. I arrived at 6:00 am and soon found a small boat, the "Arpon", being readied for departure. In my meager Spanish, I engaged the young Basque captain, Iñaki, in conversation and asked if I could accompany him on his boat for the day, explaining to him who I was, and that I wanted to observe the seabirds. To my pleasant surprise, he willingly agreed. I could come along with him and his crew of three. He said they would return in the late afternoon. Soon thereafter we left the pier and chugged out for open water, which immediately became quite rough. (In my enthusiasm, I had forgotten how readily I became seasick on a small boat.) It was a day to remember. When we reached the continental shelf, the shrimp nets were released and seabirds appeared from all directions, circling, gliding, and skimming above the water--albatrosses, petrels, shearwaters, fulmars, storm-petrels, and even a few prions (see below). What a thrilling experience! During the day (in between my frequent bouts of seasickness) I recorded 12 species of procellariiforms: 40 Wandering*, 15 Black-browed*, and 300 Shy* albatrosses, 5 Southern Fulmars*, 125 Cape Petrels*, 6 prions (unidentified to species), 750 White-chinned Petrels*, 7 Sooty Shearwaters, 60 Wilson's Storm-Petrels, and 3 Peruvian Diving-Petrels. These birds swarmed around our boat (as many as 200 at any one time) and came very close as the crew pulled in our overflowing nets of fish and shrimp. Sooty Shearwaters and prions showed somewhat less interest in the boat than did the other seabirds, and for most of the time they remained a little farther away. At the end of the day, as we returned to the harbor, we encountered several additional species--Peruvian Pelicans, Kelp Gulls, a single South American Tern, and two Chilean Skuas*.

A most extraordinary incident occurred during the course of the day. A very full net of shrimp was being pulled from the water up to the deck of the boat, and many seabirds congregated to feed on the smaller fish and shrimp which fell through the netting or spilled over the top. A Wandering Albatross (with its gigantic 11 ft. wingspan), which had followed the net to the boat, greedily reached through the netting with its head to get at some of the shrimp and fish which were inside the net. As the net was raised up out of the water the bird's head became entangled in the net and the albatross could not extract it, so the albatross was pulled up the side of the boat, hanging by its head with its body dangling outside the net. The twisting and turning of the net as it was pulled on board, and the weight of the albatross itself, broke the neck of the albatross, which was dead by the time the load of shrimp was lowered to the boat's deck. The albatross was of no use to anyone on board, so as a biologist I decided to remove the head, clean the skull, and save it as a souvenir and memoir of my day on the shrimp boat. (I didn't finalize the cleaning until a couple of weeks later.) The ten-inch skull with its massive, slightly hooked bill was then hung by a chain from the middle of the cab's roof in the front of our camping van, between Noble and me in the front seats. It was displayed there for the duration of our adventure, where it received many curious looks, comments, and questions as to what it might be. Only one or two persons ever guessed or knew that it was an albatross skull, in spite of its enormous, distinctly shaped bill, prominent tubular nostrils, and the two large depressions on top of its head, which cradled the salt glands. (I have carefully kept the skull in my treasured possessions, and it sits today in the family room of our house, on the mantle above the fireplace.)

The prions were the most exciting birds of the day. These dainty, pearly-Gray, little seabirds of cold, southern oceans (with 4-6 species in the genus *Pachyptila*, family Procellariidae) were all seen flying singly, low over the ocean surface, 50-200 yards from the boat, except on one occasion when a pair were flying together. According to *Las Aves de Chile* (volume 2, 1951), only two species of prions were known from anywhere in Chile, the Antarctic Prion and the Slender-billed Prion, with almost all the records of both species coming from the extreme southern tip of the country. The most northern documented records (at the time the volume was published) were dozens of beach washed birds after a severe storm on July 26 & 27, 1942, which left many dead birds at several sites along the coast as far north as Playa Blanca, near the town of Lota in Concepción Province, at approximately 37°S latitude. My sightings today off Valparaíso, at approximately 33°S latitude, were therefore more than 250 miles north of any previous records of prions in Chile. Unfortunately, I could not identify which species of prion(s) I observed on my June 14 boat trip, as members of this genus are so similar to one another in appearance in the field, that they are virtually impossible to separate. W.B. Alexander's *Birds of the Ocean*, 1954, was of no help. However, there is more to this story.

The very next day (Monday, June 15), Noble and I took our van for repair to the Jeep agency in Santiago (as we had done in every major city through which we traveled). As usual, we met the general manager of the

Plate 22: I spent one day birdwatching aboard a small shrimp boat off the central coast of Chile, out of Valparaíso. Highlights of the day were observing prions and salvaging the skull of a Wandering Albatross (pictured) which broke its neck in a shrimp net as it was being hauled aboard.

agency, who, in Santiago, was a man by the name of Rafael Barros, perhaps in his 60's. Guess what. By the most unbelievable coincidence imaginable, Rafael was not only the general manager for the Jeep agency, but he was a very keen amateur ornithologist! Since his youth, he had accumulated, and preserved as specimens, a very large private collection of Chilean birds in his own home. A dozen of his published sightings on various birds, dating back to 1921, were cited in the "Bibliographia Consultado" section of *Las Aves de Chile*, volume two. Even more astounding was the fact that he was one of only a very few persons who had picked up and salvaged some of the beach washed prions after the storm in July 1942, and he subsequently published a scientific paper in a university journal documenting this event. Furthermore, his name was cited in volume two under each of the detailed species accounts for both the Antarctic and Slender-billed prions! Was I dreaming? Surely this was divine intervention. It wasn't just that he was an ornithologist and an avid birdwatcher, but he was perhaps more familiar with prions in Chile than any other person. My fortuitous meeting with Rafael the day after my boat trip sightings certainly defied all odds.

I learned all these facts about Rafael very soon after Noble and I first met him on Monday, because he immediately saw my salvaged albatross skull drying on the top of our van, and of course, he inquired about it. So I told him not only about the albatross skull, but also about the prions I had seen. That's when he became very excited and told me about picking up the beach washed prions after the 1942 winter storm, and subsequently publishing a note about them. Understandably, he was at first very skeptical that I had actually seen any prions, but as I related more information to him, he became convinced enough that he said to me that he would like for the two of us to go back out to sea together in a boat on Wednesday (2 days from now), for the purpose of collecting some specimens if at all possible. He became visibly excited at this prospect (even turning red in the face), and said that he would make all the necessary arrangements and have the necessary shotguns and permits. We would travel in a small high-speed police boat with a friend of his who worked for the Coast Guard. Fortunately, in this instance, the repair work on our Jeep was estimated to require at least two more working days. How lucky could I be?

Early on Wednesday morning (June 17), Rafael picked me up at the Jeep agency and we drove to the pier for police boats in Valparaíso harbor. At 7:30 am, we departed in a very speedy little police boat, armed with shotguns for both Rafael and myself, and several boxes of shotgun cartridges. It wasn't long before we encountered our first seabirds, mostly albatrosses which glided effortlessly past our boat with hardly ever a wing beat. Rafael scattered some bait on the ocean surface to attract birds to our boat. We were about 5-7 miles off the coast and were soon surrounded by hundreds of seabirds of many kinds - - Shy Albatrosses, 2 Antarctic (Southern) Giant Petrels* (the most likely of the two giant petrels, though in 1959 only one species was recognized and I did not have to try to separate them), White-chinned Petrels (an estimated 500 individuals), Cape Petrels, Wilson's Storm-Petrels, and yes, almost a dozen prions! The prions came closer to our boat than they had on Sunday. However, collecting a few specimens by shooting was an extraordinarily challenging task in the five-foot high waves. Our little boat bounced up and down constantly, and rolled sharply from one side to the other. (Yes, I was seasick again.) The prions were also continuously in motion, twisting, darting, and turning erratically and unpredictably, just above the surface. Holding our shotguns steady was out of the question. We could only point and shoot in the general direction of a prion, hoping we might hit it. They were very small, fast moving targets. It was fortunate that Rafael had brought several boxes of cartridges with him. We needed them all. After almost three hours of effort, both Rafael and I succeeded in shooting two prions each. Rafael was ecstatic. There was no longer any doubt in his mind whether I had observed prions or not. As we returned to the pier, there were other birds - - a single Humboldt Penguin, two more giant petrels, a dozen diving-petrels, Red-legged, Guanay, and Neotropic cormorants, Peruvian Pelicans and Peruvian Boobies, hundreds of Kelp and Brown-hooded gulls, Inca and South American terns, and two sub-adult jaegers (probably Parasitic). It was a day never to be forgotten.

The next morning (June 18), Rafael took me to his house and showed me some of his extensive bird collection. We identified the prions we had collected, three of which were Antarctic Prions* (*Pachyptila desolata*) and one was a Slender-billed Prion* (*Pachyptila belcheri*). Since Rafael had to return to work, I stayed at his house alone for most of the day to prepare all four birds as scientific specimens (at the slow rate of one specimen per hour), a skill I had learned as an undergraduate university student. These specimens extended the known range of both species (at that time) approximately 250 miles northward along the coast of Chile. I left them with Rafael in his personal collection. It was an extremely gratifying experience for an aspiring ornithologist!

During our four days in Santiago Noble and I stayed in our camping van at the Jeep agency, as usual, since it was the cheapest accommodation for us. We made the decision to "turn ourselves in" to the immigration and customs offices and face whatever the penalties might be for our transgressions, even though we were not to blame. The two offices were at opposite ends of the city. Surprisingly, we received the same response at both offices, each of which told us that we would have to go back to our port of entry and have our documents stamped there! We protested, to no avail, that our border crossing was over a thousand miles back up the horrendous coastal highway, and even if we were to drive all the way back there were no officials or security posts on top of the mountain ridge where we entered, at almost 16,000 feet above sea level. They insisted this was impossible, and refused to stamp our documents. However, there did not seem to be any immediate penalty or threat of jail if we just simply walked out of their offices in Santiago and continued onward with our journey without any stamps in our passports, at least until the next inspection station. We were rather amused by the situation. Tourist travel in Latin America, by road in a privately owned vehicle, was entirely unpredictable in 1959. Patience, flexibility, and innovation were required (but we did not offer monetary bribes).

On our last day in Santiago (Friday, June 19), we paid the Jeep agency a total of $147 for all of the labor and parts involved in their many hours of work on our van. We thanked them and I said goodbye to Rafael Barros, who gave me the name of a good friend of his, Mario MacLean, in Punta Arenas, Chile, on the Straits of Magellan at the southern tip of the continent (where we were eventually headed). We were also given some tire chains as a gift from one of Noble's pilot buddies, "Bugge" (of Norwegian descent). He said they would be necessary for the winter's snow in southern Chile. Snow? What snow? Snow was not on our planned itinerary. (I suggested to Noble that perhaps we should purchase some antifreeze for our radiator, but being a warm weather native of southern California and wishing to conserve as much money as possible, he chose not to do so, a decision we would come to regret.) The two of us then went shopping for some good Chilean wine. We purchased 12 bottles of "Gran Niño Teresa", and 15 liters of "reservado" which we emptied into a large plastic bag with a spigot, and then placed it inside our canvas water bag! That evening the two of us were met again by Bugge, along with his friend, Ivan, and we all went to an exclusive restaurant where we drank wine and savored a delicious gourmet "farewell" seafood dinner. It was after midnight by the time we finished eating. Noble and I decided to drive out of town rather than spend another night at the Jeep agency. I drove only a short distance before finding a suitably quiet place to camp for the rest of the night.

The next morning, Saturday, June 20, we rejoiced in having a paved highway for some 200 miles south, all the way to the city of Chillán (our only hard surfaced highway anywhere in Chile, except between Valparaíso and Santiago). The countryside, for the first time since we entered Chile, was wet, black-soiled farmland with green pastures, woodlots, fences, and herds of dairy cows. What a pleasant change of scenery! Many of the farmers were German, having come to Chile after the war. When they answered a question with "Yes", they were just as likely to say "Ja" as they were "Sí." The most common roadside birds were Chimango Caracaras, Southern Lapwings, Chilean Mockingbirds, Chilean Swallows, Common Diuca-Finches, Grassland Yellow-Finches (the most abundant species), Long-tailed Meadowlarks, and Eared Doves. Austral Blackbirds*, Austral Thrushes, American Kestrels, and Fire-eyed Diucons were also frequently encountered. Our concrete highway ended at Chillán, and we camped 20 miles beyond. There would be no more paved roads in Chile, only horrendous, wet, slippery, muddy roads with deep ruts and water-filled holes, sometimes bordered by steep roadside ditches. Our last roadway would be an almost impassable snow and ice covered route uphill to the Argentina frontier. It would become a driving nightmare.

The next morning (June 21), we were on our way again. It was unbelievable how our road had changed from dry, dusty, and corrugated in northern Chile to wet, muddy, rutted, and slippery in southern Chile, with numerous soccer ball sized water-filled holes 6-10 inches deep. No amount of careful maneuvering could avoid them all. It was mid-winter, the rainy season, and a light rain fell all day long. I drove very slowly and cautiously, between 10 & 15 mph. The green countryside was mostly dairy farms. We stopped at a small roadside waterfall (Salto de Laja) where there was a little German restaurant, and relaxed with a cup of coffee. Noble enjoyed visiting with the owner, Herr Rodolpho von Gartzen.

We came to a major, steep-sided irrigation ditch which was an estimated 4-5 feet in depth and 8-10 feet wide, over which we had to pass. There was no permanent bridge, only six heavy wooden planks, each of which was approximately 12 feet in length, 6 inches wide, and 3 inches high. They were all lying at various angles and positions on the ground, in the bottom of or leaning against the inside of the irrigation ditch. These were the bridge. All we had to do was to lift them up out of the ditch and lay them across the top in a manner which would allow us to drive our 9,000 pound vehicle across! There was no alternative if we wanted to continue to our destination in the south of Chile. So we carefully built our bridge, which consisted of 3 wooden planks, laid side by side for each of the two wheel tracks we needed. The planks were just barely long enough to go all the way across the ditch. There was no room at all for any error in driving, which Noble was happy for me to do. One little slip on the narrow, wet planks and our Jeep would be in the bottom of the ditch (a situation I certainly did not want to contemplate). It was with considerable trepidation that I carefully drove the Jeep's front wheels onto the near end of the planks, and then very slowly continued forward until the back wheels were also now on our very narrow "bridge." Without stopping, I then steered straight across the bridge, as directly forward as possible, onto the muddy road beyond. There was a loud clatter as my back wheels drove off the bridge on the other side of the ditch. When I stopped and looked behind me all of the planks had fallen into the ditch, which was exactly where we had initially found them! We were exhilarated to have made it, and celebrated with some of our recently purchased wine. Our journey could continue.

We camped that night between Collipulli and Victoria, having driven only 95 miles all day long during 12 hours on wet, slippery, and very hazardous roads. Toward evening, very large groups of Slender-billed Parakeets* flew across the farmlands, in flocks numbering 500-1,000 birds each. In winter this species of southern temperate Araucaria and Nothofagus forests descends from the mountains and frequents the lowland agricultural areas and woodlots. Its long, pointed, sharply curved upper mandible is thought to be an adaptation for extracting the large, oval, pine-like seeds of Araucarias. Shortly after dark, two owls flew past our campsite, one of which appeared to be a Burrowing Owl and the other a Short-eared Owl.

The next morning, June 22, we drove into Temuco, center of the last Chilean Indians. Women wore colorful blankets and shawls, and large prominent silver necklaces hung low around their necks and across their chests. We asked numerous persons about the road condition between Temuco and Loncoche (30 miles south of Temuco), and they all gave us the same answer - - "Es imposible pasar, ningún vehículo ni tractor" (it is impossible to pass, no vehicle nor tractor). We were told it would be necessary to take a train for this short stretch. Noble inquired about the cost, and was told it would be $30. He said to me that he didn't want to pay that much, and, since the railroad would not bargain with him, we would have to drive, with me doing the driving! Oh dear.

Conditions could not have been worse. The wet, gooey mud on the road surface was often 1-2 feet thick, and in some of the mud-filled holes, (which were often not visible to the driver), the mud was almost three feet deep and up to the top of our wheels! Of course I was in 4WD and our lowest gear, proceeding at a snail's pace and determined not to stop, except to occasionally get out and walk through the mud with my boots on to test the depth. Fortunately, beneath the soft mud there was always a hard bottom. Sometimes the mud was above the top of my knee-length boots! When one of our front wheels dropped into an unseen mud hole, our top-heavy van would come dangerously close to tipping over. Occasional oxcarts in the middle of the road, which we had to pass, added to our hazard. At one point, I left the main road to avoid a particularly soft, sticky, muddy section, and I drove for several hundred yards through a cow pasture. People along the roadside stared at us in disbelief. On another occasion we came to the bottom of a particularly muddy hill where several trucks of varying sizes were parked in the middle of the road, waiting for it to dry out enough to become passable. All of the drivers told us "No es posible subir" (it is not possible to go up). Then a small boy saw the name "JEEP" on the front of our cab and pointed this out to the lead truck driver. The driver immediately moved his truck out of our way, waved his hand in a forward direction, and said "Suba no mas, hasta luego" (go right on up, see you later)! He was certain that a Jeep could make it to the top. I put our Jeep in low, low gear, in 4WD, and at a rate of 3 mph I started up the long, muddy incline, and I did not lose any forward progress at all as our "Roadrunner" proved it was the master of muddy hills. In ten minutes we steered all the way to the top without ever so much as a slip or a slide, mostly following in mud-filled oxcart ruts. Our heavy weight was, on this occasion, an asset and not a handicap. There were a few things it could perform better than other vehicles. The little boy had climbed up on the top of our van to ride with us, and he was as delighted as we were when we made it to the top of the hill. Noble and I got out and waved to all the truck drivers who were standing at the bottom of the hill watching us with considerable interest, and perhaps with a little envy.

Although our road was horrendous, the countryside was picturesque and scenic, with green pastures, Araucaria trees, farmhouses, two-storied or three-storied barns, many black-and-white dairy cows, and always a snow-capped volcano in the distance. Overall, the landscape reminded me somewhat of Minnesota or southern Michigan (except for the volcanoes and Araucaria trees). Oxcarts were the primary means of transportation, but they were a nuisance to us as they drove in the middle of the road and created ruts. For a pleasant change there was no rain today. Three roadside birds I had not seen previously were a Chilean Tinamou*, Austral Pygmy-Owl* (on the top of a telephone pole in mid-day), and more than a dozen Patagonian Sierra-Finches*. Chimango Caracaras were particularly abundant, and I estimated that I saw at least 750 individuals during the day. (These scavengers fill much the same niche in Chile as do American Crows throughout much of North America.) Southern Lapwings and Common Diuca-Finches were also numerous, as were Austral Thrushes and Grassland Yellow-Finches. A Short-eared Owl was foraging over a grassy pasture in the late afternoon, and a flock of 75 Slender-billed Parakeets flew past. Just after dark (at 7:30pm) we reached Loncoche, where we were told that we were the first vehicle to arrive by road from Temuco in the last two weeks. Noble had saved us the train fare with his decision, but the driving had exhausted all my energy. That night we camped a short distance beyond the town.

It was windy and raining the next morning, June 23. The road from Loncoche to Valdivia was slightly improved, but was still muddy with numerous round potholes. Valdivia was situated on the Pacific Ocean. As we were entering the city there was a pounding on the roof of our cab, above my head over the driver's seat. This was quite puzzling, so I stopped to investigate. A one-eyed man climbed down from the top of our truck, said "Muchas gracias" (many thanks), and walked off.

Both Noble and I laughed. There was no telling how long he had been riding on the top of our van. In Valdivia we stopped at a tourist hotel to ask if there was any message for us from Jean and Raymond. There wasn't. Since we were at a tourist hotel we inquired about any road that would take us from southern Chile to Argentina, as we had not yet decided how we would do this. We were given a map which showed a road from Osorno (a short distance south of our present location) to San Carlos de Bariloche, situated in the lake country of Argentina at the eastern base of the Andes. We decided on this route, which would necessitate several ferryboat rides across some of the many lakes in this region. In the meantime, we wanted to drive a little

Plate 23: In spite of horrendous deep muddy roads, S. Chile was a picturesque land of green pastures, tall snow-capped volcanos, rainbows, scenic lakes, and lush temperate rainforests with ancient "alerce" trees and Magellanic Woodpeckers.

farther south in Chile, to Puerto Montt. We left Valdivia and drove most of the way to Osorno before I chose a campsite for the night. One new addition to my bird list on this date was the Chilean Flicker*, the southernmost species in the genus *Colaptes* (which contains nine species from Alaska to the southern tip of South America). Unlike most other woodpeckers, flickers spend much time foraging on open ground outside forests, where they are particularly fond of ants.

The next morning (June 24) was cloudy but there was no rain. We were just outside the city of Osorno, the gateway to the lake country of southern Chile. This is the only region in South America where temperate rain-forests occur. In these forests is found the ancient, giant "alerce" (larch) tree, a distant relative and ecological counterpart of the North American redwood tree, filling the same niche. The alerce is an endangered species because of over-logging, and is now confined almost exclusively to a few protected areas on the continent and one or two islands in the fjord region. This coniferous tree grows to 15 feet or more in diameter and almost 200 feet tall. In 1992, a living tree was scientifically calculated to be 3,662 years old, making the alerce the second oldest tree species in the world (behind the bristlecone pine in the mountains of western North America).

We drove all the way to Puerto Montt at 42°S latitude (the same distance south of the equator as the California-Oregon border north of the equator). At a roadside stop there was a Thorn-tailed Rayadito*, a colorful, distinctly plumaged little furnariid with sharply pointed tail feathers, somewhat tit-like in its behavior. Puerto Montt was at the very southern end of the long, 1,500 mile, north-south road in Chile, over which we had been struggling ever since entering the country almost one month ago. The town was situated on the north shore of the Gulf of Ancud, where the Chilean fjord country begins. This remote region was an inhospitable land of coves, cliffs, and islands without any roads extending 750 miles south from Puerto Montt to the southern tip of the continent, on the Straits of Magellan.

Noble agreed to spend a couple of hours writing in his journal while I drove along the shoreline of the gulf to observe whatever birdlife I could find there. The relatively small diversity of avifauna included four Southern Fulmars, one of the few procellariids which foraged close enough to the coastline to be identified from shore. There were also 7 Speckled Teal, 2 Great Egrets, 15 Chimango Caracaras, 600 Whimbrels (a non-breeding visitor from North America which was almost 1/3 of the way around the world from its nesting area in arctic Alaska), 1,000 Southern Lapwings, 200 Kelp Gulls, 3,000 Brown-hooded Gulls, 5 Dark-bellied Cinclodes (my first sighting of this species since Vallenar), and 10 Correndera Pipits. Shortly before dusk we turned around in Osorno and drove 20 miles north back up the road toward Puerto Montt, on the road from which we had come, until there was a suitable side road where we could camp for the night. En route we had to stop so Noble could drag an intoxicated person from the middle of the road to a nearby fence post, where he sat him upright, propped against the post!

We returned to Osorno the next day (June 25) and then followed a road heading eastward 50 miles to a very popular hot springs and recreational area around the lake and town of Puyehue, near which the Antillanca ski lodge was located. Noble was looking forward to doing some skiing. Since the ski lodge was not going to open for another two days (on Saturday, June 27), Noble and I spent two nights in a lovely stone guesthouse on Lake Puyehue. This area was a summertime retreat and the guesthouse was almost vacant at this time of year, so we were very enthusiastically welcomed. We were at an elevation of just 1,000 feet above sea level. The view of the lake and surrounding hills was spectacular, so I took out my telescope to see what birds I could locate on the lake. I was able to identify a Flying Steamer Duck*, two White-winged Coots, two White-tufted Grebes, and six Great Grebes*. This latter species is the largest grebe in the world, magnificent even in its non-breeding plumage with its greatly elongated, rufous colored neck and long, narrow, pointed bill. Birds in the immediate vicinity of our guesthouse were Thorn-tailed Rayaditos, Austral Thrushes, Rufous-collared Sparrows, Patagonian Sierra-Finches, Common Diuca-Finches, Black-chinned Siskins, and Austral Blackbirds. Although the avifauna was not diverse, southern Chile at this latitude was a thrilling temperate environment, unlike any other on the

South American continent.

The next day (June 26), provided me with the opportunity to stroll leisurely through the surrounding tall, majestic "Andean-Patagonian" forests with their renowned Araucaria and Nothofagus ("southern beech") trees. It was overcast and cool (50°), with rain showers on and off throughout the day. The environment was paradise-like, serene and pristine. These forests were the home of the largest woodpecker in South America, the splendid Magellanic Woodpecker*, a close relative of the recently extinct and slightly larger Ivory-billed and Imperial woodpeckers in North America (all three of them in the same genus, *Campephilus*). A loud double rapping was characteristic of all members of the group. Although the male Magellanic Woodpecker was the more colorful member of a mated pair, with an entirely red head, the female was the more ornately adorned, featuring a long, prominent, thin crest that stood straight up on top of her head and curled conspicuously forward. It was a species to remember, endemic to these cool, moist forests at the southern tip of the continent, where it was now the second largest living woodpecker in the world (behind the Great Slaty Woodpecker of southeastern Asia).

Two endemic species in the family Rhinocryptidae were characteristic of the forest understory, skulking in the dense bamboo and announcing their presence with frequent loud, distinctive, far-carrying songs. These were the relatively large, dark-colored, black and chestnut hued Black-throated Huet-huet*, and the medium-sized, gorgeously colored Chucao Tapaculo*, with a bright rufous throat and breast, and a conspicuous black-and-white banded belly. Both of these species were plump, primarily terrestrial birds which often cocked their tails over their backs when running across small openings in the forest. They were frustratingly difficult to see, but once their vocalizations had been learned, they were commonly recorded and provided pleasure as I walked through the forest, under towering trees and among the patches of bamboo. It was a relaxing atmosphere. There were other birds as well, including the endemic White-throated Treerunner* (a small, arboreal, rufous and brown furnariid with a prominent white throat and white pearl-like spots on its flanks, which possessed a slightly upturned lower mandible that it used to forage on the branches and trunks of trees in a nuthatch-like manner) and the Des Murs' Wiretail* (a slender, plain-colored, little furnariid with a tail unique among passerines, consisting of only six very narrow, filamentous feathers, the two central ones being extremely elongated, twice as long as the body). It skulked in dense, low bamboo thickets.

Part of my day was spent outside the forest, where one of the most numerous birds was the ubiquitous Thorn-tailed Rayadito, of which I tallied more than 75 individuals. Some of the other species were 1 Chilean Tinamou (in the open, along a forest edge), 3 Southern Caracaras* (the southern representative of the Crested Caracara), 2 Peregrine Falcons, 2 American Kestrels, 8 Slender-billed Parakeets, 4 Chilean Flickers, 1 Tufted Tit-Tyrant, 6 Dark-bellied Cinclodes, and 1 Sedge Wren (a remarkable little passerine with a breeding range extending locally all the way from the northern edge of the great plains in central Canada to the southern tip of Tierra del Fuego, more than 7,500 miles (rivaling that of the House Wren for the New World native passerine with the most extensive north-south breeding range). The dominant scavengers, as always, were Black Vultures and Chimango Caracaras. Aquatic birds frequenting the lake were 8 Great Grebes, 4 Spectacled Ducks* (a very handsome and somewhat local species), 15 Black-faced Ibises, 5 Brown-hooded Gulls, and a single Southern Lapwing.

The ski area opened on Saturday, June 27. Noble drove the Jeep 14 miles from our lakeside guesthouse to the new wooden and stone ski lodge, and I spent six hours walking this distance, following the road from 1,000 feet elevation at Lake Puyehue to the ski lodge at 3,000 feet elevation. The weather was clear to overcast, with little wind and no rain, 40°-50° in temperature, and I frequently stopped to watch birds. The roadside habitat was mostly forest, with a few small areas of partially cleared farmland, and several small lakes. There was a light covering of snow on the ground for the last 5 miles, and the walk was delightful and invigorating. I met virtually no one else along the road.

On the small roadside lakes along the way there were 8 Silvery* and 5 White-tufted grebes (ecological counterparts of the Eared and Horned grebes in North America) and a half dozen White-winged Coots (which were distinguished with some difficulty from the other two widespread coots by the coloration and features of their bill and frontal shield, and not by the inconspicuous narrow white trailing edge of the secondary's in flight). A Plumbeous Rail walked casually along the outside, open edge of a marshy area, looking much like a dark Gray, uniformly colored King Rail with bright red legs. Seven Magellanic Woodpeckers were heard or seen, and four Striped Woodpeckers* (very reminiscent of the Downy Woodpecker) were quietly foraging in open woodland at the upper end of the road. Black-throated Huet-huets were often heard, but only rarely glimpsed, as were Chucao Tapaculos. Treerunners and wiretails were widespread (the latter identified mostly by voice). Once again the most numerous passerine species, overwhelmingly, was the Thorn-tailed Rayadito, which occurred all the way up to 3,000 feet elevation.

On Sunday (June 28), Noble spent almost all of the bright clear day, morning and afternoon, skiing on the Antillanca slopes with many other young people, most of them of German descent and from the ski club "Andino." Noble had acquired his skiing skills in southern California, and skiing was one of his most favorite recreational activities. He couldn't have been happier. Afternoon tea was served at 5:00 pm in front of one of the three fireplaces. Noble made popcorn for all 80 skiers! That evening he was given a place of honor at the table, and toasted with red wine. It was a memorable day for him. At midnight a holiday feast began celebrating the two saints, Peter and Paul. There was a large vat of hot brandy and wine mixed together. Noble provided music for everyone with our stereophonic recorder/player, powered by the electric generator on top of our van. There was much singing, dancing, and drinking far into the night. We finally stumbled into our bunk beds in the van at 4:00 am, with Noble completely exhausted from his strenuous activities of the day. While he had been skiing, I sat at a table in the ski lodge and cleaned my albatross skull, much to the wonderment of all those who came by.

After only a few hours of sleep, Noble and I got up early the next morning, (June 29), and departed the ski lodge, to follow a winding route which would take us to the Chile/Argentina border. It was necessary to pass through Osorno, one last time. From there we proceeded southeast a short distance to the village of Puerto Octay, which was situated on the northwest shore of Lago Llanquihue, the largest lake in the region with its surface only a few hundred feet above sea level. The view was magnificent, with two snow-capped volcanoes projecting above the eastern horizon -- 7,500 ft. Volcán Puntiagudo and 8,785 ft. Volcán Osorno. We followed around the northern and eastern shores of the lake and then turned inland away from the Pacific Ocean and began a gradual ascent toward the nearby Andes Mountains, which formed the border with Argentina. The entire region was part of the scenic lake country of Chile and Argentina. Our route to San Carlos de Bariloche would necessitate ferryboat crossings of three of these lakes, one in Chile and two in Argentina.

The countryside as we started upward was predominantly moist temperate Nothofagus forest. Black-throated Huet-huets flew back and forth across our gravel road, which initially was free of snow and ice. However, this changed quickly and we soon found ourselves driving through a light covering of new snow. Before very long we came to the small town of Petrohué on the west shore of Lago Todos los Santos, which was at an elevation still under 1,000 feet. Here we boarded our first ferry for a short 2½ hour ride east across the lake to the village of Peulla, which was where the Chilean immigration and customs offices were located.

Noble and I had been in Chile for 34 days, during which time we had driven nearly 2,500 miles. It was late afternoon when we walked into the border control office. A pleasant young man was at the immigration desk. When he searched our passports, he could find no Chilean entry stamp in either one. Of course he wanted to know why. So we explained to him, showed him on a map the exact site where we entered the country and told him there was no Chilean port of entry there, no barricade across the road, no control post, no police, and not even a sign saying "Bienvenidos a Chile", nothing at all. Like all the government officials with whom we talked, he said that just wasn't possible, there were no such roads entering the country. He scrutinized our bearded faces to see if we were telling the truth or not, and he went outside to look at our van. He quickly concluded that we were honest persons and he became sympathetic with our plight, and wanted to help. After pondering the situation for a minute or two, he thought of a solution. First, he stamped each of our passports with an entry stamp ("entrada") for today's date (June 29) and location (Peulla), recording that we entered the country on that day, at that site. He then took out his exit stamp (salida) and again stamped each of our passports with today's date and current location. Our passports now recorded that we entered and exited Chile on the same date, at the same place! What a clever solution. Who would ever know otherwise? We thanked the young man and gave him one of our postcards, for which he was grateful. No bribe was requested, or offered. Everyone was happy. It was now dark so we drove out of town a short distance and chose a campsite, where we ate a bowl of soup and a can of tuna each, then climbed into our warm sleeping bags for the night. We were at an elevation of less than 1,000 feet and for the moment we were not in any snow.

The weather was cool, overcast, and misty the next morning (June 30). The border and gateway into Argentina were no more than 10 miles away, at a low elevation pass in the Andes at 3,300 feet above sea level. Almost immediately we were back in the snow. An Austral Pygmy-Owl flew across the road at 1,500 ft. elevation. As we gradually climbed upward our road surface changed to ice and became very, very slippery. We stopped and put the pair of tire chains, which we had been given in Santiago, on our front wheels. Even so, all four of our wheels spun and I could not make any forward progress at all. No snowplow had been along the road, and trucks had not provided salt or gravel. No other vehicles came along to assist us. I tried several times to back down the mountain and then proceed forward again in a different track, but to no avail. We finally had to stop, get out of the van, and try to remove some of the ice from the road in front of us, with our shovels. We also used our winch to move forward bit-by-bit, one hundred feet at a time. It was very tiring work, and our travel was painstakingly slow. It took us four hours to travel the last two miles up to the border. We were both completely exhausted. At the pass was a roughly-hewn wooden arch over the road that read "Republica de Argentina, Bienvenidos." On the Argentine side of the pass the road had been cleared by a snow plow, and the surface was perfectly free of snow and ice!

In summary, Chile had the worst roads, overall, of any country through which we had yet traveled. Almost nowhere were roads adequately maintained. They were dry, dusty, and horrendously corrugated, or wet, slippery, and muddy with deep holes, or snowy, icy, and again extremely slippery. The government had failed completely in its responsibility to provide a satisfactory highway system for the country, presumably because no funds were available for road construction, maintenance, or repair. It had been an extremely arduous task in driving for me, and a formidable test of mechanical skills for Noble. Yet, by combining our efforts we succeeded, and en route we accumulated many pleasant memories, journal notes, and photographs of people, scenic seashores, snow-covered peaks, magnificent forests, and birds. Our adventure would continue!

Plate 24: We stayed two quiet nights in a lovely little guest house on Lake Puyehue. Noble enjoyed celebrating the opening day of the winter ski season at Antillanca Lodge. It required 4 hours for us to slip, slide, and push uphill the last two miles to the Argentine border.

Chapter Ten—Argentina

The Argentine gateway at the pass welcomed us into the country on Tuesday, June 30. Our gravel road, now cleared of snow and ice, took us down the forested mountain to the western shore of a pretty little lake at 2,500 ft. elevation, Lago Frías. Located here was the small village of Puerto Frías, the entry port for Argentina with immigration and customs offices. Our entry formalities were carried out pleasantly, quickly, and efficiently, and we were wished "Buen viaje" as we left. To continue our journey it was necessary to take a short distance, 30 minute ferry ride east across the lake (our second lake crossing), to the town of Puerto Alegre. The ferryboat was just big enough for our camping van to fit on board. Our route then resumed, passing for 15-20 miles through lovely temperate forests until we reached the small port town of Puerto Blest, at 2,500 ft., elevation on the southwestern shore of Lago Nahuel Huapi, in the scenic forested Andean foothills of western Argentina. The lake is of glacial origin and is the largest lake in the region, with a surface area of 204 square miles and a maximum depth of almost 1,500 feet. The lake has many islands, fjords, and long narrow arms, with a total shoreline distance of 220 miles. The lengthy Limay River originates in the lake and flows out its eastern end. The water temperature is only 45°F. It has featured in the lives and lore of the indigenous people for thousands of years, and the lake is enshrouded in mystery because of its great depth and the many persons who have drowned in the lake, their bodies never to be recovered. Legends tell of a huge snake-like creature, called "Nahuelito", which inhabits the lake.

From Puerto Blest our journey continued once again by ferry (for the third and last time). It was a 4½-hour ride east across the waters of southern Lago Nahuel Huapi. At the end of our voyage we docked at the popular tourist resort village of Llao Llao, which was located about midway on the south shore of the lake, just over a half-hour drive from San Carlos de Bariloche further east on the lake. The whole region was part of Nahuel Huapi National Park (Argentina's oldest national park, founded in 1934). The national park offered a wide range of recreational activities such as trekking, mountaineering, rafting, paragliding, mountain biking, rock climbing, snow skiing (on Cerro Catedral), kayaking, and trout fishing. The elevation in the park varied from 2,400 – 11,700 feet, and the park was home to a large diversity of wildlife in its forests and other habitats. Glaciers and rivers added to the park's great popularity. After disembarking in Llao Llao, Noble and I relaxed for a short while at a large alpine hotel overlooking the lake. Here we enjoyed a drink at the bar. As evening approached we drove out of town and found a campsite for the night on the lakeshore for our first night in Argentina.

The next morning at 10:00 am (July 1) we drove into San Carlos de Bariloche, a clean, charming little city, Swiss-like in its appearance which attracted tourists from all over the world. Considered the gateway to the Argentine lake country, it was situated at 2,500 ft. elevation on the large southeastern arm of Lago Nahuel Huapi. Nearby was a popular wintertime ski area on Cerro Catedral. An Austral Pygmy-Owl was sitting at a busy downtown street corner, on top of a light post, intently watching the traffic go by! This small owl is placed by taxonomists in the genus *Glaucidium*, a worldwide group of 32 woodland and forest "pygmy-owls" and "owlets", all of which are more diurnal in their time of foraging than most owls, and thus they are more frequently seen.

Along one of the downtown streets was a small gift and bookshop which was owned by an enterprising couple from Germany. I walked into the shop to see what kind of little souvenir I might find to purchase. Imagine my amazement when here, on a bookshelf, I discovered a small book (almost a booklet) entitled *Las Aves Argentinas, una guia de campo (Birds of Argentina, a field guide)*, by Claes Olrog, with a publication date of 1959 (that very year)! The book was published by the Miguel Lillo Institute of the National University of Tucuman and was written entirely in Spanish (except for a five page "Spanish-English Glossary" at the back). It was 6x8 inches in size with a hard cover, and contained a total of 343 numbered pages which were all rather loosely tied together with string through seven holes along the margin of each page. All 904 species of then known birds of Argentina were named, with a Spanish and Latin name, and illustrated with watercolor paintings on 48 plates (with as many as 51 very small paintings per 6x8" page). In addition, there was a brief written description for each species giving its distinguishing features and habitat, and a very small range map showing its geographic distribution in Argentina. Although the paintings were crude by today's standards (small and too bright, with overemphasis on the features that distinguished one species from another) the book was far superior to no book at all. It was a godsend for me, which I instantly purchased. I was exuberant at now possessing an authentic "field guide" for a South American country! Olrog's classification and scientific names followed almost entirely that of Peters' *Checklist of Birds of the World* (1931-1952). What a stroke of good fortune it was. In the evenings I could now stop writing regularly in my daily travel journal and instead devote this time to studying Olrog's field guide. Subsequently, it would allow me to identify some of the new birds I encountered in the field on first sight, and it would not always be necessary to write a detailed description of each one. Nevertheless, birdwatching would remain very much a challenge.

That afternoon Noble and I took a bus up to the ski area and lodge at Cerro Catedral, in the snow at 6,000 feet elevation. A ski lift took skiers and sight seekers up 300 feet higher. I squeezed in just as the door was closing and rode to the top. Noble intended to wait for the next lift, but the lift broke and there were no more rides up or down for the rest of the day! I had to walk back down, which I had intended to do anyway, observing birds and taking pictures as I descended through the light covering of snow. I was pleased when a White-throated Caracara* glided slowly along the ridge top. It was the most southern member of the three allopatric species of *Phalcoboenus* caracaras in the Andes. (A fourth species in the genus lived on Tierra del Fuego and in the Falkland Islands.) As I started my descent from the top of the ridge I encountered a group of five very handsome and distinctive Yellow-bridled Finches* feeding in the bushes and hopping in the snow. This strikingly patterned, local, southern Andean endemic is characteristic of rocky, alpine zones in summer, but descends somewhat lower in winter. The male is generally Gray or yellowish-Gray, with a prominent black throat outlined with yellow, and has a conspicuous yellow patch in the wings and the tail in flight. It became one of my favorite birds.

Noble spent the afternoon in the lodge and hostel, visiting with young people from all over the world, including a young woman who gave her name as "Princess Nogai Khan" and claimed to be a descendant of Ghengis Khan! Late in the afternoon we took the bus back to San Carlos de Bariloche, where we picked up our Jeep and then drove outside of the town a short distance to a campsite for the night.

Argentina is a vast country, the second largest in South America, extending all the way from just inside the tropics south to the Atlantic entrance of the Straits of Magellan, on the southern end of the continent. Across the straits (to the south) was the large island of Tierra del Fuego, the eastern half of which belonged to Argentina, including the town of Ushuaia which at 55°S latitude was the southernmost town in the world. It was the gateway to Antarctica, only 600 miles away.

Argentina was bordered on the east by the Atlantic Ocean (and in the NE by Brazil). On the west it was bordered by the Andes Mountains, which separated the country from Chile for their entire 2,000-mile boundary. To the north, Argentina was bordered by Bolivia and Paraguay. Within its borders Argentina could be divided into four major ecological regions characterized by climate, vegetation, and landscape features: (1) ANDEAN: the Andes mountains of the west with lush temperate forests, citrus orchards, and vineyards; (2) CHACO: the

Plate 25: Our last "lake country" ferry boat ride was across S Lago Nahuel Huapi, in the Andes of west central Argentina. Our road southward traversed the cold, barren steppes of northern Patagonia. The Upland Goose I shot was almost too tough to eat!

Plate 25a: Our road south to Punta Arenas reached the Atlantic coast, where to my amazement were wintering flamingos! There was a gorgeous orange sunset over the harbor bridge at the town of San Julian.

north-central region with dry (xerophytic) forests, thorny scrub, and cacti interspersed with savanna grasslands, cattle ranches, sugarcane and cotton fields, and plantations of quebracho trees (for tannin production); (3) PAMPA: prairies, marshes, and damp grasslands on very fertile soil in the east central part of the country, with fruit orchards, cattle and sheep ranches, wheat fields, and crops of corn, alfalfa, and flaxseed; and (4) PATAGONIA: the southern region of dry, cold, barren, windswept steppes, plateaus, and shrubby grasslands, devoted almost entirely to sheep ranches (and more recently, oil fields). A fifth land region, MESOPOTAMIA, was much smaller and situated in the far northeast of the country between the Paraná and Uruguay rivers, which originally consisted of subtropical wet forests, swamps, lagoons, reeds, palms, and wet grasslands.

We departed San Carlos de Bariloche on Thursday, July 12, heading in a mostly southerly direction which paralleled the Andean foothills to the west. Our destination was Punta Arenas, Chile, the southernmost town in continental South America, situated on the Straits of Magellan at the southern tip of the continent. It was more than 1,200 road miles away, across the cold, barren, windswept steppes of Patagonia. It would take nine days for us to reach this destination, in the middle of winter. Shortly after leaving San Carlos de Bariloche, just before we arrived at the tiny community of Mascardi, there was a prominent road sign at 2,800 feet elevation (according to our altimeter), which read in capital letters "DIVISORIA DE LAS AGUAS", "AL PACIFICO" (pointing in one direction) and "AL ATLANTICO" (pointing in the other direction). Apparently this was the continental divide between the Atlantic and Pacific oceans. I took a photograph of the sign, but I found it difficult to believe that this low elevation separated the two watersheds. If this were true, there must have been a very low valley cutting through the Andes somewhere to the west of us, leading to the Pacific.

We continued southward on a relatively good, gravel road that undulated up and down between 1,000 and 3,500 feet elevation in the forested foothills of the eastern Andes, the peaks of which were often hidden in the distant clouds. A woodchopper was cutting down a tree as we passed between the small lakes of Gutierrez and Mascardi. There were frequent cattle ranches ("estancias") with scattered trees and log fences, without any snow on the ground just yet. We were in the southwestern corner of Río Negro Province, in the ANDEAN REGION of the country, en route to the town of El Maitén. Conspicuous roadside birds included a Black-chested Buzzard-Eagle, Chimango and Southern caracaras, Austral Pygmy-Owls, a Chilean Flicker, 4 Great Shrike-Tyrants* (my first and only encounter with this large, thrush-sized, dark plumaged flycatcher, shrike-like in its behavior with a heavy hooked bill), Austral Blackbirds, and Long-tailed Meadowlarks. Two Austral Parakeets* (*Enicognathus ferruginous*) flew rapidly alongside our road, outside the edge of a forest. It was my only sighting of this species, which was very similar in its appearance to the closely allied Slender-billed Parakeet that I saw in such large wintertime flocks in the farmland of southern Chile.

At 4:30 in the afternoon we entered the far northwest corner of Chubut Province and passed through the town of El Maitén, on our route to the next town of Esquel. At a small roadside lake were a White-tufted Grebe, Speckled and Silver* teal, and two Spectacled Ducks. Raptors along our way were an American Kestrel and several Variable Hawks. Large groups (100 or 200 individuals) of Upland Geese* were standing in open pastures and grasslands, my first observations of this characteristic, widespread, abundant Patagonian goose. The species is sexually dimorphic, and males come in two color morphs. Both sexes have the same prominent black-and-white patterned wings in flight. We camped that night at an elevation of 2,500 ft. along a river north of Esquel, in open hills with scattered woodlands and trees, in northwestern Chubut Province. Our dinner menu consisted of soup and wieners. Fearing sub-freezing temperatures, we drained the water from our radiator.

Sunrise on July 3 (near mid-winter here) was at 9:30 am. We refilled our radiator with water from the river, which we carried in our little red bucket. The day was chilly, damp, and quickly became overcast. Our gravel road was firm and posed no problems for travel. In half an hour we passed through the town of Esquel, still in the northwest corner of Chubut Province, and continued south in the undulating eastern foothills of the Andes, between 2,000 and 3,000 ft., toward the next town of Tecka. The only vehicular traffic was an occasional truck.

This was predominantly cattle country with scattered trees and woodlots in the tall, dry, shrubby grasslands. The local cowboys, known as "gauchos", wore traditional berets and rode horses with fur-lined saddles. They tucked their baggy trousers into the tops of their boots, and drank "yerba mate", a tea-like beverage made from powdered leaves of an evergreen holly tree (*Ilex paraguarensis*) native to the region of Paraguay and southern Brazil. By tradition, it was drunk through a straw (called a "bombilla") from a bottle-necked gourd referred to as a "mate" (or "guampa", or "cuia" in Portuguese). The beverage was high in caffeine, vitamins, antioxidants, and minerals, and produces much the same effect as strong tea. It was the national beverage of Argentina, dating back hundreds of years, having first been used by the Guaraní Indians in Brazil prior to the arrival of Europeans. It was prepared and drunk in a time honored etiquette and custom.

As we progressed southward toward the town of Tecka, the trees and forests gradually disappeared and the countryside became a shrub-covered steppe with tall clumps of grass interspersed with sagebrush-like bushes. Sheep replaced cattle, and became almost the only livestock animal. We were transitioning from the ANDEAN region to the PATAGONIAN region of the country. Upland Geese, Variable Hawks, Southern and Chimango caracaras, and American Kestrels were common roadside birds. There was a single Black-chested Buzzard-Eagle, and 12 Silver Teal were on a small roadside pond.

We reached Tecka shortly before noon and stopped to purchase gasoline, buy bread at a bakery shop, and quickly eat a bite of lunch at a small restaurant. As we left Tecka our road climbed upward to a low pass at 3,500 ft. as we crossed the Cerro Putrachoique. Soon thereafter our road angled away from the mountains in a slightly more easterly direction, across the barren Patagonian plateau of western and southern Chubut Province, toward our destination of Lago Musters, and ultimately of Comodoro Rivadavia on the Atlantic Ocean. This would require two days of travel.

The steppes were wild country, almost devoid of people. Although sheep were the only visible livestock, there was an abundance of wildlife. Mammals included foxes, hares, and armadillos (which Noble enjoyed chasing and catching, remarking afterwards that they were surprisingly slow, and soft-shelled). Among birds, Lesser Rheas* were the most obvious, being almost 3½ feet tall. These flightless, rapid-running, "ratite" birds of open grasslands were distant relatives of Ostriches (in Africa) and Emus (in Australia). Though somewhat smaller than these other two, rheas filled the same ecological niche. They were wary and difficult to approach. (The Greater Rhea, a somewhat larger species, will be mentioned from the Chaco region.) Once again, buzzard-eagles, caracaras, kestrels, and occasional Variable Hawks were conspicuous scavengers and birds of prey. Two rare, pale-Gray morphs of the Peregrine Falcon were particularly exciting to see. These ecological counterparts of the arctic-inhabiting Gyrfalcon were at one time classed as a distinct species, the "Pallid Falcon."

Upland Geese were the most numerous species, being gregarious and widespread, and they gave me the opportunity to try out my marksmanship with the .22 caliber pistol loaned to me by the French boys. After several futile attempts, I stopped near a small group of geese standing about 30-40 yards from the road. I rolled down my window, rested my arm on the door, aimed at one of the geese, and fired. To my wonderment, the bird dropped dead, having been hit in the heart. After I retrieved the dead goose, Noble took a picture of me, pistol in hand, standing and holding it upside down by its feet! Neither of us being much of a gourmet chef, we decided the easiest way to prepare it for dinner that night would be to stew it.

An hour later, I picked out a campsite just off the road in a "quebrada" (streambed), a little out of the strong, constant, westerly wind. We were at an elevation of 2,000 feet, about 25-30 miles west of Lago Musters, in the middle of extreme southern Chubut Province. No signs of civilization could be seen in any direction. I plucked the goose and removed the edible portions (neck, gizzard, wings, breast, and drumsticks) while Noble cut up potatoes and onions. We put all these items together in a pot of boiling water. After an hour, we sampled the tenderness of the meat, which to our dismay was as tough as shoe leather. So we boiled it for another hour. The tenderness hardly changed at all. After another hour of boiling, it was still the same, so we gave up and ate most of it anyway, saving some for tomorrow. We concluded that although the goose was flavorful, with somewhat of a wild taste, it was not meant to be stewed. A glass of red wine helped considerably! That night we slept very soundly inside our sturdy van, protected from the persistent wind.

The Patagonian landscape as we approached Lago Musters the next morning (July 4) reminded me a little of Nevada or parts of Wyoming, with sagebrush-like plains. To my great delight, several Burrowing Parrots* flew across the road in front of us. These medium-sized, multi-colored parrots with long wings and a long, pointed tail were uncommon inhabitants of Patagonia, where they nested in burrows in the ground, typically on small, vertical sand banks or cliffs. They were the most southerly-distributed parrot in South America east of the Andes, and they provided me with a permanent memory of the barren, shrub-covered steppes. Another bird endemic to this area was the Patagonian Canastero*, a furtive little wren-like species which darted from one bush to another, attempting to keep out of the wind. It was distinguished from other canasteros by its dark brownish-black tail. Soaring overhead were Black-breasted Buzzard-Eagles with their distinctive broad-winged, short-tailed, bat-like shape.

We reached the south shore of Lago Musters in the late morning, at an elevation of 2,500 ft., and I stopped at a viewpoint 200 yards from the lakeshore. Here I looked through my telescope for swimming birds out on the lake, for half an hour. I was able to identify 10 White-tufted Grebes, 50 Silvery Grebes, 25 Black-necked Swans*, 2 Lake Ducks* (in the same genus, *Oxyura*, as the very similar Ruddy Duck of N. America), 200 White-winged Coots, and 3 Kelp Gulls (foraging low above the lake in the distance).

We continued onward toward Comodoro Rivadavia, arriving at this booming oil town on the Atlantic Ocean just before 5:00 pm. En route we had passed only one other vehicle (an oil company truck). The town was situated in extreme southeast Chubut Province, on the Atlantic Ocean at latitude 46°S. It was full of roughnecks, and girls to cater for them. We camped that night a short distance south of town, along the shore of Golfo San Jorge. It was cold and windy, in the middle of the austral winter. We were only about half way from San Carlos to Punta Arenas, with more than 700 miles yet to drive. The days would get colder as we went south.

Sunday, July 5, was bleak and windy all day long, with overcast skies, intermittent light rain, and a chilly temperature. The sun was never very high in the sky. Shortly after we got underway, as we were travelling along the Golfo San Jorge, I stopped to focus my telescope on a lone goose that was standing on a rock just offshore. I described it as appearing mostly dark, with some brownish-chestnut on the head and narrow white barring on the underparts. It had a yellowish bill, light ring around the eye, and yellow legs. The bird (as I later discovered) was a perfect fit for a female Kelp Goose*, a non-flocking, uncommon species of the rocky seacoast of southern South America. It was characteristic of the remote, rugged environment in which I found it, an ecological counterpart of the Emperor Goose in arctic Alaska. Birding was a never-ending adventure. It was one of my most thrilling moments. Other birds seen along the coast here were 2 Black-browed Albatrosses, a single Giant Petrel (sp?), 10 Imperial Shags* (also known as Blue-eyed Shags or King Cormorants), a single White-headed Steamerduck* (endemic to this very restricted area of Atlantic shoreline in southern Argentina), 150 Magellanic Oystercatchers*, numerous Kelp Gulls, a few Brown-hooded Gulls, South American Terns, and a single (somewhat surprising) Royal Tern.

We continued south along the coast of Golfo San Jorge and very soon we left Chubut Province and entered Santa Cruz Province (the southernmost continental province of Argentina). At the small town of Caleta Olivia our route turned inland and meandered through shrub-covered hills for 40 or 50 miles. Along this section of road we encountered our first wild Guanacos, a native mammal in the camel family and the ancestor of domesticated llamas. They inhabited the open steppes and grasslands at the southern end of the continent, and were timid animals that did not allow close approach. In this area we also saw, once again, foxes, armadillos, rabbits, and hares, plus a small pika-like rodent (not identified to species). Birds included Lesser Rheas, a single Elegant Crested-Tinamou* (a fast-running, slender, partridge-like bird with a very long narrow crest, unmistakable within its shrubby grassland habitat), Southern Caracaras, Variable Hawks, one Plain-mantled Tit-Spinetail (a small, slender furnariid with a long, pointed tail and rufous in its wings, widely distributed in shrubby areas of southern S. America). There were also scores of Mourning Sierra-Finches (which as I have previously remarked reminded me of Harris's Sparrows in coloration because of the male's black face and breast). Though relatively few in number, Patagonian birds were exciting to encounter, and to assign a name, with help from my new field guide! The cold weather was an exhilarating change from the tropics, and recalled my winter days in eastern Kansas and southern Michigan. That night we camped on the open, windy, plain, 25 miles west of Deseado. We reheated and ate the last of our goose stew, then drained all the water out of our radiator before we realized that the 12-gallon water tank inside our van was empty. We brushed our teeth with red wine!

The next morning (July 6), we drove back up the road two miles to get water from a well at a federal police check post. After filling up our water tank and radiator, Noble and I then proceeded on down the road for an hour to the small port village of Deseado, on the Atlantic coast again. Here we once more turned slightly inland and paralleled the shore in a southwest direction for approximately 150 miles, all the way to the to the coastal harbor town of San Julian. Our road surface was wet, sticky clay and required that I shift down into second gear and proceed at 15-20 mph with our little 105 hp engine working as hard as it could to pull us through the clay. We encountered numerous guanacos and Noble took moving pictures of them on several occasions. A Chilean Flamingo was standing all alone on the shore of the harbor as we entered San Julian, looking quite forlorn and out of place in this cold, desolate environment.

We drove into San Julian at 3:30 pm and stopped to purchase gasoline and to have a flat tire repaired. As we drove out of town along the waterfront I recorded 1 Giant Petrel, 4 Great Grebes, a single Rock Shag* (Magellan, or Magellanic, Cormorant), 7 Crested Ducks, 6 Magellanic Oystercatchers, and several hundred Kelp Gulls. Our road continued southwestward for two hours, until we reached the Río Chico (Small River) not far outside the town of the same name. We decided to camp there for the night, at 50°S latitude. I photographed the gorgeous, bright orange sunset. For supper, we ate bread and honey, beans, and popcorn (popped by Noble with his usual enthusiasm). We finished this simple meal with our customary glass of red wine.

Our thermometer registered a temperature of 28°F on the next morning (July 7). There was ice along the margins of the river and frost on our windshield. The sunrise (at 9:35 am) was as brilliant as the sunset last night had been. We had to break ice on the river's shore in order to gather water for our radiator (which we had drained before going to bed). As we set off on our day's adventure there was just enough frozen moisture in the roadbed to make the surface firm but not slippery, so I was able to drive more rapidly than usual. Along our route were sheep estancias with shearing areas where hides were hanging on wooden racks above the grasp of foxes, which we frequently observed in the grasslands on either side of the road. Lesser Rheas were common, as were Southern Caracaras. A new bird for my list was the Least Seedsnipe*, which we encountered in small groups on several occasions, swirling low above the steppes or foraging in the short grasslands. Also new was the Patagonian Mockingbird*, the southernmost member of the genus *Mimus* (resident here at about the same latitude south of the equator as the northernmost latitude for breeding Northern Mockingbirds in southern Canada).

As we were approaching the town of Río Gallegos, the steering bar for the left front wheel became detached, causing the Jeep to swerve into a shallow roadside ditch. While Noble held the wheel straight, I backed out of the ditch onto the level roadbed. Then I slithered under the front end of the Jeep to determine the problem. Here I could see one end of the steering bar dangling vertically, unattached, almost touching the ground. Upon inspection, I determined that the threads at the end of the bar, tying it to the inside of the wheel, were completely stripped and the end of the bar was smooth, incapable of being re-attached. It was simply worn out! With us, in our spare parts, was some bailing wire for just such an emergency. I was able to wire the end of the bar back to the wheel, not very sturdily but just firmly enough to allow us to hobble on down the road to a campsite not far outside the town of Río Gallegos, where the worn-out bar was replaced with a new one the next day. The problem never occurred again (I was not as capable a mechanic as Noble, but I could occasionally solve minor mechanical breakdowns). That night, along the Río Gallegos, there was another outstanding sunset. There had been exactly eight hours between sunrise and sunset. Twilight continued for two more hours after sunset. We camped along the river that night, and once again we drained the water from our radiator before we went to bed.

The next morning (Wednesday, July 8) there was a light covering of snow on the ground and a thin sheet of ice on our road. We refilled our radiator from the river. (Noble was still not convinced we needed to buy antifreeze, one of his few errors of judgment.) We arrived at mid-morning in the little town of Río Gallegos, situated at the mouth of the river of the same name. It was the southernmost town in Argentina on the South American continent, not far from the border with Chile. Here we took our van to the Jeep agency to have our broken steering bar replaced. This required six hours of work, much longer than we had anticipated. So we relaxed and had our first haircut in seven weeks, followed by a hot breakfast with fried eggs and potatoes (for 45 cents each). There was a loud speaker playing music on the main street downtown. I walked along the harbor with my binoculars and counted 1 Giant Petrel, 1,000 Kelp Gulls, 30 Dolphin Gulls*, 120 Brown-hooded Gulls, and 125 Magellanic Oystercatchers. The Dolphin Gull is one of the most attractive and distinctly plumaged of all the 50 species of gulls in the genus *Larus*, with its bright red bill and feet, light Gray head and body, and black wings and mantle. It is confined to the southernmost seacoasts of South America, where it is a handsome, easily recognized species, though not particularly numerous.

After stopping at the aduana on the outskirts of town, where our vehicle was briefly inspected, we departed Río Gallegos at 5:00 o'clock in the afternoon for the two-hour drive to the Chilean border. The countryside was covered with several inches of snow. At the frontier between the two countries there was an Argentine post where our passports and carnet de passage were stamped and we were wished "Buen viaje" and "Buena suerte" (good luck) as we departed. Surprisingly, there was once again no official check post of any kind on the Chilean side of the border. We were in Magallanes Province at the wide Atlantic (northeast) entrance of the Straits of Magellan, just south of 52° latitude and approximately 125 miles from the southern tip of the South American continent. The road in Chile, as we might have expected, had not been cleared of snow, and it was covered with ice (which brought back very unpleasant memories). It was about 75 miles to Punta Arenas. There was still a faint orange glow on the western horizon at 9:00 pm as we chose a campsite out of the constant wind, in the bottom of an old roadside gravel pit. We drank the last of our Chilean wine, and drained the water from our radiator before climbing into our warm sleeping bags.

Thursday, July 9, was a day of frustration. Our outside thermometer read 15°F when we arose. Even the water in our inside 12 gallon tank had some ice on top, which prevented water from being pumped out of the tank. It was necessary for us to melt some snow in a pan over our stove in order to get water for our radiator. This was a time-consuming task, but after several hours the radiator was filled. However, our problems were not over. When we started the engine it soon overheated. Water was not being circulated through the radiator (or the engine), presumably because the cooling system was blocked with ice somewhere. I was so exasperated that I uncharacteristically refused to assist Noble any longer in solving the problem, saying to him that if he had purchased antifreeze when I had suggested it to him, the problem would not have occurred. (It was one of our few disputes.) So I took my binoculars and set off walking ahead down the road by myself, leaving Noble to find a solution on his own. To his credit, he did. He took out our emergency flare, lit it, and held it under the radiator until all the ice had melted and the water began circulating once more.

We set off again for Punta Arenas. During the afternoon travel I recorded many rheas, flamingos, Upland Geese, a Black-necked Swan, Two-banded Plovers*, and a dozen Black-throated (Canary-winged) Finches*, which were a close relative of the Yellow-bridled Finches seen on Cerro Catedral (at San Carlos de Bariloche). They were equally attractive, with much more yellow in their wings, and confined geographically to a small area at the southern tip of the continent. Also present were two Common Miners and a pair of Rufous-collared Sparrows. This latter species was almost certainly the most widespread breeding passerine endemic to the Neotropical zoogeographic region. It occurred commonly all the way from southern Mexico to the very tip of South America, from sea level to 11,000 feet, in a wide variety of non-forest habitats (although it was absent from most of the Amazon rainforest). It was often found around human habitations, and its pleasant song was very well known. (Taxonomically, the Rufous-collared Sparrow is in the same genus, *Zonotrichia*, as are the White-throated, White-crowned, Harris', and Golden-crowned sparrows of N. America.)

That night (July 9) Noble and I finally arrived in Punta Arenas, at 8:30 pm in the middle of winter. The weather was clear, dark, and cold. We were at 53°S latitude, almost the extreme southern tip of the continent. It had required 165 days and 16,000 road miles from the time we left Los Angeles to reach this destination, after innumerable breakdowns and driving hazards. We purchased gasoline -- and antifreeze -- at a COPEC station, and more Chilean red wine at a small liquor store, and then drove just south of town to a campsite overlooking the Straits of Magellan. It was a magical moment for us, and we celebrated with some of our newly purchased wine. We had made it. Life was great!

We got up at sunrise (9:00 am) the next morning, Friday, July 10. Punta Arenas was a community of about 20,000 hardy people. The winter so far had been relatively mild with less than normal snowfall, and there was currently little snow on the ground. However, there were
snow-covered mountains (up to 3,500 ft. elevation) to the south of the town, on the southern end of the Brunswick Peninsula about 50 miles away. They were the southern tip of the Andes, about 54°S latitude on the north side of the Straits of Magellan, which connected the Atlantic and Pacific oceans. On the southern side of the straits was the large island of Tierra del Fuego, the "land of fire", as named by Magellan in 1520, because of the widespread fires built by the indigenous people. Politically, the island was divided between Chile (the western half) and Argentina (the eastern half). Darwin visited this island in 1834 on his historic voyage aboard the "*Beagle*." The town of Ushuaia on the southern end of the island (in Argentina) is the southernmost town in the world, at 55°S lat., and was the departure point for tourist cruises to Antarctica, the northernmost tip of which was only 680 miles away. Noble and I had hoped to ferry our Jeep over to the island and drive to Ushuaia, but in winter the ferry was unreliable and unpredictable. It came and went at the whim of the captain, depending in part on weather and sea conditions. No one could assure us of a departure date, so we were sadly unable to make a booking.

Rafael Barros (the general manager of the Jeep agency in Santiago) had given me the name of Mario MacLean, a Scotch-Irishman who worked for the Jeep agency in Punta Arenas, and said that Noble and I should introduce ourselves to him upon our arrival. We did this on our first morning there. Mario had a typically ruddy Scotch complexion, and welcomed us very warmly. We met his wife and were then introduced to an English businessman friend by the name of Mr. Arthur, and to the Frank Hammans (a very friendly middle-aged couple who owned a large sheep estancia 100 miles west of town on the Skyring Channel). It was a very congenial group of people who went out of their way during our time in Punta Arenas to entertain us with meals, afternoon teas, and alcoholic beverages. They were delighted to have some visitors in town during the middle of the cold, snowy winter. We enjoyed steaks, empanadas, mutton, roast beef, and potatoes, all washed down with wine and Scotch. There were long conversations and discussions delving into Noble's favorite topics of finance and government. Noble learned of entrepreneurship in an isolated part of the world. On one occasion, we were guests of Mario at the weekly Rotary Club meeting in the town's only hotel, where I delivered a short speech, in Spanish! We were interviewed by the local daily newspaper, "El Magallanes", for a story that would appear the next day.

That first afternoon (Friday) we drove south of town about 35 miles to the very end of the road on the South American continent. Here I was surprised at the abundance of Southern Fulmars (more than 250) and the presence of Giant Petrels and Black-browed Albatrosses, seabirds that were typically found far out at sea and only

Plate 26: S Patagonia in mid-winter was cold, bleak, snow-covered, and barren, with temperatures well below freezing. Our road ended 40 miles S of Punta Arenas (in Chile) on the Straits of Magellan at the southern tip of the South American continent, 54^0 S latitude.

occasionally seen from the shore. Some of these birds were foraging within only a few yards of the shoreline. In addition to these, I also observed 1 Great Grebe, 600 Kelp Gulls, 35 Brown-hooded Gulls, 5 Dolphin Gulls, 5 Rock Shags, 2 Flying Steamerducks, 35 Rufous-chested Dotterels* (all in winter plumage), 40 Southern Caracaras, 7 Chimango Caracaras, and 30 Black-chinned Siskins (the southernmost of 32 species in the worldwide genus *Carduelis*). We returned to Punta Arenas and camped at the same site as last night. Once again, it was clear and cold. Noble put on long underwear.

When we walked into the town's little wooden hotel for a cup of coffee the next morning (Sat., July 11) we encountered a slightly built, bearded, rather bedraggled-looking, long-haired young man of moderate height, sitting by himself and drinking a cup of coffee. He had a kind, intelligent, gentle look to his face, and friendly blue eyes. We asked if we could join him, and he readily agreed. Richard Dugdale was about our age and English by birth. Several years ago he had immigrated to Canada, where he went to work as a young man writing stories for a large newspaper in Ontario. After a year of work he wanted to travel and see South America, so he convinced the newspaper to support his travel if he would send them frequent stories of his adventures abroad, as he hitchhiked from country to country, with only a backpack. He had now been on the road for more than a year, and here he was in Punta Arenas at the southernmost end of the continent. How fortuitous that we should meet him here. He told us that he had read in the morning news and heard on the radio that the two of us were in town. The three of us shared many traits and we instantly established a friendship. When he told us his next destination was Buenos Aires we invited him to ride with us, and he happily agreed to do so. For a short period, we would all be comrades travelling together. I never tired of listening to his stories, particularly those about his experiences with an Amazon Indian tribe and his two-week visit to Easter Island. To this day, I have a small photograph of Richard, reminding me of our time together. In our cab, he sat on the crude little seat cushion on top of the engine, between Noble and me.

The southern tip of South America was Merino sheep country. The strong, westerly wind blew almost constantly, and in open areas the small trees were always bent and windblown. The scenic countryside was both rugged and beautiful, with scattered temperate Nothofagus and cypress forests on the bleak, shrub covered steppes, which in winter were cold and usually blanketed in snow. On Sunday morning, July 12, Noble and I accepted an invitation by Frank Hamman to visit his sheep ranch. We traveled with him in his Jeep over a dirt and gravel road covered with light snow and ice to his extensive property 100 miles west of town, on the Skyring Channel, with a view of distant snow-capped mountains to the west. It was difficult, awe-inspiring country, creating a very strenuous existence for the few hardy individuals who chose to live here. Frank and his wife were doing just that, and succeeding. They had to be admired. We were their guests for that afternoon, night, and the following morning.

On our first afternoon Frank took us on a short tour of the ranch buildings. He told us that rabbits were very serious pests, 1,000 having been killed on his property in the span of only one week by two men using dogs and rifles. A neighboring rancher had recently shot a very large mountain lion (*Felis concolor*, locally called a "puma"). Frank said that every year he lost 10% of his flock to natural predators. It required at least 1,000 sheep to make a living, and his flock of thick-wooled Merinos totaled 4,000. We returned to the ranch house for afternoon tea and biscuits, the latter of which were remarkably palatable even though they had been fried in sheep grease, as was the local custom. Dinner was served at 8:00 pm. It was a hot, nourishing, traditional ranch meal of mutton, potatoes, and soup. Afterwards, there was time for leisurely conversation prior to bed at 10:30 pm. Noble and I snuggled into our warm beds under thick blankets in our rather chilly bedroom. The day had been very rewarding.

I spent several hours on the ranch recording birds, the most conspicuous of which were Upland Geese, Black-necked Swans, Black-chested Buzzard-Eagles, Southern and Chimango caracaras, and Chilean Flamingos. This was certainly neither the climate nor the environment I had ever imagined for flamingos, which I had always thought of as inhabitants of warm, tropical regions of the world. How wrong I was! I also documented my first Flightless Steamerducks* and Magellanic Tapaculos* (which were common by song but so secretive that I could never see one clearly; I described their vocalizations as a "monotonous, two-syllabled, rather squeaky 'cu-tuck' with accent on last syllable, repeated 10-15 times"). Other birds in the forests and grasslands were Cinereous Harriers, Variable Hawks, American Kestrels, Magellanic Woodpeckers (heard only), and such passerine species as the White-throated Treerunner, Thorn-tailed Rayadito, Dark-bellied Cinclodes, Fire-eyed Diucon, Tufted Tit-Tyrant, Austral Thrush, Long-tailed Meadowlark, and Austral Blackbird.

On Monday afternoon, July 13, we returned to Punta Arenas and were told one more time that no definite date had yet been set for the ferry to travel across the straits to Tierra del Fuego. Reluctantly, we had to give up our attempt to visit this renowned island. Therefore, we said goodbye to all those in town who had befriended us, thanking them for their gracious hospitality, and picked up Richard at the hotel. The three of us then departed in our Jeep for Buenos Aires. I drove, Richard sat in the passenger's seat, and Noble climbed into the back of the van to sleep on his bed. It took us 4½ hours to drive the 105 miles to the border, where we arrived at 4:30 am. Not surprisingly, it was closed. Therefore, I pulled off to the edge of the road, and Richard and I joined Noble in the back for a few hours of sleep. Richard slept on the floor, inside his sleeping bag, which he placed on top of a huge pile of coats and other garments, as a mattress and for warmth. It was cold and very windy outside.

All of us slept until 10:00 o'clock the next morning, an hour after sunrise, on Tuesday, July 14. At the Chilean customs post we were informed that we lacked a necessary security check, which we should have obtained as we left Punta Arenas last night. Thus we were detained for two hours at the border until approval came via radio to allow us to proceed. It helped when we showed the customs officials a copy of the newspaper article with our pictures. An hour later we reached the Argentina "frontera", where we were cordially received and offered tea, the officials having remembered us from several days earlier. However, it still required an hour of waiting since we had arrived just behind a crowded local bus in which the documents of all 40 or 50 passengers on board were being inspected, one at a time.

It was 2:00 o'clock in the afternoon by the time we were underway for the 3-hour drive to Río Gallegos, 70 miles up the road. We stopped there for gasoline and then continued on our way, anxious to cover as many miles as possible. Noble wanted to get to Buenos Aires as quickly as we could, and volunteered to drive for a while. I agreed, confessing that I was tired and needed a rest, then I collapsed on my bed and was soon asleep. Richard kept Noble awake. At 6:00 am, he pulled off the road for a several hour nap, just outside the town of San Julian. We were in central Santa Cruz Province, on the Atlantic coast.

We arrived in San Julian at 10:00 am (July 15). There was a slight noise in the left front wheel that needed investigating. In his characteristic, capable manner, Noble soon found an Argentine-born mechanic of British descent, Magallanes MacLean, a true entrepreneur, who took us to the service station where he worked. The problem was quickly diagnosed as a broken ball bearing race. Magallanes hurried around town to find a new one for us, and soon returned with success. However, the brake drum had become scarred by the mishap and required turning on a lathe to remove the roughness. This work was time-consuming and took the rest of the day to complete.

I took the opportunity to walk around the snowy, icy coastal mudflats and sandy beach for a couple of hours, observing and recording birds. The town was situated on a large lagoon, at approximately 50°S latitude. I continued to be amazed by the presence of flamingos, and I took pictures of a large flock of 90 birds standing on the icy, lightly snow-covered mudflats. Another species that seemed completely misplaced in this environment were two Black-crowned Night-Herons. Two Southern Fulmars were foraging right next to the shoreline. Kelp Gulls were easily the most abundant species, and I estimated a total of 600 individuals. Other species that were present in relatively large numbers were 150 Flying Steamerducks, 150 Crested Ducks, 80 Brown-hooded Gulls, and 40 Magellanic Oystercatchers. Rounding out my complete list were 10 Imperial Shags, 10 Two-banded Plovers, 2 Speckled Teal, 2 Southern Caracaras, and a single White-tufted Grebe.

Our Jeep was finished by 7:30 pm and we departed for Comodoro Rivadavia, 375 miles to the north. The night was clear and cold, with a full moon and new fallen snow. Once again, Noble volunteered to drive all night, while I slept in the back and Richard helped keep him awake in the front. The next morning at 6:00 am (July 16) we switched drivers again, and both Noble and Richard got in the back to sleep. Before starting out, I chased a rabbit to help me wake up. We were in sheep country with snow on the ground, and scattered oil wells. By 1:00 pm we had reached Comodoro Rivadavia, for our second and last time. We had just crossed the border from Santa Cruz Province into Chubut Province. This was the town where we first reached the Atlantic coast on our way south from San Carlos de Bariloche, and from now on our journey would be unchartered territory for us. We were over ⅓ of the way from Punta Arenas to Buenos Aires.

In Comodoro Rivadavia, we all ate a four-course steak dinner for lunch, at a cost of 65 cents each. Then we were on our way again, all three in the cab once more (Richard still in the middle). Our route proceeded inland a short distance, away from the coast, and then climbed in elevation up to a hilly plateau at 2,000 ft., the "Meseta de Montemayor", where there were several inches of snow on the ground. Trucks were having trouble ascending the slippery hills, and some of the 18-wheelers had jackknifed, occasionally blocking the road. However, we experienced no difficulties in our 4WD. At 9:00 pm, I drove several hundred yards off the road to camp for the night, at 2,000 feet elevation. We were roughly 85 miles north of Comodoro Rivadavia, in southeastern Chubut Province, on a cold, windswept steppe 20 or 30 miles inland from the Atlantic Ocean. It was inhospitable country, with extensive sheep estancias and very few inhabitants. Our simple supper consisted only of bread and hot mushroom soup. The temperature outside was 3° above zero, Fahrenheit, when we went to bed. Richard piled dirty clothes and empty backpacks under him to insulate himself from the cold floor of the van. It was the coldest night so far of our travels (and would turn out to be the coldest campsite temperature of our entire three-year adventure).

The temperature remained the same throughout the night, and our thermometer outside still read 3°F the next morning when we got up, on Friday, July 17. All the water inside our van was frozen, but with the antifreeze we had poured into our radiator in Punta Arenas, that water did not freeze. However, there was another problem. We had not switched to lower viscosity, winter-weight engine oil, and the oil in our engine became so thick during the cold night that the starter motor could not turn the engine over rapidly enough for it to start. Luckily, we had two car batteries and could switch from one to the other as the first one gave out. It required 30 minutes of constant effort before the engine finally started. All of us breathed a huge sigh of relief. Such cold temperatures were not on our planned agenda.

Our road gradually descended to 1,000 ft. elevation and down out of the snow. Scrubby vegetation increased on the grassy steppes, and more birdlife appeared, particularly furnariids. Despite having my new little Argentina field guide, I still wrote descriptions of almost all the birds I encountered. Two of these are as follows. (1) "Furnariidae: about 5½ inches, brown, very slightly paler below, becoming whitish on the abdomen; throat prominently white; wings & tail somewhat darker; indistinct pale edges to remiges; tail rel. short, outer feathers somewhat shorter than middle but tail rather square cut; bill rather long and heavy, upper mandible somewhat decurved toward tip, bluish gray; slightly crested." This bird was subsequently identified as a White-throated Cachalote*, which was rather jay-like in its overall appearance (and was larger in size than my inaccurate estimate). It ran on the ground or perched in low scrub, and was not particularly shy. (2) "Furnariidae: medium; shape wren-like (tail held at a decided angle); grayish-brown above; whitish below (particularly posteriorly); bill dark, slender and relatively long [I drew a sketch]; tail medium, basally bright rufous-chestnut (prominent in flight) with wide black band terminally; light superciliary." This bird was later identified as a Band-tailed Earthcreeper*, an inhabitant of low shrubs and grasses where it generally kept out of sight. Its tail was characteristically carried vertically over its back when it ran. Other birds present included Elegant Crested Tinamous, which were widespread, conspicuous, and provided me with amusement as they ran, quail-like, from shrub to shrub, their long, pointed crests sticking straight up.

It was 8:00 pm when we reached Puerto Madryn, where we relaxed from our strenuous day of driving by having red wine and a steak dinner at a hotel. I went to bed while Noble drove. Within an hour of leaving Puerto Madryn we reached the border between the provinces of Chubut (to the south) and Río Negro (to the north). There was a chain barricading the road and we were required to stop for inspection. It was the middle of the night and a man was dozing inside a tiny, kerosene lighted house. After Noble aroused him, he said he wanted to look in our van, but when Noble told him that a person was sleeping inside he changed his mind, removed the chain from across the road, and motioned for Noble to continue. We were now in the extreme southeast corner of Río Negro Province (almost due east of San Carlos de Bariloche where we had first entered the country 18 days ago). A few miles up the road Noble stopped to camp for the night, in shrub-covered steppes of northeastern Patagonia. It was about 2:00 am.

Saturday, July 18, was a day we will always remember. It was a cold, mostly overcast morning. We got underway shortly after sunrise, at 9:00 o'clock. The slightly hilly countryside was barren, bleak, and desolate, with very few people or habitations of any kind. Goats now outnumbered sheep on the dry, scrubby grasslands. Widely spaced fences divided the land into large "paddocks", within which the scattered livestock roamed freely. It took many square miles to make a living here, and much of the land was owned by corporations rather than private landowners. Trucks comprised almost all the traffic on our rough, dusty, gravel road. Many of these were huge 18-wheelers that were overloaded with tons of wool, beef, wine, mercantile goods, and steel pipes or other oilfield equipment. The truck drivers mostly owned their own vehicles and traveled at all hours of the day or night, stopping frequently for roadside chats and barbecues with each other, where they drank mate, wine, or other spirits. Two persons rode together in the larger trucks, and there were one or two bunks for sleeping. They were a rough, friendly lot, always willing to help anyone out of difficulties. There was great camaraderie among them. Such companionship was necessary because of the remote, sparsely populated environment through which they mostly traveled.

It was mid-morning and I was back to my customary position behind the steering wheel, travelling about 40 mph on a wide, straight, dusty, gravel road with very little traffic, heading north toward the town of San Antonio Oeste about 75 miles ahead. I was gradually overtaking a long, 18-wheeled, heavily overloaded truck, which was producing a wide cloud of thick dust behind it. Traffic in Argentina drives on the right-hand side of the road (as in the U.S.), but on this occasion the truck ahead of me was travelling on the left side of the road because the surface was a little smoother there. If a person wanted to overtake the vehicle in front of him, he was supposed to do this on his left side (the open lane, facing any oncoming traffic) as in the U.S. However, there was no room for me to do this since the truck was driving on the left side of the road. I couldn't sound my horn to alert the truck driver to my presence because our horn was temporarily non-functioning, having lost the rubber top of the horn button that was needed for depressing it. (I almost never used the horn.) The truck driver could not see behind him because of the dust, so he was unaware of my presence. I thought that eventually the truck would return to the right hand side of the road and I could go past it. However, after several miles of dust I became impatient and decided I would speed up and go around it on his right hand side, since there was plenty of room for me to do so. There was no imminent oncoming traffic and the truck driver apparently had no immediate desire to return to the right side of the road.

My decision turned out to be the wrong one. I sped up and drove perhaps 5 mph faster than the truck, about 25 mph, to pass it on the right. The front end of our camping van advanced to a position alongside the long truck, almost halfway along its total length and 3 or 4 ft. from its right side. I was now out of the truck's rear dust and could see ahead of me down the road, which was clear of any traffic. The truck driver should have been able to see me in his right side-view mirror. Apparently he didn't look, and was completely unaware of my presence. He suddenly decided just then (for a reason I will never know) to return to the right hand side of the road. He turned rather sharply to his right, cutting in front of me, perhaps to avoid a hole in the road ahead of him. I reacted instantly when he began to crowd me off the road, and slowed as fast as I possibly could, keeping only inches between the right side of his truck and the left side of our van. Fate was not on our side today. My guardian angel had left me.

The wide roadbed on which we were travelling was built up 4 or 5 feet above the surrounding countryside, and the road had steep, very soft shoulders on both sides, which dropped off sharply at the edge of the road, with no barriers of any kind. I had almost come to a complete stop, maintaining only inches between our van and the truck, when our right front wheel went just over the edge of the road and sank several inches down into the very soft dirt at the top edge of the embankment. This immediately pulled and tilted our top-heavy van sharply to the right, and it quickly came to a stop. Then it tilted even more -- AND FELL OVER ON ITS RIGHT SIDE!

The landing was soft. I was on top of the three of us in the cab, Richard was in the middle, and Noble was on the bottom. None of us was injured. I wasn't even mad at the truck driver, only at myself for having made the wrong decision and not being more patient. The driver was apparently never aware of my presence, even after our accident, as he continued down the road in a cloud of dust. The right door of the cab was on the ground and the left door was so heavy that I couldn't push it upward and open, so I opened the left (driver's side) window and climbed out through it (reminding me of climbing out through the hatch on top of a tank). I was then able to pull the heavy door upward and hold it open for the other two to exit. The only apparent visible damage to our Jeep from the outside was a small dent at the top right corner of the van, at the point where our vehicle first hit the ground. Remarkably, no windows were broken. The right front wheel, for one small fraction of time, had supported almost the entire weight of our vehicle as it fell. Our camping van was lying forlornly on its side on the sloping embankment, nearly parallel with the road in the direction we were travelling. The front end was a little lower than the back end and all four wheels were pointing up the side of the embankment, the two left wheels six feet off the ground. Of course we took some pictures. It was almost (but not quite) amusing. The time was 10:00 am and we were in the middle of nowhere, at km marker 1248, approximately 800 miles from Buenos Aires and 350 miles from Bahía Blanca, the nearest town of any size up the road ahead.

Within ten minutes after our mishap a truck came along the road and stopped to assist us. Soon there were several other trucks, including a very large cab-over-engine Fiat. The driver of this truck fastened a heavy chain from the railing on the top of our van to his very sturdy vehicle, which was on the roadbed directly above us. He then succeeded in pulling our vehicle up to where all four of its wheels were on the surface of the sloping embankment. In this process a tremendous amount of weight was placed at an angle on our two right wheels, which were supporting the entire weight of our vehicle as it was being rotated upward. When all four wheels were on the embankment, our van was leaning very sharply to its right, and it was necessary for the Fiat to hold our Jeep up with its chain until I could start our engine and drive down off the embankment, to the level terrain below. The Fiat then tried to pull us directly backward up the steep embankment to the roadbed, but its engine was not powerful enough to do this. Noble and I assessed the situation and decided we could drive our van along the bottom of the embankment until we could find a place to drive ourselves back up onto the roadbed above. We succeeded in doing this, thanked all the truck drivers, and at 11:00 am they departed.

Back up on the road our first task was to critically assess the damage to the van, inside and out. On this occasion our solid, heavy steel frame with its tortoise-like shell was an asset rather than a hindrance, and it saved our camping van from any serious structural damage. There was the inconsequential dent at the top right corner of the van and a crack in one of the inside wooden compartments. The most serious damage was to the right front wheel and spring, which were bent in such a manner that our van tilted severely forward and to the right, even on level ground. (There will be more to say about this matter shortly.) Inside the van there was a bit of a mess from thumbtacks and staples that had spilled from their boxes and scattered across the floor. Motor oil had spilled from two opened cans and run through a cabinet, slightly soiling some electronic gear and a typewriter stowed there. In addition, there was a split at the upper front corner of our built-in food compartment. It took us several hours for us to get everything cleaned up. However, the overall damage to our camping van seemed remarkably little, considering the nature of the accident. At 5:00 pm, we finally took time out to get a bite to eat.

We then set out again for Buenos Aires, very slowly with a noticeable forward leaning starboard list (right hand tilt). I drove with extreme care at a speed of no more than 20 mph. Nevertheless, two miles after we started up the road, a most amazing event happened, which would have to be experienced to be believed. We bounced slightly over a small bump in the road and OUR ENTIRE RIGHT FRONT WHEEL CAME OFF AND WENT ROLLING DOWN THE MIDDLE OF THE ROAD AHEAD OF US! Surely, such an event had no more than one chance in a million of happening. Can you imagine the flow of adrenalin this produced in all of us? It was unimaginable! Of course the right front end of the Jeep dropped to the ground on the brake cylinder support disc, creating even more of a starboard list and causing us to veer sharply to the right and head for the shoulder -- again! I slowed as fast as I could but our van was quickly approaching the edge of the right embankment. I said to myself, "Oh no, here we go again." Luckily, we had been travelling in the middle of the road and had a little room to spare on the right side of the road. This time we were more fortunate and the van came to a stop right on the brink of the shoulder. I breathed a huge sigh of relief. Another spill had been avoided. We didn't know whether to laugh or to cry. Such an event was not on the list of possible catastrophes we had envisioned.

We retrieved our wheel and Noble determined that the spindle had been broken when the Jeep fell over on the embankment. Our Jeep had been more severely damaged in the fall than had been initially apparent. Adding to our woes was the fact that the brake cylinder support disc had become bent in the most recent catastrophe. We needed some professional help in solving our mechanical problems. Noble decided that the only solution was for him to remove the broken spindle and the bent disc and take them for repair or replacement to the nearest auto service and supply facility ahead of us, wherever that might be. He would have to hitchhike with the truck drivers. San Antonio Oeste was only 75 miles ahead, but was unlikely to be of much assistance. Bahía Blanca was 350 miles up the road. So Noble used his mechanical skills to remove these broken parts, and then he packed them in my heavy blue navy "B-4" bag. He packed another, smaller brown bag with a change of clothes, personal items, and a pouch for money, passport, and other items of value. Although he was concerned about the cost involved, Noble actually relished such adventures.

At 6:00 pm a huge Mercedes diesel, carrying 4,000 gallons of wine, came along the road heading north with two drivers (one of them originally from Italy). On top of their truck was a big wooden box full of raw, red beef for their personal barbecues. Richard flagged it to a stop. Noble explained his situation to them and they immediately made room for him and his gear in their cab. Noble said goodbye and off he went, leaving Richard and me to guard the incapacitated Jeep at the roadside. It was a bright moonlit night on the open steppe grasslands. Our evening meal consisted of vegetable-beef soup, bread, and wine. We each read or wrote notes for an hour before retiring to bed. Richard was happy to sleep on a bed for a change, rather than the hard floor of the van.

Noble's adventure began that evening when his truck pulled off the road and stopped a short distance before San Antonio Oeste, at about 8:00 pm, to prepare an "asado" (barbecue). Brush and chaparral were gathered for a fire, and within an hour the coals were ready for beefsteak. By then, five other trucks had also stopped. All the truckers shared in the festivities, drinking mate and wine, while joking and laughing in great camaraderie. Afterwards, they dozed and recuperated by the roadside before proceeding onward into San Antonio Oeste, where they arrived at midnight and checked into the Hotel Comercio for the rest of the night, as did Noble, for 40 pesos (50 cents) each.

I arose at 8:00 o'clock on a chilly, overcast to partly cloudy morning (Sunday, July 19) and ate a quick breakfast of bread, jam, an orange, and bran flakes. Grabbing my binoculars, field notebook, camera, and new little *Las Aves Argentinas*, I set off across the arid, grassy Patagonian plains with chaparral-like scrub. The livestock were mostly goats and occasional sheep, which were confined to large "paddocks" enclosed by wire fences that stretched for many miles between corner posts. A few dry gullies provided a little topographic relief. There were no houses, sheds, or livestock pens in sight. Wild animals included numerous guanacos, rabbits, and so-called "Patagonian hares." This latter mammal was actually a large rodent, superficially rabbit-like in appearance and locomotion, which ran in a bounding (cursorial) fashion. Taxonomically, it is in the genus *Dolichotus* and the family Caviidae (along with guinea pigs). It was diurnal (daytime) in its period of activity, had an antelope-like form with long slender legs and three-toed feet, weighed up to 35 pounds, and lived in colonial burrows. I saw a great many.

The passing truck drivers all stopped to offer us food and water, and to ask if we needed assistance of any kind, which we didn't. They were very friendly and concerned about our welfare, and one driver gave us some oranges. Trucks heading south sometimes stopped with a message from Noble, and we were told he was on his way to Bahía Blanca. These drivers reinforced my high esteem for them, and epitomized the comradeship that exists between persons living in isolated, hazardous regions of the earth. I admired them. Surprisingly, an ordinary passenger car came along in the afternoon and stopped. It was a 1930 Hudson with a middle-aged couple. They also inquired about our welfare and asked if we needed assistance. When we said we were OK, they immediately drove off with the friendly, polite, parting expression "Hasta siempre" (literally "Until always"). We didn't learn anything about them at all, but I thought they must have resided somewhere on the vast, barren Patagonian steppes. I was intrigued by them and wished they had remained a bit longer.

I spent an hour that afternoon scraping the last bits of dried flesh from my albatross skull. Then I consumed another hour by updating the orange line painted on the large Mercator map of the world on the outside of our van, showing the route we had traveled from California to our present location. In the evening, Richard and I tried to listen to our little radio, but there was too much static. Our cooked supper consisted of sausages, potatoes, onions, and green beans, washed down with an ample amount of red wine. As was the case last night, there was a bright, almost full moon outside when we went to bed at about 9:00 o'clock. The outside temperature was a cool 40°F. We had enjoyed a pleasant, relaxing day.

As it turned out, there were no repair facilities in San Antonio Oeste that could solve the problems with our front wheel, so it was necessary for Noble to continue north to Bahía Blanca (on Sunday, July 19). It was not until the middle of the day that he succeeded in catching a ride, along with another young hitchhiker, a soldier boy whom he had met at the hotel the night before. The truck that picked them up was the same Fiat that had pulled our fallen Jeep to an upright position yesterday, on the roadside embankment! It had been delayed because of generator difficulties. The main gravel road on which they were travelling was quite rough because of the numerous overweight trucks that followed this route, the only north-south road available along the Atlantic coast of Argentina. There were no government weigh stations. When the Fiat arrived in the town of Viedma, the driver stopped and everyone ate a steak dinner. It was about 8:00 pm and still 150 miles to Bahía Blanca. Since the Fiat was slowed by its generator problem, Noble decided to switch to another truck, a Ford 900 that was loaded with 18 tons of wool and driven by Guardo (a big, sloppy, happy fellow) and his sidekick, Negro. They drove all night, travelling at a slow rate of speed because of the heavy load, and arrived in Bahía Blanca the next morning (Monday, July 20). Guardo dropped Noble off at the IKA auto sales and service facility.

Richard and I had pancakes for breakfast that same morning (July 20). I spent the day walking leisurely across the steppes, recording the birds I encountered, taking a few pictures, and trying to shoot Patagonian Hares with Jean's .22-caliber pistol. I was not successful. A new bird for my list was the Black-and-rufous Warbling-Finch*, an attractive little bird with a distinct black mask bordered above and below by a white line. It was blackish above with a rufous breast and flanks, and a white belly. I also saw a Variable Hawk, 2 American Kestrels, 3 Band-tailed Earthcreepers, and 8 Gray-hooded Sierra-Finches. This was a rather sparse avifauna.

Noble spent the day (July 20) in Bahía Blanca acquiring a new front wheel spindle to replace the one which was broken when the Jeep fell over the embankment. The Willys parts shop did not have parts for the FC-170 model Jeep, but fortunately the spindle was interchangeable with the spindle for the big Jeep pickup they had in stock. They could not replace our bent disc, but it was straightened by hand, sufficient for use.

On Tuesday, July 21, I hurried out of bed to take a picture of the brilliant sunrise. I gulped down some oatmeal then took my binoculars and pistol to see what kind of game animal I might be able to find and shoot, so that Richard and I could vary what we had been eating for our evening meal. Sadly, I had no luck at all. Our misfortune continued when we ran out of drinking water, leaving us only red wine as a beverage! At 4:00 o'clock in the afternoon, a truck stopped to ask if we needed any assistance. The driver was a large, strongly built man perhaps in his 40's, and was wearing traditional gaucho style baggy pants tucked into his black leather boots. He said his name was "Chiquito" (which means 'the little one'). He had an "ayudante" (helper) with him, who set about preparing some "mate con bombilla" for all of us (a tea-like drink that I discussed earlier). Chiquito noticed that one of their tires was losing air, so he set about changing it. When the mate was ready, in a gourd with a metal straw, it was passed around between the three of us, from person to person all drinking out of the same straw, as was the tradition. It had a rather bitter taste and neither Richard nor I liked it very much (though of course we did not indicate such). This drink, like a great many beverages, requires a long period of consumption before the taste becomes tolerable. It was drunk as a stimulant and for its benefits to one's health. The mate taste quickly disappeared when we were offered some cognac immediately afterwards! Richard would write of this event in the next newspaper article he sent to Ontario.

There were small groups of goats roaming the countryside in the vicinity of our Jeep, within which were always a few half-grown young goats ("chivos"). They were rather wary and not easy to approach, but Richard and I decided we would try to shoot one and barbecue it for our evening meal. Our only firearm was the .22 caliber pistol. It wasn't long before we spotted two young goats with their mother, in a group of a dozen or more adults. However, they were extremely wary and would not allow us to approach them at all, certainly not within the short range of my pistol. It was decided that I would try to circle around behind the group and then Richard would slowly walk them toward me. We did this but the goats spotted me before they were in range of my pistol, and they ran away. I fired several shots as they fled, without success. Chiquito, having heard the shots, came running up with his pistol to join the hunt. Shortly thereafter, another big truck came along the road and stopped. When the one driver got out and saw what we were doing, he pulled out a long-range rifle from inside his cab, and he too began shooting at the goats, from the roadside with his rifle. Richard, Chiquito, and I were now all running after the goats, attempting to encircle them while occasionally firing at them with our pistols. The chase continued for almost an hour before Chiquito finally succeeded in shooting one of the young goats. It was now getting dark. Richard and I helped him carry the goat back to the road.

Chiquito took out a very big knife and in less than ten minutes he had the entire fleece removed, all in one piece! I watched in amazement. He took the fleece and hung it conspicuously on a large, nearby bush, explaining to me that it was OK to shoot a farmers animal to eat, when necessary, but the skin should be removed and left for the owner. Next, a pit for the fire was excavated, branches and brush were gathered, and a fire started. Using an iron tire tool and a broken truck spring, Chiquito made a vertical cross, which he firmly implanted in the ground at the edge of the fire pit. When the coals were hot, he then hung the "chivo" on the cross, and it began roasting. He had expertly removed all the inedible organs from the inside. He referred to it as an "asarito" (little roast). The others had built a barbecue pit and grill and were barbecuing huge slabs of beef that they had with them for just such an occasion. Richard and I supplied the wine, which we all drank while we were waiting for the beef and chivo to be cooked.

A lively conversation was carried on (in Spanish) while we waited. I was content to let Richard do most of our talking, as he was much more fluent in Spanish than I was. The driver of the second truck was German, having come to Argentina after the war, like so many others. He was very well informed about the world, and as did most persons we met, he expressed a strong dislike for communism. Of course our conversation included the usual masculine comments about girls, and the strong dislike between Argentines and Chileans. It was emphasized, probably correctly, that Argentina had twice as many sheep as did Chile. Finally, at midnight, 8 hours after they had arrived and after consuming a large amount of wine, Chiquito and his ayudante climbed into their truck and drove on up the road, unconcerned about their very long layover. After all, Chiquito owned the truck and was his own boss. The German driver followed them. We all wished one another "Buen viaje" as they departed. It had been another day to remember, to tell about in a book one day, or send to a newspaper in Ontario!

Almost immediately after the trucks left and headed north up the road, a 40-passenger, cross-country bus came along heading south, down the road. It stopped by our 3-wheeled Jeep, where Richard and I were standing and finishing the last drop of our wine. To our great amazement, who should step off the bus but Noble! He had boarded the bus in Bahía Blanca at 9:00 o'clock that morning, 15 hours earlier, and he had with him a new spindle for our wheel and the old, somewhat straightened disc. He, too, had stories to tell about his mini-adventure. It was another bright moonlit night and we all went immediately to bed. Richard had to sleep on the floor again.

The next morning (Wednesday, July 22) it took four hours for Noble to fit the new spindle in the right front wheel, with frequent referencing to our maintenance manual. Once again, he learned as he went along. Richard and I offered what little assistance we could, and we cleaned up the van inside to get it ready for travel once again. Finally, after four days broken down by the roadside, we were underway for Buenos Aires again. However, I had no sooner started up the road when I realized that the right front brake was not working. The brake cylinder had apparently also suffered in the fall, and had ceased to function. There was little we could do but proceed at a slower speed with extra caution, and prepare for the van to swerve to the left whenever I applied the brakes (because the left front wheel slowed but the right did not). Furthermore, we discovered yet another mechanical casualty that had resulted from our fall. The right back wheel was wobbling and overheating badly. We surmised that the most likely reason was that the right rear axle had been bent slightly by the truck and chain when all the weight of our very heavy vehicle was placed on both our right wheels as the van was rotated 50°- 60° counterclockwise, from lying on its side to a more upright position on the embankment. Our Jeep had obviously suffered more from its fall than we had originally determined, primarily because of its tremendous weight.

So we limped very slowly into the little town of San Antonio Oeste (which was 6 miles south of the main road, to the right), arriving about 6:00 pm. Here we found a small backyard repair garage, the only one in town, but it had already closed for the day. So we ate dinner at the Hotel España, went to a movie, and then drove just outside the town to a campsite near a Shell Company oilfield. The next morning (July 23), Noble explained our two problems to Jimenez, the mechanic at the town's only garage (and a young entrepreneur who enjoyed talking to Noble). He told us that it would take 1½ days to correct the bent axle, as best he could, but that he could not solve the front brake problem. All he could do was disconnect the left rear brake so that the vehicle would now stop in a straight line, but we would have only two of our four brakes working. This certainly was not good news, because even with all four of our brakes functioning there was barely enough power to stop our 4½-ton truck safely.

At noon the next day (Friday, July 24) we departed San Antonio Oeste in a severely crippled camping van and drove east toward the town of Viedma, which was situated near the mouth of the Río Negro on the Atlantic Ocean. We were in far eastern Río Negro Province, at approximately 41°S latitude. Along the roadside were flocks of Upland Geese, Patagonian Mockingbirds, and a pair of Spot-winged Pigeons (a species with a prominent white wing band in flight, which I had first seen in an Andean valley in southern Peru). I also saw my first Rufous Horneros*, Argentina's national bird, a very conspicuous and familiar bird of open areas throughout the pampas and often seen around houses. It was a thrush-sized, rather plain, brown furnariid, brighter above and paler below, with a rather short rufous tail and an indistinct cinnamon-colored wing band showing in flight. It was referred to as an "hornero" (ovenbird) because of its domed mud nest with a side entrance, which was frequently placed conspicuously on top of a fence post or a telephone pole. The nests were used only once but remained standing for several years, becoming one of the most prominent and characteristic features of the open countryside.

We reached Viedma about 5:00 o'clock, and here our highway turned north, crossed the Río Negro, and entered the very southwestern tip of Buenos Aires Province. The birdlife changed, becoming more abundant and diverse. New species included the Guira Cuckoo* (a large, untidy looking, finely streaked bird with a short, ragged crest, long rounded white-tipped tail, pale orange-yellow bill, and ochre colored crown), Campo Flicker*, and Yellow Cardinal* (a strikingly patterned olive-and-yellow emberizid with a conspicuous black throat patch and black crest, and a very popular cage bird). We camped that night along the Río Colorado, not far above its mouth on the Atlantic Ocean, in southern Buenos Aires Province 25 miles due south of the city of Bahía Blanca.

The next morning (July 25) we continued our journey toward Bahía Blanca, 50 miles north of us over a route Noble had already traveled. We were entering the PAMPA REGION of east-central Argentina, an area of fertile grasslands with marshes, lagoons, green pasturelands, fences, and windmills. Cattle were the dominant livestock on large estancias, and gauchos on horseback were a familiar sight, but many sheep were also raised. I was pleasantly surprised when a group of ten Burrowing Parrots flew past us, only my second sighting, and my last, of this handsome, multicolored Patagonian species (which migrates a little to the north and east in winter). The White Monjita* was a lovely little flycatcher of the open countryside, conspicuous in its immaculate white plumage with black primaries and a narrow black band on the end of its tail. It perched in clear view on bushes, small trees, and fences, where it was a delight to behold. Its name means "little nun", but it was also often referred to as "viudita" (little widow), and was a well known bird to the farmers and ranchers. Another characteristic and familiar bird of the region which I saw for my first time that morning was the Firewood-gatherer*, a medium-sized, slender, light brown furnariid with a long, graduated, pointed tail which in flight showed large prominent white tips to all but the central pair of feathers. It was primarily a somewhat secretive terrestrial bird of open grassy areas with shrubs and small trees, but it built a large, bulky nest of twigs which was placed conspicuously above the ground in small trees, or even occasionally on telephone poles or other man-made structures. A nest was often used for several years, becoming larger each year as new material was added. The Firewood-gatherer was easily seen in the vicinity of its nest. Other species I identified for the first time were Spotted Tinamous*, Chiloe Wigeons*, Chalk-browed Mockingbirds*, Bay-winged Cowbirds*, and White-browed Blackbirds*. I never managed to separate or describe, with confidence, any Screaming Cowbirds from the many Shiny Cowbirds, so this species is absent from my list although it was almost certainly present. There was also a single flock of 120 Grassland Yellow-Finches.

In Tres Arroyos we found an inexpensive restaurant where we ate our usual evening meal of steak and eggs. Then I drove a short distance north out of town on our main highway, which had been concrete at one time but was now badly broken into large slabs, making our travel woefully slow and a little hazardous. Soon I chose a site at which to camp, under a tree in a farmer's field on private land, but with no house in sight. We were approximately 300 miles south of Buenos Aires, in southern Buenos Aires Province, 40 miles inland from the Atlantic coast (which ran in an east-west direction in this region of Argentina). The next town up the road, north of us, was Juarez.

The next morning (Sunday, July 26), Richard decided he was tired of sitting on top of the engine and needed some exercise walking. He also wanted to write some stories about the gauchos and their life on the pampas, to send to the newspaper in Canada. So he said that Noble and I should go on to Buenos Aires without him and he would hitchhike in the pampas for a few days and then meet us later in the city. Noble had many items of business, people to meet, and some sightseeing he wanted to do in Buenos Aires, and he wanted to spend a week or longer in the city. So Richard's plan was fine with him. Noble told me if I preferred to hitchhike with Richard for several days in the countryside rather than spend it with him in the city, it was OK with him as he didn't need me for any assistance. That plan suited all three of us, and we each looked forward to our adventure. Noble would leave a message at the American Embassy where to contact him when we arrived. We waved goodbye amiably and Noble set off alone in our malfunctioning Jeep for Buenos Aires, 300 miles up the road. As always, the events that lay ahead of him could not be predicted.

Richard had with him his backpack and sleeping bag, as he normally did when he was hitchhiking. I also had my sleeping bag and backpack, in which were my binoculars, camera, pistol, notebook, field guide, rubber boots, and toothbrush. We had an assortment of food items, which included canned tuna, bread, peanut butter and jelly, canned juice, an orange, sweetened, condensed canned milk, and red wine. Each of us was carrying about 35 pounds of weight. Richard and I shared similar thinking and philosophies about most things in life. We were kindred spirits and our four days and three nights hitchhiking together (July 26-29) on the Argentine pampas were a pleasant, memorable experience for both of us. We traversed a total distance of about 200 miles, from a short distance north of Tres Arroyos to a short distance north of Las Flores (100 miles south of Buenos Aires), in eastern Buenos Aires Province.

The 250,000 square miles encompassed by the Argentine pampas were the bread basket of the nation, and produced ¾ of the country's cattle, half of its sheep, and most of its hogs, wheat, alfalfa, corn, and flaxseed. Rainfall was ample and the soil was unusually fertile. Marshes and other natural wet areas abounded, and the climate was neither too hot nor too cold. The pampas were extremely productive agriculturally, and they were also home to a great diversity and abundance of birdlife. There were bushes and scattered trees throughout. The moist air was refreshing, the temperature was mild, and the walk was exhilarating. Richard was a good companion. While he visited with farmers and gauchos in fields along the highway, I filled many pages in my notebook with descriptions of birds I had never seen before.

One of the most characteristic avian species of the pampas was the Southern Screamer*, a large, heavy-bodied, heavy-legged and big-footed "waterfowl" of marshes and wet grasslands. There are three species of screamers in the world, all of them confined to South America and placed in a single family of their own (Anhimidae). They are considered to be the most primitive living members of the Anseriformes (ducks, geese, swans, etc.), and they are the only ones which lack a web between their toes. Southern Screamers produced a very loud, far-carrying bugle-like vocalization, which reminded me somewhat of an old Model-T automobile horn. An unusual behavioral trait was their habit of occasionally soaring in circles high above the ground. Richard and I encountered them everywhere, singly, in pairs, or in small groups, and I tried several times, unsuccessfully, to shoot one with my pistol to evaluate their edibility. However, we located a floating nest in a shallow marsh late on our first afternoon, and we could see some eggs in the nest, which we thought might be edible. I lost the coin toss to see which of us would wade out to the nest to gather the eggs, up to my waist in water. There were five eggs, and I carefully placed them in my bag and returned to shore with them.

It rained that night and we sought shelter at a hacienda located several hundred yards off the main highway, knocking on the door three hours after dark, as it began to rain quite heavily. Eventually the door opened, and there stood a quite elderly man who was obviously the caretaker and not the owner. It was apparent that he was quite frightened by the appearance of two wet, bearded, and bedraggled strangers. He reluctantly picked up an oil-burning lantern and led us outside to a nearby horse stable where he said we could sleep for the night, on top of some clean straw on the ground in an empty stall. Impulsively, I then handed him the five screamer eggs and asked if he could boil them for us to eat at breakfast in the morning. Once again, with reluctance, he agreed to do this.

We thanked him and then slept comfortably out of the rain for the rest of the night. Early the next morning (Mon., July 27), the old caretaker brought us the five eggs he had just boiled. We each cracked one of them open, whereupon both of us were equally overcome by their awful, foul-smelling, rotten odor, and we immediately threw them away. Without opening them, we also threw the other three eggs away, at a nearby tree, where they broke and again produced the same horrible odor. We concluded the parent screamers had long ago abandoned the eggs, perhaps because they failed to hatch in last summer's nesting season. We were disappointed that we couldn't eat them with our bread and jam for breakfast. Before we started out for the day, Richard sat down in the stable and wrote an article about the event to include in his next mailing to the Canadian newspaper. The incident would also be implanted permanently in my memory, alongside the remembrance of the Upland Goose that Noble and I had tried to stew and eat. Culinary skills were not among our accomplishments.

Not surprisingly, almost all of the new birds that I saw in the pampas with Richard were species associated with a wet environment (wet grasslands, marshes, ponds, lakes, and reeds, or their vicinity). In the order I saw them these were: the Long-winged Harrier*, Red-winged Tinamou* (a large tinamou with conspicuous rufous primaries in flight), Hudson's Canastero*, Red-gartered* and Red-fronted* coots, Maguari Stork*, Stripe-backed Bittern* (related to the Least Bittern, in the genus *Ixobrychus*), Red Shoveler*, Rosy-billed Pochard*, Spectacled Tyrant* (an unusual looking small flycatcher of reed beds, black in color with a pale yellow bill, conspicuous pale yellow bare area around the eye, and the outer primaries entirely white, creating an eye-catching wing pattern in flight), Sooty Tyrannulet*, Scarlet-headed Blackbird* (a stunning, uncommon, local species strictly confined to reed beds, with a bright red head and upper breast which contrasted sharply with its black plumage), Great Pampa-Finch*, and Coscoroba Swan*. A non-wetland species of note was the Cattle Tyrant*, a predominantly terrestrial foraging flycatcher much like a Tropical Kingbird in overall appearance, though longer legged, browner backed, yellower below, and a little smaller. It sometimes perched on the backs of livestock.

In terms of numbers, the most astounding total was 2,500 Rosy-billed Pochards which flew overhead in a single immense flock on July 28, ten miles north of Las Flores in an area where there were many ponds, marshes, and wet fields. On this same date, in a 10-mile stretch of walking along the highway, I recorded the following numbers: 120 White-winged Coots, 200 Southern Lapwings, 400 White-faced Ibises, 60 Great Egrets, 50 Limpkins, 150 Chimango Caracaras, 100 Yellow-billed Pintails, 400 Fulvous Whistling-Ducks, 40 Southern Screamers, 500 Brown-hooded Gulls, 50 Great Kiskadees, 75 Correndera Pipits, 25 Many-colored Rush-Tyrants, 40 Firewood-gatherers, and 35 Rufous Horneros.

Noble had a few mishaps, as could be expected, on his 300-mile journey to Buenos Aires after he left Richard and me on the pampas. First, the clamp which held the right front brake hose broke and he lost all his brake fluid, and consequently all his brakes. (Remember, the right front brake and left rear brake weren't working anyway.) Second, when he pulled over to the side of the road for a rest stop, he got stuck in some soft, wet mud, and had to winch himself out. Third, he was almost run over by a train just south of Las Flores, at a crossing with no light or barricade, when the approaching train did not blow its whistle. It approached Noble from his blind spot on the right rear side of the vehicle and he did not see it until the very last moment, just in time to bring our brake-hampered Jeep to a stop. The next day (Monday, July 27) Noble drove into Buenos Aires, the first of 10 days he would spend there. On the way, he changed a flat tire in a personal best time of nine minutes (as he timed with his alarm clock)!

Buenos Aires was a vibrant city that attracted visitors from all over the world. It was a city of both cobblestone streets lined with white, tile-roofed houses, and of wide modern city boulevards, including "Avenida 9 de Julio", the widest city boulevard in the world with a width of 400 feet. Buenos Aires had a European air with sidewalk cafes and a gay nightlife. It was one of the most important cultural centers in Latin America, and had the second largest seaport in the Americas (next to New York City). The 2½ million persons who lived in Buenos Aires were referred to as "porteños" (port dwellers) to distinguish them front all the rest of the people, and they made up ⅓ of the total population in Argentina. Within the city there were many beautiful parks and plazas, statues, fountains, and subtropical shrubs and trees. Noble thoroughly enjoyed his time in the city, which he spaced between work and pleasure. Taxis, buses, streetcars, and a subway provided easy, cheap, and accessible transportation for almost everyone. An important business item was to confirm our ship reservation from Río de Janeiro to Cape Town (on the "Boissevain" of the Dutch Royal Inter-ocean Line). In his usual exuberant manner, Noble took care of the numerous tasks that needed to be done in Buenos Aires in order to keep our adventure on the road. He tried once again to obtain a permit from the Brazilian government for permission to bring our Jeep into Brazil, but as previously had no success. While I was in my realm, hiking on the pampas with

Plate 27: Our top-heavy Jeep fell gently over on its side on a steep, soft roadside embankment in Patagonia as I was attempting to pass a truck; shortly thereafter as we started out again our right front wheel came off & rolled down the road ahead of us! We eventually reached Buenos Aires.

Richard, Noble was in his realm among the entrepreneurs of Argentina, in the capital city of Buenos Aires. In between all of his business activities, Noble was wined and dined and taken to movies, nightclubs, and operas by his new acquaintances. It was a fun-filled time for him. Richard and I rejoined him at the Jeep agency at 11:00 pm on the night of Wednesday, July 29, after he had just gone to bed. We two backpackers had flagged down a bus for the last 75 miles of our travel.

On our original travel itinerary, Noble and I included all the independent nations of Latin America (which excluded the three colonial "Guianas"). The only countries remaining for us were Uruguay, Paraguay, and Brazil. Our currently planned route was to drive from Buenos Aires north into Paraguay, and from there eastward into Brazil, which would bypass Uruguay. Therefore, in order for us to visit Uruguay, we purchased a round-trip ferry ride between Buenos Aires and Montevideo, crossing the wide estuary known as the "Río de la Plata" (an expansion of the Paraná and Uruguay rivers below their confluence upstream from the Atlantic Ocean). It was an all-night ferry ride on an overcrowded boat with 400 passengers, leaving on Friday (July 31) and arriving on Saturday (Aug. 1). The ferry departed from the enormous Buenos Aires port at one of the several hundred wharfs, all of which were bustling with hard-working, dirty stevedores. During the long night, I spent much of my time attempting to engage an attractive 32-year-old grandmother in a romantic conversation, in Spanish of course. My conversation was challenging, to say the least!

When we arrived in Montevideo, which was the capital city of Uruguay, Noble telephoned a Mr. Howard Croninger, the American treasurer of General Electric/Uruguay, whose name he had recently been given as a person to contact. He invited us to his house for lunch and mid-day drinks, then took us to his private sports club for a game of tennis! Montevideo was known as the City of Roses because of its beautiful parks and gardens that attracted tourists from all over the world, who also came to enjoy the clean beaches, modern hotels, and coastal resorts.

We spent only one day in Uruguay and then returned to Buenos Aires the next day, having acquired a stamp in our passports from yet another country. I did not add any birds to my list in Uruguay. It was with great reluctance that we said goodbye to Richard Dugdale on Wednesday, August 5. He had been our good friend and travelling companion for almost four weeks. One of the most difficult aspects of our adventure was parting company with newly established friends, knowing there was little likelihood of seeing them again. Who else would I ever find to boil rotten screamer eggs with me? Our last major destinations in South America were Asunción, Iguaçu Falls, and Río de Janeiro, from which city our currently scheduled departure for South Africa was only four weeks away.

Chapter Eleven – The Road to Rio

Noble and I departed Buenos Aires on Wednesday morning, August 5, for the last leg of our South American adventure. We were headed for Río de Janeiro, approximately 2,000 miles distant by road. Our route would take us north along the eastern edge of the Argentine Chaco to Asunción, Paraguay. From there we would travel east to the Brazilian side of Iguaçu Falls, and finally eastward again across southern Brazil to Sao Paulo and Río de Janeiro

I drove over 300 miles that first day, almost nonstop, following the west bank of the Río Paraná upstream on a good paved highway. Our initial heading was northwest to Santa Fe Province and the city of Rosario, then northward to the city of Santa Fe, along the northern fringe of the PAMPAS region. The countryside was low-lying, wet, and muddy. I did not take any time to record birds. The next day (Thur., Aug. 6) we transitioned from the pampas into the southeastern corner of the CHACO REGION, a vast area in the interior of South America which extended northward from central Argentina all the way through western Paraguay into Bolivia. It was a region characterized by dry savanna grasslands, thorny scrub, cacti, scattered palms, and small patches of xerophytic woodlands. It was somewhat wetter on the eastern side of the region, which we were traversing, and there were cattle ranches, sugarcane fields, and plantations of quebracho trees (for tannin production). The eastern boundary of the Argentine Chaco was formed by the southward flowing Paraguay and Paraná rivers, and our northward travel in the region would parallel these two rivers to their west. (On the other side of the Paraná River, to the east, was the narrow, low-lying, wet, swampy, subtropical region of Argentina known as MESOPOTAMIA.) Our paved road became soft dirt, dry and dusty, or wet and muddy depending on when it last rained. We were driving north in Santa Fe Province toward the city of Reconquista. Noble remarked that the Chaco countryside reminded him of pictures from Africa, with wet savannas, palms, and thorn forest. It was an exciting environment for both of us, unlike any through which we had yet ventured in our neotropical travels. I was thrilled by the great variety of birds and kept busy recording names and descriptions in my little field notebooks.

Of many new species for my trip list, the White-banded Mockingbird* was one of the most common and conspicuous, with its dapper gray color and large white wing band. A unique furnariid was the Lark-like Brushrunner*, a mostly terrestrial, somewhat social species of open woodland and scattered trees. It was distinct in appearance with its heavily streaked plumage, long pointed crest, pale pinkish bill, and relatively long, orange legs. It ran along the ground in a very lark-like fashion. Another furnariid was the Stripe-crowned Spinetail* (genus *Cranioleuca*), an arboreal bird of woodlands and scrub which foraged in a chickadee-like manner, often hanging upside down. It was distinctly plumaged, being brownish above (like most furnariids) with a prominently black-and-white striped crown, a conspicuous white superciliary, a narrow black eye-line, and noticeable rufous shoulders and tail. Red-crested Cardinals* (genus *Paroaria*, family Emberizidae) brightened the landscape with their brilliant red-crested head and red bib, Picazuro Pigeons* sat in the treetops, small groups of Monk Parakeets* flew back and forth across the fields and savannas, and several Roadside Hawks perched conspicuously in trees or on telephone poles. Small numbers of White-browed Blackbirds* (genus *Sturnella*) foraged in pastures, where they were easily located by their bright red breasts. The short crest, reddish-orange eye, white-tipped tail, and heavy body streaking of another new bird enabled me to identify it as a female White-tipped Plantcutter* (an aberrant cotingid). My favorite bird of the day was the Whistling Heron*, which was seen in both wet and dry grasslands, singly or in small groups. It was one of the most attractively colored and appealing of all the ardeids, being pale buff-colored with blue-gray back and wings, white rump, light blue eye ring, dark bluish crown, and pinkish, black-tipped bill. More than a dozen individuals of this lovely little heron were recorded. I documented a total of 44 species for the day (some of which, as usual, were not identified until I had time to compare notes with my new little field guide, *Las Aves Argentinas*).

We camped a short distance after passing through Reconquista. The weather was becoming warmer as we approached the tropics. The next day (Fri., Aug. 7) we drove northward toward the town of Resistencia (in Chaco Province), still paralleling the Paraná River. The flat terrain was utilized for cattle ranches and sugarcane fields. People (Mennorites?) along the road often traveled in 2-person, 4-wheeled buggies pulled by a team of two horses. The sugarcane was cut by hand, bundled and stacked on oxcarts, and then transported to tall wooden and rope hoists at the edges of the fields. Here the bundles were hoisted onto trucks for transport to the nearest sugarcane factories. Sugar was one of Argentina's principal exports. We drove at a leisurely speed and I stopped frequently for brief periods to search for birds in the Chaco woodlands while Noble wrote in his journal. It was one of our more pleasant mornings.

In the early afternoon, just after we had crossed the provincial boundary between Santa Fe Province (to the south) and Chaco Province (to the north), we came to a sugarcane field in an arid area of scrub and cactus. I pulled off to the side of the road to identify some foraging birds in the scrub. Very soon, a woman who was cutting cane about 100 yards away saw me and was frightened by my bearded appearance, binoculars, and strange behavior. She called to a group of other women working nearby, and they all looked at me warily. Then they called a young girl over to them, and after a moment of conversation with the girl, she went running off through the trees bordering the cane field. I walked back to our Jeep to get my camera so I could take pictures of some of the cacti, and then I returned to the edge of the cane field near where I had previously been. Within a few minutes, a toothless old man carrying an ancient single-barreled shotgun came running up to me rapidly, completely out of breath. I almost laughed at the sight of him, but instead I greeted him with a cheerful "Buenas tardes." His concern immediately disappeared and he broke into a broad, toothless smile. It was apparent to him that in spite of my beard and strange appearance I had no bad intentions. I attempted to converse with him, in Spanish, saying I wanted to take some pictures of the cacti. He responded in a somewhat disgusted tone, and with hand gesticulations he said there were lots of cacti everywhere and that they weren't good for anything at all. He then motioned for me to follow him and he took me to a particularly large individual cactus and said I should take its picture, which I did, and thanked him. In the meantime, several of the women and the girl had sneaked up through the trees to have a closer look at me, but they kept their distance and continued to peer at me with suspicion. The old man said I could take all the pictures I wanted, wished me "Buen viaje" (pleasant journey), and then left. I smiled to myself, waved goodbye to the women, and walked back to the Jeep. Life was an adventure!

Many of the 69 avian species that I documented today were associated with water, such as the Brown-and-yellow Marshbird*, Snowy Egret, Neotropic Cormorant, Wattled Jacana, Ringed Kingfisher, White-faced Ibis, Cocoi Heron, Brazilian Teal*, Ringed Teal*, Giant Wood-Rail* (the largest New World rail), Chestnut-capped Blackbird*, Saffron-cowled Blackbird*, Limpkin, Snail Kite, Common Moorhen, and S. Am. Snipe. The Ringed Teal was a particularly attractive little duck that preferred the smaller ponds and marshes. Giant Wood-Rails walked conspicuously along the edges of water and wet woodlands, where they were more readily observed than most other rails. Non-aquatic species seen included a Golden-winged Cacique* (which was building a typical, long-hanging nest in a group of trees), a Cattle Tyrant (feeding on the back of a horse), and a Rufous-browed Peppershrike (practicing its spring song, which I described as something like that of a House Finch). Several groups of Blue-fronted Parrots* flew noisily overhead, with members of a pair customarily next to one another. This parrot is the most southerly distributed of all the 31 species in the genus *Amazona*. While I was searching for birds in a patch of dry forest, I suddenly gasped at the sight of a foraging Red-billed Scythebill (a woodcreeper). It was inspecting epiphytes and probing into crevices and under the bark of trees with its amazingly long, sickle-shaped bill. I watched it for quite some time, wondering how such a bill could possibly evolve. In this same woodland were my first views of Plush-crested Jays* (genus *Cyanocorax*), to which I was attracted by their beautiful bell-like vocalizations. (One day I would make an effort to see all the corvids in the world, to keep me traveling to new places.)

Noble and I arrived at dusk in the town of Resistencia and found a small restaurant in which to eat some supper. Men stood around in the streets conversing with each other and drinking brandy made from sugarcane. Resistencia was located strategically at the junction of the Paraguay River (flowing from the north) and the Paraná River (flowing from the east then curving south here at Resistencia). The southwest corner of Paraguay was just across these two rivers, to the east and north of their junction. As we headed north out of town we were now paralleling the Paraguay River (rather than the Paraná River). It formed the boundary between Argentina and Paraguay, and Asunción (the capital city of Paraguay) was just 200 miles upstream. I drove only a short distance before choosing a campsite for the night. During our 2nd day in the Chaco, I added 39 birds to my trip list, one of my best days so far. We both slept well that night.

The next day (Sat., Aug. 8) Noble and I continued northward and within several hours we crossed the border from Chaco Province into Formosa Province, the northeasternmost province of the Argentine Chaco. We started out on a flat, dry, roadbed that was built several feet above the surrounding wet, low-lying marshes and grasslands. However, the countryside soon became drier, the vegetation changed to scrub and scattered xerophytic woodlands. We were proceeding slowly, generally at a speed between 15 and 25 mph. It allowed me the opportunity to observe some of the more conspicuous roadside birds. These included Greater Rheas* (see comments below), Southern Screamers, Whistling Herons, Great Egrets, Maguari Storks, a single Jabiru* (see comments below)*, Snail Kites, Savanna, Roadside, and Black-collared* hawks (see comments below), a Long-winged Harrier, Southern & Chimango caracaras, Black & Yellow-headed vultures, Southern Lapwings, Wattled Jacanas, Giant Wood-Rails, Picui Ground-Doves, Monk Parakeets, Rufous Horneros, White & Black-crowned monjitas (the latter being only a wintering bird in the Chaco from its breeding range in western and southern Argentina), Cattle Tyrants, Great Kiskadees, and as always Rufous-collared Sparrows. Remarkably, no Turkey Vultures were observed, nor in fact had I recorded this usually conspicuous species from anywhere in Argentina. Does it disperse somewhere else in winter?

The Greater Rhea is the bigger relative of the more southerly distributed Lesser Rhea. Like its smaller relative, the Greater Rhea was a tall, flightless, fast-running (almost 40 mph) "ratite" of open grasslands. At a height of 4½-ft. and a weight of up to 75 pounds (for males), it was by far the largest of all New World birds (though somewhat smaller than its Old World ecological counterparts, the Ostrich in Africa and the Emu in Australia). The Jabiru was another very large bird, the largest stork in the New World. It was mostly white, with a bare black head and neck bordered below by a red collar. With its 3½-ft. overall height, immense, slightly upturned bill, and somewhat commanding appearance, it was a bird that sometimes frightened small children! The Black-collared Hawk* is a very handsome, moderately sized, broad-winged raptor which is predominantly rufous in color with a conspicuous white head, black upper breast patch, black flight feathers, and a black band on the end of its tail. It was a bird found in low-lying wet areas where it fed partly on fish, having strong, curved toes which were spiked underneath, as in the Osprey. It had a wide geographical range extending all the way from Mexico to Argentina, but this was my first sighting. It often soared, and I attempted to photograph a soaring individual with my 35mm wide-angle lens, but it was not much bigger than a spot in my viewfinder, so I also drew a sketch of it in my notebook. I considered it one of my favorite raptors.

During one of our brief stops in an area of dry scrubby thickets I was provided with one of my few good looks of a caprimulgid. I discovered it resting quietly on the ground in a small opening, almost hidden by its cryptic coloration. It was possible for me to write a long, detailed description of its plumage pattern and color. Finally, I flushed the bird to observe its wing and tail patterns in flight, which were without any white. My field notes enabled me to eventually identify this bird as a female Little Nightjar*, a fairly common, widely distributed species in South America.

In the middle of the afternoon when we were about 15 miles south of the town of Formosa I stopped to birdwatch in an area of moist woodland with scattered palms. As I got out of the cab, I was happy to hear the familiar "hah-hah" vocalization of a Laughing Falcon, which I quickly located near the top of a tall tree. The site was good for birds and I stayed here more than an hour. Included among the variety of birds present were both a Greater and a Little Thornbird* (furnariids in the genus *Phacellodomus*), a Green-backed Becard*, and a White-bellied Tyrannulet*. My most exciting observation of the day, however, was that of a Pale-crested Woodpecker*, a Chaco specialty and one

Plate 28: The Chaco region of inland, lower South America extends from N Argentina northward to Paraguay and Bolivia. It was characterized by dry thorn scrub and cacti. Mennonites drove horse-drawn carriages.

Plate 29: Sugarcane was a chief export of Argentina. It was cut by hand, loaded onto oxcarts, and transported to processing factories.

of 11 species in the neotropical genus *Celeus*. All members of this genus possess a long pointed crest and feed largely on arboreal ants, quietly and mostly unnoticed in the lower and middle forest, sometimes descending to the ground. The Pale-crested Woodpecker was an unusually handsome woodpecker with its pale yellowish-blonde head, wide red malar streak, dark chestnut-black underparts, and heavily barred upperparts (wings and back, with black, rufous, or pale yellow bars of varying width). The red malar streak identified it as a male. This sighting became a lasting memory and the species was added to my list of favorites. (I was particularly fond of barred plumages, and of buff, ochraceous, cinnamon, rufous, chestnut, and russet colors.)

In the late afternoon we drove into the town of Formosa. There were several police check posts since this was the last community of any size before we reached the border of Paraguay. Somewhat amusing to us was the fact that the local policemen rode bicycles, and carried both a .45 caliber pistol and a sheathed bayonet in their belts. We were generally just motioned on past the check posts, without being asked to stop, but as we headed out of the town we were told to follow a policeman on a bicycle, at 10 mph! That night we camped just 5 miles north of Formosa, less than 100 miles from the Paraguayan border. For me, it had been another excellent day in the Chaco and I wrote descriptions of, or otherwise documented, 74 species of birds (which was a slight increase in number from each of the two previous days). It was one of the most prolific single days of my travel to date.

Sunday, Aug. 9, was our last day in Argentina. We drove at a slow to moderate rate, stopping regularly for picture-taking, writing notes, or observing birds. The weather was pleasantly mild and sunny, and we both enjoyed the ride and leisurely pace. There were many oxcarts on the road, frequently driven by indigenous people of Guarani descent, some of whom spoke no Spanish. Cane brandy was the beverage of choice by almost everyone. The most numerous roadside birds were Black Vultures, Southern Lapwings, Monk Parakeets, Rufous Horneros, and Red-crested Cardinals. Several species were new for my trip list: Short-billed Canastero*, Wedge-tailed Grass-Finch*, Strange-tailed Tyrant* (see comments below), and Red-crested Finch* (which not only had a small red crest but also was almost entirely reddish in color). Bare-faced Ibises were identified for only my second time (the first occasion having been in Colombia).

The Strange-tailed Tyrant was a local, uncommon, resident little flycatcher of marshes and wet grasslands. A pair were seen together, which enabled me to study their sexual dimorphism, particularly in the length and shape of the male's unusual tail. The male perched conspicuously on top of a long grass stem, giving me ample time to write a careful description. It was mostly black and white with an orange-yellow bill and a long, twisted tail (twice the length of the body). The two outer tail feathers were greatly lengthened and rotated 90°, with a bare shaft for the proximal ⅓ of their length and then with a wide, broadened inner web for the terminal ⅔ of their length (the outer web being absent). The overall appearance of the tail was remarkably similar to that of some of the African whydahs that live in the same kind of grassland environment. This phenomenon has been termed "convergence" by evolutionary biologists. How had natural selection arrived at the same answer in two populations that were very widely separated in both space and genetic ancestry?

A particularly memorable event occurred at one of our roadside stops. I was walking through some Chaco woodland, rather dense and with a closed canopy, carefully searching for birds. Suddenly my eyes focused on a monkey resting quietly in the middle of a tree, about 15 feet above the ground in a tangle of vines. It had a rather small round face and was watching me keenly, but did not seem overly alarmed, and it did not move. So I took out my little lined field notebook and sketched its face and wrote a description of its overall appearance (as best as I could see through the foliage). My notes said, "monkey: medium sized; very long heavy tail; generally gray above and light fulvous below; face small, patterned black and white; tail tawny russet bordered with black; eyes brown." Years later, I was able to identify the species, quite conclusively, as an Azara's Owl Monkey (also called Southern Night Monkey) in the very distinct genus *Aotus* (which is placed by some authorities in its own family, Aotidae).

Members of the genus *Aotus* occur only in the Neotropics, from Panama south throughout much of South America, in tropical forests and woodlands all the way upward into temperate elevations. Traditionally, only a single species was recognized in the genus throughout its entire range, but more recently taxonomists have split the genus into as many as 4-8 species, of which the Azara's Owl Monkey is the most southerly distributed. Owl monkeys are unique among primates in their principal time of foraging, which is at night rather than in the daytime. They are also distinct in being omnivorous in their diet, eating a wide assortment of not only plant material, but also animals and animal products (honey, insects of all kinds, spiders, other arthropods, eggs, and small vertebrates such as lizards, frogs, and even bats). No other New World (or Old World?) primate is as carnivorous. They snatch or grab small animals with their hands. Unlike most New World monkeys their tail is not prehensile and is not used in locomotion. They move through the trees in a quadrupedal fashion (walking on the top of branches with all four limbs), and sometimes by leaping, since their legs are longer than their arms. They are said to be most active on moonlit nights. In the daytime they characteristically sleep in dense foliage, vines, or in tree hollows. The Owl Monkey was a most fortuitous and thrilling find for me.

Noble and I crossed the rather small Pilcomayo River, which formed the border between Argentina and Paraguay, in the late afternoon (Aug. 9). Border formalities were simple on both sides of the river, with very little delay. A few miles inside Paraguay our road turned right and came to the west bank of the broad Paraguay River, at the insignificant little community of Puerto Pilcomayo. Here we crossed the river on a very large ferry, to the capital city of Asunción. It was 7:00 pm and just before sunset when we ferried across. I stood on the deck with my binoculars in readiness for any birds that might come by, and was soon rewarded for my effort when a group of six parrots came flying rapidly low overhead. I studied them as they quickly flew past and then wrote a careful description in my notebook. I described them as an *Aratinga*-sized parrot with pointed wings and a long pointed tail, primarily green in coloration, with a dark head and blue in the primaries. They proved to be Nanday Parakeets, in the monotypic genus *Nandayus*. This species is endemic to the Chaco and the Pantanal, from Argentina to Bolivia and Brazil, with a rather long, narrow geographic distribution in the southern interior of the continent, remarkably similar to that of the Pale-crested Woodpecker. It was a great finish to the day.

My total list of birds seen during our four days in the Chaco would eventually reach 118 species (after many years of sleuthing out the last few descriptions in my field notebook). It was a fun time. Noble and I had spent over five weeks traversing Argentina in the wintertime, and had driven a total distance of more than 4,000 miles in the west, south, east, and north of the country, for the greatest cumulative distance we traveled in any Latin American country.

Paraguay was one of the poorest and least populated countries in South America, and one of only two without a seacoast (along with Bolivia). It was divided by the Paraguay River into an eastern region and a western region, which are ecologically distinct from each other. There was virtually no tourism in this land-locked country. Most of the people were "mestizos" (persons of mixed European and indigenous ancestry), and two thirds of the population lived in rural areas. It was the only country in South America where the native language (Guarani) was used more frequently than was Spanish. Poorer women were shabbily dressed, frequently smoked cigars, and were often barefoot. Catholicism was overwhelmingly the dominant religion (as elsewhere in Latin America) and played a prominent role in almost everyone's life.

As in Argentina, yerba matte was a very popular beverage (and indeed the leaves for this tea-like drink came from a holly tree native to Paraguay). The bark of native quebracho trees was harvested as a source of tannin, and the very hard wood was used for railroad ties. Cotton and sugarcane were the two principal crops, and there was a booming cattle industry, with large estancias (ranches). As in Argentina, the local gauchos (cowboys) rode horses, carried whips, and wore wide-brimmed hats, baggy shirts and pants, boots, and chaps. Oxcarts and horse-drawn carriages were widely used for transportation.

Asunción had very little of interest. There were a few attractive tree-lined streets, green parks, and a well-kept botanical garden. Houses in the central part of the city were constructed of brick covered with plaster. Women on the streets wore neat cotton dresses and lace shawls. Local offices and foreign embassies carried on the business of government. Although it was not our original intention to do so, it became necessary for us to spend three full days (Mon.-Wed., Aug. 10-12) in Asunción, while Noble worked aggressively to obtain

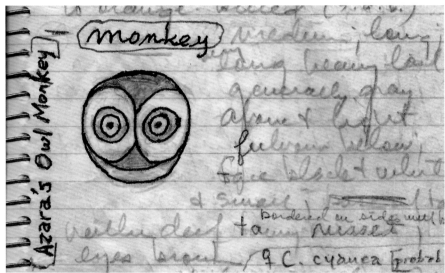

Plate 30: Palms grew in the more moist areas of the Chaco. A thrilling daytime discovery was an omnivorous, nocturnally active, little known Azara's Owl Monkey, which I sketched. A huge, grotesgue-looking stork, the Jabiru, flew overhead.

a permit from the Brazilian government that would allow us to take our Jeep into that country without paying an enormous "tax." All of his attempts were in vain, and the U.S. Consul in Asunción (a Mr. Wilson) was not sympathetic and of no help at all. This was a situation we had tried to solve, without success, before we left California (by correspondence to and from the Brazilian Embassy in Washington). The problem originated from the fact that our international carnet de passage (which was issued in Belgium and delivered to Noble in Los Angeles before we departed) had a large red stamp on every one of its 40-50 pages which said, in English, "NOT VALID IN BRAZIL." Brazil was the only country in the whole world that did not honor this document! The Brazilian government did not want any foreigner to sell his vehicle for profit inside the country, and so they gave no permission to anyone to bring a vehicle across the border without first paying a very large sum of money (in the neighborhood of 500 dollars for a vehicle such as ours), as a guarantee that it would not be sold. Of course, Noble and I could not afford such a cost. No amount of reasoning with the officials at the Brazilian Embassy in Asunción could persuade them to give us permission, without the exorbitant payment. Our road to Rio had a major obstacle.

Noble finally exhausted all his arguments and pleadings, which were to no avail, and we left Asunción on Thursday morning, August 13, heading due east on a brand new dirt road for 225 miles to the border with Brazil, on the Paraná River. On the outskirts of the city were many groups of Nanday Parakeets, which continuously flew back and forth overhead. Several prehistoric looking Guira Cuckoos sneaked through the thickets, a Tataupa Tinamou* flushed from the edge of a woodlot, and a pair of brightly colored Red-crested Finches foraged on the ground in a vegetable patch overgrown with weeds. We stopped to fill up our 40-gallon gas tank as we passed through the village of Coronel Oviedo. The countryside was subtropical grasslands, just below the Tropic of Capricorn, with scattered trees, orange groves, and woodlands. Cotton and sugarcane were the two principal crops. Oxcarts and horse-drawn carriages were widely used for transportation. Many rural houses were built of mud and sticks.

Birdlife along the way included two species of *Cyanocorax* jays that flew across the road--the Purplish Jay* and the somewhat more numerous Plush-crested Jay. Also, my heart skipped a beat at the sight of several White Woodpeckers*, a very striking two-toned species with an all white head, rump, and underparts, which contrasted sharply with the solid black back, wings, and tail. Its stunning appearance was enhanced by a conspicuous, yellow, bare area surrounding its white eye. This social species is placed taxonomically in the genus *Melanerpes* (which includes the Red-headed Woodpecker of N. Am.). It had a less undulating flight than most other woodpeckers and frequented dead trees in open areas. This would be my only encounter with this striking woodpecker.

Other noticeable roadside birds included a handsome, ferruginous-colored Savannah Hawk, which was gliding low above the grasslands as it searched for large insects, snakes, lizards, amphibians, or the occasional unwary bird, and two Burrowing Owls that sat separately on grassy mounds, keeping a watch for any predators. Red-breasted Toucans* were conspicuous in the canopy at the edge of woodlands, where they were easily located by their red and orange-yellow underparts and their large, pale, lime-green bills, and several Scaly-headed Parrots* (genus *Pionus*) flew by just above the treetops. I identified them by their moderate size, rounded wings, short tail, generally overall green coloration, reddish undertail coverts, and narrow white ring around the eye. We stopped at a wooded stream to bathe and wash a few clothes, and a Guarani man came by and told us, in Spanish, that a "plataforma" (a barge) was available at the river to take small vehicles across into Brazil. We continued on our way.

Noble and I had been told that if it rained our new road would be barricaded and closed to traffic until it dried out, to avoid destroying the road surface. This thought motivated us to get to the river as soon as possible, before it rained. However, not wishing to drive in the dark we stopped for the night at a location somewhere east of the settlement of Caaguazú, about 40 miles from our destination at the river. As we climbed into our bunks for the night, it was mild, calm, and peaceful, with no indication of rain. Then disaster struck. About midnight a very heavy torrential rain commenced. Noble immediately jumped up, hurriedly got dressed, and climbed into the cab to see if he could drive ahead far enough along the road to get past the last barricade before traffic was stopped. He didn't make it. Within a very few miles he came to a barricade which prevented any further travel. We were imprisoned in an area of mixed forest, farmland, and ranchland, and stuck there until the road dried out.

The next morning (Friday, Aug. 14), the rain had stopped. However, the temperature had gone down, the humidity had gone up, the sky was cloudy, and there were patches of mist and fog. It was not a good day for the road to dry, but for me it was not so bad. The forest road was virtually devoid of people and the area was quite suitable for birding. Noble commenced writing in his journal and I set off on foot to see what birds I might find.

My day got off to an exciting start with one of the first species I encountered. I peered at a little bird in the forest understory through a small opening where I could see only a small fraction of the whole bird, out of any sunlight. My brief, inadequate description said, "small: black with red head and narrowly black around the base of the bill; flicked wings." I tried for a long time afterwards to find a bird that matched this woefully incomplete description. A manakin seemed most likely, but no illustration or description seemed to fit to my satisfaction. More than fifty years later when I was comparing all the colored illustrations of manakins in *A Field Guide to the Birds of Brazil* (Ber van Perlo, 2009), studying their geographic distributions as shown on the range maps, and reading the habitat and behavioral descriptions, I came to the conclusion that the bird in Paraguay must have been a male Helmeted Manakin*. The site where I recorded it was only just outside the southwest edge of the documented range for this species, and the "narrow black area around the base of the bill" was not a feature of any other small black bird with a "red head." At the time I described the bird, with my very restricted view in dim light, it was not evident that the conspicuous red color was only on the upper half of the head, and the fact that I did not observe a "helmet" could have been because it was lying flat or obscured by foliage. After a great many years, I had finally arrived at an identification that I found satisfying. As my early mentor said, "Birds are where you find them." My list increased by one.

I struggled throughout the day with the many small flycatchers and eventually recorded the following: Euler's Flycatcher*, Sepia-capped Flycatcher*, Southern Bristle-Tyrant*, Bay-ringed Tyrannulet*, Yellow Tyrannulet*, Eared Pygmy-Tyrant*, and Ochre-faced Tody-Flycatcher*. Easier to identify were Plovercrests* (a hummingbird with a long, thin, wispy crest), Blue Manakins*, Red-rumped Caciques*, and dazzling Green-headed Tanagers* (a multicolored member of the genus *Tangara*). In the canopy were Chestnut-eared Aracaris* and Maroon-bellied Parakeets*. Toco Toucans*, the largest member of the family Ramphastidae, were conspicuous in tops of the taller trees in more open areas. I saw countless other birds - - hummingbirds, woodpeckers, antbirds, foliage-gleaners, woodcreepers, flycatchers, tanagers, etc. - - for which I was unable to write a description or assign a name. Neotropical birding without a field guide (or any other kind of bird book) was a daunting task. However, my field descriptions eventually allowed me to identify 55 species for the day. It was a challenging and fun-filled day in a Paraguayan forest, far off the beaten path.

The next morning (Sat., Aug. 15), I again went off with my binoculars for several more hours of observing. The temperature was a remarkably cool 44°F and there was even a little frost on the open ground! A few additional birds that I encountered were Blue-fronted Parrots, Blond-crested Woodpecker* (another eye-catching member of the genus *Celeus*, a close relative of the Pale-crested Woodpecker), Blue-winged Macaw* (one of the large *Ara* macaws), Sharp-billed Treehunter *(a rather small, trunk-foraging furnariid), Southern Antpipit* (a confusingly named, small, terrestrial forest flycatcher with a behavior and superficial appearance not unlike that of the N. American Ovenbird), White-throated Woodcreeper* (one of the largest woodcreepers, with a relatively heavy bill), and Red-crowned Ant-Tanagers* (which surprisingly were my first identifications of this wide-ranging, gregarious, uniformly reddish tanager of the lower and middle levels of neotropical forests).

The sun came out, the day warmed, the road dried rapidly, and by mid-morning the highway authorities came along and removed our barricade! Noble had spent his time writing letters and journal notes, and talking to Mennonite farmers who came along the road. We were underway again just after 9:30am. Not far ahead of us, an hour away, was the wide Paraná River that formed the border with Brazil. We were quite apprehensive about our reception there and of entering the country with our Jeep, since we had no official written permission to do so. We had no backup plan available (without a huge financial expenditure), and we were scheduled to leave Rio de Janeiro in less than 2 weeks on a ship for Cape Town. It was not a good situation for us.

We reached the bank of the river at an insignificant little settlement that was called Puerto Franco at that time. We were about 600 ft. above sea level in elevation. (Years later this small settlement would be a thriving city with a new name, an international bridge, and a large hydroelectric dam just upriver). As we had been told, there were several rafts or barges, of varying sizes, to ferry goods, people, and small vehicles back and forth across the river. There was considerable traffic here between the two countries. Obviously, we would need the largest barge we could find for our 4½-ton camping van. But first, it was necessary for us to go across the river, without the Jeep, and talk to the custom authorities on the other side. We parked the "Roadrunner" on the riverbank and climbed aboard one of the ferries, along with other passengers. It was a slow 10-15 minute ride, powered by a diesel engine. The rather fast-flowing, moderately shallow river was approximately 300-400 yards across.

Providence was with us. On board the ferry with us was a 30-year-old, slender, moderately tall Chief Petty Officer in the Brazilian Navy who was stationed at a small naval base a short distance down the river. He walked up to us and introduced himself, in perfect English, as Jose Pimentel and said "Where are you guys from?" When we told him we were from the U.S.A., he said that he had once spent a year's duty in New Orleans. We told him what we were doing and that we were having difficulty obtaining permission from the government to bring our Jeep into Brazil. He said that maybe he could help, as he knew the junior Customs official at the entry post on the other side of the river. When we arrived at the landing and disembarked, Jose walked with us to the customs post and introduced us to his friend (whom I will call "Paulo"). As luck would have it, he was the only person working at the post that weekend, as the senior officer in command had gone to Rio de Janeiro, leaving him in charge. Jose explained to Paulo what we were doing and said that we only wanted to bring our camping van into the country.

Paulo turned his attention to us and said, in fairly good English, "Do you have permission?" We tried a bluff, confidently said "Yes", and showed him our carnet de passage. He looked at it for only a couple of minutes and then said to us, correctly, "This does not give you permission." (Of course every page said in bold red letters "Not valid in Brazil.") So Noble pulled from his briefcase one of his letters from the Brazilian Embassy in Washington, D.C, which he had received prior to our departure from California. Paulo read the letter carefully and then remarked once again, "This does not give you permission." As before, he was correct. If ever there was a time to offer a monetary bribe this was it, but Noble and I had agreed, at the start of our adventure, that we would never do this, and we didn't. We had no more bluffs. Paulo studied us intently for several minutes without saying anything, apparently giving careful thought to our situation. It appeared to me that he wanted to help us if he could, without getting himself into trouble from his superiors. Finally, he surprised us by saying, "Let me think about this; come back this afternoon and I will give you an answer." He had not said "No" but neither had he said "Yes." There was nothing more Noble and I could do, so we thanked Jose for his assistance, said goodbye to Paulo, and went back across the river to our van, to wait for Paulo's decision that afternoon. Jose returned to his naval base.

Early that afternoon, Noble and I ferried across the river again and walked into the Customs post, where Paulo was waiting for us. As we walked in, he looked up at us and immediately said, "I have thought it over and it will be OK." We couldn't believe our ears! It was difficult to hold back tears of joy. Without any hesitation, Paulo put a stamp in our passports and carnet de passage, and wished us well. We thanked him emphatically and gave him one of the postcards that pictured the two of us standing together next to our Jeep (taken just before we left California). We caught the next ferry back across the river to our waiting van, which we were anxious to take into Brazil as quickly as we could, before Paulo changed his mind or his superior returned.

Plate 31: We ferried across the Pilcomaya River from Argentina into Paraguay. Lower income women frequently smoked cigars! Paraguayan gauchos (cowboys) were much like their Argentine counterparts.

We located the largest barge we could find and told the captain we wanted to cross the river. He motioned for us to drive on board, and then he positioned our Jeep in the middle of the barge. He started his motor and slowly began edging out into the river. Almost immediately disaster struck, as it had in so many other similar situations. The barge could not support the 4½-ton weight of our van and it very slowly began to sink down into the river. The alarmed captain reversed the engine, which was slow in responding. We were saved from submerging by a group of military recruits who came jogging along the beach just at that time. Seeing our impending doom they came to our rescue by fastening some ropes to our barge and pulling it safely back to shore before we sank! Our Jeep had survived yet another brush with death. However, we had not yet entered Brazil.

Noble and I were extremely disheartened. All of our joy at having obtained permission to take our Jeep into Brazil vanished. Our adventure was a roller coaster ride. Then, at the height of our despair, who should appear once again, out of nowhere, but our guardian angel, Jose Pimentel. He came up to us and said in his usual good-natured manner, "How's it going today?" So we told him both the very good news, that we had received permission from his friend to bring our Jeep into Brazil, and the very bad news, that it was too heavy to be transported across the river by barge. In his customary calm way, Jose smiled and remarked that he could solve that problem for us. He would simply go back down river to his naval base, commandeer an old Brazilian WWII landing craft, and bring it back upriver tomorrow morning to transport us across the river. No problem. Our despair turned to hope. It was like the wheel of fortune. First you win and then you lose, then you win again. What would we have done without Jose?

As promised, Jose returned the next morning (Sunday, Aug. 16), chugging slowly up the river in a decrepit little landing craft. After he arrived with the boat, Noble could foresee no more difficulties in getting across the river so he said to me that if I preferred, I could go on ahead of him, on a somewhat faster ferry, and have a short time to search for birds on the other side before he got there with Jose. So I did. By now, Noble should have known better.

He drove the Jeep on board, and to everyone's surprise it was discovered that our van was slightly too long to fit completely, lengthwise, on the small craft. The Jeep was just a few inches too long to allow the front landing ramp to be raised completely and fastened securely in a normally watertight manner. Jose decided the problem was not that serious since there were four motor-driven water pumps installed on board to remove what little water would leak inside. So the low-powered craft set off across the river, with water slowly leaking in around the edges of the unsecured landing ramp. The pumps were turned on, at which time it was discovered that only one of the four was working! Oh dear! Jose told Noble not to worry, that the craft would not sink in the 15 minutes it would take to cross the river. However, as usual, things got worse. Not only did the boat begin to very slowly fill with water, making it more difficult to steer the unwieldy craft as the water sloshed from side to side, but the underpowered engine could not propel the boat, with its very heavy load, in a straight line directly across the fast flowing river. The current slowly pushed the boat downstream, toward a sandy beach on the opposite riverbank that was well downriver from the firm, gravel, well-used ferryboat landing. Noble told me later that he had grave concern about whether the craft was actually going to make it all the way across the river. But it did - - barely - - without sinking.

Instead of jubilation at having successfully made it, Noble experienced great consternation because it was necessary for him to back the Jeep off the landing craft onto a wide beach of very soft, dry sand. There was no alternative. So he backed off (having driven forward onto the craft), and all four wheels immediately sank in the soft sand, all the way down to their axles. Four-wheel drive in our lowest gear ratio did not cause the Jeep to budge at all. The front of the Jeep was pointing toward the river, where there was nothing to which our winch could be attached. The van was very thoroughly bogged and our winch could not be used to pull it out. Would Noble's problems ever end? Once again, Jose came to our rescue. In a voice without panic, he told Noble not to worry, that he could again solve the problem. Saying to Noble that he should wait there, he set off up the beach and soon disappeared into the trees. Noble waited patiently, not knowing what to expect. In about an hour or so, Jose returned - - with a team of eight mules!

The mules were hitched to the back of our van, where they tugged and tugged, and, sure enough the Jeep began slowly to move backward out of the soft sand, up to firm ground at the top of the beach, 50 ft. away. Noble's problems, however, were not quite over yet. He discovered that sand had penetrated the wheels, interfering with brake function. Before he could drive anywhere he had to take off all the wheels, one at a time, and dust out the sand. It was a time-consuming task, but Jose helped him. Finally, Noble was able to drive the van along the river shore, up to the ferry landing where I had been waiting patiently several hours for his arrival, unaware of the cause for the delay. It was now about mid-day. For Noble, the morning's activities were several more pages to write in a book one day. (He might tell it a little differently from my own remembrance of what he told me.) We celebrated with a glass of wine. We had entered Brazil without a written government permit, with the help of a Brazilian naval chief petty officer and a junior Brazilian border officer, both of whom were sympathetic with our adventure. We would not have succeeded without their assistance (or that of the military recruits who helped retrieve our slowly sinking ferry barge before it settled on the bottom of the river). As it turned out, we would be able to continue on our road to Rio, and catch our ship for Africa. Again, we celebrated with a glass of wine.

Early that afternoon we drove south in Brazil a few short miles downstream from our ferry landing to the Brazilian town of Foz do Iguaçu (Iguaçu Falls), which was located at the junction of the Iguaçu River (coming from the east between Brazil and Argentina) and the Paraná River (coming from the north between Brazil and Paraguay). The Falls were located 12 miles upstream on the Iguaçu River, which was one of Brazil's principal rivers, with its source 800 miles to the east in the Serra do Mar range of mountains not far from the Atlantic Ocean. The Iguaçu River flowed westward across the Brazilian Plateau. Along its lower stretches, including the falls themselves, the river formed the border between Brazil (to the north) and Argentina (to the south). Both of these countries have tourist facilities for visitors to view the falls, each from their own side of the river. A person could travel by road back and forth between Argentina and Brazil across the river on a bridge below the falls, and thus easily see the falls from both sides. The falls are near a point where the three countries of Brazil, Argentina, and Paraguay all came together, and tours to the falls could be arranged from any of these three countries.

Iguaçu Falls are considered one of the three "Great Waterfalls of the World", along with Niagara Falls in North America and Victoria Falls in Africa, and they attract visitors from many different countries. They are a UNESCO World Heritage Site and are one of the "New Seven Wonders of Nature." The surrounding forests were preserved in national parks, both in Brazil and in Argentina. Noble and I spent all or part of three days in this magnificent area, viewing the falls, taking photographs, hiking, birding, and just relaxing. Of course I spent most of my time looking for birds, and one of my most vivid and pleasant memories is of watching Great Dusky Swifts* swooping in and out of the mists rising above the falls. Rainbows created in the mists were gorgeous. We drove back and forth from one side of the falls to the other, between the towns of Foz do Iguaçu in Brazil and Puerto Iguazú in Argentina. The falls extended for a width of 1¾ miles from one side to the other, in a long arc that was broken into many separate falls of varying heights and widths. The maximum height of unbroken water was 270 ft., at the "Devil's Throat" on the main channel of the river (forming the international boundary between Brazil and Argentina).

I documented 90 species of birds in the vicinity of the falls (Aug. 16-18). Heading the family lists in number of species were flycatchers (12), tanagers (9), furnariids (8), and woodpeckers (7). Dusky-legged Guans* flew noisily from treetop to treetop, and a Black Hawk-Eagle soared high above the forest, directing my attention to it by its far-carrying "eeowww" vocalization. Surucua* and Blue-crowned trogons called regularly from inside the forests, the latter with a long, rapid, mellow trill which I laboriously tracked down through the forest (and recorded in my notes that it sounded like an Eastern Screech-Owl). Swallow-Tanagers*, an unusual member of the family Thraupidae, perched conspicuously on high bare branches of trees, from which they sallied out after flying insects. They also fed on fruits. Males were conspicuous with their blue-and-white coloration and black faces. Unlike other members of the family, Swallow-Tanagers nested in holes in a bank (or even man-made structures) and gathered in flocks outside the nesting season.

I wrote the following description of an unknown woodpecker: "woodpecker: med-large; head with a medium long crest from forehead to nape; head mostly red with a dull reddish-brown face; bill pale; a short wide white stripe on each side of the neck; upperparts blackish with white lower back and rump; underparts from chin to undertail coverts rather narrowly barred dark and light." In an effort to identify this bird I referred to the only "field guide" I had, Orog's *Las Aves Argentinas*. As inadequate as this publication was (by today's standards) I concluded, to my satisfaction, that this sighting was of *Dryocopus galeatus* (Helmeted Woodpecker*), a species occurring in Argentina only in Misiones Province. The wide white shoulder stripe and the brownish face seemed conclusive to me. There was no mention in Olrog of the abundance or overall range of this species (and it provided me with only a Latin or Spanish name, not an English name). It was many years later that I read in both Short (*Woodpeckers of the World,* 1982) and Sick (*Birds in Brazil,* 1993) that this was a rare species, with few records and a small geographic range. Sick wrote (on page 45) that " The Helmeted Woodpecker…is a vanishing species." Modern illustrations of the Helmeted Woodpecker in these two publications (and in *A Filed Guide to the Birds of Brazil* by Ber van Perlo, 2009) confirmed my conclusion that the bird I saw was indeed this species. Sick commented that Short told him in a personal communication, in 1989, that vocalizations of this woodpecker had recently been recorded by Ted Parker, and that information on its nesting had been obtained at Iguaçu!

It was almost 1,000 miles from the falls to Rio, across the southern Brazilian tableland and then northward along the Atlantic coast. Noble and I would see only a very small part of this immense country, the largest in South America and the 5th largest in the world. Our journey started out on Wednesday (Aug. 19) from the falls at only 500 ft. above sea level, but we soon climbed up to the tableland, at elevations between 2,000 and 4,000 feet. We would be in the state of Paraná for three full days (Aug. 19-21), as we traveled east across this plateau region all the way to Curitiba, 400 miles away, following a dirt or gravel road through the towns or little municipalities of Cascavel, Guaraniaçu, Laranjeiras, Guarapuava, Guará, Irati, and Palmeira, in that order.

The countryside was vegetated primarily with a mixture of grasslands, tropical seasonal forests, and remnant Araucaria woodlands, just south of the Brazilian "cerrado." It was a rather poor area agriculturally, sparsely populated, and utilized mostly for raising cattle and goats, or for farming. Its signature feature was the tall, stately, distinctly-shaped Araucaria tree (*Araucaria angustifolia*), known locally as the "Brazilian Pine", "Paraná Pine", or "Candelabra Tree." (It is not closely related to "true" pines, in the genus *Pinus*.) These giant coniferous trees grew with a tall straight trunk up to 250 ft. tall, with their long, dense needles covering upper branches in a wide-spreading whorl. They were once the dominant tree on the plateau. Araucarias are a very ancient group of conifers, dating all the way back to the early Mesozoic Era, some 200 million years ago. (They may have comprised the principal food of herbivorous sauropod dinosaurs in Jurassic times.) There were 19 living species in the genus, occurring in a "Gondwanaland" distribution in the world: southern South America (with just 2 species), Australia, New Guinea, Norfolk Island, and New Caledonia (with 13 endemic species). There was one Brazilian bird, the Araucaria Tit-Spinetail (*Leptasthenura setaria*), which was confined exclusively to Araucaria trees.

Barefooted children walked along our road to Curitiba, sometimes following several goats, which they kept on rope leashes. A conspicuous roadside bird was the White-eared Puffbird*, rather kingfisher-like in its appearance with a big orange bill, barred rufous upperparts, white underparts, and distinctly patterned black-and-white head. It sat in the open on trees or telephone wires. Other noticeable birds along the way, in the order in which I first saw them, were Plush-crested Jays, Guira Cuckoos, Blue-and-white Swallows, a Picazuro Pigeon, Burrowing Owls, a White-tailed Kite, S. Rough-winged Swallows, an Aplomado Falcon, Smooth-billed Anis, Roadside Hawks, White-collared Swifts, Great Kiskadees, Yellow-rumped Marshbirds, Great Pampa Finches, American Kestrels, Campo Flickers, a Ruddy Gournd-Dove, Yellow-headed Caracaras, and ever present Black Vultures. Turkey vultures reappeared, having been inexplicably absent throughout Argentina and Paraguay. On several occasions, one or two beautiful Diademed Tanagers* flew across the road, looking very similar to the Blue Grosbeak in shape, size, and color, except for their rather inconspicuous white crown (narrowly red on

Plate 32: Our new dirt road eastward through the forest in Paraguay was barricaded when it rained. Iguaçu Falls were one of the three "Great Waterfalls" of the world, a UNESCO world heritage site. Children on the southern Brazilian Tableland were often impoverished and barefoot.

the front) and the small black area around the base of their bill. Twice when we stopped at a grove of roadside Araucaria trees there was a pair of Araucaria Tit-Spinetails* foraging like chickadees in the tall canopies, where they remained mostly out of my sight in the dense foliage. We also made several stops at patches of woodland next to the road, where I added the Blue-naped Chlorophonia*, White-throated Spadebill* (recently split from the more northerly Stub-tailed Spadebill), and Red-rumped Warbling-Finch* to my trip list.

We reached Curitiba late Friday afternoon (Aug. 21). Here we were happy to discover that there was now, at last, a good highway to follow all the way to Rio, 600 miles away. We camped that night at 3,000 ft., just northeast of the city. The next morning (Aug. 22), at daybreak, we headed north to the town of Rabeira, on the border between the states of Paraná and Sao Paulo. The scheduled date of our departure for Cape Town was one week away. Our scenic highway wound up and down between 2,500 ft. and 3,000 ft., through green hills paralleling the Atlantic coast along the eastern edge of the Brazilian Plateau. It was a very pleasant drive during which I added the Chestnut-headed Tanager* to my bird list. We camped that night about 90 miles west of Sao Paulo, in a forest at 1,500 ft. elevation. Pauraques and a Long-tufted Screech-Owl* called throughout the night. As usual, I could not locate the owl with my flashlight.

There were two hours available for me to look for birds before we got underway at 9:00am the next morning (Sun., August 23). New birds for my Brazilian list were Shiny Cowbird, Chalk-browed Mockingbird, Bran-colored Flycatcher, Hangnest Tody-Tyrant* (a plainly colored little flycatcher endemic to SE Brazil), Chicli Spinetail*, Rufous-winged Antshrike*, Cinnamon Tanager*, Swallow-tailed Hummingbird*, Blackish-blue Seedeater*, Grassland Sparrow, Blue-winged Parrotlet, Planalto Hermit*, Pale-breasted Thrush, Olive-green Tanager*, Masked Yellowthroat, and Striped Cuckoo.

Noble and I reached São Paulo that afternoon. The city was situated among hills at an elevation of 2,800 ft., inland about 35 miles from the Atlantic Ocean. With a population somewhere between six and seven million people it was the largest city in South America -- and one of the five largest cities in the world. It was Brazil's leading industrial and manufacturing center. Since most of the businesses were closed on Sunday, we continued on through the city and drove 40 miles southward down through the hills to the port city of Santos, on the Atlantic Ocean. Here, along the beach, 2 Magnificent Frigatebirds glided effortlessly past us, and 2 Royal Terns stood with 30 Sandwich (Cayenne) Terns on a sand spit. As the sun went down below the mountains behind us we chose a campsite on the beach where we could watch the moon come up above the bay, over a glass of Argentine wine.

We spent all of Monday in São Paulo at the enormous Willys agency. Here we had our Jeep maintained and serviced while Noble talked at length with the service manager and inspected their large assembly plant. The next morning (Tuesday, Aug. 25), we departed for the very last leg of our "road to Rio." The service manager at the Willys agency admired the adventure that Noble and I were undertaking and he did not charge us any money for his service or parts.

As we left São Paulo our highway dropped in elevation and descended down to the Paraíba River valley, which we then followed northeastward (downstream) toward Rio. The coastal Serra do Mar range was on our right and the Serra da Mantiqueira range of mountains was on our left. Rio was only 275 miles ahead. We arrived late that afternoon. Since our departure from California in January, Noble and I had been on the road exactly 7 months and had driven 20,000 miles. We had successfully traversed Central and South America over unimaginably difficult roads, some of which we had been told were impassable (and they almost were). Although we had suffered numerous vehicle breakdowns and spiritually disheartening delays, we had never given up. Noble and I were both determined to succeed in pursuit of our adventure, and we had completed touring the first two of our planned six continents. That night we once again celebrated with a bottle of wine.

Rio de Janeiro was the capital city of Brazil in 1959, but the government shifted to the newly built city of Brasilia, further inland, the very next year. Rio, with its many tall-forested hills, gleaming white beaches, green parks and gardens, lively nightlife, and colorful festivals, was one of the most popular destinations for tourists from all over the world. Its two most famous landmarks were: (1) Sugarloaf Mountain, a huge, vertical rock monolith rising abruptly to 1,299 ft. above sea level at the end of a peninsula jutting out into Guanabara Bay; and (2) the 125 ft.-tall statue of "Christ the Redeemer", overlooking the city from the top of Corcovado Mountain. This 700-ton statue was built between 1922 and 1931, of reinforced concrete and soapstone, with funds donated by Brazilian Catholics. It depicts Christ with outspread open arms, symbolizing peace, and is surrounded below by the Tijuca Forest. Noble and I drove to the top of the mountain one day to photograph the statue and marvel at the spectacular view of the city below, extending all the way to the ocean.

Rio, with four to five million people, was somewhat smaller than São Paulo but it was more densely crowded. It was a chief seaport and served importantly for finance, trade, and transportation. It was a city of great contrasts, and although there were many elegant hotels lining Copacabana Beach, there were also large slum districts on the steep hillsides, and along the swampy shores of the bay. Streetcars provided cheap, easy public transportation. Noble and I met many new friends and were hosted by them for a dinner or two. One of the persons with whom I became acquainted was a U.S. Navy petty officer who was preparing to return to the United States after a tour of duty in Rio. He was selling many of his personal belongings to save on the cost of taking them back with him. He asked me if I would like to purchase a rather old double-barreled 12-gauge shotgun, for $15. I really had no use for such a firearm, but I couldn't resist the bargain price, so I did. (This shotgun would feature in two future events in my forthcoming travels in Asia.)

On Wednesday morning (Aug. 26, our first full day in Rio) Noble and I walked together into the office of the Royal Interocean Lines, with whom Noble had several months earlier (in Lima) booked passage for us (and our "Roadrunner") from Rio to Cape Town on the Dutch ship "Boissevain." He had received confirmation of this booking in Buenos Aires, saying that the departure date would be August 29th. We were met in the office by a very pleasant young Dutchman by the name of Derek, who spoke perfect English (as did most Dutch). He was particularly friendly and very much interested in our adventure. Noble inquired how much the cost was going to be to freight our Jeep across the Atlantic. Derek asked us how much it WEIGHED. Oh goodness! We told him 4,000kg (approximately 9,000 lbs, or 4½ tons). So he immediately went to a table for freight costs, all of which were calculated entirely by the weight of the goods being shipped. After checking this table, Derek looked up at us and said the cost would be approximately US $800 (rounded out after conversion).

We were very dismayed. That was a sum of money we could scarcely afford. We pleaded with Derek, explaining our situation to him, and asked if there was any other alternative. He genuinely wanted to help us if he could, so he gave the matter careful thought. Suddenly he asked us if we were going to be traveling as passengers on the ship, along with our Jeep (as we had not mentioned this fact to him). We said yes we were. That gave him an idea. He said that passengers were allowed to take with them a certain maximum amount of luggage at no extra cost to their passenger ticket, but if their luggage exceeded this amount there was an "excess baggage fee." This fee was based entirely on VOLUME of the excess luggage. He asked us for the outside dimensions of our Jeep, which we knew exactly, and told him 15'x 8'x 7½'. These numbers were converted to meters and the volume of our Jeep was then calculated, in cubic meters. Derek then retrieved a different table for shipping costs, one for excess baggage. After inspecting this, he said that to take our Jeep with us as "excess baggage" the cost would be US $300 (again rounded out after conversion). He told us that would be fine with him! This was less than 40% of the cost as calculated by weight. Our elation could not have been greater. A sympathetic young Dutch shipping agent had come to our rescue. We thanked him sincerely, and of course gave him one of our postcards. He also informed us that the date of our departure had been delayed a few days, from Aug. 29 to Sept. 2. That was OK with us. It gave us some time for relaxation and leisure. Not surprisingly, I opted for some birdwatching, along with getting a badly needed haircut (but not a beard reduction). Here in Brazil we no longer heard the whisper of "Fidel Castro" as we walked by people on the street.

Itatiaia National Park was situated about 125 miles by road from Rio, via the town of Resende on the highway to São Paulo. This park was the oldest national park in Brazil, having been established in 1937. It was in the

Mantiqueira Range of mountains, on the border between the states of Rio de Janeiro and Minas Gerais. The park covered almost 700 square miles and was a ruggedly beautiful region of lush forests, lakes, rivers, waterfalls, and alpine meadows, varying in elevation from 1,000 ft. at the entrance to 9,500 ft. at the top of Pico Agulhas Negras (Black Needles), the third highest mountain in all of Brazil. A good dirt road went from the entrance to the visitor's center at 3,500 ft., which was the altitudinal division in the park between the lower elevation Atlantic Tropical Rainforest and the higher elevation Montane Seasonal Forest. A separate road, some distance away, took adventurous visitors all the way to the top of Pico Agulhas Negras. In the vicinity of the visitor's center were many hiking trails and several accommodations for staying overnight. Waterfalls were one of the most popular attractions. Naturalists and birdwatchers came to the park from countries around the world, and over 300 species of birds had been documented from within its boundaries.

I drove to the park in our Jeep on Thursday (Aug. 27), arriving at 6:00 am for a full day of birding. It was partly cloudy to overcast, but without any rain. The temperature was mild, but at higher elevations in the afternoon the weather was cool and occasionally misty. Orchids and Fuchsias added color to the forests. Bird songs came from all direction. It was a delightful, awe-inspiring environment. I hardly knew where to begin. The woodland edge, as always, was a particularly productive area in which to see birds. I spent the morning in the lower park and much of the afternoon in the higher park, which necessitated quite a bit of driving (thus subtracting from my time for birding). I drove all the way up to 8,000 feet, on the road to Pico Agulhas Negras.

Flycatchers and tanagers were the two families with the greatest number of species, a dozen in each. Tanagers came in all colors of the spectrum. Particularly attractive were the jewel-like, multi-colored Green-headed Tanagers, in their gorgeous aqua-blue, green, black, lime, and orange colors. Red-crowned Ant-Tanagers were suffused red all over, Diademed Tanagers were uniformly vivid blue, Chestnut-headed Tanagers were soft gray with a bright orange-chestnut hood, Ruby-crowned Tanagers (males) were glossy black with a concealed ruby-red crown patch, Brassy-breasted Tanagers* (one of the many *Tangara* tanagers) were lovely green with an orange breast and black-streaked upperparts, Golden-chevroned Tanagers* were purplish-blue with olive wings and a conspicuous yellow shoulder patch, and Magpie Tanagers were handsomely patterned black and white, remarkably like a magpie in color as well as in size and shape.

Three kinds of foliage-gleaners, each in a different genus, foraged in all the vertical levels of the forest, probing in palm fronds and epiphytes. These were the Buff-fronted (*Philydor*), Buff-browed (*Syndactyla*), and White-browed* (*Anabacerthia*) foliage-gleaners. Two kinds of woodcreepers, the small Olivaceous Woodcreeper and the much larger White-throated Woodcreeper, both searched forest tree trunks and branches for beetles and other arthropods. Joining these passerine trunk foragers were five kinds of woodpeckers: Yellow-fronted, Green-barred, White-spotted, Yellow-browed, and ANOTHER HELMETED WOODPECKER. This last species was described very briefly in my field notes as: "woodpecker: large; pileated type; large crest, top of head red; face reddish-brown; looked like *Dryocopus galeatus*." (I had recorded the name from Olrog's field guide when I identified this species at Iguaçu Falls.) The face color was once again a defining feature. Maybe in 1959 this species was not as rare as it was subsequently said to be in Short's and Sick's later publications (see my earlier comments).

Sooty* and Gray-rumped* swifts swooped and circled above the forest. Yellow-legged Thrushes*, black with a contrasting bright yellow bill and legs, frequented the middle and upper layers of the forest, where they fed in fruiting trees, as did Plumbeous Pigeons*, Golden-winged Caciques, and Red-rumped Caciques. In the forest understory were stunning Blue Manakins, conspicuous gems in their vivid blue and black coloration, with a bright red crown. Sadly, they were not displaying at this time of year. Much more difficult to locate in the understory, and to identify, was a plain, somberly colored Serra Tyrant-Manakin* (included with the Wied's Tyrant-Manakin in 1959). It required much scrutiny and a long description before I could confidently assign a name to this sighting. (It may not be a "manakin" at all.) One of my favorite birds of the day was the Bay-chested Warbling-Finch*, with its immaculate gray upperparts, white underparts, and wide chestnut-brown breast band and flanks, somewhat resembling a Bay-breasted Warbler.

Brown Tanagers* in the middle and upper levels of the forest were unusual, for a tanager, in their overall tawny and cinnamon coloration, with a rather heavy, short, black bill. The sexes were alike. Green-winged* and Thick-billed* saltators were remarkably similar in appearance and color, but separated by slight differences in throat color and bill size. The distinctive Black-throated Grosbeak* was entirely black in color with a bright orange-red, heavy, grosbeak-like bill. However, it was really a "saltator" (in the genus *Saltator*, family Cardinalidae). The Brazilian Ruby* was one of several hummingbirds, easily identified by its shiny green foreparts, bronzy-rufous lower back and tail, and (in the right sunlight) brilliant ruby-red throat. Males of this species were once killed by the thousands (in the late 1800's and early 1900's) for use in making "feather-flowers" which were widely produced by nuns at convents and then sold to distributors for use in women's millinery. The handsome, wispy-crested Plovercrest was frequent at flowering shrubs along the forest edge. Among the many tyrannids (flycatchers) were the Long-tailed Tyrant, Shear-tailed Gray Tyrant*, Mottle-cheeked Tyrannulet, Velvety* and Blue-billed* black-tyrants, Gray-hooded Flycatcher, Yellow Tyrannulet, and two Chestnut-crowned Becards* (genus *Pachyramphus*, once placed with the cotingas).

In the afternoon I drove up the poorly maintained road that ascended to the top of Pico Agulhas Negras, but I did not go quite all the way to the top. I stopped in an area of grass and shrubs at about 8,000 ft. elevation. It was damp and cool. Here I observed a bird foraging in the low vegetation which I observed carefully for several minutes and then wrote the following description: "spinetail: medium; tail rel. long, graduated, & feathers narrow & pointed terminally; rather plain; generally rather brownish, somewhat paler below; small throat patch yellow-buffish." I sketched the slender pointed bill. Eventually I was able to identify this sighting as an Itatiaia Thistletail*, a species with a very restricted range in the Mantiqueira mountains of southeastern Brazil, in shrubby grasslands at higher elevations. (I read much later, in Ridgely, 1994, that it could easily be found along the upper Agulhas Negras road in Itatiaia Nat. Park!) The Itatiaia Thistletail (formerly called Itatiaia Spinetail) was probably the most iconic bird of the park, and I was happy to have found it, without knowing this fact at the time. It was a perfect way to finish my day of birding, during which I documented 65 species in the park (most of which were not identified until a later date). As always, a great many birds were seen or heard that were never identified. My day in Itatiaia National Park came to an end much too soon, and I vowed to return one day. As always, it had been fun and challenging.

Back in Rio the days passed rapidly. Noble and I spent a few hours on Copacabana beach one afternoon, and on another occasion (Aug. 28th, my birthday) we drove along the shore of Guanabara Bay. Here I marveled at a Magnificent Frigatebird, which conjured up the image of an ancient pterodactyl in my mind. This species was widely present along the entire tropical coasts of both the Pacific and Atlantic oceans, breeding as far north as Baja California, central Texas, and Florida. Also present along the bay were several Brown Boobies, South American and Sandwich terns, and a sub-adult wintering jaeger from the Arctic (probably a Parasitic). On two different days I visited the Zoo and the adjoining National Museum, where I studied living and mounted birds, and scientific specimens, in an effort to identify some of the birds described in my field notebooks. I was moderately successful. I also browsed in the extensive scientific library for further help.

On Sunday, Aug. 30, Noble and I drove 40 miles north of Rio to the city of Petropolis, where we had been invited to a luncheon and afternoon party with some American college students who were studying in Rio. We arrived early in the morning and I had a couple of hours to find birds before festivities began. Serra dos Orgãos National Park was nearby, a magnificent scenic area encompassing 25,000 acres of peaks, fantastically shaped rocks, and steep hillsides covered with scrub, grasslands, and dense Atlantic Tropical Rainforest. Lianas and climbing cacti were prevalent, as well as colorful yellow Cassias and magenta-flowered *Tibouchina* trees (a melastome). Many hiking trails were available. Because of time limitations I had to devote all my birding to lower elevations in non-forested areas, such as parks, gardens, and woodland edge. In these habitats the most active and conspicuous birds were hummingbirds and tanagers. I identified Amethyst Woodstars*, White-throated Hummingbirds*, Swallow-tailed Hummingbirds, and Glittering-bellied Hummingbirds. Tanagers present were

Sayaca Tanagers, Cinnamon Tanagers (very handsome in their cinnamon colored foreparts, black mask, blue-gray back, whitish lower underparts, and cinnamon vent), Golden-chevroned Tanagers, Burnished-buff Tanagers (a *Tangara*), and two plainly colored, ochre-brown, female Ruby-crowned Tanagers. Other species were the ubiquitous House Wren and Rufous-collared Sparrow, along with Great Kiskadees, Rufous Horneros, Blue-and-white Swallows, two Bran-colored Flycatchers, a Chicli Spinetail (*Synallaxis*), and a Rufous-browed Peppershrike. I could easily have spent several days in this lovely reserve. It was one more place for me to return.

On Tuesday, Sept. 1, Noble drove our Jeep to the waterfront to have it loaded on board the "Boissevain" for our departure the next day. I took a streetcar to the Parque da Cidade (City Park) on the outskirts of the city, for one last morning of birdwatching. I arrived at 8:30 am and spent 4 hours in this delightful park with its gardens, ponds, trees, fast-flowing stream, tiny waterfalls, and forested hillsides. There were many footpaths and all areas were easily accessible. It was a pleasantly mild and sunny morning and a most relaxing way to enjoy my last full day in South America.

I documented 36 species in the park, 10 of them that I had not previously seen. Brazilian Tanagers*, a species endemic to the SE coastal region of Brazil, were the most brightly colored of all the birds I encountered. Males, with their brilliant scarlet color, black wings and tail, and shiny silver bill, frequented the forest edge and interior, particularly along the stream. Females were more somberly plumaged. Another species endemic to this coastal region was the handsome Gray-hooded Attila*, a large flycatcher of the middle forest, with a lengthened bill. It was more readily heard than seen, but two separate individuals were tracked down and easily identified by the sharp contrast between their all-gray head and bright ochraceous color everywhere else. Salvadori's Antwrens*, yet another endemic to SE Brazil, were diminutive, active, wren-like formicariids of the woodland edge and forest interior. Still another formicariid restricted to this region was the strikingly patterned Scaled Antbird*, which called attention to itself by a frequently sounded double note (perhaps used in pair contact). A female obligingly allowed me to obtain a good description of her distinctive buff-spotted, dark-brown upperparts, prominently banded tail, and streaked black-and-white underparts.

One of my most satisfying finds of the morning was a Sharp-tailed Streamcreeper*, a brown, short-tailed, waterthrush-like furnariid with bold white spotting below. It was hard to observe as it foraged secretively in the streamside undergrowth. Throughout most of its wide, disjunct distribution in South America it was a difficult species to find, but apparently it was encountered more frequently here in southeastern Brazil than elsewhere. I was happy to observe it along the stream in Parque da Cidade.

Three male hummingbirds in the park were particularly attractive. In order of increasing size these were the Violet-capped Woodnymph* (4" long), Black Jacobin (5" long), and Swallow-tailed Hummingbird (6⅓" long). The diminutive woodnymph was bright emerald-green except for a beautiful violet-blue crown. The Swallow-tailed Hummingbird was shiny dark-green with a purplish-blue hood and a long, forked, blue tail, almost twice as long as the body. It was one of the easiest hummingbirds to recognize, and was said to be one of the most bellicose. The Black Jacobin* was entirely black except for its striking white tail (all but the middle pair of feathers, which were black). The tail was flashed open and shut in visual displays, creating a breath-taking spectacle. It became another of my favorites. My morning in the park was a perfect ending to my seven months of neotropical birding. A complete list of all the 1,101 avian species that I documented in Central and South America is provided in Appendix A.

The next day (Wednesday, September 2, 1959) Noble and I departed for Africa. I looked forward to pursuing birds with illustrated books and field guides!

Plate 33: A new, scenic paved highway meandered through the Serra do Mar Mountains between Sao Paulo and Rio de Janeiro. The seacoast city of Santos served as the port for Sao Paulo. Our Jeep was hoisted aboard the "Boissevain" for a ten day voyage from South America to Africa.

Chapter Twelve – South Africa

Noble and I arrived with our Jeep in Cape Town on September 11, 1959, after a 10 day voyage from Rio de Janeiro across the South Atlantic Ocean on the "Boissevain." It was a cool (55° F), overcast, windy, rainy day. We were at the southern end of the African continent, on the northwest shore of the Cape of Good Hope. As we approached Cape Town from 15 miles to the west there were numerous seabirds. I counted the following species: 4 Wandering Albatrosses, 18 Shy Albatrosses, 14 Black-browed Albatrosses, 5 Giant Petrels, 80 White-chinned Petrels, 2 Sooty Shearwaters, 35 Cape Petrels, 25 Cape Gannets*, 1 Bank Cormorant*, 4 Brown Skuas, 75 Kelp Gulls, 50 Hartlaub's Gulls*, and 30 unidentified terns (probably Common). As we approached Cape Town I photographed the renowned Table Mountain just south of the city, a flat topped granite mountain two miles wide and 3,500 feet high on the upper end of Cape of Good Hope. We were viewing a new continent. It was an exciting moment.

Africa was a wildlife paradise. Nowhere else on earth was there such a diversity of large mammals. Foremost among these were the "ungulates", or hoofed mammals, inhabitants primarily of grassland savannas where they often occurred spectacularly in very large groups. These included more than 80 different kinds, among them such popular and well known animals as wildebeests, gazelles, zebras, antelopes, impalas, buffalos, reedbucks, kudus, rhinoceroses, giraffes, hippopotamuses, waterbucks, hartebeests, duikers, warthogs, elands, gerenuks, bushbucks, dik-diks, and nyalas. Other African mammals were elephants (the world's largest land animal), primates (gorillas, chimpanzees, baboons, monkeys, and others), carnivores (lions, leopards, cheetahs, caracals, servals, hyenas, jackals, hunting dogs, foxes, mongooses, otters, and honey badgers), pangolins, aardvarks, hares, squirrels, hedgehogs, elephant-shrews, hyraxes, and bats. All of these could best be seen in the game reserves and national parks scattered around the continent. Sadly, many species were declining, endangered, or on the verge of extinction in the wild.

I had only one major reference with me to help identify the mammals we saw: *The Mammals of South Africa* by Austin Roberts, second ed., 1954. This monstrous, technical, scientific document contained 700 pages and weighed 6½ pounds, but it had color plates of the mammals. I had transported this book with me all the way from California, and in it I checked off all the mammals I identified as we saw them. When we reached East Africa I purchased a small booklet, *Game Animals of Eastern Africa*, by C. A. W. Guggisberg, 7th ed., published in 1959, which had black and white sketches of the better known game animals, including a variety of species which were not included in Roberts. At various locations along the way I was also able to acquire checklists of the local mammals.

Birdlife in Africa, though less diverse than in South America, was almost as spectacular, thrilling, and challenging to pursue. Best of all, I now had some published references with color illustrations to assist me in field recognition, and I studied these intently almost every evening in our van. My primary reference for southern Africa was *Birds of South Africa* (which I brought from California), first published in 1940 by Austin Roberts (the same author as for mammals) but revised in 1956 by McLachlan & Liversidge. I also had with me a little guide called *A First Guide to South African Birds*, (first published in 1936 and last revised in 1956), by E. L. Gill. It contained plates with colored illustrations of most of the species.

Our travel routine in Africa was much the same as it was during our seven months in Latin America. The driving was primarily my task and Noble took care of vehicle maintenance and repairs (as well as handling almost all financial transactions). It was a system that worked well. Happily, the dirt roads over which we traveled in Africa were for the most part less demanding and hazardous than in Central and South America, largely because the terrain was not as mountainous. To help cope with the overburdening weight of our van we had heavier duty custom wheel rims built in Cape Town (which required eight days to complete). However, our excessive weight continued to be an albatross around our neck. Luckily, much of our travel in Africa coincided with the local dry seasons, and mud was not something with which we usually had to contend. Our ambitious plan was to drive from Cape Town to Algiers, crossing the middle of the Sahara Desert. It was an adventure few had attempted, and fewer had accomplished.

South Africa was a picturesque country of contrasts. Situated below the tropics at the southern end of the continent it had a coastline with both the Atlantic Ocean (to the south and west) and the Indian Ocean (to the south and east). There were lush hardwood forests along the southern coast that were bordered inland by plateaus, ridges, escarpments, steppes, and mountains as high as 11,000 feet. Semi-arid savannas and woodlands covered much of the western, central and northern regions of the country. The country was situated almost entirely below the Tropic of Capricorn, in the southern temperate region of the continent. There were botanical remnants of Africa's once Gondwanaland connection.

In Cape Town we attracted our usual attention because of our beards (which no longer elicited the whispers of "Fidel Castro") and our unique camping van, with its large Mercator map of the world on the left side. I kept an updated painted orange line on this map which indicated the route we had traveled to date. Of course a journalist soon found us and wanted a story and picture for the newspaper. On our very first day in town, a middle-aged man of British ancestry happened to walk by our van as we were climbing in, a Mr. Gant, and he stopped to visit with us. He owned a 9,000 acre fruit farm outside Cape Town, near the town of Somerset West about 20 miles east. The farm produced a great variety of fruits, including peaches, apricots, plums, and grapes for wine production. After several minutes of talking with us, Mr. Gant invited us to be his guests for two days at his farm, where he lived with his wife. We gladly accepted his invitation, and the chance for a hot bath. It was an opportunity for Noble to learn about the life of an enterprising fruit producer and a chance for me to see my first African birds, a dream come true.

We passed through fields of yellow mustard on the way to the fruit farm, which was named "Lourensford." When we arrived at the farm, we were amazed to see snow on the tops of the distant Hottentot Mountains to the east, at about 5,000 ft. elevation! In the two days on the fruit farm, I tallied 63 species of birds (see Appendix B, Sept. 12 & 13, 1959). Colonial hanging nests of the Cape Weaver* hung from the ends of branches in small trees at the edge of streams and ponds. This species was one of 56 African weavers in the genus *Ploceus*, members of which were characteristic, well known, and widespread birds everywhere on the continent. Nests within the genus were among the most complex of all birds, intricately woven into pouches as long as two feet in length, suspended from their top. An entrance was at the top, side, or bottom of the nest, sometimes through a long neck.

Other birds on the farm included Bokmakieries* (a colorful yellow, olive-green, and black "bushshrike" which sang from inside the understory with lovely onomatopoeic flute-like notes), Laughing Doves* (which called throughout the day from yards and gardens with a quiet soft series of bubbling "coos"), and such memorable birds as the Hammerkop*, Egyptian Goose*, Crowned Lapwing* (a noisy, conspicuous species of Vanellus), Eurasian/African Hoopoe*, Speckled Mousebird*, Water Thick-knee*, Ground Woodpecker* (see my comments on Sept. 22), Cape Sugarbird*, Jackal Buzzard* (a *Buteo*), Rufous Rockjumper* (one of two endemic South African birds in the genus *Chaetops*, of uncertain taxonomic affinity), White-necked Raven*, Malachite Sunbird*, and Tinkling Cisticola* (one of 51 African species in the genus *Cisticola*, which were small, brownish, mostly streaked, grassland "warblers" which were difficult to separate from one another in the field by appearance but were recognizable by their songs, as indicated by such English names as "Tinkling", "Chattering", "Singing", "Trilling", "Whistling", "Bubbling", "Piping", "Siffling", "Rattling", "Churring", "Wailing", "Chirping", "Winding", and "Croaking")!

On Sept. 15 & 16 we drove to Cape Point at the end of Cape of Good Hope, which was the southwest extremity of the continent. (Cape Agulhas, 100 miles to the ESE, was actually the southernmost tip of Africa.) Along the rocky seacoast were four species of cormorants - - the Great Cormorant, Bank Cormorant, Cape Cormorant*, and Reed Cormorant. A Rufous-chested Sparrowhawk* (one of 49 species in the worldwide genus *Accipiter*), 2 Gray-winged Francolins* (a South African endemic), and an African Stonechat* were also added to my incipient African bird list.

From the tip of the cape, we drove back north to the Rondevlei Bird Sanctuary just outside the southern suburbs of Cape Town, at the upper end of False Bay. The sanctuary was a small area of wetlands with a shallow lake ("vlei"). Here there was a wide assortment of aquatic birds, the most exciting of which were both Greater* and Lesser* flamingos. I estimated that 200 individuals of each species were present. Like all of the six species of flamingos in the world, adults were predominantly pink in overall color. (I have already commented on the rather amusing fact, to me, that in mentioning flamingos almost everyone always says "pink flamingos", as if there were alternative colors, which there aren't.) Other water birds present were: Great White Pelican*, Gray Heron, Black-headed Heron*, Red-knobbed Coot*, Little Grebe, Egyptian Goose* (a handsome bird), Red-billed Teal*, Yellow-billed Duck*, African Marsh Harrier*, Blacksmith Plover* (a strikingly patterned black, white, and Gray *Vanellus* species, so named because its loud "tink" call sounds like a blacksmith hammering on steel), Kittlitz's (Sand) Plover*, Three-banded Plover*, and Water Thick-knee*.

After eight days in Cape Town our new custom wheel rims were finally installed by the Jeep agency and our van was ready for departure, on Sept. 19. Our first destination was the vast, semi-arid, inland region of grasslands, savannas, scrub, and mountains, known as the "Karroo", which was situated in western Cape Province on an elevated, undulating plateau 3,000-4,000 feet above sea level. Hottentots once lived here. The larger, more northern part of the region was called the "Great Karroo" and the narrower more southern and mountainous part was referred to as the "Little Karroo." We camped that first night in the fertile Oliphants River valley just north of the town of Citrusdal, about 120 road miles north of Cape Town at an elevation somewhat below 2,000 feet. Fruit farming was the principal source of livelihood in the valley, which was on the very western outskirts of the Karroo. Roadside birding during the day produced 19 new species for my trip list, including the endemic Karroo Scrub-Robin* (Appendix B).

The next morning (Sept. 20) I awakened at dawn and went birdwatching for an hour before Noble arose, as was our normal routine. Laughing Doves called repeatedly and Cape Weavers flew back and forth to a nesting colony. A Cape Sugarbird performed its aerial display, clapping its wings while holding its greatly elongated tail over its back. We got underway at 7:30am and followed the Oliphants River northward in southern Namaqualand, through the town of Clanwilliam and on to the town of Klawer. Here we left the river valley and veered slightly eastward along the base of some hills to our right, and began ascending slowly to the town of Vanrhynsdorp, which was only a short distance ahead.

I stopped in a rocky area to see what birds I might find. There were Karroo Scrub-Robins, Piping Cisticolas (locally called "Neddickys"), a Layard's Warbler* (a Karroo endemic formerly referred to as a Tit-Babbler, in the genus *Parisoma*), Speckled (Rock) Pigeons*, Karroo Chats* (endemic to SW Africa), and African Pipits* (a member of the widespread "Richard's" pipit assemblage). In this area, I also came across a Rock Elephant Shrew scurrying among the rocks. This cute little mouse-like mammal with a long narrow snout feeds on ants, termites, and other insects. It is one of 19 species in the family Macroscelidae, which is endemic to Africa (and so distinct that it is sometimes placed in a taxonomic order of its own).

As I was clambering over more rocks in this same area, my eyes suddenly fell on a sizeable snake sunning itself on the rocks. It was about five or six feet long and entirely black in color, without any prominent pattern. I thought to myself "here's a chance for a good close-up picture." Removing my little Canon camera from my shoulder bag, I crouched and slowly approached the snake to get as close as possible with my wide angle, 35mm lens. The snake remained motionless and looked harmless enough stretched out over the rocks. When I was seven or eight feet from the front of the snake, I slowly raised my camera to take its picture. The snake had not moved. However, as I was manually focusing my camera the snake, without any warning, suddenly raised its head and front body vertically almost 12 inches above the ground, looked directly at me, and fanned a very wide cape at the sides of its neck. I instantly jumped back, snapping my shutter as I did so. (The eventual picture was blurred.) The snake immediately crawled hastily under some rocks and disappeared. My heart was beating much faster than normal. When I relayed this experience to a young park ranger in the Drakensberg Mountains ten days later he said to me, "You were lucky, that was a Black Spitting Cobra." He went on to explain that their aim (with venom ejected through their fangs) is very accurate within ten feet, and that the venom can cause temporary or permanent blindness, rarely even death. I took a deep breath and marveled that natural selection had not long ago eliminated my genes from the gene pool.

From Vanrhynsdorp we left our main north road and turned off northeastward on a well-used, graded, secondary gravel road. We began climbing up to Nieuwoudtville, a town of 2,000 people situated on the Karroo plateau at an elevation of 3,000 feet. Additional birds that I identified along this 30 mile stretch of road included three more Karroo endemics - - Pale-winged Starling*, Karroo Lark*, and Namaqua Prinia* - - plus my first sightings of the Pin-tailed Whydah*, Southern Anteater-Chat*, Cape Lark*, Black Crow*, Stanley's Bustard*, Namaqua Sandgrouse*, Black Bustard (Koorhan)*, Spike-heeled Lark*, and Mountain Wheatear* (genus *Oenanthe*). Along with the Capped Wheatear* which I recorded the day before, it was one of the two southernmost of the 20 species of wheatears in the world, most of which were Palearctic breeding. We camped that night east of Nieuwoudtville, on the open, sparsely scrubbed plains. During the day I added 23 species to my bird list (Appendix B). As dusk approached several hares appeared (genus *Lepus*) looking very much like North American jackrabbits, but I could not distinguish between Cape and Scrub hares (although the latter is somewhat larger).

The next morning, Sept. 21, I again arose at dawn and went walking for an hour prior to our departure. Spring-time flowers were in bloom and the hillsides were patchily covered with a lovely blanket of lavender flowers, even though the normally wet winter had been unusually dry. Cape Province of South Africa makes up the geographically smallest of the six "Floral Kingdoms" of the world recognized by botanists. The Cape Kingdom has more than 1,000 species of indigenous plants, including those of the Karroo. Some of these had their origin as far back as the Jurassic Period, 150 million years ago at the breakup of Gondwanaland. Botanists and others come here from all over the world to document the plant diversity and witness the springtime floral spectacle.

During my morning walk I crossed a low grassy hill and on the other side, to my excitement, was a mixed co_lony of Suricates and Bushy-tailed Meerkats. The two species often live in a mutually compatible group, working together cooperatively. Both are members of the Old World carnivorous family Viverridae (along with civets, genets, and mongooses). (Viverrids are closely related to weasels, otters, badgers, skunks, polecats, and others in the worldwide family Mustelidae.) Suricates are well known and often featured in African movies because of their behavior of posting sentinels that stand motionless for a long period of time in an upright position on an elevated lookout post, watching for potential predators on the colony (eagles, jackals, servals, ratels, Kaffir cats, and others). A vocal alarm call is sounded when a predator is spotted and all the members of the colony (including the sentinel) scurry for the nearest burrow or other safe cover. Evolutionary biologists ponder how such "cooperative behavior" evolves.

We drove eastward from our campsite for 25 miles to the town of Calvinia, and from there we continued eastward for another 50 miles to the town of Williston. This was the heart of the Great Karroo with dry grasslands and scrubby hillsides on a plateau 3,000-4,000 feet in elevation. Most of the streambeds, including the Sak River west of Williston, were dry. Sheep and cattle ranches were scattered far apart. Young native boys herded the livestock on foot with the aid of long whips. The countryside reminded me of our own Southwest, with buttes and flat arid plains covered with vegetation similar in outward appearance to sagebrush and catclaw. Even the small towns were reminiscent of the Old West. New birds for my list included the Sacred Ibis*, Karroo Bustard (Koorhaan)*, Black-eared Sparrow-Lark*, Southern Masked-Weaver*, Yellow-bellied* and Yellow-rumped* eremomelas (family Sylviidae), and three species of birds endemic to the Karroo: Chat Flycatcher* (genus *Bradornis*), Tractrac (Layard's) Chat*, and Ferruginous (Red) Lark*, a species with a very small geographic range. It was springtime and many birds were singing, particularly the half dozen species of larks. The open savannas and plains of the Karoo were a pleasant, inspiring environment.

One of the most iconic and delightful mammals of the Karroo was the Springbok, a very handsome mid-sized gazelle (antelope) weighing 75-100 lbs. and standing 2½ ft. tall at the shoulder. Like most gazelles it was a gregarious species, living in groups of a dozen or more individuals. It possessed a distinctive, wide, black horizontal band on its sides which contrasted with its white underparts, and there was also a narrow black band running from the back to the front of its white face, through its eyes. The thick, rounded, heavily ridged horns were 15-18 inches long in the male, but somewhat shorter in the female. They were lyre-shaped and lightly curving backwards and upwards, and at

Plate 34: Table Mountain is a world famous landmark on the peninsula just south of Cape Town. Our African adventure began in the "Karroo", a semi-arid plateau covered with grass, shrubs, and springtime wildflowers, north of Cape Town.

the same time outwards and then inwards toward the tips. (I confess this description is difficult to envision, but the horns were indeed very attractively shaped and sculptured.) Like other gazelles, Springboks were slender legged, fast running (up to 60 mph), and capable of taking leaps through the air of more than 12 feet. Its name comes from the habit of stiff legged trotting interspersed with great upward springs and arching of the back, a characteristic behavior of antelopes referred to by mammalogists as "stotting." Such behavior in Springboks displayed the white rump patch and was performed by males in courtship, or to threaten predators. Similar to most other antelopes, the Springbok was polygynous in its mating behavior (one male to many females). Males defended a territory during the breeding season, protecting their harem and fending off rival males. In its geographic distribution the Springbok was confined to southwestern Africa, where it was a free spirit of the open plains and was endeared by many. The national rugby team of South Africa is nicknamed the "Springboks."

At Williston Noble and I left our good E-W gravel road and turned SE on a less well maintained and less traveled gravel road heading toward Fraserburg, 60 miles away. As we left Fraserburg our road began climbing slowly up to a low ridge of mountains, and I chose a campsite at 4,500 ft. elevation for the night. We were in the middle of the Great Karroo. The next morning (Sept. 22) I was delighted to find two Rufous-eared Warblers* (family Cisticolidae) foraging actively in the scrubby bushes near our van. This handsome, long- tailed little warbler was prinia-like in its size and shape, and in coloration it was buffy brown above with prominent dark streaking on the crown and back. It possessed a wide, conspicuous, rufous ear patch (for which it was named), and was white below with a narrow, sharply outlined black band on its chest. The abdomen was tinged light buff. Overall, I found its appearance and coloration to be particularly appealing. It was confined to arid and semi-arid Southwestern Africa, with almost the exact same distribution and habitat as the Springbok. Together they provided me with pleasant memories of our short time in this enchanting environment.

The Swartberg Mountains, which rise as high as 7,500 ft. above sea level in an E-W direction, separate the Great Karroo from the more southerly, much smaller, narrower, wetter, and more rugged Little Karroo. Our road south to the coast crossed over these mountains at Swartberg Pass, which was at an elevation of 5,500 ft. between the towns of Prince Albert and Oudtshoorn. I stopped a mile just below the pass for a short hike, pursuing birds, and picture taking, while Noble wrote in his journal. Low, rocky cliffs were in the immediate vicinity. It was a cool, cloudy day, which was quite a pleasant, invigorating environment for a brisk, 45-minute walk 5,000 ft. above sea level.

Shortly after starting out among some rocks I encountered several Cape Rock Hyraxes (locally referred to as "Dassies", in the family Procaviidae, order Hyracoidea). These peculiar, pika-like, rabbit-sized, herbivorous, hoofed animals are so distinct from other mammals that taxonomists place them in an order of their own. Hyraxes are nimble climbers and jumpers. Some populations (tree hyraxes) live in forest trees, and others live in open grasslands where their homes are in terrestrial termite hills. The bottoms of their feet have rubbery pads which prevent them from slipping on rocks or tree branches. I was amused by their running and jumping among the rocks, appearing and behaving much like pikas in the Rocky Mountains of North America (although they were somewhat larger in size).

Another mammal which I was fortunate to observe here was the Vaal (Gray) Rhebok, a dainty, mid-sized antelope usually placed in the reedbuck/bushbuck subfamily. A small group of this gregarious species was seen at a distance of approximately 150-200 yards. Rheboks feed not only on grasses (as a "grazer") but also on the leaves of woody shrubs (as a "browser"). With their grayish-brown overall color (lighter on the underparts) and moderate size, they were very similar in general appearance to the Mountain Reedbuck. They could be distinguished from this species by their horns, which were virtually straight and more slender and pointed than the heavier, ringed, forward curving horns of the reedbuck. Like many ungulates they were territorial and aggressive during the breeding season. The name "Reebok" of the worldwide British company specializing in athletic shoes (and other sporting goods) is the Afrikaans name for this fast-running little antelope.

Birds which I encountered during my walk at the pass included a Hoopoe, Ground Woodpecker (first seen on Sept. 12), Yellow-tufted (African Rock) Pipit*, and Karroo Long-billed Lark*, all of the last three being endemic to southern Africa. The enigmatic Ground Woodpecker preferred rocky, hilly areas and was unique among woodpeckers in being entirely terrestrial in its foraging and breeding, feeding on ants and other insects and nesting in a burrow in the ground or in a sandbank. In its captivating appearance, it reminded me somewhat of a Lewis's Woodpecker, being pinkish below and dark colored above.

After I returned to the Jeep, Noble and I descended southward in the mountains for 40 miles, all the way down to the sizeable town of Oudtshoorn in the upper Olifants River valley at 1,000 feet elevation. This town was the largest in the Little Karroo and was typical of many in South Africa. The population of 20,000-30,000 inhabitants was ethnically 15% "white", 12% "black", and over 70% "coloured" (a mixture of the two). Afrikaans was the predominant language, resulting from the fact that the Dutch were the first European settlers (in 1839, followed 20 years later by the first British settlers). Bushmen lived here prior to the arrival of Europeans. In the late 1800's and the early 1900's the principal source of income for the town was from ostrich farming, and this activity was still very important, along with wine production, tourism, and other activities. The town called itself the "ostrich capital of the world."

From Oudtshoorn we proceeded southward for another 40 miles, leaving the Karroo behind us as we crossed over the low Outeniqua mountain range, via the Outeniqua Pass at 2,500 ft. elevation. We were now entering the lush evergreen Tsitsikamma Forest, which was interspersed with areas of heath-like, stiff twigged, leathery-leafed (sclerophyllus) vegetation known locally as "fynbos" (see below). On the hillsides, I took pictures of "yucca-like plants with stalks of red flowers at the top" (probably *Aloe ferox*) and of small purple-flowered herbs. Our good gravel road descended into the town of George, which was situated on the "Garden Route" highway, a good E-W asphalt surfaced road. This was a very popular tourist route extending eastward along the Indian Ocean all the way from Cape Town to Port Elizabeth. We turned east on this route and followed it a very short distance to the town of Wilderness, where shortly thereafter I found a suitably remote pull-off to camp for the night. It had been a very productive, exciting day.

The Tsitsikamma Forest extended in a very narrow E-W belt for over 100 miles along the middle of South Africa's southern coast from George eastward to beyond Storms River, in an area where the average rainfall was 25-50 inches per year. The Garden Route traveled along its southern edge. The dominant tree in this lush, evergreen, coniferous-hardwood forest was the Outeniqua Yellowwood (*Podocarpus falcata*), a giant conifer of Gondwanaland origin which occasionally attained a height of 200 feet, a diameter of 10 feet, and an age of 1,000 years. Interspersed with this forest were areas of "fynbos" vegetation, which occurred along the entire southern coast of South Africa. Such botanical communities were typical of worldwide temperate coastal regions with mild wet winters and hot dry summers (as I have previously mentioned from my travels in Chile). These ecosystems were called "fynbos" in South Africa, "matorral" in Spain and Chile, "maquis" in France, "macchia" in Italy, "chaparral" in California and Portugal, and "kwongan" in SW Australia. The widely used general term "heath" was often used to refer to such vegetation. In South Africa the botanical genus *Erica* (in the family Ericaceae, the heath family) was a widespread component of the fynbos, as was also the indigenous genus *Protea* in the family Proteaceae. The diversity of the flowering plants in the fynbos was an important factor in designating the Cape Province of South Africa as a distinct floristic Kingdom of the world, which had its origins in Gondwanaland.

We took 2½ days (Sept. 23-25) to drive the 210 miles from Wilderness to Port Elizabeth, stopping for a day at Plettenberg Bay to spend time with a young couple we had met in Cape Town, now on their honeymoon. Wine sampling at a few of the vineyards along our route was a pleasant diversion. We stopped to photograph the deep, scenic gorge below the Paul Sauer highway bridge at Storms River. This concrete arch bridge was completed only a few years earlier, in 1956, and spanned 328 feet above the gorge, which was 400 feet below the bridge. Not far from Port Elizabeth we encountered our first troop of Baboons. These terrestrial quadruped primates stood 2½ feet tall and weighed between 55 and 65 pounds, making them the largest primates on the continent behind gorillas and chimpanzees. (Mandrills were only very slightly smaller.) Males were larger than females. Baboons were found almost everywhere in Africa below the Sahara except in the central rainforests. They lived in troops of 20-80 individuals and spent the day on the ground, but ascended into trees (or among rocks) to sleep for the night. They were known to live 30 or more years in captivity. Baboons varied greatly in color across the continent and taxonomists were not in agreement on how many "species" there were, with most authors recognizing from one to seven. The population along the coast of South Africa was referred to as the "Cape Chacma Baboon." I did not find baboons a very appealing animal, as individuals were constantly squabbling and fighting among themselves.

It was only a short drive from Wilderness to Plettenberg Bay, so on Sept. 23 I opted to hitchhike and search for roadside birds in the Tsitsikamma Forest and fynbos for most of this distance, while Noble drove on ahead to join our friends from Cape town. Rameron Pigeons* called from the forests with a low, hoarse "coo" repeated up to three times. Their bright yellow eye ring, bill, and feet contrasted with their overall dark purplish coloration. African Paradise-Fly-catchers* (genus *Terpsiphone*) flitted through the middle and lower levels of the forest with their greatly elongated central tail feathers fluttering behind them. They also attracted attention by their striking bluish-black, rufous, and white plumage. Related species are distributed throughout the tropics and subtropics of Africa, Asia, and the South-west Pacific. The signature bird of the forest, however, was the Knysna Turaco* (Loerie), an extraordinarily attractive chachalaca-sized arboreal frugivore of the forest canopy, in the endemic African family Musophagidae. It was vivid green in color with a prominent pointed crest and with bright crimson primaries (outer wing feathers) which were very conspicuous in flight. I heard them calling frequently from the dense forest, but to my dismay, I was never able to locate one visually. Their "kow, kow, kow" vocalizations reminded me somewhat of Neotropical trogons. Nectar-feeding sugarbirds and sunbirds were important flower pollinators in the fynbos areas. I finished the day with 21 new species for my trip list (Appendix B).

I documented birds around Plettenberg Bay on the morning of Sept. 24. Foraging along the rocky coast was a pair of African (Black) Oystercatchers*, eye-catching in their all black plumage with a bright red bill, legs, and eye ring. As their name implies, the 11 species of oystercatchers in the world all feed predominantly on shellfish such as oysters, mussels, and chitons, which they pry open with their long, stout, laterally compressed bill. This is the only oystercatcher that breeds in Africa. Perched conspicuously in a tree was a very handsome African Fish-Eagle*, strikingly patterned in rufous, black, and white. To my eye it was one of the most attractive raptors anywhere, and its loud, far-carrying yelps were one of the most characteristic and memorable sounds of Africa. It was one of eight species in the genus *Haliaetus* (along with the Bald Eagle of North America), and I placed it near the top of my list of favorite birds. In the forest I encountered a Green Woodhoopoe* (locally called a Red-billed Hoopoe) which was clambering in creeper-like fashion on the branches and trunk of a large tree. Also present was a Scaly-throated Honeyguide*, one of 17 species in the unique family Indicatoridae, members of which feed predominantly on beeswax. Some species in this family (particularly the Greater Honeyguide) are renowned for their uncanny ability to lead a ratel (honey badger) or other predatory mammals (including a human) to a beehive where it can exploit the beeswax after the hive has been ripped open by the predator. Evolutionary biologists have difficulty explaining how such behavior can evolve by natural selection. All honeyguides are brood parasites, laying their eggs in the nests of hole-nesting birds such as barbets and woodpeckers. That afternoon we said goodbye to our honeymooning friends, Leicester and Rosemary Dicey, and continued our travel eastward along the Garden Route. We drove only 50 miles before I chose a campsite just beyond Storms River, on the eastern outskirts of the Tsitsikamma Forest.

Prior to our departure the next morning (Sept. 25) I strolled through the forest briefly for one last time. I was pleased to track down a Narina Trogon* (one of only three species of trogons in Africa) and several Chorister Rob-in-Chats* (which in coloration reminded me a little of an American Robin except for the rufous sides to its tail). This latter species sulked in the understory but announced its presence by a loud song which consisted mostly of mimicked phrases of other birds, much like a mockingbird. We then climbed into the cab of our camping van, left the unique Tsitsikamma Forest, and headed east toward Port Elizabeth, two hours away. En route I saw three Spur-winged Geese* and two Eurasian Buzzards (a migrant from Eurasia). The goose was a very large, rather unattractive waterfowl (to my thinking), endemic to Africa. It was black and white in color with a bare pinkish-red face, bill, and legs.

While we were in Port Elizabeth, Noble and I stopped by at the Goodyear Tire Company, one of the largest of the company's many sites throughout the world. We had purchased Goodyear tires throughout our travel, and now needed two new ones. Due to Noble's entrepreneurial skills, the company offered to give us 2 new tires and to supply us with new tires every 20,000 miles for the rest of our trip, at no cost to us, simply as a token of good will! After leav-

Plate 35: Native Xhosa people lived in round adobe houses on the undulating "Transkei" grasslands in East Cape Province. The scenic, rugged Drakensburg Mountains, home to majestic Lammergeiers, rose to more than 10,000 ft. elevation.

ing Port Elizabeth, we drove for two more hours before I found a roadside campsite 25 miles west of Grahamstown, in open forest. Fiery-necked Nightjars* called repeatedly throughout the night - - "Good Lord, deliver us" - - a well known nocturnal sound of the African bush.

The next morning (Sept. 26) I awoke to the pleasant vocalizations of the Emerald-spotted Wood-Dove*. This song - - several, soft, sad series of notes, the last longer and descending in pitch - - became one of my most fond memories of Africa. It wasn't long after getting underway that we drove into Grahamstown, a mid-sized country town. In the middle of the main street was a long team of 16 oxen that were pulling a heavy wagon loaded with firewood. The slender, middle-aged native man who was directing the team with a very long whip was dressed in a wide-brimmed, badly worn felt hat, a heavy ragged jacket, and well worn faded brown trousers. He was selling the firewood at the equivalent of 35 cents for a full gunnysack.

We left Grahamstown and continued on to the sizeable town of East London, where our highway veered to the left, away from the coast and headed north, climbing upward to an inland grass-covered rolling plateau known as the "Transkei", in far eastern Cape Province. It was an area with hot summers and mild, dry winters. Among the birds which I recorded along the way were: Buff-streaked Bushchat* (a handsome, very local endemic, wheatear-like grassland bird) and a Long-tailed Widowbird* (which caused me to gasp as it flew across the grasslands with its remarkable long wide tail fluttering behind it). It was a member of the weaver family, Ploceidae. We camped that night at the Great Kei River highway bridge. At daybreak the next morning (Sept. 27) a Giant Kingfisher* flew by. It was about the same size and general overall appearance as the Ringed Kingfisher of the Neotropics. Another bird in the vicinity of our campsite was the Black-collared Barbet*. I tracked down a pair which was vocalizing loudly in a duet. Their ringing song has been described as a harsh "krrr" followed by a ringing "too poodly, too poodly", with the first note higher. It was one of the most frequently heard and characteristic sounds of the African bush, and one which I remember with nostalgia.

The native tribal people of the Transkei were the Xhosa. They lived in a manner little changed over the centuries. Their circular adobe houses with a pointed thatched roof were known as "rondavels", and were sometimes painted white. Many of the women, girls, and young boys wore traditional bright ochre-colored blankets draped over or around them, although girls and women were sometimes bare-breasted. Women carried goods of all kinds in a large cloth bundle on the top of their head. Men and boys tended cattle. Many of the words in the Xhosa language were spoken with a distinct clicking sound, and the Xhosa were known as "the people with the clicking tongue." Because of their recent exposure to tourism, some of the people held out their hand for money if we asked permission to take a close-up picture of them, so we either gave them a few coins or just didn't take their picture. (Elsewhere in Africa we were only very rarely asked for money if we requested a person's photograph.) It was our first encounter in Africa where people were living and dressing in a traditional (non-European) manner.

Our journey now took us northeastward toward the towns of Umtata and Kokstad, still on the green, rolling, grassy plateau of the Transkei. The Indian Ocean was 50-70 miles away, to our right. On our left were the distant foothills of the Drakensberg Mountains. Along our route were small colonies of Village (Black-headed) Weavers*, many Cape Griffons* (large, tawny colored vultures), and a pair of Gray Crowned-Cranes*. These unique cranes, icons of Africa, were distinguished from all other cranes by the gold colored, fan-shaped crown adorning the top of their head. (They were a popular zoo bird around the world.) Largely gray, black, white, chestnut, and golden-orange in plumage coloration, this handsome crane was characterized by its white face contrasting with its black crown. Like all cranes it engaged in elaborate courtship displays and two individuals mated for life. Also on this elevated plateau was another African icon - - several 5½ ft. tall, arrogant-looking, slender, long-legged Secretary-birds* which were striding majestically across the grasslands. They were dapple gray in color with a bare red face and long plumes hanging from the back of their head. As a terrestrial raptor, they fed on a great variety of invertebrates and vertebrates, including snakes which were killed with their feet. Secretary-birds could run rapidly but they were also strong fliers, and sometimes engaged in aerial displays. For me, they epitomized the "real Africa."

In Kokstad we paid 42 cents per gallon for gasoline, more than twice as much as the average price that we paid in South America. Fuel for our 10 miles per gallon, overweight camping van was one of the major costs of our adventure.

We spent that night near Kokstad, at 4,500 ft. elevation. The next day (Sept. 28) was rather tortuous driving all day long as we finally left Cape Province, after 17 days, and entered Natal Province on our way to Pietermaritzburg. Our poorly maintained gravel road wound back and forth, and up and down along the eastern base of the Drakensberg Mountains. A noteworthy roadside bird sighting was that of an early arriving Pallid Harrier, which visited South Africa from the Palearctic during the northern winter.

Giants Castle Game Reserve was 40 miles northwest of Pietermaritzburg, in the rugged Drakensberg Mountains. It was one of the oldest game reserves in the country, having been established in 1903 to protect the Eland. It was a scenic region of high grasslands and rocky escarpments, coves, and cliffs rising to elevations over 10,000 ft. above sea level. Noble and I spent two days in the reserve, hiking, photographing, and searching for wildlife. On one of the two mornings, Sept. 30, I walked with an elderly park ranger for several hours up to the 10,000 ft. level, where there were patches of snow on the ground. We walked across the international boundary (marked only with a crude, simple wooden sign) into Basutoland, a small, independent, geographically isolated mountainous nation now called Lesotho. Bushmen once lived here. The ranger and I proceeded only a short distance before retracing our steps back into South Africa.

I took pictures of red-flowered heath-like shrubs, and of 3 Black Wildebeests ("Gnus"), a very distinct looking antelope shaped somewhat like an American Bison. Wildebeests were related to Topis and hartebeests, and were characterized by thick based horns which were directed laterally from the side of the head and then curved sharply upward. Even more unusual were the erect or long, hanging manes on their head, upper back, neck, chin, and chest. It was a species which was almost extinct outside of protected areas but its closest kin, the Blue Wildebeest, was still widespread and abundant in eastern and southern Africa. Two other kinds of antelopes which I encountered on this high elevation walk were the Mountain Reedbuck (a rather plain, mid-sized species with forward curving horns) and the Klipspringer. This latter species was a small, stocky, rock-inhabiting antelope which stood on the tips of its toes, like a ballet dancer, motionless as it scanned the surrounding mountainside for potential predators, such as leopards, caracals, and crowned eagles.

My short trek in the high Drakensbergs gave me a chance to experience the remoteness of this harsh environment, and the wildlife adapted to it. In addition to the above antelopes, I was thrilled to observe, on 2 separate occasions, a soaring Lammergeier* (Bearded Vulture). This large, majestic, predatory scavenger was very local in its distribution in Africa (but more widespread and numerous in Palearctic Asia). It was easily recognized by its long wings and distinctive long, wedge-shaped tail. Other birds which I observed during our time in the reserve, at slightly lower elevations, were the Blue Crane*, Martial Eagle*, Verreaux's (Black) Eagle*, Lanner Falcon*, Gurney's Sugarbird*, and Orange-breasted Rockjumper* (in the genus *Chaetops*). Like the closely related Rufous Rockjumper (which I mentioned from the Capetown area) this latter species was a handsomely colored, babbler-like bird of uncertain taxonomic affinity, endemic to South Africa. I saw a pair of individuals, and it was exciting to watch them as they hopped and jumped from boulder to boulder. It was sudden, unpredictable encounters such as this that motivated me to pursue birds. Was I really crazy?

Noble and I departed the Drakensbergs on Oct. 1 and drove down from the mountains 100 miles to a campsite just outside Durban. Early the next morning I pursued a rising and falling series of bubbling "doo's" coming from a dense wet thicket and was able to locate my first White-browed Coucal* (the Burchell's subspecies), a member of the cuckoo family. The pleasant, almost flute-like notes became a familiar and haunting sound to me as it emanated from wet patches in the bush throughout the next several months of our travel in Africa.

We got underway soon after sunup and drove into the large commercial city of Durban, an important trading port, manufacturing center, and tourist destination on the eastern seaboard of South Africa in the province of Natal. A mixture of peoples from Europe, Asia, and Africa provided great cultural diversity. We were there for 2 days taking care of our usual travel chores which were necessary to keep us on the road. One of these was to purchase, or acquire, some new clothing. So we ventured into a clothing outfitter store and Noble asked to talk to the manager. He explained what we were doing and convinced the man, in his usual persuasive manner, that he could benefit by taking a photo of us buying clothes in his shop, which could then be displayed for advertising purposes. In return, the manager would allow us to pick out a few garments to take with us, without charge. Only Noble could succeed with such an argument, but the manager agreed and we both walked out with a new wardrobe worth about $75 each. I obtained a new pair of trousers, 2 pairs of khaki field shorts (which were an absolute necessity in Africa), and some socks! Back on the street outside the store we were unable to avoid a newspaper journalist who wanted a story and photograph for the local newspaper, having spotted our Jeep parked on a downtown street. Such interviews were impossible to escape.

From Durban we drove north on a well-used dirt road through Zululand to the Hluhluwe Game Reserve in northern Natal Province. Along the way there were fields of pineapples and sugarcane. Some of the Zulu people wore traditional garments, and rural women and girls were occasionally bare-breasted. I photographed a young Zulu woman (with her permission) who was wearing a single piece of cloth wrapped around her waist as a skirt, and nothing above. She was not embarrassed, nor did she ask for money, and was planting corn by hand, 2 or 3 kernels at a time from a handheld gourd container. Her small baby sat nearby on the ground, unattended, at the edge of the little corn plot. The young woman seemed happy and satisfied with her life. Was anything more important than this?

Hluhluwe was a popular nature reserve, both for its variety of mammals and its abundance of birdlife. It had been established in 1895 to protect the White Rhinoceros. The reserve was spread out over 370 square miles of rolling green grasslands ("veld") and dry woodlands. Characteristic of African savannas everywhere were small, flat topped acacia trees scattered across the open grasslands. We were met at the entrance to the reserve by a pleasant, courteous young game ranger who explained to us that visitors were not allowed to take their own vehicles into the reserve, but that he would take us around in his truck. We spent 3 days there, arriving on Oct. 3 and departing on Oct. 5, camping at night in a designated area. We had many opportunities for good close-up pictures of the game animals since they were unafraid of our truck and often allowed us very close approach.

Noble was very excited by the many large mammals we encountered, as was I, so he decided to count them while I kept my usual tally of the birds I identified. His numbers of the game animals were: 100 Warthogs, 100 Impalas (a medium sized, slender, elegant antelope with lyre-shaped horns, which was common and well known throughout much of eastern and southern Africa), 10 Waterbucks (a large, stout, plainly colored and rather undistinguished looking antelope with long, upward curved horns), 10 Burchell's Zebras (the best known and the only widely distributed of the 3 species of zebras in Africa), 10 Baboons, 6 Giraffes (which were up to 18 ft. tall, occasionally even more), 5 White Rhinoceroses, 2 Greater Kudus (a large, stately appearing antelope, approximately the same size as an American Elk, with long, picturesque horns winding upward in a corkscrew spiral, considered by many to be the most magnificent of all the African antelopes), and 1 Black Rhinoceros. This latter species was distinguished from the White Rhinoceros by being slightly smaller and darker, by lacking a hump at its shoulders, by a narrower, more pointed and grasping (prehensile) upper lip, and by carrying its head higher. It was the more widely distributed and usually the more numerous of these two sympatric (geographically overlapping) species.

Birdwatching in Hluhluwe was as rewarding as were our sightings of game animals. During our time there I added over 40 species to my trip list (Appendix B, Oct. 3-5, 1959). Among the more memorable species, in the order I recorded them, were: Bateleur* (a remarkable bat-shaped raptor with wide wings and an extremely short tail, a gliding teetering flight from side to side, and handsome black, chestnut, and white plumage coloration), Trumpeter Hornbill* (a large, guan-sized, black and white arboreal forest bird with a bulky, casqued bill, filling somewhat the same ecological niche in Africa as do toucans in tropical America), Purple-crested Turaco* (my first sighting in this memorable family), Crowned Eagle* (a magnificent, heavily barred and banded species which, along with the Martial Eagle, was one of the two largest, non-vulture, African raptors), Lilac-breasted Roller* (a conspicuous "fly-catching", jay-sized bird which immediately became one of my most favorite birds with its narrow elongated outer tail feathers and splendid colors of aqua-blue, violet, chestnut-brown, and emerald green), Marabou Stork* (a huge black and white stork, another African icon with its enormous bill, bare red head, and long pendulous gular pouch, which fed with vultures on carrion at carcasses of large game animals, and roosted gregariously at night in the tops of tall dead trees), and the Red-billed Oxpecker* (one of two unique African species which were related to starlings and had the remarkable habit of feeding on the backs of large mammals, carefully removing and devouring ticks and other ectoparasites). My complete list of birds from Hluhluwe, as recorded in my field notebook, includes 10 species of hawks, eagles, kites, and Old World vultures (family Accipitridae). One of these was the Black Kite (a common, widespread Old World species with many subspecies, some of which, such as the yellow-billed race in South Africa, were considered specifically distinct by some authors). On Oct. 5, prior to our departure, I found the nest of one of these kites 25 ft. above the ground in an acacia tree alongside a dry streambed. I climbed up the tree with my camera (against park regulations) to look at the nest, and discovered that it contained 2 eggs, so I took a picture of the nest and eggs.

We left Hluhluwe (pronounced "shlooshloowee") on Oct. 5 and continued our adventure in Natal northward for about 75-80 miles to the tiny village of Ingwavuma, traversing the extreme southeast corner of Swaziland enroute. This was very dry, thickly wooded "bush" country. We were on our way to the small Ndumu Wildlife Sanctuary on the border of Natal and southern Mozambique. Our dry, dusty road was designed for small jeeps and was so narrow in places that both sides of our overweight camping van simultaneously scraped against the stiff roadside scrub (reminding me of our "bypass" road in the mountains of western Panama). Driving demanded all of my concentration to keep in the shallow established wheel ruts. I was happy the road was not wet and muddy. We camped that night a short distance along this road, in the middle of the bush. Supper consisted of soup and canned tuna. Fiery-necked Nightjars and Southern White-faced Owls ("Owlets": genus *Ptilopsis*, formerly *Otus*) called regularly throughout the night. (In Appendix B this owl is recorded for Oct. 8, 1959, from Swaziland where I first <u>saw</u> it.)

Ndumu was a small reserve of only 250 sq. mi. of wetlands, rivers, savannas, and woodlands, situated approximately 250 miles north of Durban where the Pongola River joined the Usutu River. As a result of its subtropical climate, high rainfall, and diversity of vegetation the reserve supported 430 species of birds, the greatest variety of any single locality in South Africa. Insects were also in great numbers and 66 different species of mosquitoes were known to occur! Since some of these were known to carry malaria, Noble and I took a preventative tablet daily. A characteristic tree here (and elsewhere) was an acacia tree with smooth yellowish-green bark and spiny nodules, popularly referred to as a "fever tree" (*Acacia xanthophloea*) because it grew in the more moist patches of savanna where there were malaria carrying mosquitoes, and one of the symptoms of this disease is a high fever. These 50-80 ft. tall trees were unusual in that photosynthesis occurred in the bark, the compound leaves having very small leaflets. Noble and I wore our new hiking shorts almost everywhere we went, and we continued to do so throughout the rest of our time in Africa. My short-sleeved T-shirts gradually became more and more ripped and ragged because of the many hours I spent in the bush, and eventually they looked just like what the natives wore so I affectionately referred to them as my "African" shirts.

African Hippopotamuses ("hippos") were widespread in the rivers and lakes at Ndumu, our first encounter with this icon. Next to elephants and rhinos it was the largest land animal in the world, weighing up to 5,000 pounds. Hippos spent most of the daylight hours floating lazily under the surface of lakes and rivers, with only their eyes and nostrils above water. In times of danger they submerged entirely for periods of 5 minutes or more. Hippos are 2-toed, barrel shaped ungulates, almost completely hairless. At night they emerge from the water and venture on shore to graze on grasses, sometimes wandering for several miles. They are quite territorial and mean-tempered and will readily charge a person if one gets too close, chasing him at speeds of 20 mph or more, faster than most people can run. Therefore they were responsible for more deaths annually than any other animals in Africa except crocodiles and buffalos! Hippos mate, give birth, and even sleep under water (automatically surfacing to breathe). Noble took a color slide with my camera of me sitting on the shore of a lake looking through my telescope studying the hippos. Another mammal which we saw in Ndumu was the Red Duiker, one of the smallest antelopes. It was about 15-18 inches tall at the shoulders with tiny pointed horns 3-4 inches high which stuck straight up on the top of its head between equally high ears. Its overall dark reddish-brown color helped conceal it in the scrub. Two conspicuous reptiles in the park were crocodiles (our first encounters with these large, well known inhabitants of aquatic shorelines) and 4-5 ft. long monitor lizards (family Agamidae).

Birds were abundant during our 2 days at Ndumu (Oct. 6 & 7) and I recorded 38 species new for my trip list (Appendix B). These included the following birds of aquatic habitats: Goliath Heron* (the world's tallest heron at 5 ft. in height), Yellow-billed Stork*, African Openbill* (a stork), Glossy Ibis* (a worldwide species but my first ever sighting), African Spoonbill*, Squacco Heron*, Black Heron*, Intermediate Egret, African Pygmy-goose*, White-faced Whis-

Plate 36: Flat-topped Acacia trees in Natal Province were characteristic of grassland savannas throughout Africa. A Zulu woman was planting corn by hand. Wide-buttressed, grotesgue-looking Baobab trees were another African icon.

tling-Duck* (occurring also in the Neotropics, where I did not encounter it), Hottentot Teal*, White-backed Duck*, Lesser* and African* jacanas, Black Crake*, and Purple Swamphen*. In addition, three bush-inhabiting species were particularly memorable: (1) the Arrow-marked Babbler* (in the genus *Turdoides*, a group of captivating, noisy, gregarious, thrush-like birds colored in appealing patterns and shades of brown, chestnut, gray, black, and white), (2) the Four-colored Bushshrike* ("Gorgeous Bushshrike" of some authors, which was both thrilling to view, with its wide black necklace and bright red, yellow, and olive plumage coloration, and pleasant to hear, with its melodious, ringing "kong, kong, koit" vocalization), and (3) the Pink-throated Twinspot* (a strikingly patterned little estrildid finch with a brown back, black breast with large white spots, and bright pinkish-red face and throat). The possibility of encountering, by chance, such delightful species as these was the single most important factor which motivated my birdwatching. I remember the profound comment of an early birding mentor of mine - - "Birds are where you find them."

Noble and I departed Ndumu on Oct. 8 and retraced our route back through the thick bush to Swaziland. This was the shortest route to our next primary destination, which was Kruger National Park in the northeastern corner of South Africa (in Transvaal Province). We spent most of the day in Swaziland. I stopped several times to search for birds in the Brachystegia woodlands. Two of the most exciting species which I encountered were the Wahlberg's Eagle* (a small, polymorphic member of the genus *Aquila*, most frequently all dark in color) and Gray Go-away-bird* (a subdued all gray member of the turaco family, an inhabitant of dry bush and open woodland, and named for its familiar, oft- repeated, nasal "go-away" vocalization, almost always sounded whenever a bird first encountered a person). Other birds were the Bearded Scrub-Robin*, S. White-faced Owl* (my first sighting), Neergaard's Sunbird* (a local endemic), and African Broadbill* (in the family Eurylaimidae). Unlike its colorful Asian relatives the broadbill was rather drab in appearance, being brown above with a black cap and concealed fluffy white rump patch, and white below with prominent black streaking. It was the most common and widely distributed of the four, mostly very local, African broadbills. I didn't encounter it again until Tanganyika.

We left Swaziland shortly before sunset and crossed the border eastward into the Portuguese colony of Mozambique, a long irregularly shaped country extending north-south along the Indian Ocean. Soon thereafter I chose a campsite in a relatively secluded spot just off the road. This diversion, in and out of Mozambique, was the shortest route for us to Kruger National Park.

The next morning at daybreak, Oct. 9, a small number of villagers had gathered around our Jeep to stare at it, as usual, waiting for anyone to step out. Since the bush was our only toilet this meant that we had to walk some distance away to perform our early morning necessities as discretely as possible. In the vicinity of our campsite were African Palm-Swifts* and Senegal (Lesser Black-winged) Lapwings*. After a quick breakfast and tidying up the van we were on our way for the short, 45 minute drive eastward into Lourenco Marques, the capital city of Mozambique at the extreme southern end of the country, on the Indian Ocean. We stayed in the city only 6 hours, long enough for a welcome hot meal, and picture taking along the waterfront of fishermen, boats, and fruit markets. We couldn't escape our customary interview with a young reporter on the street who wanted a story for the local newspaper, "Noticias." By late afternoon we were once again on the road, heading back to the border of South Africa. Thirty minutes later as the sun was going down I picked out a campsite in an area of bushed savanna and woodland, 15 miles northwest of Lourenco Marques (still just inside Mozambique). Another eventful day had ended.

Noble wanted to bring his journal up to date the next morning for a couple of hours (Oct. 10), so this time was available for me to pursue birds in the surrounding countryside. I located 4 species I had not yet seen: Lizard Buzzard*, Black Cuckoo*, Small Buttonquail*, and Gabar Goshawk* (another of the 49 species of accipiters in the world). By mid-morning we were underway again, on an excellent paved highway connecting Lourenco Marques with Johannesburg. We crossed the border back into South Africa at the Mozambique town of Ressano Garcia. Custom formalities were very simple and required only a few minutes on both sides of the border. We were now in the South African province of Transvaal, a few miles south of the southern entrance to Kruger National Park. It was an exciting moment for both of us.

Kruger was famous throughout the world for its diversity of large game animals, and one of the most visited national parks in all of Africa. It was a top priority for both of us. In 1959, the number of visitors was considerably less than it was in subsequent decades. Never-the-less, there were strict rules which applied to all visitors bringing their own vehicle into the park, most specifically that one must never get out of their vehicle and must camp only in designated, fenced campsites. (Such stipulations applied to most national parks in Africa.) Wild animals always had the right-of-way, and feeding them was strictly prohibited. Because of these rules most animals were relatively unafraid of vehicles and allowed visitors to approach them closely in their cars, affording the opportunity for close-up pictures of such large animals as elephants, rhinoceroses, giraffes, and even lions. A car was like a zoo cage where wild animals could come close to view the people inside! We were fortunate to be in the park during the dry season when the roads were passable, and at a time of the year when temperatures were mild.

Kruger was a very large reserve with an area of 7,500 sq. mi. stretching N-S in a 30-50 mile wide band for over 200 miles, all the way north to the border of Southern Rhodesia. On the east it was bordered by Mozambique for its entire length, and the Tropic of Capricorn crossed the upper middle so that the northern third of the reserve was inside the tropics. Several rivers ran through or bordered the park, which was mostly flat or gently rolling country below 2,000 ft. in elevation. There were numerous seasonal shallow lakes ("pans"). The principal vegetation was grassland savanna with scattered Baobab trees and acacias (including fever trees), but riparian (riverside) forests and Brachystegia ("miombo") woodlands provided habitat diversity for the great variety of wildlife. Terrestrial termite mounds up to 12 ft. or more in height dotted the plains and open woodlands.

Baobab trees (genus *Adansonia*) were an African icon and were one of the park's most characteristic features in the grassland savannas. These grotesque looking, bottle-shaped trees were 15-100 ft. tall with an enormous 20-35 ft. diameter trunk. The wide branches, leafless much of the year, stuck up into the air at all angles much like the waving tentacles of an octopus. The extremely thick cork-like bark of the trunk sometimes held as much as 30,000 gallons of water, which could be used by wildlife and people in times of water shortage. Baobabs were hollow inside, providing shelter and homes for barbets, hornbills, swifts, bats, squirrels, galagos, bees, and many other kinds of wildlife. There was only one species of baobab in Africa but 6 species occurred in Madagascar, and one species was native to northwestern Australia. Near the outskirts of the park I drove our camping van off the road and up to a baobab tree and parked in front of it, facing it just long enough for me to get out and quickly take a photograph before a park ranger could come along and notice us off the road.

Noble and I spent 5 days in Kruger (Oct. 10-14), slowly navigating the park's roads from south to north for the entire length of the reserve. We had no fear that an elephant would be able to push over our 4½ ton truck, although they regularly pushed over trees 1-2 ft. in diameter, and rhinos, we thought, would have difficulty putting much of a dent in the van's 18 gauge steel perimeter. In spite of its negative features our Jeep was a very sturdy, almost impenetrable fort on wheels. We were prepared for the African wildlife, and our wide, wrap- around front windshield afforded us great viewing over an extensive area. On several occasions a park ranger, armed with a rifle, guided us on foot to a river or shallow lake for wildlife viewing and photographing. We saw all the hunter's "big five": African Elephant, Lion, Leopard, "rhino" (both species), and African Buffalo. All but the rhinos were first sightings for us of these iconic mammals. I took as many pictures as my limited budget would permit, but I had very little money for purchasing film and was thus restricted in my picture taking. Sadly, many of my pictures were underexposed because my camera was not automatic and I had no way of measuring light. Therefore, I always had to guess where to manually set my shutter speed and my aperture opening for the existing light conditions, and I had a tendency (unknowingly) to underexpose my pictures. I mailed my exposed film home for developing and didn't see any of my pictures until after I returned home to California more than 2 years later, although my father had the exposed film developed when he received it and he would periodically send me a note to say that my camera was still functioning OK, without critiquing my pictures!

Mammals which we saw for the first time (in addition to the above) were Blue Wildebeest (quite distinct in color and appearance from the Black Wildebeest, with which it did not overlap in geographic distribution), Bushbuck, Nyala (a close relative of the Bushbuck with slate gray coloration, narrow vertical whitish stripes across its back, and distinctive dorsal and ventral manes running from head to tail), Side-striped Jackal (filling the Coyote niche), and

Vervet Monkey. This latter species was common throughout the park and made a nuisance of its self by climbing on vehicles and grabbing food, as it did many places elsewhere in Africa. Other mammals which we encountered were species we had seen previously (and would see many more times): hippopotamuses, giraffes (which would stick their head in our window if we left it open), zebras, kudus, waterbucks, impalas, and baboons. We were unable to identify (to the species level) some of the smaller antelopes we observed, as well as the squirrels and mongooses. Crocodiles and monitor lizards were common along the river banks.

All of the 41 new trip birds which I identified in the park are listed in Appendix B (Oct. 10-14, 1959). Ostriches* were frequently encountered, walking sedately across the open grasslands or running down the middle of our road in front of us, sometimes with 10-12 little chicks. Both parents looked after their offspring, which could run as soon as they were hatched. Ostriches were flightless ratite birds (without a keel on their sternum) and were the largest of all living birds, standing 7-9 ft. tall, weighing up to 300 lbs., and running as fast as 40 mph (considerably faster than a man). Males were a little larger than females. They are the only bird with just 2 functional toes on each foot. Their rudimentary wings provide evidence to ornithologists that they evolved from a flying ancestor, as did Emus in Australia and Rheas in South America. These species are distant relatives of one another, filling the same ecological niche on each of the three continents. Their common ancestor probably lived as far back as the breakup of Gondwanaland at the end of the Mesozoic, in late Jurassic or Cretaceous times 60-120 million years ago (unless one assumes that their flightless behavior and other similar features evolved independently on each of these 3 continents, which is an unlikely scenario). Ostriches were mostly vegetarian in their diet, eating seeds, fruits, grasses, flowers, and shrubs. They now lived only in Africa but formerly their range also extended to the deserts of Syria and Arabia. They were an African icon.

A large, terrestrial-foraging, turkey-like bird of the open woodlands in Kruger was the Southern Ground-Hornbill*. It was completely black in plumage coloration except for its all white primaries, which were very conspicuous in flight. Other distinguishing features were its bare red face and throat pouch, and a black bill with a small basal casque. It fed on large insects and small terrestrial vertebrates (frogs, lizards, snakes, rats, and mice), and utilized trees for roosting at night and for nesting, in large holes. Almost 3½ ft. tall, these birds (along with the geographically separate Northern Ground-Hornbill) were the largest of all 29 species of African hornbills. I was happy to obtain a clear, though distant, picture of this species.

Perhaps the most thrilling bird in Kruger for me was the 5 ft. tall Saddle-billed Stork, arguably the most handsome of the world's 19 species of storks. This majestic bird was strikingly patterned in black and white, conspicuous both while standing and in flight. Its large, pointed, slightly upturned bill and its long slender legs were prominently colored red and black. My first observation of this species was of a single bird standing among some palm trees at the edge of a small stream, perhaps 200 or 300 yards from our road. I was so excited that I decided to risk life in jail by walking out across the savanna to get close enough for a picture. So I did. (Of course the bird flew just as I was focusing, and the resultant picture was out of focus, again). No, I was not observed by the rangers during the time I ventured across the open veld, away from my car.

Some of the pictures I was successful in obtaining at Kruger were the following: Impalas drinking at a waterhole, Nyalas in an open grassy woodland, an Elephant digging with its feet and trunk for water in a dry sandy streambed, a pair of Ostriches with ten very small chicks running down the road ahead of us, Waterbucks in an open woodland, 2 crocodiles on a flat, bare bank of the Luvuhu River, a Fork-tailed Drongo harassing a Wahlberg's Eagle in the top of a small leafless tree, a tall slender termite mound of red dirt, a Verreaux's (Giant) Eagle-Owl* standing conspicuously in the daytime among the dense leafless mid-story branches of a small roadside tree, Vervet Monkeys sitting on the ground next to our camping van, a Knob-billed (Comb) Duck standing in the top of a small tree in which there was also a nesting colony of Red-billed Buffalo-Weavers*, several yellow-barked "fever trees" at the edge of a dry pan, a Marabou Stork standing in the top of a bare tree, a Greater Kudu standing in open woodlands, a young park guide with his rifle leading us to a hippo pool on the Limpopo River where there was a "BEWARE OF CROCODILES" sign on the trunk of a large fever tree, a distant Blue Wildebeest on the savanna, and finally, a standing White-backed Vulture in a tree silhouetted against the sunset. Kruger was an unparalleled experience!

We spent our last day in the park, Oct. 14, at the far north end of the reserve, between Pafuri and Punda Maria Camp. I was able to find another species of *Vanellus* for my favorite list, the boldly patterned White-headed (White-crowned) Lapwing*, a bird of river sandbanks, and I also saw my first African Finfoot (one of 3 species in the pan-tropical family Heliornithidae, members of which were often referred to as "sungrebes"). Additional species which I saw for the first time were 5 Crested Guineafowls* and a Brown Snake-Eagle*.

We finally left the park in mid-afternoon and headed west toward Makhado, stopping to camp for the night shortly thereafter about 40 miles east of this town, in the far northeast of Transvaal Province. The next morning (Oct. 15) we continued our journey on a southwestward bearing toward Pretoria and Johannesburg, travelling back across the Tropic of Capricorn and once again inside the southern temperate region of the continent. Roadside birds included my first sightings of the Marsh Owl* (a close relative of the Short-eared Owl), Temminck's Courser*, and White-browed Sparrow-Weaver*. We camped that night near Nylstroom, at an elevation of 4,500 ft. on the Transvaal plateau. On the next day, Oct. 16, we continued to Johannesburg, via Pretoria. Additional roadside birds were Spotted Thick-knee* ("Cape Dikkop"), White-bellied Bustard*, Latakoo Lark* (a very local endemic of the grasslands here), African Quailfinch*, and my first White-quilled Bustard* (a taxonomic split from the Black Bustard).

We devoted the next 8 days in Johannesburg to Jeep maintenance, banking, visa applications, laundry, re-supplying, sightseeing at the nearby gold mines, writing letters home, and dining with new acquaintances. Noble found many persons to enlighten him on local government and economics. Time went by quickly. At a dinner party in Pretoria to which we were invited we met and visited for over an hour with a middle-aged woman (Mrs. Cilliers) who had a degree in astrology from a European university. Our conversation was enhanced by "amarula", a special South African liqueur. Afterwards, Mrs. Cilliers said to us that if we would each give her our date, place, and time of birth she would be happy to send us, in the mail, a several page horoscope describing our past and predicting the future. Of course we did this. When I finally read my horoscope 3 months later in Nairobi I was astounded at the accuracy in which she referred to my past, and to the events she predicted for the future which eventually turned out to be surprisingly correct. My conclusion - - don't underestimate the power of the stars!

We left Johannesburg on Oct. 26 and proceeded toward the border of Bechuanaland, the neighboring country to the west (now called Botswana). Our 2 days of travel to the border took us across southwestern Transvaal Province, in a semi-arid grassland region at an elevation between 5,000 & 6,000 feet. This was wheat country, where many acres of golden fields were being harvested by workers with hand-held sickles. The severed wheat stalks were pitched with a hay fork onto ox-drawn carts in the fields for transport to the nearest location where the grain could be thrashed. Women working in the fields took time out to nurse their babies nearby. Small boys on foot tended cattle across the unfenced grasslands. Lanner Falcons pursued and caught large flying insects with their feet. Greater Kestrels* sat on the highest viewpoints, scanning for small birds, rodents, and lizards. They were easily distinguished from the non-resident, migrant Lesser Kestrels by their larger size and uniformly barred upperparts. Untidy looking nests of the White-browed Sparrow-Weaver were widespread in the sparsely distributed roadside trees. Southern Ant-eater Chats performed conspicuous aerial flights on fluttering wings, displaying their prominent white wing patch. These burrow-nesting, ant-eating, mottled brown birds of the plains often perched on the top of termite mounds. (I had first encountered them in the Karroo.) Chestnut-backed Sparrow-Larks*, Pied (Jacobin) Cuckoos*, Red-headed Finches*, and Violet-eared Waxbills* (one of my favorite estrildids) were all recorded for the first time. Two Spotted Thick-knees were seen, and later heard near our campsite at dusk on the evening of Oct.26, west of Ventersdorp. According to Roberts their "melancholy whistling notes cause misgivings in the minds of the superstitious." Again, I thought happily to myself that this was the real Africa.

We reached the border of Bechuanaland the next day, on the afternoon of Oct. 27. After 6½ weeks in South Africa my list of new trip birds recorded from the country was almost 400 species (Appendix C), the greatest number for any single country during our entire 3 years of travel. Having a good illustrated field guide eliminated the need for writing time consuming field descriptions, and enabled me to recognize on first sight many of the birds I encountered. South Africa was a great beginning to my African wanderings!

Plate 37: Elephants had the right-of-way in Kruger Natural Park. Stately giraffes and food-snatching Vervet Monkeys were among the scores of wildlife for which Kruger is renowned around the world.

Chapter Thirteen -- Bechuanaland to Nyasaland

Noble and I crossed the border between South Africa and southeastern Bechuanaland (Botswana) on Oct. 27, shortly after noon at the S. Afr. town of Mafeking. Customs and immigration were quick and without incident, owing to the fact that Bechuanaland was a British Protectorate and everyone on both sides of the border spoke English. The country was a little larger than California in size. Our entry point was on the SE edge of the vast Kalahari Basin, an area of almost one million sq. miles which covered most of the country, generally between the elevations of 2,500 and 4,000 ft. above sea level. Average annual rainfall in the basin varied from a low of 4" in the southwest to a maximum of 20" inside the northern and eastern edges. The large interior of the basin received less than 10 inches of rainfall and thus qualified as a true desert, the Kalahari Desert. Typically there was a long dry cold season (March through October) and a short wet, hot season between November and February. We were arriving at the end of the dry season, but the whole region had been in a severe drought for the past two years and many of the pans had been without water for this extended period of time. Wildlife had either perished or migrated elsewhere. Unfortunately, it was not a good time for viewing large game animals, or very many of the birds.

The basin was covered primarily by grasses and herbs, which varied in the amount of ground cover dependent on the rainfall. However, acacia trees were widely distributed, and along the northern and eastern borders there were dry "mopane" woodlands characterized by Mopane trees (*Colophospermum mopane*) and African Teak trees (*Baikaea plurijuga*). Mopane trees grew on poorly drained alkaline soils and attained heights of 15-60 feet. They were a member of the legume family and were distinguished by butterfly-shaped (2-winged) leaves. A small variety of birds occurred almost exclusively in mopane woodlands. The San people ("Bushmen") still sparsely inhabited the basin, following their traditional lifestyle as nomadic hunter-gatherers and pursuing game with bows and poisoned arrows.

In 1961, the Central Kalahari Game Reserve was established in the middle of the basin, an enormous area of over 20,000 sq. miles, making it one of the largest protected land areas in the world. In the far northern part of the basin was the world famous Okavango Delta, a vast permanently flooded wetland situated in a tectonic trough into which the Okavango River emptied (from the north). It was a mecca for wildlife and attracted tourists from all over the world, and was designated a UNESCO World Heritage Site in 2014. The delta had no outlet for the water that came in from the river, and the water eventually either evaporated or was transpired by the vegetation. Unfortunately, the delta was too far north from our planned route for us to include it in our adventure.

Noble and I only skirted the eastern edge of the Kalahari Basin in our Bechuanaland trek as we traveled north for three days (Oct. 27-29) from Ramatlabama (across the border from Mafeking) to Gaborone, Francistown, and Nata, a total distance of 500-600 road miles. En route we crossed the Tropic of Capricorn once again, and re-entered the tropics. Our dirt and gravel road was dry and dusty but generally well maintained, and not hazardous. I continued to do the driving. The weather was very hot on some of the days, although it was not quite yet summer. One afternoon the temperature reached 103° inside our van, so I took time out to paint the top of our van's metal roof with a reflective silver color in an effort to keep the van a little cooler on the inside.

On Oct. 29, we were detained for several hours in Francistown while our front axle was welded back together after it had, once again, developed a crack. Such problems never went away.

After we arrived in Francistown, Noble and I decided that we wanted to view one of the larger salt pans in the extensive Makgadikgadi group of salt pans, with the hope of seeing some large game animals. According to our road map, this group of pans was situated just southwest of the small town of Nata, 120 miles northwest of our present location in Francistown. So after getting our jeep welded, instead of driving straight out of town on the main road to Bulawayo we turned left onto a less traveled road and headed northwest toward Nata. We arrived as it was getting dark and camped off the road just south of the town, in an area of thorn veld and Brachystegia woodland at an elevation of 3,000 feet.

During our first three days of travel in Bechuanaland (between Ramatlabama and Nata), I recorded 16 new species for my trip list (Appendix B). I photographed two of these, the Social Weaver* with its huge communal nests, and a Three-banded Courser* as it walked cautiously across a woodland opening (one of my most exciting memories). Burchell's Sandgrouse* flew back and forth to water in the early mornings (as did almost all members of the desert inhabiting family Pteroclidae). Black-cheeked Waxbills* (considered conspecific with the allopatric Black-faced Waxbill by Clements) were one of the most attractive estrildids, adorned in colors of black, gray, and pink. Crimson-breasted Gonoleks* (a magnificent bushshrike with black upperparts, scarlet underparts, and a long white bar along the inner wing) sang pleasantly back and forth in duets. Southern Pied-Babblers* (genus *Turdoides*) foraged in small groups through the dry understory, conspicuous in their all white body plumage and solid black wings and tail. Flying above the woodlands were flocks of migrant (or maybe resident) European Bee-eaters* and a handsome, easily recognized Black-breasted Snake-Eagle* (genus *Circaetus*), a widespread African raptor. Another memorable raptor was the Dark Chanting-Goshawk*, a rather uniformly dark gray bird with a bright red cere and long, bright red legs, 1 of 3 species in the African genus *Melierax*. A pair of brown and green colored Meyer's* (Brown) Parrots flew through the trees, screeching loudly. This parrot is one of nine species in the African genus *Poicephalus*, and the most widespread member of this group in East Africa. In spite of the dryness, there were some exciting birds in the Bechuanaland bush.

On the following morning, Oct. 30, I arose at dawn and went for a short walk in the surrounding scrub. I was thrilled to spot a Bat-eared Fox sunning itself in a grassy woodland opening. Its huge ears were standing upright in a "V" on top of its head, providing me with yet another vivid memory of wild Africa. Soon thereafter we got underway and drove into the town of Nata, where we turned west on a good gravel road. Fifteen miles later we came to the northern edge of the Makarikari Pan (one of many pans in the Makgadikadi complex, according to our map). It was an extensive flat grassy area with scattered patches of shallow water of varying sizes. Many of the smaller "vleis" were crusted with salt around the edges. There were carcasses of dead wildebeests scattered about. The drought had taken its toll. Nowhere were there any concentrations of large game animals, and all that we encountered were small groups of zebras and wildebeests. We were quite disappointed. To console us, Noble decided to get out his 16mm movie camera, climb up to the roof of our van, and take action movies of zebras for 5-10 minutes as we chased them in our Jeep back & forth across the dry, grassy, flat edges of the pan. We drove only fast enough to keep the zebras just in front of us as they zigged-zagged back & forth. It was a great highlight for Noble!

We then retraced our route back to Nata, at which point we decided to follow a secondary road heading east out of the town. Our map indicated this would take us across the border into Southern Rhodesia and eventually intersect with the main highway between Francistown and Bulawayo. It would save us many miles of travel if we did not drive all the way back to Francistown. We were willing to attempt this route, so we set out. This narrow, little used road traversed many miles of open grassland, thorn scrub, and Brachystegia woodland, but it was quite passable, without any hazards. It was a good choice. We camped that night in the bush 80 miles east of Nata, just before the village of Maitengwe on the border of Southern Rhodesia (Zimbabwe).

Plate 38: From South Africa our route headed west briefly into Bechuanaland, on the eastern edge of the vast Kalahari Desert, where the country was in a severe drought, and many animals were dying. We chased zebras in our Jeep across the dry Makarikari Pan.

It was our last night in Bechuanaland. There were many birds along our track, and I particularly enjoyed the Southern Pied-Babblers, flocks of European Bee-eaters, Yellow-billed Hornbills, Long-tailed Shrikes, Gray Go-away-birds, Meyer's Parrots, and flocks of Quailfinches. I saw my first Double-banded Sandgrouse*, Abdim's (White-bellied) Storks*, Swallow-tailed Bee-eater*, and Violet-backed (Plum-colored) Starlings*. Other notable species included Blacksmith Plover, Arrow-marked Babbler, Swainson's Francolin, White-crowned Shrike, Lilac-breasted Roller, White-backed and Lappet-faced vultures, Secretary-bird, Tawny Eagle, White-quilled Bustard, Ostrich, Gray Crowned-Crane, Kittlitz's Plover ("Sandplover"), and Capped Wheatear. Scops-Owls called repeatedly throughout the night.

We entered Southern Rhodesia the next day, Oct. 31, and proceeded 100 miles east to the city of Bulawayo, joining the main road not far out of the city. Of the 140,000 inhabitants in the city, 40% were white settlers and 60% were non-white. We were invited to have a rare hot bath - - at the railroad station! Twenty-five miles south of the city were the "Matopos Hills", a rocky, hilly area in which Matopos National Park was located. The park was a world famous archaeological site containing many ancient rock and cave paintings by Bushmen, dating back 2,000 years. It also was the location of Cecil Rhodes grave, which Noble wanted to visit. So we drove to the park and camped there that night.

We spent all of the next morning in the park (Nov. 1), allowing me this time to hike and record birds in the wooded valleys and among the granite "kopjes" (small hills), at elevations up to 5,000 feet. I found nine birds that were new for my list (Appendix B), the most exciting of which was the African Harrier-Hawk (genus *Polyboroides*), a predominantly gray and black bird with a yellow face and a prominent white band in the middle of its tail. It was a large raptor with broad wings and a long tail, but a relatively small head, and it flew in a peculiar loose, floppy manner. Its most unusual feature was its manner of feeding. It clung with one of its long, double jointed legs to a tree trunk, using its wings for support, and with the other leg it reached inside holes, cracks, and crevices to extract lizards, birds, large insects, and other vertebrate or invertebrate prey. Harrier-hawks also regularly clung, with flailing wings, to the hanging nests of weaver birds to remove young birds from the nests. Additional birds which I added to my list in the park included the attractive black & chestnut colored Harlequin Quail* (a *Coturnix*) and the Wattled Lapwing (the largest *Vanellus*, with a brown body, white crown, long yellow facial wattle, and the usual white wing band and white rump characteristic of the genus). Two species of small mammals which I was able to identify were the Southern Dwarf Mongoose and the Smith's (Yellow-footed) Bush Squirrel (genus *Paraxerus*). Although I didn't see any, Leopards were said to be quite numerous in the park, where more than half their prey items were the locally abundant Rock Hyraxes.

We left the park at noon, drove back into Bulawayo and then continued toward our next destination, Wankie Game Reserve, 125 miles northwest across "Matabeleland." Both lanes of our highway were initially asphalted, then after a short distance only one lane was asphalted, and then sometime later there were simply two narrow asphalted strips, one for the right pair of tires and one for the left! Eventually, some 20 or 30 miles from Bulawayo, the road became entirely gravel and we were once again on a dusty road. Along the way, next to the roadside, was a pair of lovely Southern Carmine Bee-eaters*, certainly one of the most beautiful of all birds with their reddish-maroon, aqua, and blue plumage. As in many bee-eaters the central pair of tail feathers was narrow, pointed, and greatly elongated. Of the 26 species of bee-eaters in the world (family Meropidae) the majority lived in Africa, where they were one of the most colorful and conspicuous components of the avifauna, and one of my most favorite groups of birds. These tropical and subtropical Old World birds were conspicuous with their "fly-catching" behavior, characteristically sallying out after passing insects from exposed high or low perches. The bee-eaters were a thrilling finish to the day for me. The unexpected encountering of a species was one of the most exciting aspects of pursuing birds, and not knowing what I would find was a primary stimulus for my avid searching.

Wankie National Park was situated in the NW corner of Southern Rhodesia and was the country's largest game reserve, with an area of 5,600 sq. miles. It was also the oldest, having been established in 1928. The reserve was named after a local Nhanzwa chief, Hwange Rosumbani, and it claimed to have a greater diversity of mammals – 108 species – than any other national park or game reserve in the world! Nineteen kinds of large herbivores and 8 species of large carnivores were known to inhabit the park. The mammalian fauna included such species as the Lion, Leopard, Cheetah, Hyena, Hunting dog, Jackal, Bat-eared Fox, Baboon, Elephant, Black & White rhinos, Giraffe, Eland, Kudu, Buffalo, Wildebeest, Waterbuck, Zebra, Sable Antelope, Impala, Warthog, and many smaller species. The elephant herd was the largest in the world with tens of thousands of individuals which migrated back and forth between Wankie and the nearby Chobe ecosystem across the neighboring border of Bechuanaland. With more than 400 species of birds in the park, including 50 kinds of raptors, Wankie was truly a wildlife paradise. Noble and I would encounter only a tiny fraction of the total wildlife here during our 3½ days in the park (Nov. 2-5). Noble said the park was even more exciting for him than was Kruger.

The habitat of Wankie was a mixture of grasslands and woodlands, interspersed with frequent natural and man-made sources of water. It was situated at an average elevation of 3,500 ft. above sea level, and received 20-25" of rainfall annually, quite a dry environment. Baobab trees were scattered across the savannas, and mopane, teak, and Brachystegia trees dominated the woodlands. The reserve had over 300 miles of tracks and roads. We met two of the parks young rangers, Tim Braybrooke and Harry Castle, and they took us around the park in an uncovered Land Rover, often off the designated roads. On one occasion Harry insisted on driving up to within 50 yards of a very big bull elephant, face to face, causing it to threaten us by fanning its huge wide ears. I was more than a little frightened since our open Land Rover would have been easily crushed if the elephant chose to charge us. Harry wanted to prove to us that he could bluff the elephant and it would back down, which fortunately it did! On another occasion we drove up to within 35 feet of a lioness lying on the ground and took a close-up picture as she concentrated on eating a freshly killed buffalo calf.

We encountered three new mammals for our list: Eland, Sable Antelope, and Steenbok (also spelled Steinbok). The Eland was the largest antelope in the world, males of the East African subspecies weighing up to over 2,000 lbs. and standing an average of 5 ft. tall at the shoulders. It was somewhat ox-like in shape with a prominent dewlap and heavy, posterior directed 3 ft. long horns with spirally twisted bases. (A slightly larger subspecies, the "Giant Eland", inhabited West Africa.) The Sable Antelope was a relatively large, somewhat undistinguished looking antelope except for its spectacular, heavy, 5 ft. long horns that curved in a long arc backwards from the top of its head. Males weighed up to 550 lbs. and stood an average of 4½ ft. tall. They were predominantly dark chestnut or black in color, and were placed in the same subfamily as the Roan Antelope and the Oryx. The Steenbok was one of the numerous small reddish-brown antelopes, being only a little larger than a Red Duiker, from which it also differed by its slightly longer more pointed and slender horns. Noble again kept a count of the number of each game animal we saw, and I took pictures of many, particularly those drinking at waterholes.

During our 3½ days in the reserve, I tallied a total of 171 species of birds (40% of the total list for the park). Eighteen of these species were new additions to my trip list (Appendix B, Nov. 2-5), and 17 species were diurnal raptors. I particularly enjoyed watching the aerial courtship display of the male Red-crested Bustard (Korhaan)*. This involved an individual flying straight up into the air, 100 feet or more, then suddenly tumbling vertically downward almost to the ground with its wings and crest displayed, before pulling out at the last second and gliding a short distance to a landing. Other birds of particular note were large flocks of Pin-tailed Whydahs in the savannas (mostly in non-breeding plumage), and a mixed group of three kinds of falcons -- Lanner Falcons, Eurasian Hobbies (a non-breeding visitor), and Lesser Kestrels – which were all feeding aerially on flying insects. Also there was a noisy flock of 50 migrating European Bee-eaters high overhead and a flock of 500 migrating African or Common Swifts foraging low over a vlei. I took photos of a fish-eagle perched in a tree, a nesting colony of White-browed Weavers, a Long-tailed Shrike, a Dark Chanting-Goshawk sitting on a nest, and White-bellied Storks flying overhead in the late afternoon with the golden sun reflecting from their white bellies. The African Spring was an exciting time of the year.

On the afternoon of Nov. 5, we left Wankie and drove the short 30-40 mile distance northwest to Victoria Falls, on the upper reaches of the mighty Zambezi River. The small town of Victoria Falls was on the south side of the river in Southern Rhodesia, and the town of Livingstone was on the north side in Northern Rhodesia (Zambia). It was here that the Scottish explorer David Livingston was probably the first European to ever view the falls, in 1855. We camped near the falls that night. A half dozen Rock Pratincoles* (family Glareolidae) were on the rocks along the river. These small dark birds with a bright red bill were strikingly patterned in black, gray, and white. The species was locally distributed in Middle and West Africa on large rocky rivers. Victoria Falls was an ideal place to find them.

We devoted all of the next day, Nov. 6, to sightseeing and picture taking at the falls. Victoria Falls are considered to be one of the three "Great Waterfalls" of the world, along with Niagara Falls and Iguaçu Falls. They were designated as a UNESCO World Heritage Site in 1989. The falls are 354 ft. high and more than a mile wide, and are said to have the largest single sheet of falling water in the world, at maximum flow. The spray created by the falls rises as high as 1,300 ft., producing magnificent rainbows when the sun (or moon) is at the right angle. The falls drop over the edge of a wide basalt plateau and drop into a narrow chasm which is only 200-400 ft. wide. Below the falls the river continues through a long series of gorges. At the time of our visit the amount of water coming over the falls was almost at an all time low, resulting from the current long drought. Never-the-less, the falls were a spectacle we would never forget. Among several new birds for my list was the Red-billed Quelea*, a tiny estrildid finch distributed over much of the African continent, which sometimes occurred in flocks of several millions of birds! I saw only a small group of a dozen individuals. We camped again at the falls.

The next morning, Nov. 7, we drove into the town of Victoria Falls, not far below the falls. The streets were lined with beautiful red-flowering, wide-spreading Royal Poinciana trees (also called Flamboyant trees or Flame trees). This tree, *Delonix regia*, was native to Madagascar but was grown all over the warmer parts of the world as an ornamental. It was a member of the legume family, with long hanging pods and fern-like leaves. A new bird for my list was the Grosbeak (Thick-billed) Weaver*, a reed nesting member of the Ploceidae with a massive bill, which in overall appearance reminded me of the South American seed-finches (*Oryzoborus*).

Between the towns of Victoria Falls (in S. Rhodesia) and Livingstone (in N. Rhodesia) there was a 650 ft. long, single arch steel bridge over the Zambezi River spanning the second gorge below the falls, at a height of 420 ft. above the river. The bridge was designed and built in England in 1905, and was then shipped to Beira (Mozambique), from where it was transported by rail to the town of Victoria Falls. The bridge carried road, rail, and foot traffic across the river. It was officially opened on Sept. 12, 1905, by Prof. George Darwin, son of Charles Darwin! Both N. and S. Rhodesia were at one time united under the single name of "Rhodesia", after Cecil Rhodes, the influential British businessman, politician, and mining magnate who lived in South Africa most of his rather short lifetime, working to expand the British Empire throughout the region. When Noble and I drove across southern Africa in 1959, this part of the world was still under British rule, but it would be only a very few years later that independence from European imperialism would occur throughout almost all of Africa.

We left the falls that morning (Nov. 7) and headed NE from Livingstone toward the Northern Rhodesian capital city of Lusaka, 290 miles away. The gravel road was in excellent condition and we were able to travel between 40 and 45 mph, an extremely rare event for us anywhere in Africa. Brachystegia woodlands dominated most of the countryside, but I did not stop for birding. There were almost half a dozen dead Pennant-winged Nightjars lying along a short stretch of road where they had apparently been killed by traffic the previous night. I stopped to identify them. Several were males in breeding plumage with greatly elongated, white inner primaries. I salvaged a wing for a souvenir (which I still have in my possession today). After it got dark there were several of these spectacular nightjars flying back and forth across the road, catching insects. I was curious how abundant this species might be, as well as what other nocturnal birds might occur along the road, so I decided to continue driving after dark. Not far before we got all the way to Lusaka there was a major

Plate 39: Wankie National Park in Southern Rhodesia was perhaps our most favorite park in all of Africa. We spent three and a half days there observing the tremendous variety of wildlife, particularly the elephants and lions. I recorded 171 species of birds (40% of the total park list).

Plate 40: The Eastward flowing Zambesi River drops 354 ft. over the edge of the Zambesi escarpment, forming Victoria Falls. With a one mile width, the Falls can produce the largest curtain of falling water in the world. Also pictured is a ten-foot high termite mound.

road junction south of the city, and a highway to the right carried traffic back toward Southern Rhodesia and the Zambezi River. I took this turn (after dark) and soon thereafter I counted 16 Pennant-winged Nightjars in the span of thirty minutes along a ten mile stretch of this road, between 10:00 and 10:30 pm. Their long, conspicuous white middle wing feather trailed behind them, fluttering like a pennant. A wide white wing band added to the distinction of males. This nocturnal experience was another unforgettable memory for me of the African bush. A few minutes later we turned left off this main road and drove several miles to a small tourist park overlooking the scenic Kafue River Gorge, where we camped for the night, at an elevation of 3,000 feet. It had been a long, rewarding, adventuresome day.

We spent the next morning, Nov. 8, in the undisturbed visitor campgrounds. While Noble slept in I had a few hours in which to search for birds. We were surrounded by Brachystegia woodland, and there was also riparian vegetation along the river gorge. My most notable sightings were the Pale-billed Hornbill* (an uncommon endemic to this region), Gray Tit-Flycatcher* (genus *Myioparus*), Peter's (Red-throated) Twinspot* (an attractive little estrildid finch), Greencap Eremomela*, and Rufous-bellied Tit* (another endemic to this region). After a leisurely morning, we packed up and drove the short 45 mile distance south to the Zambezi River. Here we crossed a bridge and once more entered Southern Rhodesia, having spent only 2 days in Northern Rhodesia. A small settlement had been established around the 3,000 acre Chirundu sugar plantation on the south side of the river, at an elevation of 1,500 feet. Noble was so enthralled with the enormity of the operation that we stopped and he introduced himself to the manager, an Englishman by the name of Graham Lester. Immediately they were engaged in warm conversation and Mr. Lester invited us to spend the rest of the day with him, while he showed Noble around the plantation and explained the economics of his immense operation. I happily studied my bird books. We camped near the main house that night.

Birds in the vicinity of the plantation homestead the next morning (Nov. 9) were such abundant and widespread species as Common Bulbul, Ring-necked Dove, Long-tailed Starling, Southern Masked-Weaver, and Red-billed Quelea. A flock of about 500 weavers (*Ploceus* sp) in an overgrown field were all in non-breeding plumage, and went unidentified by me. It was not until mid-afternoon that we got underway to visit the nearby Kariba Dam, about 35 miles upriver. We arrived an hour before sunset and I picked out a campsite for the night. I saw my first Mosque Swallow*, and a migrating flock of 100 Southern Carmine Bee-eaters circled overhead. An African Scops-Owl and a Square-tailed (Mozambique) Nightjar called occasionally during the night.

We spent the next morning (Nov. 10) viewing and learning about the dam. It had been finally completed only a few months before our arrival, and became one of the largest dams in the world, 420 ft. high and almost 2,000 ft. long. It created Kariba Lake, which was 175 miles long, up to 20 miles wide, and contained more water by volume (44 cubic miles) than any other man-made lake in the world. It averaged 95 ft. deep and had a maximum depth of 320 feet, and was considered a great engineering accomplishment for its time. As a hydroelectric dam it generated 6400 GWh of electricity per year and 184,000 kilowatt hours daily. During its construction it was necessary to resettle 57,000 indigenous people, and to capture 6,000 large animals (plus a great many smaller ones) and move them to other areas, a project that came to be known as "Noah's Ark." We departed from the dam in mid-afternoon and drove back to the main Lusaka - Salisbury highway, where we turned right and headed southeast toward Salisbury, approximately 180 miles down the road. I chose a campsite shortly thereafter, at 4,000 ft. elevation in Brachystegia (miombo) woodland.

I was awakened the next morning (Nov. 11) by a pair of Black-collared Barbets vocalizing loudly in a duet outside our van. Here I saw my first Miombo (Stierling's) Wren-Warbler*, an eye-catching little bird with pronounced black and white barring on its throat, breast, and belly. Also present were several Cardinal Woodpeckers (filling the Downy Woodpecker niche), paradise-flycatchers, Scarlet-chested Sunbirds, and a Scaly-throated Honeyguide. Ever present Fork-tailed Drongos swooped aggressively at bush-squirrels. We ate a quick breakfast of fruit and corn flakes and got underway for Salisbury, at 7:00 o'clock.

Salisbury was the capital city of S. Rhodesia, situated at 5,000 ft. on the Highveld Plateau. We spent the next 4½ days here, engaged in our usual variety of activities, which as always included some maintenance and repair work on our Jeep. The right wheel bearing needed replacing and a brake booster system was installed to provide us more safety and quicker stopping capability. We visited several embassies to obtain visas for some of the countries on our agenda ahead of us, and one evening we were invited for dinner with the mayor of the city! Time went by quickly. At our campsite north of the city I recorded, on Nov. 14 & 15, several birds new for my trip list: W. Violet-backed Sunbird*, Spotted Creeper* (a certhiid in the monotypic genus *Salpornis*, with a wide, disjunct distribution in Africa and India), Miombo Rock-Thrush*, White-breasted Cuckoo-shrike*, and Afr. Pygmy-Kingfisher* (a non-aquatic species which was characteristic of dry woodland, and was slightly smaller than the Am. Pygmy Kingfisher).

We departed Salisbury on Nov. 16 in the late afternoon and drove a short distance out of town to a campsite just 5 miles south of the city. I birded until late the next morning at this site, on Nov. 17, while Noble wrote in his journal. Species I had not seen previously included the Locustfinch* (a tiny, uncommon estrildid, black in color with a red face and breast), Rosy-throated Longclaw* (perhaps the most colorful of all the species in the worldwide pipit family), and 2 kinds of cisticolas – Wing-snapping* and Pectoral-patch*. At noon we left our campsite and drove southeast on a well maintained, rather curving gravel highway for 105 miles, to the small town of Rusape. Here we turned left onto a secondary road and began climbing upward into the Eastern Highlands of Southern Rhodesia, situated along the western border of Mozambique. At 5,500 ft. elevation we reached the mountain resort town of Nyanga, which was the northern gateway to Nyanga (Inyanga) National Park. We camped at 6,000 ft. just inside the park.

Nov. 18 was another day to remember. It began pleasantly on a mild sunny morning in the exhilarating mountain air of Nyanga National Park. Noble and I decided we would go for a short hike in the mountains, following one of the well used hiking trails to the top of Mt. Inyanga, which at 8,500 ft. was the highest mountain in Southern Rhodesia. Since we were starting at 6,000 ft., this was not a stressful hike. A park brochure said "the peak can be reached in 1-3 hours by anyone of average fitness." The trail was moderately steep as we started upward through a nice forest with relatively little understory, on the south side of the mountain. I took time out occasionally to identify the birds we encountered. Noble walked on slowly ahead and I would soon catch back up. One of the birds I discovered was my first Wahlberg's (Sharp-billed or Brown-backed) Honeyguide*, a very plain brownish little bird (except for its white outer tail feathers) which was easily confused with several other species. After 1½ hours of walking we left the forest, at 7,500 ft., and came out onto a heath and grass covered moor. We were on a rolling, rocky plateau which was gradually climbing upward to the bare, rounded summit of Mt. Inyanga, clearly visible some distance in front of us. On our right was Pungwe Gorge, which dropped off very steeply into the forest below us, through which we had just walked. One could see far into the distance. I remember thinking to myself that if either Noble or I should fall over the edge of the escarpment no one would ever find us in the forest below!

Adding to the splendor of our environment was a pair of truly majestic Wattled Cranes* which were foraging on the moor. I viewed them with awe. With a height of just over 4 feet they were one of the tallest cranes in the world, only slightly less high than the Whooping Crane. They were immaculate white, gray, and black in color, with long white facial wattles, a bare red face, black crown, and black legs. They fitted perfectly in this remote, mountain ecosystem. What a thrill they were for me, and another permanent memory of Africa. Also along the edge of the forest and grasslands here was the attractive Bronze Sunbird*, with an overall coloration of shining bronze-green. It was one of numerous species in the genus *Nectarinia*, and like many in the genus it was characterized by narrow, pointed, greatly elongated central tail feathers. The flowers of *Loranthus* and *Erythrina* were among its favorite feeding sources.

We walked up the gently sloping plateau for another hour, toward the poorly defined summit. Clouds were beginning to roll up the mountainside from below. Noble wanted to go a bit farther, saying that we were not quite all the way to the top of the mountain yet. My altimeter (which I always carried with me) read 8,500 feet.

That was close enough for me, so we agreed that he would continue a little while longer, all the way the last 100 ft. or so in elevation to the top while I started slowly back down the mountain to search for birds as I went. Our footpath was quite well defined, and Noble could eventually catch up with me. The descent back to the road at 6,000 ft. shouldn't require more than 1 or 2 hours. It was now just past noon.

Within an hour after I started back down, the mountain became enveloped in clouds. I could still find birds in the forest so I continued to walk slowly, thinking that Noble would soon catch up with me. The clouds quickly became denser and the forest disappeared in a thick fog. I could barely see the trail. There was no more birdwatching, but I was not far now from the road where we had parked our Jeep at the trailhead. The time was about 3:30 in the afternoon. I arrived shortly thereafter at the van and relaxed inside to finish writing up my bird notes, thinking that Noble would soon return.

Time went by and Noble did not appear, and the dense fog did not go away. I began to get a little worried, thinking that in the fog maybe Noble had stumbled over a rock or tree root and had twisted his ankle, or worse. As darkness set in I turned on our headlights and regularly sounded the horn loudly. The little park brochure we had picked up at the park entrance was not encouraging. It read: "Hazards to climbing are bewildering fast weather changes that can switch from sunny skies to thick fog. Under these conditions hikers have lost their way, fallen into a ravine, and died." Oh my! There was nothing more I could do but go to bed and wait until morning, hoping for the best.

I got up before daybreak the next morning, Nov. 19, and filled my backpack with everything I could think of that might be needed to find and rescue Noble – ropes, walking sticks, knee and arm braces, tape and bandages, pain pills, a loud whistle, and trail food. At the crack of first light I started back up the mountain on the same trail we had hiked together yesterday. Within 45 minutes after I began climbing there was a very loud "hello" from just up the trail ahead of me. Down came a happily smiling Noble. Nothing in my backpack was needed! He explained that he simply lost the trail in the fog, and after stumbling and falling down the mountainside without knowing or being able to see where he was going, he decided to just sit down under a big tree in the forest and wait for daylight. The night was not particularly cold but he said that he constantly worried all night long that he would be eaten alive by some wild animal (even thinking he could hear one not far away). As soon as there was enough light the next morning and there was no more fog, he found his way back to the trail. We walked together down to the van and I prepared a big pan of hot oatmeal for the two of us. It was a story Noble could tell to his grandchildren one day. As the years went by his rendition of this event became more and more enhanced!

We spent the remainder of the day in the national park. I pursued more birds and Noble took time to record the events of yesterday in his journal. I followed some of the forest walking trails in the upper Pungwe River valley, between 6,000 & 7,000 ft. elevation. In addition to 4 species of birds which were new for me (Appendix B), I also encountered 2 new kinds of mammals, the Common Reedbuck (very similar to the Mountain Reed-buck) and a small group of Samango Monkeys (usually classed as a subspecies of the White-throated Guenon, *Cercopithecus albogularis samango*). These delightful long-tailed monkeys of the forest mid-story and canopy were easily recognized by their long white beard and whiskers which encircled their lower face from ear to ear. They peered at me inquisitively from between the leaves and branches, or ran and jumped through the trees using their arms and legs to swing from branch to branch. (Unlike many of the monkeys in the New World there are no monkeys in the Old World which have a prehensile, grasping tail, and thus they did not swing or hang by their tail, contrary to many Tarzan movies.) That night we camped in the national park for our last time.

On our way south out of Nyanga National Park the next day (Nov. 20), I stopped to look at a Long-crested Eagle* perched at the forest edge alongside our road. It was easily recognized by its long floppy crest and its all black plumage. In flight it showed a large, conspicuous white patch in the outer wings, and the inner wings and tail were noticeably barred. Soon thereafter we discovered that our front axle, once again, had developed a crack which required some welding. We limped down the road 20 miles to the sizeable town of Umtali (now called

Mutare), near the Mozambique border. We arrived in the afternoon and found a welding shop, but it was not until the next day that the Jeep could be repaired. So we found a wooded campsite not far out of the town, at 4,000 ft. elevation. Prior to leaving our campsite the next morning (Nov. 21) I found my first African Wood-Owl* (a congeneric relative of the N. Am. Barred Owl) and Whyte's (Yellow-fronted) Barbet*. The barbet was a distinctively colored bird, mostly dark olive-brown with a yellow forecrown and large white patch in the wings. It had a very small geographic range in Africa.

Our Jeep was ready to leave at noon, and we headed east out of town for the nearby Mozambique border (our second entry into this Portuguese colony). We wanted to visit the Gorongosa Game Reserve, which was approximately 150 miles away, east of us then a short distance north. As we began our travel the road was paved for one lane, but soon it became entirely gravel. An Impala ran ahead of us down the middle of the road, at 30 mph, before it finally veered off into the adjacent bush. We camped that night near the tiny village of Gondola. We were in lowland, open, hot tropical forest at 2,000 ft. elevation, about 75 miles east of the border. I was excited when 2 Silvery-cheeked (Crested) Hornbills* flew overhead, with very noisy wing beats. This species was one of the largest African hornbills and was easily recognized by its black & white plumage pattern and its huge, cream-colored casque. Also at our campsite I was able to track down a calling African Barred Owlet*, in the genus *Glaucidium*. Like other members of this worldwide genus, birds often hunted and vocalized in the daytime.

One hour after we got underway the next morning (Nov. 22), when we were not far from the entrance to Gorongosa Reserve, we came to a barricade across the road which prohibited traffic from going any further. The road ahead was completely washed out from a recent rainstorm and even our Jeep was not allowed to pass. We were extremely disappointed, of course, but we had no choice but to turn around and go almost 50 miles back from the way we had come, to just south of the town of Nova Vanduzi. At a major road junction here there was a main gravel road going north for 223 miles to the city of Tete, on the Zambezi River. Shortly after we reached this junction and headed north we crossed the Pungwe River, on a raft. The river was flowing west to east from its source in Nyanga National Park. Luckily, the single vehicle raft was large enough to float our 9,000 lb. camping van (unlike several similar situations in Central & South America). The raft was hand-winched along a steel cable which stretched across the river.

We continued north from the river, and just prior to the town of Vila Gouveia (Catandica) I stopped to carefully observe a Dickinson's Kestrel*, which was perched conspicuously in an isolated savanna tree. This handsome little falcon was gray plumaged with a whitish head and rump, and with narrowly barred wings and tail in flight. It was an uncommon species of East Africa which was said to favor areas with scattered palms and baobabs. While I was watching, it flew a short distance and then alighted again. It was my only sighting of this species, and I remember it fondly. That night we camped 5 miles N. of the town of Vila Gouveia, in a grove of trees just off the road.

On Nov. 23 & 24, Noble and I tried to follow a local, primitive bypass road from the town of Mungari for 50 miles north to the town of Mandie, on the south bank of the Luenha River. This diversion was recommended in an A.A. club pamphlet we had picked up at a travel office in Salisbury, saying that it would allow a visitor in Mozambique to "experience more of the local culture." Fifty miles on a bypass road was not a very long distance. Unfortunately, things did not work out as described in the booklet. When we arrived in Mandie we discovered that the pontoon bridge said to cross the river here had recently been washed away (probably in the same storm that washed out our road to Gorongosa). There was no way for us to cross the river and continue northwatd. We would have to retrace our 50 mile dusty bypass road back to the main road. We camped on the river bank that night. One of the birds along the river was a Banded Snake-Eagle (*Circaetus cinerascens*). Roberts wrote that this species "is usually found in the neighborhood of rivers, perched conspicuously in large trees." This is exactly where I found it, and it was my only encounter with this species in my African travels.

We returned to our main road the next day, Nov. 24, and drove north once again to the Luenha River, where there was now a bridge. We camped here along the river for a second time. Tete was our next destination, just up the road. We drove there the next morning. This sizeable city was situated on the south side of the lower Zambezi River at 500 ft. elevation, 400 miles below Kariba Dam. In 1959 it had a population of approximately 100,000 people, 10% white (mostly of Portuguese ancestry) and 90% native or of mixed ancestry. We stopped only long enough to purchase gasoline and fill up our inside 12 gallon water tank (to which we added a bottle of 50 Halazone tablets for purification).

Immediately after leaving Tete we ferried across the wide Zambezi River on a large, motorized raft. (Unlike similar situations in South America, the raft did not sink.) A new bird for my list here as we crossed the river was the Collared Pratincole*, a graceful bird with a swallow-like flight, a deeply forked tail, and a conspicuous white rump. This species was widespread in Africa (and Eurasia) along the margins of rivers and lakes. It was 1 of 8 "pratincoles" in the family Glareolidae (along with 8 "coursers" and the enigmatic "Egyptian Plover").

It was not quite 50 miles from Tete to the southwestern border of Nyasaland (Malawi). We entered the country at the border post of Zobue. Nyasaland was a long narrow country bordering the west and southern shore of Lake Nyasa for the entire 365 mile length of the lake. The lake was the southernmost, and the second deepest, of the 3 large lakes in the "Great Rift System" of East Africa (from N to S these were Lakes Victoria, Tanganyika, and Nyasa). This system (also known as the "Great Rift Valley"), was a series of trenches being slowly formed by the splitting of the African tectonic plate into an eastern and a western plate. Lake Nyasa had a maximum depth of 2,300 ft., a maximum width of 47 miles, and an average width of just 25-30 miles. The lake was further distinguished by having more species of fish (mostly 1,000 different kinds of cichlids) than any other freshwater lake in the world! Many of these were endemic to the lake. The surface elevation of the lake was 1,650 ft. above sea level, and the surface area was 11,500 sq. miles.

The warm, wet season (Dec.-Apr.) was just beginning but it was already hot in the southern lowlands of Nyasaland. On our way north we therefore decided to leave the main road and follow a somewhat longer route which took us on a southeast loop through the cooler Mlanje and Zomba mountains at elevations as high as 5,000 ft., via the towns of Mlanje and Zomba.

At this point in our African journey I switched my major bird reference from Roberts to Mackworth-Praed & Grant, *Birds of Eastern and North Eastern Africa*, a two volume set of books published in 1951 & 1955. Volume 1, Non-passerines, had 806 pages (plus an introduction and a long index of scientific names, but no English names), and Volume 2, Passerines, had 1099 pages (plus an index of scientific names and an "Index of English Group Names in Volumes One and Two"). What this meant was that nowhere in either volume was there an index of English names for the species, passerines or non-passerines. However, taking up a lot of space in these two volumes was a written description and discussion of every SUBSPECIES occurring in East Africa. The English names used for species (in the text) were often different from the names used by Roberts, which considerably complicated my efforts to determine whether I had seen a species previously or not. The two volumes collectively weighed 6½ pounds! Therefore, they could hardly be described as "field guides", but they were my major reference for East Africa. Their most welcome feature was a colored illustration of virtually every species. I spent long hours every evening studying these two volumes. They were a great improvement over having no bird book at all, but my birdwatching continued to be a demanding challenge. I couldn't have been happier.

On Nov. 28, there were many tea estates as we drove at elevations between 2,500 ft. and 3,000 ft. through the Mlanje Mountains. I encountered 6 species new for my list (Appendix B). One of these was the White-eared Barbet*, and I not only saw 5 individuals of this species but I also found a nest with almost fully grown young, which were poking their heads outside a tree cavity in which the nest was located. This species inhabited a very narrow geographic zone along the coast of East Africa. On the next day (Nov. 29) we traveled at elevations as high as 5,000 ft. in the Zomba Mountains, where I observed my first Augur Buzzard* (a buteo hawk which was separated by Clements from the "Jackal Buzzard" of others). New additions to my trip list for our two days of travel in these low elevation mountains can be found in Appendix B.

Early on the morning of Nov. 30 we arrived in Lilongwe, the capital of Nyasaland and the largest city in the country. On the outskirts of the city was a large, crowded, colorful market (which reminded me of Sunday markets in the Peruvian Andes). We stopped to purchase some fruits and take pictures. My first photo was of a native policeman standing next to his bicycle at the edge of the large market plaza. He was wearing a neatly pressed uniform of dress shorts, short-sleeved shirt, and wide brimmed hat, and he was the only person I could see wearing shoes of any kind at all. He grinned when I took his picture. Everyone in the market who was selling goods, both men and women, had their wares either spread out on a straw mat on the ground in front of them, or on the top of a long wooden table covered with a tarpaulin in case of rain. Another picture I took was of a small two-wheeled cart loaded with firewood being pulled across the plaza by 2 donkeys. All kinds of items were being sold or bartered in the market: tobacco, tea, coffee, rice, sorghum, cotton, sandals, hats and garments, flour, potatoes, corn, beans, rice, sugar, salt, oil, fish, poultry, mangoes, avocados, bananas, pineapples, guavas, blankets, rugs, mats, glassware, pottery, eating utensils, scissors, pharmaceuticals, toiletries, soap, hardware, firewood, etc. Vendors were all very simply dressed. Women usually wore a head covering or scarf and were most frequently dressed in a colorful or patterned "chitenje", a sarong-like one piece garment. They regularly carried a baby on their back, supported in a blanket (or any other piece of cloth). Hardly anyone wore shoes or sandals. People were generally polite, friendly, and curious. Here in the market on the outskirts of the city there were no Europeans and the native traders and buyers were largely uneducated. They were too timid or self conscious to attempt a conversation with us, although a few of them understood a little English (since Nyasaland was a British Protectorate). People were equally divided in their religious upbringing between Islam and Christianity.

Noble and I spent more than 2 hours in the teeming marketplace before we climbed back in our cab and headed north out of the city toward the distant town of Mzimba, 182 miles away. We arrived at this town in the late afternoon and turned east off the main road onto a secondary road which took us 5-6 miles up onto the cooler Vipya Plateau at 6,000 ft. elevation. We camped here that night in a grassy area of Brachystegia woodland. The next morning, Dec. 1, there were 3 first time bird sightings for me in the vicinity of our van: (1) Red-capped Crombec* (a very small, extremely short tailed, canopy foraging, gray & white sylviid warbler with a diffused reddish crown, chestnut ear patch, and wide chestnut breast band), (2) Siffling [Short-winged] Cisticola* (a plain-colored cisticola without any streaking, rufous-brown above and pale buff-brown below, which was singing from rank grass), and (3) White-headed Sawwing* (an eye-catching, all black swallow with an almost entirely white head and a deeply forked tail).

After breakfast we left the plateau and drove back down to the main highway, then proceeded northward. Before long we came to a road junction where we turned east on a very winding road that took us 57 miles through hills to the town of Nkata Bay, on the west shore of Lake Nyasa at 1,650 ft. elevation. The water in the lake was clear, cool, and the most refreshing of anywhere in our African travel, so we spent the rest of the day swimming in the lake and relaxing on the shore. I was surprised to see some Gray-headed Gulls, the same species which I had first seen in South America! Then we found a secluded site to camp where I spent 2 hours in the van studying my Praed & Grant before retiring for the night, in my narrow little bunk just under Noble. Life was an adventure.

I awoke at dawn the next morning (5:00 am on Dec. 2), got dressed, and went birding in the woods along the lake shore. Some of the birds I encountered were: 1 Bateleur, 1 Purple-crested Turaco, 5 Dideric (Diederick) Cuckoos (a distinctive little cuckoo in the genus *Chrysococcyx*, which vocalized with an oft-repeated, loud, plaintive "dee, dee, deederik"), 2 Little Bee-eaters, and three kinds of kingfishers - - Malachite, African Pygmy, and Brown-hooded. Unlike in the New World, many of the African kingfishers were woodland birds and not associated with water, or catching fish. It was a very pleasant, relaxing 3 hours of tracking down birds. Again, I thought of the oft-repeated comment by my early mentor, L. B. Carson (the president of the Topeka Audubon Society in Kansas, in 1947-48): "Birds are where you find them." This comment motivated my birdwatching throughout my entire life.

We packed up and departed at 8:00 am, with our next destination being the Nyika Plateau in the northern tip of Nyasaland. Our road was slow and winding so we didn't arrive at the town of Katumbi (via the town of Rumphi) until late in the afternoon, at an elevation of approximately 6,000 feet. Here we once again left the main road and turned right (east) to climb in altitude up to the top of a plateau at 7,500 feet. Here I found a campsite in isolated, grassy miombo woodland (for yet another time). Before sunset I located a pair of Souza's Shrikes* (*Lanius souzae*), my first sighting of this local, uncommon shrike associated with miombo woodlands. In appearance, it was much like other gray, black, and white shrikes of the genus, but it had a distinctive wide white bar on its shoulder. It turned out to be my only sighting of this species. Other birds present were a Black-bellied Bustard, 2 Fiery-necked Nightjars (seen flying at dusk), 2 unidentified young eagle-owls, 2 White-headed Black-Chats (Arnott's Chat), and an African Stonechat.

Noble wanted to write in his journal the next morning (Dec. 3), so I had my first real opportunity to bird in a patch of lush East African montane forest, and I was excited at the chance to do so, until noon time. This area was not a national park in 1959, but it became one a few years later (in 1966). Avian highlights for me were the following: 2 Livingstone's Turacos (seen first here but not recorded in Appendix B until Dec. 18), 3 Blue Swallows (an all dark glossy blue swallow with extremely long outer tail feathers, very local in Africa), 4 Spotted Creepers, 2 Cameroon Scrub-Warblers* (*Bradypterus lopezi*, also called Evergreen Forest Warbler, a plainly colored dark brown, skulking warbler which was challenging to identify), 1 White-chested Alethe* (a medium-sized, striking, very localized member of the thrush family, with a gray crown, rich chestnut back, and immaculate white underparts), 2 Mountain Yellow Warblers* (a relatively large, local sylviid warbler with bright yellow underparts, in the genus *Iduna*), 1 African Green Pigeon (the most widespread African species in the genus *Treron*, a group of tropical Old World fruit-eating pigeons), 1 Speckled Mousebird (a peculiar, mostly brownish bird which resembled a parakeet in size and shape, and which clambered or crept about like a mouse in small groups in thick, low bushes - - then flew rapidly to the next clump), 1 Chapin's Apalis* (also called Bamenda Apalis; a local, long-tailed, attractive cisticolid warbler, dark grayish-black above and very pale Gray below, with a diagnostic deep red face and breast), and 1 Olive-flanked Robin-Chat* (*Cossypha anolama*, a very local SE African endemic of which I recorded a long description in my field notebook and then devoted considerable time to finding in my Praed & Grant, vol.2, on pp. 302-304 and plate 66). The Latin name of "*anomala*" was very appropriate. Thank goodness I had a published reference available, even if it did have 1099 numbered pages and no English names for species in the index! It was a challenging, fun-filled morning which provided me a glimpse of the hard to find, endemic birds of E. African mountain forests.

During my morning on the plateau I also encountered several groups of zebras and one small group of 3 Roan Antelopes, a species I had not previously seen. It was one of the largest antelopes, closely related to the Sable Antelope and the Oryx. Males weighed 550-650 lbs. and stood an avg. of 5 ft. tall at the shoulders. The 3 ft. long horns of males were ringed at their broad base, and they rose vertically from the top of the head, and then curved sharply backwards (as in the Sable Antelope). Roan Antelopes varied geographically in color, but here in East Africa they were usually reddish-brown with a whitish belly. They had a dark face with a narrow vertical white stripe, and a white nose. This species was locally distributed in east-central, north-central, and western Africa, but was not often encountered.

Noble and I began our descent westward down from the top of the plateau at noon, back to our main north-south road. On our way down we came across a migrating flock of about 50 Amur (E. Red-footed) Falcons*, a Palearctic migrant from the steppes of NE Asia. These long distance migrants made the world seem a bit smaller. We reached the main road at 5,000 ft. elevation and turned north, toward the border of Tanganyika 120 miles away. We arrived in the town of Chisenga as the sun went down, and found a campsite shortly thereafter. Along the roadside was a dead Square-tailed (Mozambique) Nightjar and an injured Fiery-necked Nightjar, both probably having been hit by a passing vehicle. I prepared the dead bird as a "scientific study skin" and eventually gave it to John G. Williams at the Coryndon Museum in Nairobi.

The next day at 2:00 pm, Dec. 4, we crossed a tiny portion of Northern Rhodesia, via Fort Hill, and arrived at the border of Tanganyika (Tanzania). The town of Tunduma was on the Tanganyika side of the border. We had traveled for 38 days since leaving South Africa and I had recorded another 144 species for my trip list. The last 2 species were seen this morning, a Woodland Kingfisher* and a Black-winged Lapwing* (Appendix B). Our African adventure would continue.

Chapter Fourteen -- Tanganyika and SE Kenya

Tanganyika covered an area of 364,000 sq. mi. (three times the size of New Mexico) just below the equator in the heart of British East Africa. It was situated between the three largest Rift Valley lakes: Lake Victoria to the north, Lake Tanganyika to the west, and Lake Nyasa to the south. To the east was a long coastline on the Indian Ocean. The two most important destinations on our planned itinerary in Tanganyika were the vast Serengeti National Park and Mt. Kilimanjaro, the tallest mountain on the continent, both of them in the northern part of the country. We were now in the land where Swahili was spoken, a hybrid language used for communication between native Bantu peoples, Europeans, and Arab traders. I learned to say "yambo" (or "chyambo" depending on your local accent), which meant good morning or good day, and "ndegi" which meant bird. This was about the extent of my Swahili. We were always greeted politely by native Africans with "Yambo, bwana" – good morning, sir. When I was out walking by myself, or birdwatching, I was always queried with "Safari a whapi, bwana?" This translated as "Where are you going, sir?"

On the cloudy afternoon of Dec. 4, Noble and I headed NE from the southern border town of Tunduma, at 4,000 ft. elevation. Our main gravel road was badly corrugated, the worst we had experienced anywhere in Africa so far. Noble wrote in his journal that the road was so rough that the second hand vibrated off my wristwatch! It was 70 miles to the large town of Mbeya, an important agricultural hub for the country situated at 5,500 ft in a narrow highland valley surrounded by mountains. Mt. Rungwe, a volcanic mountain to the south, was almost 10,000 ft. high, the tallest mountain in the southern half of Tanganyika. Tea estates were situated on the mountain slopes of this scenic region. That night we camped along the road 5 mi. E of Mbeya, at 5,000 ft. elevation. A surprising bird sighting was that of a Northern Wheatear, a migrant from the Palearctic winter. (This is the only wheatear that breeds anywhere outside the Old World, with a breeding population in arctic North America.) I carefully observed and wrote a description of a Spotted Eagle-Owl which was on a low hilltop in an open area of mixed grassland and cultivated weedy fields, standing on top of a small rock.

For the next three days (Dec. 5, 6, and 7) we drove northward through the cities of Iringa and Morogoro to the Tanganyikan capital city of Dar es Salaam, on the coast of the Indian Ocean. The 21 birds which I added to my trip list during these travel days are listed in Appendix B. On Dec. 7, we traveled east from Morogoro through a sparsely populated area of dry woodland where two young boys wearing sandals and carrying long slender spears were tending cattle. They were short-haired and had elongated, punctured ear lobes with wide rings and ornaments dangling from them, and wore necklaces of red beads. They wore red or black capes wrapped around their body from their shoulders to their knees, and had a dagger tucked in their waistband. In body build they were very slender and moderate in height. They must have belonged to one of the several subgroups of the wide-ranging Masai (Maasai), although I was surprised to encounter them at this location, which was at the extreme SE edge of the Masai Steppe. They smiled cooperatively when I asked to take their picture.

With a population of ½ million, Dar es Salaam was by far the largest city in Tanganyika. Early on the morning of Dec. 8, Noble wanted to look around the city so I decided to catch a local bus and see what birds I could find along the Indian Ocean coast northwest of the city, toward the town of Bagamoyo. We agreed to meet in Dar es Salaam at the end of the day, back at the campground where we spent last night, but set no definite time. The morning was hot and sunny, and fishermen with nets were walking along the beach. Small outrigger fishing boats with sails were just offshore. Thatched houses were scattered among the mangroves. As I began my birdwatching I was happy to see and hear the lovely flute-like notes of a Four-colored (Gorgeous) Bushshrike singing in the coastal thickets. During seven hours of walking throughout the day, on the beach and in the adjacent trees and scrub, I recorded a total of 60 different species, 11 of which were trip additions (Appendix B). Perhaps the most notable of these was the Crab Plover*, an enigmatic species placed in its own family, Dromadidae. This unusual shorebird, which was about the same size and shape as an oystercatcher, was endemic to the Red Sea and NW Indian Ocean coasts. It was black and white in color, with Gray legs and a thick, pointed black bill that it used for feeding predominantly on crabs in the coastal mangroves. Two other species of particular interest here were 40 Sooty Gulls* (with a geographic range approximating that of the Crab Plover) and a single Palm-nut Vulture* (a small, short-tailed, black and white vulture which fed extensively on the skin of Oil Palm nuts, as well as such other plant and animal items as fish, carrion, mollusks, and Raphia Palm fruits).

A very memorable experience on my walk along the beach was an encounter with a group of 6-8 native people performing or watching a witchcraft ("traditional healing") ceremony under a large, wide spreading mangrove tree. A male "witch doctor" was attempting to heal, through magic and spiritualism, a sick woman who was sitting on a small thatched bench in front of another woman who was sitting on the ground behind her massaging her bare back and shoulders. The scene was mostly in the dark shade and it was a little difficult for me to see, or photograph, exactly what was going on. I did not want to approach too closely, so as not to anger or offend anyone. A man wearing a string of bells around his bare torso sat on the ground next to the sick woman, constantly jingling his bells. A young boy continuously beat on a large drum. The witch doctor danced around and around the sick woman, waving a wand, jiggling a large rattle, and producing smoke from a kudu horn. A bamboo table nearby with plates of different foods and bowls of broth had unknown significance. Before very long I was approached by a neatly dressed young African boy who said to me, in perfect English, "I think you should go; there are people who do not want you here." Of course I left immediately but the event remains a permanent memory, even though my resultant pictures were mostly too dark to show exactly what was happening. Witchcraft and spiritualism were widespread in Tanganyika at this time (but sacrificing albino Africans was strictly prohibited).

Noble wanted to visit the nearby offshore island of Zanzibar, a small, politically distinct nation whose inhabitants were almost entirely Moslem, from Africa or Arabia. So while I was hiking along the beach Noble drove down to the port of Dar es Salaam to find the most inexpensive way we could do this. He enjoyed hassling and bartering, and explained to the harbor master what it was that he wanted. The harbor master was sympathetic and arranged for us to travel as passengers on a small Arab dhow which was transporting coconut oil and soap to the island. Our cost would be only one British pound ($2.80) each. That suited Noble just fine, though he had no idea what the boat would be like. Our scheduled departure was 7:00 pm. Noble drove back to our campground to wait for my return, though of course I was unaware of the arrangement he had made. It was 6:30 pm by the time I arrived, and we hastily grabbed our shoulder bags and drove to the port, fearing we would miss the boat's departure. Although we were late, the boat had not left on time (as usual) and it was not until 8:00pm that we got underway. Noble and I were the only passengers on this small, 43 ft. long cargo boat, the "St. George", which was captained by a man named Tamadawas from the Seychelle Islands. As the only guests on board we were given bunks in the captain's quarters. Although it was just a 50 mile trip, the boat required seven hours to make the trip because it traveled at a maximum speed of only seven knots. Therefore, it was 3:00 am (Dec. 9) when we finally arrived.

We were greeted by the harbor police, who wanted to stamp our passports. Guess what? In our haste to get to the boat on time we had neglected to put our passports in our shoulder bags. After much pleading with the authorities they agreed to let us stay on the island for only one day without a visa, but we were required to depart before midnight on that same day! To insure that we would do this a policeman accompanied us to the airport and waited while we purchased a flight back to Dar es Salaam for that evening. This gave us one whole day for sightseeing, which was really all we wanted. Our spirits were brightened by a beautiful rainbow over the harbor as the sun came up, shining though the mist of a local rain shower. It was a good start to our day of sightseeing.

In addition to the African and Arabian inhabitants on the island there were also a significant number of Hindu merchants from India. Zanzibar was the world's number one producer of cloves, and there were mats along the sides of many streets where cloves were drying in the sun. The narrow, crowded streets were lined with shops of all kinds, and a large market sold the usual wide variety of fruit; bananas, coconuts, pineapples, mangos, and jackfruits (a native fruit of western India, the largest tree borne fruit in the world, up to 80 lbs. in weight and 3x2½ ft. in dimension). There were more bicycles than cars on the city streets. Grownups were typically dressed in white gowns, both men and women, with traditional Muslim caps. Children were mostly barefoot and raggedly dressed. Everyone was polite and courteous. Children happily let us take their pictures. Of course we were a curiosity with our long beards and western style field shirts, shorts, and shoes. In our sightseeing we walked past the Sultan's palace, where a uniformed policeman in khaki shorts was keeping watch on the pedestrians and cyclists. We enjoyed sipping tea at a little sidewalk table outside a tea parlor. The day was relaxing and photographically rewarding. It passed by quickly and pleasantly, with no new birds for me on the island. That evening we caught our obligatory flight back to Tanganyika on a small British airplane, and returned to our campground on the outskirts of the city.

The next morning, Dec. 10, Noble took our Willys camping van to the Jeep Agency for servicing, and I went strolling in the countryside south of the city to see what birds I might encounter. While our vehicle was being inspected and maintained, Noble again used his people skills to convince the agency owner that it would be good advertizing for him to have a picture in the local newspaper showing our "Roadrunner" being serviced at his company. In return, he would not charge us for the servicing. The owner willingly agreed and Noble contacted a young newspaper journalist, Tony Hughes, who came to the agency to interview Noble and take a picture of our Jeep. Such person to person interactions gave Noble as much pleasure as I derived from encountering new birds. In the meantime, I tracked down a total of 36 species on the outskirts of the city. These included two trip additions, the Black-backed Puffback* (a malaconotid shrike) and the Green Coucal* (family Cuculidae). Also, I was surprised to encounter my second African Broadbill. We returned to our campground again that night.

The newspaper story appeared early the next day (Dec. 11) in the morning edition of the "Tanganyika Standard." It brought us more rewards than those for which Noble had bargained. We drove into the city early that morning and parked along a city street. Almost immediately a tall, good-looking American man in his early 40's, Dusty Rhodes, spotted our van and walked over to us and introduced himself. He had just read our article and seen the picture in the newspaper of Noble and our Jeep camping van. What a coincidence! By a remarkable stroke of good fortune he was one of the two co-owners of a successful safari company, the East African Game Safaris, Inc. After several minutes of conversation he asked us if we would like to be his guests for two weeks at his main safari camp south of Dar es Salaam, six or seven hours away by road. He would provide us with everything we needed, including guns, except we would have to purchase our own "small game" licenses (at a cost of only $6 each). These permits would allow each of us to shoot at least 1 individual of 32 different kinds of game animals, including warthogs, many small to mid-sized antelopes (up to the size of hartebeests), and zebras. (Elephants, rhinos, buffalos, kudus, big cats, and others were excluded.) The only other expense for us would be a per mile cost for the times we were travelling on safari in one of his Land Rovers. Dusty was going to be at the camp for some of this time, with his wife who was temporarily visiting him from the United States. Otherwise his partner and principal game guide, Clary Palmer-Wilson, would be in charge of our activities.

It sounded too good to be true. Neither of us could quite believe the invitation we had just received. Here was the opportunity of a lifetime, to participate on a real African safari! Noble had dreamed of shooting a zebra since his childhood days. Of course we quickly said yes. One of the best features of our world travel was that we were not on an absolutely rigid itinerary and we could choose to alter it in time or space whenever we wanted. For me, it would also provide a great opportunity for pursuing more birds. Truly, some great spirit was on our side.

Noble and I got underway almost immediately, at 9:00 am, to drive the 120 miles to Morogoro, where Dusty told us we could stop at a wildlife office to purchase our permits. He had provided us with a map and directions to his camp, which was located not far from the small town of Mikumi, about 200 miles SW of Dar es Salaam. On the way to Morogoro, I stopped to identify a Fasciated Snake-Eagle* which was flying low alongside the road, then perching in a tree. It could be distinguished from its close relative, the Banded Snake-Eagle, by the barring on its belly and under wing coverts and by the pattern of banding on its tail in flight. After purchasing our permits in Morogoro, we continued SW toward Mikumi. (This was the same road we had followed from the Tanganyika border to Dar es Salaam.) At Mikumi we turned south off the main road and headed toward the Great Ruaha River. Dusty's camp was located on the west side of this road, a short distance before we reached the river, in a rugged area of rolling savannas and woodlands varying mostly between 1,000 and 2,000 ft. in elevation.

The average annual rainfall here was 30-35 inches, and the wet season was just beginning. The savannas were dotted with acacias, baobabs, palms, and a few tamarinds (a native legume with edible fruit, which was widely cultivated around the world). Terrestrial termite mounds of varying heights (up to 15 ft.) added to the inspiring landscape. The area was sparsely populated with native huts and very small villages. Dusty's camp was well equipped with dining and sleeping quarters for guests and workers, and maintenance shops for his many vehicles. It was a large and successful operation. We met Dusty's wife, Ricki, and Clary Palmer-Wilson, his partner and principal guide who was officially licensed as a "White Hunter." Clary was a native of Nairobi, a man in his late 50's or early 60's whose father was British and his mother was from India. He was a pleasant, knowledgeable, quiet spoken man with a good sense of humor and many stories to tell of his 20-30 years in the African bush on safaris. He was the principal hunting guide for the company and the person in charge of all field activities, while Dusty managed the company's finances and business operations.

Ricki (Dusty's wife) had come to Tanganyika to shoot a buffalo, which she wanted as a trophy for their new home in the states. Also at the camp were Clary's 13-year-old son, Mike, and a rather elderly native African guide by the name of "Timatayo" who accompanied us on most of our hunting expeditions. We were a very compatible group of safari-goers. Except for Ricki's buffalo (and Noble's zebra) no one much cared whether we shot very many animals or not. Simply participating on an African safari and experiencing the rugged bush gave Noble and me a thrill and provided us with unlimited memories and a lifetime of stories. I devoted more time to tracking down birds by myself than I did with the group tracking down game animals. We spent 15 full days at the camp, arriving on the evening of Dec. 11 and departing on the morning of Dec. 27.

Almost all of East Africa's game animals, large and small, inhabited the extensive area over which we hunted, but stalking and approaching them close enough to shoot was very different from driving up to them in a national park or game reserve, where they were not wary or frightened of people and vehicles. It was a challenge, but trophy hunting was not a major priority for either Noble or me. Therefore, the total list of animals which we shot during our 2 weeks on safari was very meager: one hartebeest and two warthogs each and a duiker by Noble and a reedbuck by me. We ate all of these, and I saved some warthog tusks and the horns of my Lichtenstein's Hartebeest (even though I did not find them very attractive since they reminded me more of a cow than anything else)! Only one time did I get close enough to shoot at any of the zebras we saw, and I missed, nor did Noble succeed in getting the zebra he wanted, but the safari was fun.

All night long we could hear the loud, blood-curdling screams of hyenas outside our sleeping quarters, as

Plate 41: Noble and I were invited guests for 17 days of "East African Game Safaris, Inc.", headquartered in the forested hills of eastern Tanganyika. We each shot a warthog and a medium-sized antelope, which we ate and salvaged their tusks as trophies.

they hunted and killed their prey. One evening there were several "bushbabies" (Lesser Galagos) outside, not far away in the trees. These primitive, cute little kitten-sized primates looked at us with enormous round eyes, which were positioned on the front of their equally round faces. They were characterized by a very long tail and by larger hind limbs than fore limbs. Bushbabies were well known for their ability to take enormous leaps, either through the branches of a tree or in kangaroo-like hops on the ground. Even though their total length (front of the head to end of the tail) was only a little more than 1 ft., these little mammals regularly took leaps of 12 feet (and rarely even up to 20 feet, fifteen times their total length)! They ate a wide variety of both plant and animal foods, but insects were most preferred, which were captured by grabbing them with their prehensile hands and feet. (As in almost all primates the thumb and big toe were opposable to the other digits, which enabled them to grasp and pick up objects with their hands and feet.) During the daytime, bushbabies most often took refuge in a hole in a tree, often in baobabs.

Clary told us that elephants were the most unpredictable of all game animals, and that a person should never trust them and always treat them warily. In spite of their great bulk they could charge for short distances at speeds up to 25 mph, faster than most persons could run. Both Noble and I, on separate occasions when we were walking alone on a trail through thick bush, suddenly encountered an elephant a short distance ahead of us on the trail. We each responded as we had been instructed, which was to stop and slowly back away, continuing to face the elephant until it was out of sight, and then turn around and quickly walk away, constantly checking over your shoulder to see if the elephant had decided to chase you or not. Luckily, neither of us was pursued, although we were both a little frightened by our encounter. (I never decided what action I would take if an elephant did decide to charge me, though I would probably attempt to quickly climb a large tree, hoping that one was readily available and big enough that the elephant could not push it over.)

A very memorable day was the occasion when four of us (Noble, Timatayo, Mike, and I) took the company's old Ford pickup truck and went hunting together. After an hour or so we arrived at an extensive, largely dry grassy swamp where there were elephants, lions, buffalos, zebras, sable antelopes, and reedbucks, mostly far in the distance. Only zebras and reedbucks were on our permit. One of the reedbucks, an animal smaller than a deer, was about 200 yards away. It was a long shot for a novice, inexperienced hunter like me, but I supported my .300 Winchester rifle on a tree, located the buck in my telescope, and fired. The buck immediately ran off, away from us, and then after a short distance it suddenly fell into the very tall grass, either dead or seriously wounded. Noble was watching it in his borrowed binoculars. Not too far away was a group of buffalo, one of the most feared of African game animals (along with lions and elephants). Noble and I wanted to go look for the fallen reedbuck, but both Timatayo and Mike refused to come with us because they were frightened of the buffalo. So Noble and I walked out into the tall grass by ourselves, and Mike climbed a tree to watch for lions and to warn us if any came our way. (In reality it was more for his safety than ours.) Eventually, Timatayo joined Noble and me in the search when it appeared it was safe for him to do so. It was very difficult for any of us to see anything in the tall grass, so after 20 or 30 minutes of searching without success, we finally gave up. In the meantime, Mike shouted at us repeatedly to say that both lions and elephants were coming in our direction. Actually, they were still almost a mile away. At one point during our searching my binocular neck strap broke, unnoticed at first by me, and my binoculars fell into the tall grass. I was extremely fortunate to find them 10 minutes later, by retracing my path through the grass.

It was now almost noon and the sunny day had become quite hot. Noble and Mike both decided they had hunted enough for one day, so they took the truck into a nearby small village for lunch and a cool drink, arranging to come back for Timatayo and me at the end of the day. After they left, Timatayo and I located a native hut where we solicited some drinking water from the owner, and then Timatayo sat in the shade of a tree and talked with him for half an hour. (The elderly Timatayo spoke almost no English.) Eventually it clouded up and I set out for the swamp again, with Timatayo reluctantly following after me. (After all, he was supposed to be the guide.) Before very long, about mid afternoon, a heavy thunder shower suddenly came over us, the temperature dropped considerably, and it poured heavy rain for more than 30 minutes. Both Timatayo and I were soaked, and I began shivering slightly from the sudden drop in temperature. The day had ceased to be much fun.

We were standing under a small tree, at the edge of a vast grassland in front of us, and I couldn't stop my slight shivering. Several elephants from a great distance off started to come our way and Timatayo, who was deathly afraid of them, wanted to leave. I indicated to him through sign language that we could climb the tree if they got too close, but he gesticulated that they would then pick up a log with their trunk and throw it at us in the tree to dislodge us! I was quite amused, but he was very serious. To make matters worse, some buffalos also appeared off in the distance and began to come our way. Poor old Timatayo, if it wasn't one thing it was another. It had started to rain again so we moved to another tree for better shelter, while he constantly peered around the trunk to see if the buffalos or elephants were getting any closer. It was now late in the afternoon and a small group of zebras suddenly appeared in the grasslands not too far away. I wanted to stalk them to get within shooting range. When Timatayo refused to come with me I set off by myself. Eventually I approached to within 150 yards of them but all I could see was their heads above the top of the grass. When they began to show alarm I tried to stop shivering long enough to hold my rifle steady for a shot. I took aim through my mist-clouded scope and fired. All the zebras disappeared at once, running off through the grass. I searched briefly to see if one had fallen, but since Timatayo refused to join me in the search and it was now getting dark, I soon gave up, deciding that I had probably missed anyway. That was the only shot I fired at a zebra on our safari.

During our two weeks on safari I documented a total of 227 avian species, 34 of which were trip additions (Appendix B: Dec. 12-26, 1959). My birding covered a large area around the camp, and extended up to 4,000 ft. into the Uluguru Mountains, which were part of a chain of isolated mountains in eastern Africa known as the "Eastern Arc Mountains." Included on my total list were 19 species of diurnal raptores, 7 kinds each of kingfishers, barbets, and bulbuls, 6 different cuckoos, 5 species of hornbills, 4 kinds of bee-eaters (Little, European, Swallow-tailed, and Boehm's*), and 3 species each of parrots (Brown-headed, Meyer's, and Brown-necked*), musophagids (Purple-crested & Livingstone's* turacos, and Bare-faced Go-away-bird*), and rollers (Broad-billed, European, and Lilac-breasted).

A dozen or so species on my safari list were endemic, widely or locally, to East Africa. These included the Hildebrandt's Francolin, Boehm's Bee-eater, Green Tinkerbird* (a small barbet), Green Barbet, Yellow-streaked Bulbul* (see comments below), Shelley's Greenbul*, Livingstone's Flycatcher*, Kretschmer's Longbill* (see comments below), Chestnut-fronted Helmetshrike*, Long-tailed Fiscal* (a *Lanius* shrike), Uluguru Violet-backed Sunbird* (extremely local), and Lesser Seedcracker*. The longbill was a moderately large (6½") plain olive-colored sylviid with the general appearance of a small bulbul, possessing a long slender bill and a rather short tail, which foraged mostly out of sight in the forest understory. The Yellow-streaked Bulbul was unusual in its creeper-like manner of foraging on the branches of trees, while constantly flicking its wings open and shut, one after the other (as did Shelley's Greenbul).

Our two week safari in the East African hinterland was a once in a lifetime adventure which we would always remember. However, Noble still had a strong desire to take home a zebra skin trophy while he currently possessed a hunting permit, so he asked Clary how he might accomplish this before we left Tanganyika. Clary gave him the name and address of a long time acquaintance and good friend of his, Thorkild ("Andy") Andersen, a Danish man who managed a sisal estate near the small town of Lembeni, not far south of Mt. Kilimanjaro. Andy was an avid hunter of many years who could provide him with a gun, a guide, and transportation. I, too, was happy to have another attempt at procuring a zebra skin. This arrangement was perfect for us, as we were planning to climb the mountain and could easily stop by the sisal estate. Clary wrote a letter of introduction for us to give to Andy.

On the morning of Dec. 27, Noble and I said goodbye to Clary and all the others at the safari camp, thanked them for their generosity and companionship, and departed northward in our camping van for whatever new adventures might lie ahead. Our proposed travel was constantly being modified and rearranged, both in time and in space. At our current rate of progress we might never get back to California! Fortunately, our time delays were adding hardly any extra expense to our travel, and there was still sufficient credit left in Noble's $20,000 loan to accomplish what we set out to do. Daily costs for food were averaging $1.50 per person and we rarely paid anything for lodging. Our major concern was how much longer our greatly overloaded Jeep might survive. Gasoline, repairs, and vehicle maintenance were 80% of all our expenditures. After Africa, there were still 3 more continents to go!

From our safari camp on Dec. 27 we headed north on secondary roads for the district capital of Kilosa, 45 miles away at an elevation of only 1,500 feet. Beyond this capital city was the town of Korogwe, and further yet was the principal town of Tanga on the Indian Ocean. We chose this route because we had not previously traveled this way. Along the road were small native villages, sisal estates, coconut palms, and ceiba (kapok) trees. We were going to stop in Kilosa only long enough to have a front shock absorber welded.

However, as so often happened, this simple repair required considerably more time than it should have because a necessary replacement part had to be ordered from Dar es Salaam and then transported to Kilosa by truck. Therefore, it was not until Jan. 3 that we finally departed Kilosa. In the meantime I looked for birds just south of the town at a small, reed-lined lake, and Noble, in his customary good natured manner, found lots of people with whom he could interact while we waited. A new experience for him was to be invited to play in a rugby match when the local district team came up one player short in an important match with a visiting team from the British cruiser HMS Gamia, which was docked temporarily in Dar es Salaam. Noble quickly read the rule book and was jubilant when "his team" won 16-3! On New Year's Eve we were invited to a black tie dinner at a local private club. Our champagne glasses were constantly refilled and there were many young ladies with whom to dance. Of course we each found a partner willing to sing "Auld Lang Syne" with us and then participate in the traditional kiss afterwards. What more could two young bearded adventurers in borrowed suits and black ties ask for?

We finally left Kilosa at noon on Jan. 3 and drove 55 miles NE to a wooded campsite along the roadside, at an elevation of 2,500 ft. about halfway to Korogwe. Along the way I added the Von der Decken's Hornbill* to my trip list, a species widely endemic to NE Africa, 1 of 15 species in the African genus *Tockus*. It was my first "life bird" of 1960, a characteristic bird of the dry savannas and woodlands which called attention by oft-repeated low, monotonous "wuk, wuk, wuk, wuk" vocalizations (like other hornbills in the same genus). In a curious way, I found this a pleasant, contenting sound of the open bush.

The next morning (Jan. 4) we drove into Korogwe, arriving at noon. This moderate-sized town was situated on the Pangani River not far above its mouth on the Indian Ocean. A new addition to my trip list along the way was the Grasshopper Buzzard*, a rather small, long-winged, long-tailed raptor in the Old World genus *Butastur*. Like others in the genus it possessed a large rufous patch in the outer wings, visible from above. It characteristically foraged low over the savanna in search of insects, which comprised the major portion of its diet. We did not stop in Korogwe but continued onward, following a twisting highway which ascended to the town of Amani at 3,500 ft. in the East Usambara Mountains. Like the Ulugurus, these mountains were part of the Eastern Arc chain. We camped in a mountain rainforest here for 2 nights, Jan. 4 & 5. I chased down birds in the forest and surrounding areas from daybreak to dusk on Jan. 5, and from 0630-1000 on Jan. 6, between the elevations of 3,500 and 4,000 feet.

The Usambara Mountains were best known for their botanical endemism and biodiversity, but during my day and a half there I documented a total of 56 species of birds, 9 of which were trip additions (Appendix B). Of these, 4 were endemic to these mountains: Green-headed Oriole*, Long-billed Tailorbird* (or "Apalis", an endangered species in the genus *Orthotomus*), African Tailorbird* (or "Red-capped Forest Warbler", also in *Orthotomus*), and Sharpe's Akalat* (genus *Sheppardia*, family Muscicapidae). All but the oriole were inhabitants of the dense forest undergrowth. These endemic forest species were geographically isolated in the Usumbara mountains in the same manner oceanic species were isolated on islands. A bird I saw here but never again anywhere else was the Fischer's Turaco*.

We departed Amani at 10:00 am on Jan. 6 and drove down the mountains on a good paved road, then turned east toward the principal coastal city of Tanga, on the Indian Ocean in the far NE of Tanganyika. We stopped here only long enough to fill up our 40 gallon gas tank, and then headed north toward the border of Kenya 50 miles away. Along this last stretch of road in Tanganyika I was delighted to see 2 new species of bee-eaters, the N. Carmine* and the White-throated*. Both were intra-African non-breeding migrants to this area, and they often occurred in large, high-flying, vocal flocks. Carmine bee-eaters were frequently attracted to grass fires where they fed on the disturbed insects. They also regularly perched on the backs of large mammals, and sometimes even bustards, to pick off arthropod parasites. In their breeding areas to the north and west of Kenya they excavated nests in vertical river banks, or sometimes even on almost flat ground, nesting in small to large colonies, as do many bee-eaters. This species was very similar in coloration to the S. Carmine Bee-eater except the whole head was turquoise-green, not just the crown. (The two species were considered conspecific by some authors.) It was a marvelously colored bird and a thrill to observe. Such encounters reinforced my enthusiasm for birdwatching.

Noble and I crossed the border in the middle of the afternoon, from Tanganyika into Kenya. Since both were still British colonies in 1960, the formalities were very simple and there was very little delay. Our passports and vehicle carnet de passage were all stamped and we were on our way again. Freedom, or "Uhuru", was only a year ahead. We were in arid scrub country not far from the Indian Ocean coast, heading NE toward the major port of Mombasa, 70 miles away. Here in the hot, dry lowlands women and girls generally wore only a skirt, and no top. They carried water from wells to their houses in ceramic jugs on top of their head, often with a baby tied in a sheet on their back. On this well-traveled road they wanted money if we stopped to take a picture. We journeyed only about 20 miles before finding a campsite for the night, just off the road in an abandoned patch of coconut palms.

Mombasa was the principal seaport for Kenya and the second largest city in the country (behind Nairobi). We stopped there the next morning (Jan. 7) only long enough to buy some fresh fruit for our cupboard - - bananas, oranges, mangos, pineapples, and plums. Then we continued north along the coast for 60 miles to the popular resort town of Malindi, where Noble had the address of three Kruger brothers whom he had met along the road in southern Tanganyika. They were spending some time there with their mother and a girl friend, and they had invited us to spend a day or two with them at their modest holiday cottage in the nearby suburban village of Watamu - - snorkeling, swimming, and relaxing in the sun on a lovely white sand beach. It was a delightful spot, comparable with the best beaches anywhere in the world. We arrived in mid-afternoon and Noble immediately went snorkeling on a coral reef, and was thrilled to closely observe many colorful tropical fishes of all sizes and shapes. This area would one day be a marine preserve. I went beachcombing and had a very long, cool, refreshing swim in the surf. That night everyone attended a dance in the town. For us, it was an almost cost free life of luxury.

The next morning, Jan. 8, the Krugers took us to view some ancient Arab ruins along the seacoast south of town. Afterwards we expressed our gratitude to them for their hospitality, said goodbye, and retraced our route back to Mombasa. From there we headed NW on a good hard-surfaced "macadam" highway connecting Mombasa to Nairobi, about 300 miles away. Noble had some business which needed attention in Nairobi before we returned to Tanganyika. A railroad track ran alongside our highway, paralleling it almost all the way to Nairobi. That night we camped approximately 20 miles NW of Mombasa in a dry wooded savanna at 1,000 ft.

elevation, near the tiny village of Mariakani. There was a beautiful orange and crimson sunset.

We had a leisurely day of travel the next day (Jan. 9) as I stopped frequently to pursue birds in the dry bushed savanna on both sides of the road, while Noble read or wrote in his journal. We progressed only 25 miles for the day and stopped just before we reached the tiny village of Mackinnon Roads at the SE corner of Tsavo National Park, Kenya's largest game reserve with approximately 8,000 sq. miles. We were at an elevation of 1,500 feet. For the day, I recorded a respectable number of 93 species of birds, 13 of which were trip additions (Appendix B). I included "Dodson's Bulbul" on this list, following a consensus of recent authors, even though Clements had this form as a subspecies of the Common Bulbul. Of the new additions, 9 species were endemic to NE Africa: (1) Golden Palm Weaver* (genus *Ploceus*, a brightly colored golden-orange and rufous bird which nested colonially under the large fronds of palm trees), (2) Dodson's (Common) Bulbul*, (3) Pangani Longclaw* (in the pipit family), (4) African Bare-eyed Thrush* (1 of 64 species in the worldwide genus *Turdus*), (5) Eastern Chanting-Goshawk*, (6) White-bellied Go-away-bird*, (7) Golden-breasted Starling* (genus *Lamprotornis*, perhaps the most beautiful of all African starlings with its golden-yellow underparts and long, graduated tail), (8) Fischer's Starling*, and (9) Tiny Cisticola* (which at 3½" long was barely larger than Africa's smallest birds, which were several penduline-tits).

As we proceeded onward toward Nairobi the next day (Jan. 10), our highway (and the railroad track next to it) formed the SW boundary of Tsavo East National Park. This unfenced park was on our right and I frequently stopped to view birds within the park. Our highway entered the park at the village of Manyami, and divided it into two main sections, Tsavo East to our right and Tsavo West to our left. At several points we left the highway and followed a side road into the park for a short distance, both east and west. Before going to bed that night I counted all the birds which I had tallied during the day. The total came to 72 species, another respectable endeavor, of which 16 were species new to my adventure (Appendix B), and 11 of these were NE African endemics: (1) Scaly Chatterer* (a special bird for me), (2) Black-throated Barbet*, (3) Pale Prinia*, (4) Pink-breasted Lark*, (5) E. Yellow-billed Hornbill*, (6) Crimson-rumped Waxbill*, (7) White-breasted White-eye*, (8) White-headed Buffalo-Weaver* (a handsome weaver), (9) Yellow-necked Francolin* (Spurfowl), (10) Superb Starling*, and (11) Taita Fiscal* (a *Lanius* shrike).

We stopped at an information center in the park on that same day and picked up pamphlets and brochures pertaining to the park and camping facilities. Noble was anxious to get to Nairobi where he wanted to spend 4 or 5 days with business activities, visas, and various money enterprises, but I preferred to spend this time in the park to leisurely search for birds. (Noble not only did not need my assistance in Nairobi, but he was happier carrying out these activities by himself.) One of the park brochures I had obtained described an inexpensive lodge and campsite called "Bushwhackers", where a person could spend the night at very little cost if he had his own food and sleeping bag. It was recommended by birdwatchers and was situated adjacent to Tsavo East Park, on the west bank of the Athi River as it flowed south along the NW boundary of the park. Bushwhackers' location was only 12 miles north of the Nairobi highway, so Noble readily agreed to drop me off there, with several days of clean sox and underwear, and some food. Four days later (on Jan. 14) I would hitchhike to Nairobi and rejoin him. We settled on a time and place to meet in Nairobi, and at 7:30 pm that evening (Jan. 10) we said goodbye to each other at Bushwhackers and temporarily parted company. This was not the first time that we had followed our own pursuits independently for a short period of time.

I spent the next 3 days wandering around in the grasslands, thorn bush, and dry riverine woodlands, chasing birds, taking pictures, and simply enjoying life. I couldn't have been happier. The days were mostly dry, sunny, and hot at this low elevation of 2,000 ft. above sea level. Not only did I look for birds but I was also always on the lookout for mammals and other wildlife. I could wade across the river, being alert for the abundant crocodiles, and meander around in the park. Monitor lizards were numerous, as they were almost everywhere in Africa (but surprisingly, I never came across a python). Almost all the East African game animals were present in the park, including the "big five." However, since I was on foot the larger mammals were generally wary of me and did not allow me to approach them closely enough for a picture. Some of the smaller mammals which I saw were jackals, Spotted Hyenas, a Bat-eared Fox, baboons, mongooses, squirrels, and Vervet Monkeys. Around my campsite at dusk were Greater Galagos, a larger relative of the bushbabies which I saw at our safari camp in southern Tanganyika. I could hear them vocalizing outside my hut at night, as I could the hyenas and the African Scops-Owls.

Three species of ungulates which I saw for the first time were Kirk's Dikdiks, Gerenuks, and Oryx Antelopes. These certainly spanned the spectrum in sizes and shapes of "antelopes." Dikdiks were only 16-17" tall at the shoulder and weighed just 12-14 pounds. Gerenuks were medium-sized, graceful, gazelle-like antelopes with long slender legs, a small head, and a longer neck (for their size) than any other African mammal except the Giraffe. Their horns, which were present only in males, were strongly ridged and S-shaped, curving backward from the top of the head and then upward at their tip. Gerenuks were strictly a browsing antelope and ate only leaves from bushes and trees. By standing vertically, straight up on their back legs, they could reach branches which were almost 7½ ft. above the ground - - more than twice their shoulder height! They were fascinating to watch and became one of my favorite mammals.

The Oryx was one of Africa's most spectacular large antelopes, with extremely long, pointed, rapier-like, straight or slightly backward curving horns, up to 4 ft. in length. Horns were present in both sexes. They were used effectively to ward off predators, and in territorial disputes. In addition to their inspiring horns, Oryx were distinctly patterned with a black & white face, a black throat, a horizontal black flank stripe, and a wide black band around their upper leg. The race in eastern Africa (including Tsavo) had long, stiff black hairs on the tips of its ears and was known as the "Fringe-eared Oryx." (The race in South Africa was called the "Gemsbok.") The Oryx was a species of open dry woodland and bush country, and at one time was distributed locally over much of the African continent, in suitable habitat, but as a result of both legal hunting and illegal poaching its numbers were much reduced almost everywhere. It was an African icon. I saw several, at moderate to long distances away, and they are a lasting memory for me of wild Africa.

During my 3½ days of meandering on foot in the park, on both sides of the river, I tallied 163 species of birds, 16 of which were additions to my trip list (Appendix B: Jan. 11-13, 1960). Of these, the following 10 species were endemic to NE Africa: Violet Woodhoopoe* (a species in which Clements lumps the 2 geographically isolated subspecies, "*damarensis*" and "*grantii*", often separated specifically by others as the "Violet" and "Grant's" woodhoopoes; the form at Tsavo was Grant's), Red-bellied Parrot* (also called the African Orange-bellied Parrot, a name which I prefer), Rueppell's Glossy-Starling*, Parrot-billed Sparrow* (1 of 22 species in the Old World genus *Passer*, which includes the non-native House Sparrow in North America), Black-bellied Sunbird* (with an extraordinarily small geographic distribution along several rivers in SE Kenya), Nubian Woodpecker*, Abyssinian Scimitarbill*, Red-and-yellow Barbet*, Black-faced Sandgrouse*, and Purple Indigobird* (a small, dark purplish-black, finch-like bird in the genus *Vidua* - - indigobirds, whydahs and paradise-whydahs - - within which all 19 species were brood parasites, each species having its own 1 or 2 host species whose songs they mimicked and in whose nests they placed their eggs).

My total list of 163 kinds of birds from Bushwhackers was roughly ⅓ of all the species documented for the entire 8,000 sq. mi. of the combined Tsavo East and Tsavo West parks. In addition to the birds which I have mentioned, I was particularly happy to encounter a White-headed Barbet*, a pair of Bat-like (Boehm's) Spinetails*, and a White-backed Night-Heron* (genus *Gorsachius*). The spinetail was a tiny, relatively broad-winged, very short-tailed swift with a fast, erratic, bat-like flight. It preferred to nest in baobabs, and became a favorite of mine. The night-heron was perched motionless among the branches of a densely foliaged riverside tree, mostly hidden from sight until I found just the right place from which to view it, where I had a good look. It was distinctly and attractively colored with a black head, white eye ring and lores, lengthy rufous neck, white belly, and black upperparts with a mostly hidden narrow white back. It was the last addition to my trip list at Bushwhackers. I felt extremely fortunate to have located this secretive nocturnal species.

As planned, I hitch-hiked to Nairobi on Jan. 14, departing Bushwhackers about 10:00 am and meeting Noble in the middle of that afternoon. His 3 days had been successful for him and we celebrated in the Jeep with some good South African wine. One of his accomplishments was to purchase a new spring for our Jeep, replacing one which had finally given out from the over burdening demands we had placed on it (as did all our springs eventually).

On Jan. 16, we headed south from Nairobi on a paved highway back toward the border of Tanganyika, at the Kenyan town of Namanga 120 miles away. We drove only 40 miles this day before I found a campsite 10 miles north of the large town of Kajiado, on the Athi Plains at 5,500 ft. elevation. En route I saw my first Kenya Rufous Sparrows* (a local endemic in the genus *Passer*) and my first Pied Wheatears* (*Oenanthe pleschanka*, a non-breeding visitor from the Palearctic).

In several hours of birdwatching on the Athi Plains the next morning (Jan. 17) between our campsite and Kajiado, I recorded 3 trip list additions: 2 Kori Bustards* (Africa's largest bustard, a stately and inspiring bird), 4 Short-tailed Larks* (my only encounter with this NE African endemic), and 1 Golden-winged Sunbird* (a medium-sized, handsome, dark bronzy-black sunbird with a large golden wing patch, golden-edged rectrices, long decurved bill, and narrow elongated central tail feathers, in the monotypic genus *Drepanorhynchus*). The moderate elevation grasslands here, with mild temperatures and wet marshy patches, were a pleasant change from the hot, dry, lower elevation grasslands of SE Kenya. This was the home of the Masai.

We continued southward that afternoon on the plateau between Kajiado and the Kenya border, where we arrived in the early afternoon at the town of Namanga. Again, the border formalities were quick and efficient, with stamps placed in our passports and in our carnet de passage. From the border we continued south, toward the important Tanganyika provincial capital city of Arusha, at 4,500 ft. elevation, where we arrived at 5:00 pm.

Between Kajiado and Arusha I recorded the remarkable total of 88 species of roadside birds, of which 14 were trip additions (Appendix B). My most enjoyable of these was the Rosy-patched Bushshrike* with its sandy-brown coloration, wide black necklace, rosy-red breast, and white corners to the tail which were conspicuous in flight. It was an endearing bird of dry scrub and thickets. Also among my favorites were the Double-banded Courser*, Chestnut Sparrow*, Straw-tailed Whydah*, Speckle-fronted Weaver*, and Gray-headed Social Weaver*, which I photographed at a colony of its hanging nests in a flat-topped acacia tree. This color slide is one of my most favorite pictures from Africa, reminding me of the vast, open African savannas with their flat-topped acacia trees and nesting colonies of weavers.

That evening we ate dinner at the Safari Hotel in Arusha, and then found a campground on the outskirts of the city for the night. We awoke the next morning, Jan. 18, with a distant view of Mt. Meru rising 15,000 ft. above sea level to the NE of Arusha. This popular, thriving city had a population of about 40,000 people in 1960, the 3rd largest city in Tanganyika and the capital of Arusha Region. It was the gateway to many tourist attractions, including Serengeti National Park to the west and Mt. Kilimanjaro to the east. Noble was excited to spend the morning walking around the town, so I had the opportunity to take the Jeep and go birdwatching. It was a morning I will never forget.

I drove west out of the city on a good hard surfaced road, in a sparsely populated area of grassland savanna with scattered trees and thorn scrub, at an elevation of almost 5,000 feet. This was Masai domain. Drying corn shocks and hollow logs for beehives hung in trees, and cattle were tended by young warriors draped in an orange blanket and carrying shields and long slender spears, mostly as protection from wild animals. The Masai people lived in small domed huts made of mud and cow dung, as they had done for centuries. I passed one small European-owned restaurant and bar, and there was an occasional vehicle on the highway. I drove only about 10 miles, to a suitably remote, unspoiled area in which to birdwatch, and parked the Jeep just off the highway. It was

Plate 42: The Masai people in East Africa lived in primitive, domed huts made of cow dung. Unluckily, a non-conventional family group that I stumbled upon near Arusha in northern Tanganyika threatened to kill me with spears because they thought I had taken their pictures inside their huts. .

7:00 am, about an hour after sunrise, and there were no signs of people or houses anywhere around. Wearing my usual hiking shoes and khaki shorts, I took my small shoulder bag with binoculars, camera, compass, and notebook, and set off on foot through the open woodlands directly away from the road to see what birds I could find. It was a calm, clear, cool morning and I was thrilled to be wandering alone in the "real Africa."

I was in no hurry as I meandered back and forth in pursuit of the many birds I could glimpse or hear around me, in all directions. It was very exciting. Flappet and Rufous-naped larks were conspicuous as they sang on the wing while performing aerial courtship displays. Crowned Lapwings screamed loudly at my presence, and 4 Yellow-collared Lovebirds* screeched as they darted just above the treetops. This was one of 9 species in the genus *Agapornis*, in which species were unusual among parrots by their habit of carrying nesting materials tucked under their body feathers, from outside the nest site to the cavity or hollow in which they were nesting. Lovebirds were a popular cage bird, so named because of their constant mutual preening, and like all parrots they kept the same mate for their lifetime. This was my only sighting of this diminutive species which was almost entirely confined, very locally, to Tanganyika. I took time to write a long description in my field notebook to help me identify several Mariqua Sunbirds, and I recorded Hunter's* and Desert* cisticolas for the first time. I was happy to once again encounter a pair of Rosy-patched Bushshrikes, a species I never tired of seeing. After several hours of leisurely wandering through the open bush I had recorded a very satisfying list of over 70 different kinds of birds, and I decided it was time to turn around and return to Arusha. It was about 2 miles back to the highway. I had encountered no persons or human habitations.

At that point, not far ahead of me through the trees, I could see a clearing in which there was a "krall" - - a few houses enclosed within a circular fence of thorn branches. I was intrigued and decided to have a closer look. The krall was perhaps 80-100 yards across. The 4 igloo-shaped, one room "houses" were built of mud and cow dung plastered on a framework of small branches, and each was approximately 10-15 ft. in diameter and 4½-5½ ft. high. Each of them had a single oval opening through which a person had to stoop for entering. Because there were no windows and no lighting, the houses were very dark inside. I had never before come across anything quite like this in my travels. Not seeing anyone around, I walked up to the thorn branch fence surrounding the krall, at the gateway, took out my camera and from the outside of the fence I took a picture of the overall scene inside, not focusing on any particular house.

To my great astonishment there was immediately a very loud scream from inside one of the houses. I did not know which house. Then there was shouting back and forth between all the houses. Very quickly a woman came running out of each of the 4 houses, and up to where I was standing just outside the entrance gate. Of course I could not talk to any of them. They were all very angry, and the oldest one took command of the group and motioned for me to hand over my camera to her. I tried to maintain a smiling, friendly manner and pleasant composure and appease them by explaining, as best I could in sign language, that I did not take a picture of any person. (From their viewpoint, I suppose it is possible they thought because they could see me standing outside in the sunlight that I could also see them just inside their dark houses, which I couldn't.) They were genuinely furious, perhaps because taking their picture also took away their soul. Destroying my camera would preserve their soul. As you can imagine, I did not want to lose my camera, particularly because I knew I had not really taken a picture of anyone, at least one that I could see. However, my explanation was not understood or accepted, and they were adamant that I must hand over my camera to them. I offered them a few coins in the palm of my hand but the lead woman simply slapped my hand and the coins fell to the ground, where nobody bothered to pick them up. When I continued to refuse, the woman in charge of the situation picked up a very wicked looking thorn branch lying nearby on the ground and threatened to hit me over the head with it, as did the other women. I would have been completely outmatched in such a physical confrontation, and I quickly came to the conclusion that the best solution to my dangerous situation was simply to run away (forgetting my pride). I had run distance races quite successfully all through high school and I knew how fast I could run and how to pace myself, and I thought I could almost certainly keep ahead of the women if they chose to chase me. I was young, lean, and very fit.

The women were momentarily surprised when I suddenly broke away from them and began to run away, but they all instantly began shouting loudly and running after me, as fast as they could. (The three youngest were probably wives of young warriors.) As I had predicted, I was able to run fast enough to keep just ahead of all of them, without tiring or breathing too heavily. Astonishingly, they did not immediately give up their pursuit but continued to chase me even as they slowly dropped behind, continuing to shout angrily and probably cursing me. Before very long, however, there was a change in their language and their shouts, which did not seem to be directed at me any longer. In response to this change, I noticed some movement through the trees far to my left, and far off to my right was more movement. Quickly I became aware of the fact that young boys who had been tending cattle, with their usual spears, were running toward me from almost all directions. I knew there was no chance that I could outrun them, so I just stopped running. Three boys with spears, and one older man with a long knife, quickly surrounded me. Shortly thereafter the women caught up with our group. They jabbered loudly at the males, undoubtedly telling them of the picture I took. The oldest person in the group of males then assumed command, and motioned for the women to return to the krall, which they obligingly did.

All the males were as angry as the women. Again, the oldest person in the group, the man with the knife, took charge (as almost certainly was a traditional rule). Like the woman, he motioned for me to give him my camera. When I refused, in as friendly a manner as possible, he threatened me with his knife. My irrational thinking at this point was that he must know what the penalty would be for killing a white person, but I decided to try and appease him by offering him my binoculars rather my camera, as I could more easily and cheaply replace them. (I should have realized this wouldn't work.) As soon as I gave him my binoculars he immediately handed them over to one of the boys, and motioned for him to take them to the krall. The boy obliged and began slowly walking toward the distant enclosure with my binoculars. The man then turned back to me and once again indicated that I must give him my camera. At this point, it was certainly the only sensible thing for me to do, if I wanted to walk away alive and unharmed. This was a time for rational reasoning, not for obstinacy. My life was worth much more than my camera.

So what did I do? I made an instantaneous decision and suddenly changed my friendly behavior, looked directly at the man, and glared at him as fiercely as I could, face to face. Then, without hesitating, I turned and quickly walked away from the group, after the boy who was taking my binoculars to the krall. I soon caught up with him, grabbed him by the shoulder from behind, turned him around, and forcibly took my binoculars back from him. He was quite surprised (perhaps a little frightened) and did not resist. Then I turned and began walking away from the group, at a fast pace but not running, back toward the highway where my Jeep was parked, about 1½ mile away. I had always heard that chills run up and down a person's spine when he is truly frightened, and that is exactly what I felt as I began walking away, shaking in my footsteps. I could feel a spear hitting me in the middle of my back at every step I took. Before very long I could not resist glancing back over my shoulder to see what the group of males was doing. They were all standing quietly and motionless exactly where I had left them, watching me as I walked away, undoubtedly trying to decide whether to throw a spear at me or run after me. They did nothing at all but watch.

I was still quivering as I arrived at the Jeep and climbed into the cab. On my drive back to Arusha I once again passed the small restaurant with a bar, and on the spur of the moment I stopped and went inside. It was almost noon time. I walked up to the vacant bar and rang the bell for the bartender. An older man appeared who was the owner of the restaurant and still a British citizen after many years in Tanganyika. He said to me, "What can I get for you?" Remembering the best cowboy movies I had seen in my youth, and still shaking from my harrowing experience, I said to him, "I'll have a double whiskey." When he gave it to me I immediately downed it all in one gulp, in proper cowboy fashion. The old man looked at me and said, "Are you all right?" I responded that I was going to be. Noticing the camera which I had over my shoulder, he asked if I was taking pictures around there. I answered that I was. He then said, "Let me give you some advice." "Don't take any pictures of the natives in this immediate area." "It violates their local beliefs and several years ago they killed a young tourist from Britain, a lad who was taking their pictures." My stunned response was, "Give me another double whiskey." My life had been spared by the spirits in the sky yet another time, but would my luck run out some day? Could I ever moderate my irrational, impulsive, behavior and innate obstinacy? Natural selection would surely eliminate the genes which compelled such behavior before I lived long enough to pass them on to the next generation. My father's words came to me again, in a different vein, "You must be crazy."

Upon returning to Arusha I found Noble, who had thoroughly enjoyed his morning talking with a fellow Trojan (USC graduate) he chanced to run into, Don Higley, who had been living in Tanganyika for the past two years. Don helped Noble arrange an itinerary for our imminent travel to Ngorongoro Crater and Serengeti National Park. Shortly after lunch the two of us set off for the crater, situated on the eastern edge of the Great Rift Valley approximately 110 road miles (in those days) from Arusha, on a winding, narrow dirt road in the newly established Ngorongoro Conservation Area (NCA). The road climbed in elevation from 4,500 ft. in Arusha up through montane forest to the rim of the crater at 7,500 ft. on the eastern side of the mountain. Because of trade winds coming from the east, this was the wetter side of the mountain, with more forest. The oval, grassland floor of this long extinct volcano was 2,000 ft. below the rim, and was a protected area for wildlife. Not far away from the crater was Olduvai Gorge, also located in the conservation area and famous archaeologically for some of the most ancient fossils assigned to the genus *Homo* ("humans"), who lived here approximately 2 million years ago, but we did not visit Olduvai. We had intended to drive down into the crater on a dirt tourist road to view the wildlife there, but recent rains had made this road temporarily impassable. Since we were going to return to Arusha along this same route, we would try again on our return. That night we camped in scrubby grasslands on the rim of the crater.

During the first several hours around our campsite the next morning, on a cloudy, cool day (Jan. 19), several small groups of unidentified sandgrouse flew past, and I saw my 1st Streaky Seedeaters* and Jackson's Widowbirds*. There was also an Augur Buzzard, a few Speckled and Rameron pigeons, 2 Egyptian Geese, a Pallid Harrier, Tawny Eagle, Lanner Falcon, and 15 White Storks (a non-breeding visitor from Europe and western Asia). At 10:00 o'clock we put our van in order and headed down from the crater's rim onto the vast Serengeti Plains, at an elevation of 4,500 feet, where we turned NW for the short drive to the entrance of Serengeti National Park. This was Masai country.

Almost immediately we encountered a small group of 6-10 Masai moving from one location to another across the plains, the first of several such nomadic groups we would meet in the next 10 miles. Young and old men and women (with or without babies on their backs) and boys and girls of all ages were travelling in families and genetically related groups along the road, bringing with them their cattle, goats, sheep, and other livestock as they moved to greener pastures at the beginning of the rainy season. The Masai had followed such a semi-nomadic, pastoral lifestyle for centuries. Unlike the small, isolated group which I was most unlucky to encounter yesterday, all the Masai we met today were cordial and friendly, and waved happily at us as we drove past them. They did not object to being photographed (without any payment) and smilingly posed beside our Jeep. They were also more colorfully and traditionally dressed than the ones I had encountered yesterday, and were highly decorated with beads, bracelets, necklaces, and dangling ornaments of all kinds - - around their neck, in the top and elongated, slit lobes of their ears, around their wrists and entire length of their arms, and around their ankles. Men were occasionally as highly decorated as were the women, and both men and women sometimes had their head entirely shaved. Warriors wore braided, ochre-dyed hair.

Body garments were very simple for both men and women, consisting usually of a single ochre colored piece of cloth or cowhide draped loosely over one shoulder or wrapped around their body. Women were more

Plate 43: While driving across the Serengeti Plain, Noble and I encountered a semi-no-madic group of Masai moving across the grasslands at the onset of the rainy season, as was their tradition. They followed centuries old customs of dress and behavior.

conscientious about keeping covered than were men, who sometimes let their blanket hang open in front, exposing their nakedness underneath! The majority of people were barefooted, but a few wore primitive cowhide sandals. Warriors and boys tending cattle carried shields and spears. The Masai lived in small, very primitive dome-shaped huts of sticks plastered with mud and cow dung (as the ones I encountered yesterday). Their traditional diet consisted of large amounts of raw meat, milk, and blood from cows. There were many rituals for different rites of passage for both girls and boys, and both were circumcised at the time they reached puberty.

Two features of the Masai which both Noble and I quickly noted were: (1) a very strong body odor which neither of us found very pleasant, and (2) constant swarms of little moisture-seeking bush flies which almost completely covered their bare shoulders and back, and much of their face (particularly around their eyes, nose, and mouth, wherever moisture could be found). These persistent little creatures did not bite but we found them extremely annoying on us, and they could not be shooed away no matter how many times you waved your hand at them. The Masai seemed to have evolved complete tolerance to them. Fortunately, Noble and I rarely had to contend with them except when we were near a group of Masai. In 1960, this well known, widely publicized and much photographed African tribe (with distinct subgroups) lived over an extensive region in southern Kenya and northern Tanganyika, resisting attempts by the governments to change their traditional lifestyle.

Shortly thereafter (on Jan. 19) we entered the vast Serengeti National Park, covering 5,700 sq. mi. east of Lake Victoria in NW Tanganyika, on the border with Kenya (where the Masai Mara Game Reserve would soon be established). In the far north of the park the Mara River flowed westward into Lake Victoria. Serengeti N. P. was said to be the most widely known game reserve in the world, and it claimed to have a greater concentration of plains inhabiting game animals than any other location on earth, with up to 2 million Blue Wildebeests (also called Brindled or White-bearded Gnus), ½ million Thompson's Gazelles, and ¼ million Burchell's Zebras. These 3 species, particularly the wildebeests, were renowned for their annual mass migrations north and west out of the central plains region of the park in the months of May, June, and July as the dry season commenced and the grassy forage withered and died. Many animals drowned while attempting to swim across the Mara River, while the weak, elderly, and very young fell prey to the lions, cheetahs, hyenas, and hunting dogs, all of which closely followed the migrating herds. It was an unparalleled spectacle, and the focus of many photographic documentaries. At the onset of the rainy season in December the animals returned to the central steppe for reproduction. In addition to the three most numerous ungulate species there were many other grassland and woodland mammals in the park, with up to 3,000 lions and 1,000 leopards. More than 500 kinds of birds had been documented. Serengeti was truly a wildlife paradise.

Noble and I visited just a tiny fraction of the park, following only the main SE to NW road for 75 miles from the SE park entrance to the park headquarters situated in the very small village of Seronera, and then returning over the same route. We were in open grassy savannah for almost all of our 5 days in the park, at elevations between 4,500 and 5,500 feet. The narrow park road on which we traveled was not hard surfaced and consisted of fine black clay soil, which was very dusty when it was dry. However, very soon after we started out through the park (on Jan. 19), a severe thunderstorm developed and heavy rain fell for more than an hour. (It was, after all, the beginning of the rainy season.) Our road quickly became a quagmire of black sticky mud several inches thick. Even when I put our van in 4 wheel drive, low transfer case, and low gear, it was almost impossible to make any forward progress. We crept along at 2 or 3 miles an hour, frequently becoming temporarily stuck in the mud.

Driving was very laborious (recalling our difficulties in the Chilean mud) but I stopped briefly on numerous occasions to identify birds. It was almost dark when we finally reached "Lion Hill', a low, rocky hill ("kopje") with scraggly comiphora trees, where I succeeded in leaving the road and driving a short distance up the hillside to a level hard rocky surface where we camped for the night (illegally), at 5,000 ft. elevation. Lion Hill was about halfway between the park entrance and Seronera. During our day in the park we had not encountered any other vehicles on the road. In spite of the difficult driving, I never-the-less had managed to see a good variety of birds along the way. These included the following: 30 Lappet-faced Vultures, 2 Marabou Storks, 2 Augur Buzzards, 15 African Pipits, 60 Fischer's Sparrow-Larks, 8 Rufous-naped Larks, 75 White Storks, 5 Northern Anteater-Chats*, 20 Capped Wheatears, 4 European Rollers, 30 House Martins (a non-breeding Palearctic visitor), 5 Hildebrandt's Starlings, 25 Red-capped Larks, 40 Crowned Lapwings, 15 Kori Bustards (wow), 25 Yellow-throated Sandgrouse* (my only record), 25 Chestnut-bellied Sandgrouse*, 2 Two-banded Coursers, 2 Pallid Harriers, 1 kestrel (not identified to species), 5 White-backed Vultures, 10 swifts (probably White-rumped), and 25 Caspian Plovers.

The next morning (Jan. 20) the weather was still cloudy, with scattered rain showers. The average annual rainfall here was 20-25 inches. The road had not dried out at all, so Noble and I made the decision not to attempt driving any further yet. Therefore, against park regulations I spent the day strolling back and forth on the open plains, from one acacia tree to another, adding to my bird list from yesterday. Noble was content to sit in the van and write letters, or in his journal. I finished the day with 57 different species. Additions to yesterday's list included (in part) 5 Meyer's Parrots, 18 Red-fronted Barbets (a remarkable number), 1 Jacobin Cuckoo, 10 Rattling Cisticolas, 1 Pied Wheatear, 4 Abyssinian Scimitarbills, 3 Scarlet-chested Sunbirds, 15 Rufous-tailed Weavers* (a large speckled bird with a whitish iris and rufous-edged wings and tail), 2 Tawny Eagles, 1 Hoopoe, 3 Bateleurs, 1 Saker Falcon* (a tough identification but I am sticking with it; I saw this species again in Turkey), and 7 Helmeted Guineafowls. I took a picture at sunset of the exquisitely patterned clouds in the western sky.

The road condition was no better the following morning, Jan. 21, so once again we chose not to attempt the still very muddy road. I was enjoying my leisurely wandering across the open savanna, between the widely scattered, small umbrella-shaped acacia trees. All around me were wildebeests, zebras, gazelles, soaring vultures, striding bustards, noisy lapwings, passing sandgrouse, and aerial displaying larks and pipits. I added a Montagu's Harrier (a Palearctic visitor) and a Black-shouldered Kite to my park list. Wildlife was conspicuous and abundant. It was a very exciting environment. Although we were camped on "Lion Hill", I saw no lions anywhere around.

Very shortly after I started out on this morning, when I was perhaps ½ mile from our van, I noticed a dark shape near the top of a distant, flat-topped acacia tree. My initial thinking was that it was probably an *Aquila* eagle (which were common here), and I approached it for a closer look and possible identification. It was on the other side of the tree from me, partially obscured behind the foliage. When I had approached to within 40 or 50 yards from the tree, I discerned that the object on the other side of the tree, partially hidden in the treetop foliage, was not an eagle at all but a LEOPARD! I was astounded, and delighted. It was very intent on eating a White Stork, which it no doubt had caught in the treetop before daybreak while the stork was roosting there. Under the tree on the ground were some scattered remains from the stork - - feathers and its entire orange-red bill - - remnants the leopard had chosen not to eat. (I salvaged the bill as a souvenir and I still have it at my house today, in my collection of miscellaneous plant and animal bits and pieces from all over the world.) Amazingly, the leopard was so engrossed in gleaning every last bit of flesh from the stork that it apparently didn't notice me at all, standing not far away on the open plain. I thought to myself what a wonderful opportunity this was to get an action picture with Noble's movie camera of a leopard. So I turned around and walked rapidly back to our van to get the camera. What were the chances that the leopard would still be there when I returned?

Remarkably, it was, and so concentrated on eating the stork that it apparently still had not yet noticed my presence, as unlikely as this might seem. My obscured view of the leopard was not a good moving picture, but never-the-less I sat the camera on its tripod and focused on the leopard, no more than 30 yards away in the treetop, still partially hidden in the canopy. Just as I was about to start shooting, an event occurred that had less than one chance in a million of happening. At that very moment, in the early morning, a park ranger came across the plains, flying a small, lightweight single engine aircraft, and flew very noisily 200 feet directly above the leopard in the tree! The airplane was heading for our Jeep on the hilltop, which almost certainly was what

Plate 44: Happily, this particular group was inquisitive, friendly, and did not object to being photographed. The Masai resisted attempts by the government to change their lifestyle and primitive traditions.

Plate 45: While in Serengeti National park we encountered a heavy rainstorm. Our black clay road became impassable for three days. During our delay, I discovered a Leopard eating a White Stork in the top of a small, isolated, flat-topped Acacia tree. I salvaged the bill.

221

the pilot was focused on. I doubt very much whether the pilot ever saw the leopard in the treetop as he flew just above it, or me, standing on the open grasslands as he looked directly into the early morning sun. As you can imagine, the leopard experienced the fright of its lifetime and in a single bound it leaped all the way to the ground from 20 feet up in the tree, and ran as rapidly as it could across the open plains, directly away from me. I don't think the leopard ever saw me, or that the pilot ever saw either one of us. However, the next day when we finally made it to the park headquarters Noble and I were immediately fined $15 for "camping illegally in the park." Our explanation that we could not drive on the muddy road was not considered an acceptable excuse.

We finally left our Lion Hill "campsite" on the morning of Jan. 22. I had listed a total of 73 different bird species during our 2 days there. During our 4 hour drive to Seronera I saw my first (and only) Fischer's Lovebirds*, a group of 3 which flew past our van. Other birds I encountered along the way, in part, were 1 Three-banded Plover, 4 Abdim's Storks, 1 Kori Bustard, 1 Lanner Falcon, 4 Double-banded Coursers, 1 Black-winged Stilt, 3 Egyptian Geese, 1 Secretary-bird, 2 Blacksmith's Plovers, 1 Greater Honeyguide, 2 Yellow-throated Longclaws, 1 Northern Wheatear, 5 Rueppell's Glossy-Starlings, and 5 White-headed Buffalo-Weavers. Also, I finally identified, to my best ability, a Lesser Spotted Eagle* (a non-breeding visitor from the Palearctic). Like almost all the individuals of *Aquila* here on the savanna, it was first seen perched in a treetop, but when I stopped the van it flew and soared above me, low over my head, allowing me to discern the few features which helped in distinguishing it from the resident Tawny Eagle, which was a common, widespread species. (Also possible were the Steppe Eagle, a Palearctic visitor which I had not yet identified with certainty, and the Greater Spotted Eagle, a very unlikely visitor which I also had not yet identified).

We arrived at Seronera in mid-afternoon and spent one night there before departing at 9:00 o'clock the next morning (Jan.23). This small community was spread out in the middle of Serengeti and contained the park headquarters, tourist lodges, campgrounds, and a little airport (from where the ranger had flown). At the park headquarters I purchased for $5 a painted, raw cowhide Masai shield which had been confiscated from poachers. (I still have this souvenir in my den at home.) There was a large, rocky kopje nearby where I had a close encounter with 2 lions during my early morning walk! One of them growled at me but then turned and walked away. Trees and ponds around the town provided a habitat for birds which was distinct from the open plains. In 6 hours of birdwatching here (on Jan. 22 & 23) I recorded 7 species which were additions to my trip list: Great Spotted Cuckoo*, Woolly-necked Stork*, Greater Painted-snipe* (a new family for me), Black Bishop*, Swahili Sparrow* (genus *Passer*), White-tailed Lark*, and Gray-breasted Spurfowl* (Francolin). Other species included Wire-tailed Swallows, Gray-headed Social-Weavers, Gray Crowned-Cranes, Yellow-billed Storks, Black-headed Herons, Comb (Knob-billed) Ducks, an African Jacana, White-crowned Shrikes, Lilac-breasted Rollers, Pin-tailed Whydahs, and a conspicuous, mottle-plumaged D'Arnaud's Barbet. An African Scops-Owl was heard after dark, giving its soft frog-like trill.

Noble and I departed Seronera at 9:00 am on the morning of Jan. 23 to retrace our route back across the Serengeti plains to Ngorongoro Crater and Arusha. During our 5 days in this world famous game reserve we had observed many of the more common and well-known mammals, including the leopard in the tree, lions at Seronera, hyenas eating a freshly-killed buffalo calf (after having killed its mother), "Coke's" Hartebeests (a race of Lichtenstein's), buffalos, zebras, waterbucks, and various dik-diks. Four species which were new for my mammal list were Thompson's and Grant's gazelles, Topi (also called Sassaby or Tsesseby, a relative of hartebeests), and Oribi (a medium-small, plainly colored antelope very similar in size and appearance to the Steenbuck). Considering the fact that we were limited by the muddy roads, we were quite well satisfied with our wildlife encounters in Serengeti. The leopard was a lifetime story. On our way out of the park, prior to our arrival at the exit, we became stuck in the mud again, when our rear axle sank well under the road surface. We were in the center of one of the parallel dirt tracks leading across the open plains, and had to wait 6 hours until a big truck came along, at midnight, to pull us out!

The next morning, Jan. 24, it required 4 hours for us to travel the last 10 miles out of the park through the three inches of sticky black mud on the road, in 4 wheel drive, low, low gear. It demanded all my concentration to prevent us from becoming bogged again. We saw no other tourists in the park. From the park entrance to the rim of the crater necessitated another 4 hours of driving. During this span of time I encountered one new trip species, the Schalow's Wheatear* (*Oenanthe schalowi*), a bird which was locally endemic to the highlands of southern Kenya and northern Tanganyika. That night, Noble and I camped on the eastern side of the Ngorongoro Crater rim, in a montane forest at 7,500 ft. elevation.

The next morning (Jan. 25, the first anniversary of the date we departed Los Angeles) Noble wanted to drive down into the 2-3 million-year old crater. There were many game animals on the grassland covered floor and he particularly wanted to see a cheetah (as did I), which surprisingly neither of us had yet observed anywhere. It would be an all day excursion, and I opted to remain behind on the rim, to search for birds in the montane forest there at 7,500 ft. elevation. Occasions for me to bird in mountain forests were rare, and I wanted to take advantage of the opportunity. Thus we parted company for the day, shortly after daybreak on a mild, overcast morning.

My day of searching produced 6 new trip birds: Tacazze Sunbird* (a purplish-black *Nectarinia* with elongated central tail feathers), Scaly Francolin*, Cinnamon-chested Bee-eater* (my 10th species of bee-eater), Gray-headed Negrofinch* (a handsome, dapper Gray & black estrildid with a long white eyebrow), Abyssinian Crimson-wing* (an elusive estrildid of the forest understory), and Thick-billed Seedeater* (a large, mostly drab brown fringillid finch with a whitish forehead and belly, and a huge pale grosbeak-like bill). Noble returned with the Jeep at 4:00 pm, sadly without having sighted a cheetah. We then drove all the way into Arusha for a beer at the Safari Hotel bar, where we met a young British chap about our age who was temporarily working in the region. The three of us were soon engaged in a lively conversation.

Noble mentioned that he and I were planning to climb to the top of Mt. Kilimanjaro in the near future, as one of our priorities while we were in Tanganyika. He said we had studied a tourist brochure describing the normal procedure for doing this. The brochure said that the usual way was to hire a single guide for the two of us and a porter each to carry our gear, and to spend 5 days going up and down the mountain, staying at night in 3 climber's huts at elevations of 9,000, 12,500, and 15,500 feet. Such a trip would cost the two of us a total of $60. We would bring our own food and sleeping bags, and that was what Noble was planning for us to do. The round trip was approximately 60 miles of walking, 30 miles up and 30 miles back down along the same route. The trail started at a hotel situated at 4,500 ft. elevation, and ascended all the way to the top of the mountain at 19,340 feet, the highest mountain on the African continent. It was a vertical climb of almost 15,000 feet. The young man listened intently to our plan and then said to us, in a knowledgeable manner, "if you are really fit you can do the climb in just 3 days, and if you are really fit you don't need a porter to carry your gear for you, and you don't really need a guide because the trail is very well defined and you can't get lost, and it won't cost you any money at all." He concluded by saying that he himself had done that. These comments were not only a challenge to Noble's competitive spirit and his physical fitness, but they provided us the opportunity to save some money. So Noble turned to me and declared emphatically, "that's what we're going to do"! As dumbfounded as I was by this proclamation, my own thinking was that if Noble could accomplish such a feat, so could I. We both enjoyed challenges. Without any hesitation I said "let's do it." The climb was still 3 weeks ahead of us.

We had one other destination on our agenda prior to climbing the mountain, and that was to visit the 5,000 acre Kisangara Sisal Estate, where we hoped to get some help in obtaining a zebra hide for each of us, as a trophy of our African adventure (a long time desire of Noble's). You may remember that one month ago our safari guide in southern Tanganyika, Clary Palmer-Wilson, had given Noble a letter of introduction to a good Danish friend of his, Thorkild ("Andy") Andersen, who had been managing this estate in NE Tanganyika for almost 30 years. Andy was a keen trophy hunter and Clary told us he could provide us with guns and a guide in our efforts to shoot a zebra each, as we were still allowed to do with our small game hunting licenses. Kisangara was British owned and was located between 3,000 & 3,500 ft. elevation, 40 miles south of the equator and 5 miles north of the small town of Lembeni (south of Moshi). Mt. Kilimanjaro's snow-capped peak was clearly visible from the estate, 65 miles away in the distance, looking NNW.

Tanganyika was one of the world's principal producers of sisal in the 1950's and 1960's. Sisal (*Agave sislana*, family Asparagaceae) was a native plant of Mexico and elsewhere in Central America but was widely cultivated in warm moist climates around the world. The plant grew to a height of 4-6 ft. with a whorl of thick, dark green, lustrous, spine-tipped leaves, each 2-6 ft. long and 5-7 inches wide. Cultivated plants were spaced evenly 4-8 ft. apart in large sunny fields. Young plants matured in 3-5 years, at which time they sent up a central flower stalk 15-20 ft. tall. After reaching maturity, a plant produced harvestable leaves for 7-8 years, about 70 in their first year and then 20-30 annually thereafter, for a total lifetime production of approximately 300 leaves. Leaves were cut for processing from around the outside base of the plant once annually. After being cut, the leaves were crushed between heavy rollers and pulp was scraped off the fibers. The creamy-white, coarse fibers were then washed and dried in the sun, producing strands 40-50 inches long. These strands were used in making rope and twine which was widely employed for marine, agricultural, shipping, and industrial use. Sisal was also used in making mats, rugs, millinery, and brushes. It was an intricate process.

We arrived at Kisangara in the late afternoon on Jan. 27, and were very warmly greeted by Andy, who had been told we might be coming. He was a man in his 50's or early 60's and lived with a much younger, very attractive wife, Ruth, who had only just recently married him, having come from Denmark to Kisangara to live with him (along with her cute, blonde-headed 10 year old daughter, Nina, from a 1st marriage). On the wide front veranda of their neat, spacious manager's house were mounted trophy heads of Andy's many years of game hunting: eland, sable, kudu, buffalo, oryx, impala, Grant's and Thompson's gazelles, and numerous others. In spite of our untrimmed beards, Noble and I immediately became part of their household, although we slept in our camping van which we parked in the side yard. Whenever we were present, we were always invited to eat our meals with them. Of course they had lots of household help at the house from African cooks, maids, workers, and laborers of all sorts. When he wasn't out in the bush attempting to shoot a zebra, Noble enjoyed reading the many books in Andy's extensive library.

I was astounded to discover that Andy had a permit to collect bird specimens for scientific study, and was quite keen on acquiring birds for various museums around the world (for which he received handsome payment). He was personally acquainted with John G. Williams at the Coryndon Museum in Nairobi. What a miraculous coincidence (rivaling that of William Mille and Rafael Barros in Chile) that I should chance to become acquainted with Andy. Of course he was equally surprised at my knowledge of local birds and experience in collecting and preserving specimens. Therefore, he gave me a long list of species which museums had requested of him, and he loaned me his shotgun for collecting some of these birds. By the time I left Kisangara I had collected a total of more than 200 individuals for him, representing 40 or 50 of the species on his list, several of which were species he himself had never located in the field. He gave all the birds I collected to a young African boy whom he had trained to prepare specimens, which he did extremely well and rapidly. What were the odds that in 1960 either Andy or I would encounter, entirely by chance, another person with similar knowledge and expertise, in the middle of East Africa?

Noble and I spent 3 weeks at Kisangara, Jan. 28-Feb. 17. My major birding areas were: (A) the estate itself (between 3,000 & 3,500 ft. in elevation), on Jan. 28, 29, & 30; (B) the nearby Pangani/Ruvu River (at 2,500 ft. elevation), on Jan. 31 and Feb. 1,2,5,6, & 14; (C) the South Pare Mts. (between 6,000 & 7,000 ft. elevation), on Feb. 2 & 3; (D) the North Pare Mts. (between 6,000 & 7,000 ft. elevation), on Feb. 8 & 9; (E) dry thorn scrub at the eastern base of the North Pare Mts. (at 2,500 ft. elevation), on Feb. 11; and (F) Lake Jipe (at 2,000 ft. elevation), on Feb. 15. I documented a total of over 200 species of birds, of which 34 were trip additions (Appendix B, beginning with the Vulturine Guineafowl* on Jan. 28, and finishing with the African Emerald Cuckoo* on Feb. 17). A remarkable total of 161 species was documented just along the Pangani/Ruvu River during 6 days and 41 hours of birdwatching there (area B above).

The Pare Mountains are part of the chain of isolated mountains in eastern Africa known collectively as the Eastern Arc Mountains, which from north to south are the Taita Hills in SE Kenya, and the Pare Mts., Usambara Mts., Uluguru Mts., and Udzungwa Mts. in eastern Tanganyika. The highest peaks in this chain are between 8,000 and 8,600 feet. Wet montane forests and dry woodlands cover much of these mountains. Local endemism in the plants and vertebrates illustrates the importance of geographic isolation in the process of speciation. I noticed differences in the numbers and kinds of birds even between the North and South Pare Mountains, in my brief visits to these two areas. The taxonomy of many of the birds is still being worked out, with DNA playing an increasingly important role.

One of the museums where Andy was sending bird specimens had requested that he send them population samples of the local nightjars, so he asked if I could collect some for him. He suggested an area of dry thorn scrub and grasslands at the eastern base of the North Pare Mountains, at an elevation of 2,500 feet. So I drove his Land Rover to this location on Feb. 11 and birded there in the afternoon. When the sun went down I began patrolling the roads looking for nightjars along the edges. As it got dark I had to depend mostly on eye shine to locate them visually before they flew. I persevered at this task for a total of five hours, from 7:00pm until midnight. It was a remarkable experience. I finished with the following list of species identified, all of which are nocturnal or mostly so: 3 Spotted Thick-knees, 2 Three-banded Coursers (only my 2nd encounter with this attractively patterned species, the 1st being the bird I photographed in daytime in Bechuanaland), 7 Bronze-winged (Violet-tipped) Coursers, 12 Sombre Nightjars* (7 coll.), 10 Plain Nightjars* (9 coll.), and 15 Slender-tailed Nightjars* (5 coll.). I also had a close-up look at a Verreaux's (Giant) Eagle-Owl. It was a very exciting and memorable night!

Lake Jipe was a small, shallow lake east of Mt. Kilimanjaro, on the border of Kenya (and Tsavo National Park). The lake was 7½ miles long and 2½ miles wide, with a surface area of 12 sq. mi. and an avg. depth of just 10 feet. It was fed by a river coming from Kilimanjaro, and water left the lake to form the Ruvu (and eventually the Pangani) River. It was rich in biodiversity, particularly birds, and was important to local fishermen and for irrigation of the surrounding countryside. Extensive swamps and marshes existed around the edges.

I drove the 20 miles from Kisangara to the lake on Feb. 15, for a day of birdwatching. When I arrived I quickly found a boy who was willing to paddle me around the lake in his long, narrow, hollowed out log boat, for a total fee of 50 cents for the day. This turned out to be more difficult than I had envisioned, because of the great instability of the boat. If a person leaned to his left or right at all the boat rolled precariously and could be prevented from tipping over only if the other person quickly leaned in the opposite direction. The boy sat in the back, on top of the gunnels, and I sat in the bottom of the boat in the front. It took much effort for me to sit upright while looking at birds through my binoculars, or taking pictures with my camera. The boy paddled me anywhere I wanted to go, and I gradually became more adept at keeping us stable.

I tallied 30 different birds during my 6½ hour boat ride. Three of these were additions to my trip list: 30 Pink-backed Pelicans* (a breeding pelican which was endemic to Africa), 4 Garganeys* (a common, non-breeding duck from the Palearctic), and 7 Whiskered Terns* (a breeding bird not only in Africa but also in Europe, Asia, and Australia). Other noteworthy birds were 12 Purple Herons, 50 White-backed Ducks, 15 Fulvous Whistling-Ducks, 4 Fish-Eagles, 5 marsh harriers (not identified to species), 5 Black Crakes, 6 Malachite Kingfishers, 10 Blue-cheeked Bee-eaters, 3 Taveta Golden-Weavers, and 2 Zanzibar (Red) Bishops.

To assist Noble and me in shooting a zebra, Andy loaned us the use of his ancient, roofless, dilapidated Land Rover, and he gave each of us a gun to use. He only went with us himself on one occasion, but he had two young boys whom he sent with us as "guides" (one at a time). Neither could speak much English. He gave us maps and told us where to go. The gun he gave to me was a telescope equipped, high-powered 9.3mm Czech made rifle which produced such a severe kick to my shoulder whenever I fired it that it pushed me backward, much to my discomfort and displeasure. Noble and I usually went hunting independently of the other, thinking we might have more success than if we were together. Therefore, we were pretty much on our own in our hunting expeditions.

On my second outing into the bush to shoot a zebra, the young guide with me and I were walking through open grassy woodland when we spotted a group of 4 or 5 zebras standing 400 or 500 hundred yards ahead of us, mostly obscured between the trees. We crouched down and slowly approached to within perhaps 250 yards of them, at which point there was a very large log lying on the ground behind which we could hide. I rested my rifle on top of the log for support, took careful aim through the scope at the most visible zebra, and pulled the trigger. The resultant jolt to my shoulder pushed me backward and all I could see was a cloud of dust as the zebras ran off. I asked the boy if I hit one, and he excitedly said, "yes, bwana."

Then a most amazing thing happened. The zebras were frightened by the shot but they had not seen us and they didn't seem to know where the shot had come from. In their panic they ran straight away from us through the trees for a short distance, but then for some unexplained reason they curved around in a big circle and headed back toward us, where we were still crouched out of sight behind the log. They came straight at us and then suddenly stopped when they were about 200 yards away, closer to us than when I first shot at them. Looking through the trees I couldn't see exactly how many there were so I was unable to know if one zebra was missing or not. My little guide was very excited and said "shoot again, bwana." I pondered this for a moment, knowing that my permit allowed me only one zebra. I thought to myself, what if I had missed and really didn't shoot a zebra the first time. I was in a quandary about what I should do. It was definitely a closer shot this time, with more chance for success. So I rested my rifle on top of the log for a second time, took aim at a zebra, and fired again. The result was a replica of my first shot. I got pushed backward, the zebras disappeared in a cloud of dust, and when I asked the boy if I had hit one he again responded excitedly, "yes, bwana."

The zebras ran off through the trees away from us and then for a second time, even more astounding than the first time, they once again circled around and came back toward us. They were still unaware of where the shot had come from. This time when they stopped they were no more than 100 yards from us, only a stone's throw away! I really couldn't believe my eyes. I was flabbergasted, and was still unable to get an accurate count of the zebras so I couldn't be absolutely certain whether I had shot one or not. My ecstatic young guide could not control himself and he shouted at me once more to "shoot again, bwana." This time, however, I refused to shoot again, and I stood up from behind the log. It was the first time the zebras had seen either one of us, and they instantly ran off through the trees, this time without circling back.

At my direction, the boy and I headed off toward the spot where the zebras were standing when I fired my first shot. A careful inspection all around found no zebra on the ground, or any evidence that one had been shot. Not only was I disappointed but I was a little annoyed at the boy for telling me I had shot one. We walked through the trees in the direction they had run, but found no evidence of having injured one. So we proceeded to the location where they were standing the second time I shot. Once again, there was no dead zebra, or any sign that one had been shot. Now I was truly irritated. What if I had missed on both occasions? I could hardly control my annoyance, or frustration, with the boy. On a last hope we set off walking through the trees again, in the direction the zebras had fled on this occasion. I became angrier with every step I took. Then suddenly, after we had walked 300 or 400 yards through the open dry bush, there on the ground right in front of us, miraculously, was a dead zebra! I couldn't believe my good fortune. My anger at the boy quickly disappeared. The zebra had been shot in one of its shoulders, right where I had aimed. My zebra hunt had come to a successful end.

Now I was confronted with the problem of what to do with the zebra, a dilemma I hadn't thought about. At that point another miracle happened. From out of nowhere a middle-aged African man appeared, carrying a long, sharp knife. My first thought was "oh no, not again." But it turned out that he was friendly and all he wanted was for me to give him the meat for his family to eat, and in return he would skin the hide off the zebra and give it to me. It was certainly a win-win situation. With amazing expertise this man had the entire skin off in one piece within a matter of 15 or 20 minutes. He then folded it up carefully and carried it all the way back to our Land Rover for us. I shook his hand, thanked him, and then drove with my guide back to Kisangara, where I happily related the events of the morning. My zebra hunting expedition in the East African bush was a story I could tell to my grandchildren one day.

Plate 46: Andy and Ruth Andersen greeted us at the Kisangara sisal estate. On Andy's front porch were exhibited some of his many African game trophies. During our 5-week stay at the estate, I shot a zebra and saved the hide as a "souvenir."

At the estate house, I stretched out the skin and tacked it to the plywood roof of our van, then salted it and left it to dry in the sun. Several weeks later, in Nairobi, I took the skin to a taxidermist for tanning, and left my home address in California where it could be shipped. (When I eventually settled down in Nacogdoches, Texas, in 1971, it decorated the wall of a hallway in my house for many years, until it finally fell apart and like all things came to an end.)

On Feb. 17, Noble decided it was time for the two of us to go climb Mt. Kilimanjaro, even though he had not yet succeeded in shooting a zebra. As we were gathering our gear together outside the house that morning I heard a familiar song from the top of a tall tree in the yard. It was a bird I had never succeeded in locating visually and therefore had not yet identified, or listed. After many minutes of critically peering into the foliage at the top of the tree I finally found the bird, a beautiful emerald green little cuckoo aptly named the African Emerald Cuckoo*. I was delighted to at last be able to put a name to this sweet, far-carrying, whistling song (rendered as "teeu-tu-tui") which I had heard many times in my African travel. Like all Old World cuckoos, it was a brood parasite and laid its eggs in the nests of other birds. This sighting was a good omen for our forthcoming ascent of the mountain!

Both Noble and I had looked forward to climbing Mt. Kilimanjaro from the start of our adventure together, a year ago in California. Trekking to the tops of mountains not only was stimulating exercise, it gave one a sense of accomplishment, which we both enjoyed. We departed Kisangara in mid-afternoon and drove the short distance north of the estate to the base of the mountain, to the rustic Kibo Hotel at an elevation of 4,500 feet. The hotel was owned by a rather elderly German woman, and in 1960 this was the starting point for the "Marangu" route to the top of the mountain. We were given permission to leave our camping van in the hotel parking lot during our planned three days on the mountain.

Thinking that if we ate a big meal that evening in the hotel we would have to carry less food with us on the mountain, we walked into the dining room about 6:00pm and sat down at a dinner table with the intention of eating as much as we could afford. Prices were reasonable, and one of the five-course meals on the menu was advertised as "all you can eat" for the equivalent of about 3½ U.S. dollars. Although this was considerably more than we usually spent for eating we decided to splurge, with the plan that we would indeed eat all that we could. Along with several mugs of beer each, we then managed to eat three, 5-course meals for the price of one, much to the chagrin of the German owner who protested to us that we couldn't do that (even after we pointed out to her that the menu clearly said we could). We won the argument, and at 10:00pm we waddled out to our camping van in the parking lot and tumbled into bed for the night, having indeed eaten three 5-course meals!

The challenge for me on Kilimanjaro was not to find birds but to get to the top and back down in 3 days, which was the goal Noble had boldly set for both of us. There would be very little time for me to birdwatch. We would have no guide, no porters, and no cost. I was very fit, weighing only 128 lbs., but my pack weighed 25-30 pounds. It contained food, water, warm clothes for the higher altitudes, first aid supplies, compass, altimeter, and miscellaneous other items. My sleeping bag was tied on top. We started up in just hiking shorts, each of us with a walking stick. In my shoulder bag, as always, were my camera, binoculars, and field notebook, but I did not take with me my 2 heavy volumes of Praed & Grant. To reach the top of the mountain did not require ropes or any special equipment for snow and ice (which was present only on the very top of the mountain). To succeed required only strong legs, strong lungs, and an indefatigable spirit. We would ascend upward in elevation for 14,840 ft. in 30 miles of walking, for an average grade of almost 10%. Noble and I were both 26 years old. Noble was confident he could make it since he had been told that others had accomplished such a feat "if they were fit" (and he certainly regarded himself in this category).

We started out at the crack of dawn on Feb. 18, at an elevation of 4,500 ft. from the hotel parking lot. There was a wooden bar across a 4-wheel drive jeep track with a sign that said "Kilimanjaro summit road." There were no entrance gates, information booths, ticket kiosks, or any other persons. There were just the 2 of us as we began our long anticipated trek up Kilimanjaro. It was an exciting moment. For the next 3 days we would not encounter a single other person anywhere on the mountain (except one native African man in the forest after dark on the very final night, five miles above the hotel). Noble and I agreed that it was not essential for the two of us to walk at exactly the same rate and stay next to each other, so we each walked individually at the pace which we found the most comfortable. Noble was bigger than I was in both height and weight, though he was trim and moderately fit, and he walked at a slightly slower rate than I did. However, when I stopped briefly for a picture or to momentarily look at a bird he quickly caught up. We walked steadily and were never far apart. Initially our vehicle track took us upward at a moderate slope through mixed farmland and woodland, but after several miles the jeep track ended and at 6,000 ft. elevation we entered a lower montane forest with only a well defined walking trail.

There were small waterfalls alongside our trail and I could hear birds and monkeys in the forest, off to the sides of the trail, but I did not stop to pursue them. By noon we had reached the first of the 3 climber's huts, at 9,000 feet. We were on schedule, ascending at a rate of almost 1,000 ft. in elevation per hour. My breathing rate was only slightly elevated and I was only a little fatigued. (Noble was a little more out of breath than I was and slightly more fatigued.) We were in a botanical zone with dense bamboo. The few birds which I had identified in the forest included various greenbuls, several kinds of warblers, Starred Robins, Double-collared and Olive sunbirds, Montane White-eyes (the most numerous bird), Olive Thrushes, Hartlaub's Turacos, Lemon Doves, Cinnamon-chested Bee-eaters, and vocalizing Emerald Cuckoos (now that I could recognize their calls).

The going became more difficult in the afternoon as we gained in altitude, and our rate of progress slowed. The forest changed, becoming thinner and less high, with lichens and mosses on the branches. It was just before dusk when we arrived at the 2nd hut, about 6:00 pm at tree line, 12,500 feet. We were both tired, but on schedule. In the open areas on the elevated rocky moorlands were giant senecios, heaths, and lobelias. In this environment I found my first Moorland (Hill) Chats* and Red-tufted (Scarlet-tufted Malachite) Sunbirds*, both species confined to this high ecosystem. The shining, dark, emerald green sunbirds with elongated central rectrices and a red tuft of feathers under their shoulder were particularly thrilling to see. Other species here were 15 White-necked Ravens, 40 Barn Swallows, 30 Hunter's Cisticolas, and 5 Streaky Seedeaters. Noble livened up our spirits by preparing some very welcome hot popcorn for us on his little Primus gas stove, which he always took with him everywhere we went, to everyone's delight. We ate a cold supper of canned tuna and fruit, along with a sip of brandy. Very soon we climbed into our warm sleeping bags on the hard wooden bunk beds in the chilly, unheated and unlighted hut. We had completed our first day, having ascended 8,000 ft. in elevation (from 4,500-12,500 ft.) but tomorrow was planned to be the most difficult on our 3 day agenda.

Daybreak the next morning (Feb.19) came much too soon. I had recovered quite well from our strenuous climb yesterday, but this was not true for Noble. He ached all over, groaned as he struggled out of bed, and quickly announced that he was much too stiff and sore to climb to the top of the mountain that day, as we had planned. He said for me to go ahead and we could meet each other at the end of the day at the 3rd hut, situated at an elevation 15,500 ft., after I had ascended to the top of the mountain and come back down to this hut, as was our plan. He would proceed more slowly and walk up only from the 2nd hut to the 3rd hut that day, where we could spend the night together. It would give him a chance to recover from yesterday, and he could then decide what he wanted to do on our planned 3rd day. This solution seemed satisfactory to both of us. Therefore, I quickly ate some bread & honey and an apple, and set off alone for the peak, which was clearly visible almost 7,000 ft. above me, about 12 miles walking distance away. The day began bright and sunny, with no wind and a chilly temperature. As I would soon discover, it had snowed lightly higher up on the mountain last night, not far above me. I was now wearing long trousers, a cap, and a jacket.

Soon I was on the high elevation "saddle" between Kilimanjaro to my west and 16,900 ft. Mawenzi Peak to my east. The vegetation was mostly grassy moorland with giant lobelias and yellow-flowering heath-like shrubs, between 13,000 and 14,000 ft. elevation. I did not slow down to identify the few birds which I encountered, but I did stop briefly to look at several Common (Gray) Duikers, a small antelope weighing only 25-40 lbs. and standing just 1½-2 ft high at the shoulder. I had not positively identified this little ungulate previously, even though it was very widespread and common throughout much of Africa. Because of its abundance and wide geographic and altitudinal distribution, this species was preyed upon by a long list of predators: lions, leopards, cheetahs, servals, caracals, kaffir cats, genets, civets, hyenas, jackals, hunting dogs, ratels, baboons, crocodiles, pythons, monitor lizards, eagles, large owls - - and others!

As I slowly gained in elevation along the saddle the grass covering became less dense, and the ground was blanketed with a light covering of snow which had fallen just last night. I reached the 3rd hut in late morning, situated at the western end of the saddle at the base of Kilimanjaro's steep volcanic cone, at 15,500 ft. elevation. It was still a clear, sunny day. Since I would be coming back down to this hut after I reached the peak, I was able to take almost everything out of my pack and leave it here. I was relieved to not have this weight on my back any longer, during my steep climb up the cone. I rested briefly at the hut, drank some water, and changed into my warmest clothing. There would be wind, snow, and ice on the very top.

I put on my snow cap and sun goggles and started up the steep, barren, windswept cone for the last 3,840 ft. of ascent. This was by far the most difficult part of the entire climb. The cone had a grade of 20-25% and was entirely covered with loose, rocky volcanic ash known as "scree." It was a climber's worst nightmare. For every hard-earned step upward, one slipped halfway back down. It was terribly frustrating. With the lower oxygen content of the atmosphere I was breathing more heavily than I had ever done. I began to question my wisdom in trying to reach the very top of the mountain on this 2nd day of climbing. My rate of ascent slowed to a snail's pace as I stopped frequently to catch my breath, and I worried if there was enough time left in the day to succeed. I asked myself if I were having fun. (For the moment the answer was clearly no.) Furthermore, clouds were now rapidly moving in, as they did almost every afternoon, obscuring most of my view. However, I told myself this was not a time for panic, and I continued to persevere.

With great joy and relief I reached the top of the mountain by mid-afternoon, quite exhausted by my strenuous climb. I had challenged the mountain and won! It was a wonderful feeling, and sufficient reward for all my effort. My success in reaching the rim provided me with the sense of achievement for which I was striving. From now on it would be all downhill! On the rim there was considerable snow and thick ice, as there was all year long even though Kilimanjaro was just a few degrees south of the equator. Footing was precarious and there was no visible trail. It was not possible to walk anywhere safely without crampons, which I did not have with me. I waited 10 or 15 minutes for the clouds to clear enough that I could take a picture looking down over the icy rim a short distance into the crater. On the top of the rim, firmly secured to a rock, was a large glass container covered with a heavy lid, within which was a book and pencil where a hiker could write his name, date, and address. I did this, in spite of the numbness in my fingers from the cold wind. Unfortunately, clouds prevented me from getting any pictures of the distant scenery (which is why climbers who take 5 days arrive at the top of the mountain in the very early morning).

Going down the steep, scree-covered cone was actually fun. The lava ashes to the side of the trail were so loose that one could take giant steps and then slide great distances downward before coming to a stop. Progress going down was 8-10 times faster than it was going up. It took me less than an hour to descend to the 3rd (top) hut, well before dusk. Noble was there to greet me and take a picture of me looking like a man from the moon, with snowcap, goggles, and frost in my beard. His day of rest had reinvigorated him. With my success of having reached the summit today, he was more determined than ever to start out tomorrow morning very early, 3 or 4 hours before sunup, in order to reach the top soon after daybreak. He said that he would then catch up with me on my way down to the bottom! It would be a tremendous achievement for him if he could succeed with this very ambitious plan. I told him I would start down slowly at first to allow him the opportunity to try and catch up. He encouraged me to stick to our planned agenda and not wait for him. (We were both very independent persons.)

As he had planned, Noble got up several hours before daybreak the next morning (Feb. 20), at 3:00 am, and began the very steep climb through the terrible scree to the top of Kilimanjaro. Moonlight allowed him to see where he was going. When I got up at sunrise I could clearly see through my binoculars that he was almost to the top. What determination! I waited an hour before I started down, at a very moderate pace. It was 27 miles to the bottom for me, and 30 miles for Noble (from the top). He would certainly descend through the scree faster than I was walking, and it seemed quite likely that he could indeed catch up with me. I hoped he would succeed. In spite of our independence and competitiveness, we were comrades. As I arrived at the 13,000 ft. level, just above the 2nd hut at treeline, I stopped for a few minutes to watch some birds, including 2 soaring Lammergeiers, which were always an inspiring sight. Again, there were White-necked Ravens, Moorland Chats, and Hunter's Cisticolas. I looked at my watch and decided I would have to pick up my pace in order to reach the hotel before dark, so I did. I told myself that if Noble didn't catch up with me before dark he could always safely spend the night inside the bottom hut, and finish his descent early tomorrow morning.

I arrived at the hotel at 5:00 pm, well before sunset. Noble had not yet caught up with me but there was still more than an hour of daylight left. I put my backpack in our van and went inside the hotel for a leisurely meal and a beer or two, or three, or four, while I waited for Noble to appear. Finally, when he hadn't come into the hotel by 11:00pm and the doors were being locked for the night, I went outside to sleep in the Jeep for the night.

When I got to the van I could not believe my eyes. There was Noble, lying horizontally on the hard asphalt parking lot right next to our van. I instantly asked him, "Are you all right?" He replied in a very low, husky whisper, "not really." I asked "why are you lying there?" His response was "I don't have the energy to find my key." The truth was that after 33 miles and 20 hours of walking Noble had completely exhausted every ounce of energy he had, like the marathon runner who "hits the wall." He had absolutely no strength left. He went on to say that it got dark just as he arrived at the bottom hut, but he was absolutely determined to keep going all the way to the hotel (as we had planned) rather than stay in the hut that night, to fulfill his challenge. Not long thereafter he lost his way in the dark and began stumbling down the mountainside through the forest, off the trail, without direction. He had almost completely exhausted all his strength when, miraculously, he saw a dim orange light off through the forest. Noble told me that he shouted as loudly as he could, over and over, until the light finally came toward him and an African man, perhaps in his fifties, appeared in front of him, with an oil burning lamp. Noble did not know what the man was doing in the forest with his lamp, but upon seeing Noble's plight he immediately offered to help him walk the last 3 or 4 miles down the mountain through the forest, all the way to the hotel parking lot, supporting him and constantly telling him (in Swahili) to "walk slowly." Noble admits that without his help he would surely have collapsed in the forest for the night, waiting for the hyenas to find him. So there he was, without enough energy to find his key, open the door, and climb into the van. I helped him into the van but Noble's indefatigable spirit had allowed him to do what he said he would - - hike all the way to the top of the mountain and walk back down again in 3 days. What determination! For several days afterward he walked very slowly and struggled to climb up into his passenger's seat in the Jeep's cab, complaining mostly that his feet hurt.

Noble wanted to make one more attempt at procuring a zebra skin, so we returned again to Kisangara, as Andy and Ruth had invited us to do. After 2 more unsuccessful hunting days, when Noble missed on several shots at a zebra, he reluctantly acknowledged that he needed a bit more target practice! We finally said goodbye to Andy and Ruth on the morning of Feb.24, and thanked them profusely for their 4 week hospitality and tolerance of two bearded young adventurers. (Ruth was going to be in Copenhagen in 6 month's time and she provided me with her address, saying I should visit her there if I had the opportunity to do so - - which I did!) Andy was very grateful for all the museum specimens I had collected for him. Before departing we took pictures of everyone (and my stiff zebra skin) next to our Jeep. Tanganyika was one of the most memorable countries of our 3-year adventure. Our next destination would be Kenya, for a second time.

Plate 47: Our trail began at 4,500 ft. elevation as a Jeep track, initially climbing gradually upward through mixed forest and farmland. No other persons were present. The 19,340 ft. summit was 30 miles ahead and almost 15,000 ft. above us.

Plate 48: I started the 2nd day from the middle hut at 12,500 ft., just above treeline in a zone of yellow-flowered heath. (Noble remained behind.) Soon I reached the "saddle" between Mawenzi (middle) and Kilimanjaro (bottom) peaks, and discovered new light snow covering the mountain.

Plate 49: From the top hut at 15,500 ft. it was a grueling, steep, 6-hour climb in volcanic ash ("scree") to the permanently snow-covered summit at 19,340 feet. Clouds almost completely obscured the view. After a short stay, I slid down rapidly and met Noble at the top hut, with ice crystals in my beard. The next day he climbed to the top (starting at 3:00am) & then we both descended, separately, to the bottom - - see text!

Chapter Fifteen -- SW Kenya, Uganda, and the Congo

On the afternoon of Feb. 24, Noble and I once again crossed the frontier between Tanganyika and Kenya, at the Kenyan border town of Namanga. Our immediate destination was Amboseli National Park, just a short distance east of Namanga. We were still hoping to see a cheetah in Africa, and this was our best opportunity left to find one. Amboseli encompassed 150 sq. mi. of grassland plains, yellow-barked acacia woodland, rocky, lava strewn, thorn bush country, swamps, and marshes. Most of the park was between 4,000 and 5,000 ft. elevation. Kilimanjaro was plainly visible only 20 miles south of the park. We camped that night just outside the park's southwestern boundary.

Although we saw large numbers of wildlife in Amboseli on Feb. 25, we did not manage to locate a Cheetah, which was certainly a nemesis for us. I took photographs of herds of zebras on the plains, with Kilimanjaro behind them on the distant horizon. I also photographed Marabou Storks standing in the top of a fever tree, a very tall giraffe not far from us along the edge of an open plain, two Gray Crowned-Cranes at the side of a marsh, a large flock of Collared Pratincoles flying over this same marsh, and a huge bull elephant just 20 yards away challenging us in an angry mood with his ears fanned out widely to the sides of its head. Even inside the cab of our van we did not feel entirely secure, in spite of the total 9,000 lbs. of our vehicle weight. On another occasion, a Black Rhinoceros charged at our van when we approached it too closely, but it stopped at the last moment. Large mammals in the game reserve were not always friendly! Other mammals included buffalos, wildebeests, hartebeests, and a variety of other antelopes such as Impalas, Grant's and Thompson's gazelles, and several Sunis (a tiny dwarf antelope even smaller than dik-diks). It was my first encounter with this species.

My total bird list for our day in the park was 63 species, of which only 2 were additions to my trip list, the Long-toed Lapwing* (a *Vanellus*) and the African Silverbill* (a plainly colored estrildid finch with a large silver-Gray bill). The lapwing was one of the most spectacular of the genus, with its white face, throat, fore neck, and belly, and its black hind crown and nape, grayish back, and broad black breast band. In flight, the wings were almost entirely white and the tail was black terminally and white basally. It was a very strikingly patterned bird that walked on floating vegetation like a jacana, with its long legs and greatly elongated toes. Other birds included Ostriches, Egyptian Geese, Black-bellied and White-bellied bustards, White-bellied Go-away-birds, Lilac-breasted Rollers, Fischer's Sparrow-Larks, White-headed Buffalo-Weavers, and Gray-headed Social-Weavers. These species were among the most characteristic and distinctive birds of East Africa.

Amboseli was the heart of Masai country and there were many Masai kralls within the park and the surrounding countryside. The local people here were friendly and happy to have us take their photograph (unlike the small local group I encountered near Arusha). Following their long tradition, they still tended cattle, goats, and donkeys, and lived in small, round, dung huts. I did not request permission to take pictures inside their houses, as I would very much have liked.

That night we camped again just outside the park. It was so hot inside our van that I decided to sleep on top of the van, on the flat plywood roof, on a thin mattress under a mosquito net. It was necessary to remove my large, one-piece, stiff zebra skin from the rooftop for the night and stand it upright on the ground resting against the side of the van. When I got up the next morning, to my utter amazement, it was gone! A quick inspection of the ground indicated that it had been dragged off through the dry open woodland. The culprit(s) had to have been one or more hyenas. I set off in pursuit, following the conspicuous drag marks in the dirt. It was an easy trail to follow. I feared the worst for my treasured trophy, and I was very angry at the hyenas. (I was not a seasoned African bushman and could not have envisioned such an event happening.) Sure enough, after perhaps ½ or ¾ of a mile I discovered my zebra skin lying in the dirt on the ground. Remarkably, it was still in one piece, mostly unscathed, and had not been chewed on. I was greatly relieved. The hyena (or hyenas) apparently just got tired of dragging it, dropped it, and proceeded elsewhere for something easier and more suitable to eat. I was able to drive our Jeep through the open woodland to the location where the large, unwieldy skin had been abandoned, and thus did not have to drag it or carry it back. The incident was one more memory for me of the African bush.

At mid-morning, we departed for Nairobi, the capital city and commercial center of Kenya, which was situated on a 5,500-ft. plateau. Noble was enthralled with this bustling city and looked forward to spending four or five days there. At the top of his priority list, as always, was Jeep maintenance and repair. This included having the valves ground in an effort to increase the performance of our underpowered engine. Two other priorities were to eliminate as much nonessential weight as possible, and to find additional funding to help with our diminishing funds. All of these activities challenged his entrepreneurial skills and he preferred to attend to them without any assistance from me. Before he left Nairobi, Noble succeeded in (1) selling our heavy Ampex stereophonic tape recorder for $500, (2) bargaining with Bardahl Oil to advertize their product with a painted logo on the back of our van until we left Africa, in return for $250, and (3) obtaining a voucher for free gasoline from Mobil Oil, in return for placing their well-known Pegasus sticker on the side of our van. No one was better at accomplishing these tasks than was Noble. Now we needed to find buyers for our heavy electric generator and our heavy refrigerator, both of which were mostly non-essential.

The first thing I did upon arriving in Nairobi on Feb. 26 was to take my zebra skin to Zimmerman's Taxidermy shop to have it tanned, and afterwards to have it shipped to my home address in Long Beach, California. The next morning (Feb. 27) Noble dropped me off at the Coryndon Natural History Museum, where I introduced myself to John G. Williams, the foremost East African ornithologist, mammalogist, and naturalist, who was 46 years old at the time. I conveyed greetings to him from Thorkild (Andy) Andersen, and told him of our four weeks at the sisal estate, and my activities in collecting birds there. John was a very friendly person and we visited together for more than hour, at the end of which he gave me permission to examine bird specimens in his collection to help me identify some of my unnamed descriptions in my field notes. Among his many research activities John also collected butterflies, mostly as a hobby, and when I mentioned to him that I also collected butterflies in my travels he gave me his mailing address and asked if I would send him some specimens from the Congo and Central Africa when Noble and I traveled through these regions in the near future – which I did! When I inquired where I should birdwatch in Kenya along our proposed route of travel to Uganda, he told me about Lake Nakuru (of which I had already read) and the Kakamega Forest (with which I was unfamiliar). My morning at the museum with John G. Williams was an extremely pleasant and rewarding experience. John was one of the leaders in field studies of East African birds and mammals. He had few peers.

Since Noble didn't need or request my assistance with his activities in the city for the next several days, I decided to take my backpack and sleeping bag and hitchhike to the Kakamega Forest to birdwatch there for this period of time, stopping at Lake Nakuru on the way. Noble was quite happy with this plan. So I set off the next day, Feb. 28, for Lake Nakuru, with about $25 of spending money. It was the beginning of a very memorable five-week adventure for both of us. The initial plan was for Noble to pick me up at the Kakamega Forest in about 4 days time. Noble would have to do all the cross-country driving, a task he had done for more than one day only once before (in Argentina when I hitchhiked for several days across the pampas with Richard Dugdale).

From Nairobi to Lake Nakuru was just 100 miles, on a good hard surfaced asphalt ("bitumen" or "tarmac") road. In my primitive Boy Scout rucksack I had a sheet, a very lightweight jacket, several changes of sox & underwear, another short-sleeved shirt, food, water, binoculars, camera, pocket knife, compass, altimeter, snakebite kit, simple first aid kit, small bar of soap, butterfly net, cyanide jar, two volumes of Praed & Grant, field

Plate 50: Amboseli National Park is on the southern boundary of Kenya. I photographed a Hooded Vulture feeding on an Elephant carcass, an Ostrich, and Buffalos at a waterhole. One night my zebra skin was dragged, unscathed, a short distance away by hyenas.

notebook, paper envelopes (for butterflies) and several pencils. Only the two bird books, and my one quart of water, were particularly heavy. I wore just hiking shoes, hiking shorts, a short-sleeved shirt, underwear, and sox. I didn't have a hat, and I never used sunscreen in spite of the often hot and sunny days. I was distinct in my sun-tanned appearance with my shaggy hair, long black beard, and a blond moustache! If it rained I simply got wet (then soon dried out). Daytime temperatures were mild to hot, but it cooled off somewhat at night. I could always find a rest house, guest house, or some other kind of shelter for the night, at little or no cost. It was a fun, carefree life.

Hitchhiking was remarkably easy. Almost every vehicle that came my way offered me a ride, both Africans and Europeans. Africans always asked politely "Safari a whapi, bwana?" (Where are you going, sir?) Most people were not going very far. Some vehicles were in very poor condition. One old flatbed truck with an ancient African driver had a top speed of 20 mph. The driver stopped every five miles to add water to the radiator, and every ten miles to fasten the accelerator cable together with a rubber band. Fortunately at one of his stops another vehicle came along, to which I quickly transferred. I arrived in the little town of Nakuru that night and found a cheap rest house – 35 cents for the night with my own food and sleeping bag. (This was a popular tourist town and there were, however, several very nice hotels available, which were much more costly.)

Lake Nakuru was one of many shallow "soda" lakes in the central Rift Valley, at 5,500 ft. elevation. It had a surface area of 30 sq. mi., a maximum depth of only 6 ft. (in the middle), and an average depth of less than 1 foot. The amount of water in the lake fluctuated from year to year and season to season, depending principally on rainfall (annual average of 34") and temperature. Its high salt content was mostly sodium carbonate (from diatomite in the soil). Around the margin of the lake was a wide band of yellow-barked "Fever Trees" (an acacia) which attracted a variety of non-aquatic birds.

The lake was widely known and publicized throughout the world as an "ornithological wonder" because of the 1-2 million flamingos that frequented the lake for feeding throughout most of the year. Of these, 95% were Lesser Flamingos which fed on microscopic blue-green algae and diatoms, and 5% were Greater Flamingos which were ecologically separated by feeding on bottom, mud-dwelling invertebrates. Both species were filter-feeders which fed (singly or in dense groups) by standing in shallow water, bending their neck down and forward, submerging their head upside down under the water and moving their lamellate (filter-edged) bill back and forth sideways along the bottom of the shallow lake, filtering food items from the water as they did so. Flamingos did not breed on Lake Nakuru, but did breed on several of the other soda lakes in the area, principally Lakes Natron and Elmenteita. Roger Tory Peterson visited Lake Nakuru in August 1958 (one and a half years prior to my visit) and after viewing the one million or more flamingos around the edges of the lake, he wrote, "I witnessed the most staggering bird spectacle in my thirty-eight years of bird watching."

I spent all morning (0700-1200) at Lake Nakuru on Mar. 1. During these 5 hours I tallied 52 species of birds, 2 of which were trip additions, the White-eyed Slaty-Flycatcher* (easily recognized by its dark to light Gray coloration and very prominent dark eye surrounded by a wide white ring), and the Gray-capped Warbler* (distinctly colored with an olive back, whitish underparts, Gray cap, chestnut throat patch, and prominent black band through the eye to around the nape). The warbler's loud, variable song directed me to it in the middle of thick bushes. Both of these were exciting finds. The shoreline was entirely covered in pink by the flamingos, which in my field notebook I estimated their number to be "the whole lake." Among the hordes of Lesser Flamingos I could find only ten Greater Flamingos. It was truly a remarkable sight. (Sadly, my telephoto camera lens was being repaired in Nairobi.) Other birds of note were 50 Marabou Storks, 10 Sacred Ibises, 25 Black-winged Stilts, 120 Ruffs, 25 Speckled Mousebirds, 5 Arrow-marked Babblers, 3 Malachite Sunbirds, and 9 Baglafecht Weavers.

I left Lake Nakuru at noon and continued my hitchhiking north and west toward the Kakamega Forest. The road soon began ascending the west side of the Rift Valley, into the scenic "White Highlands" region of western

Kenya where many Europeans had settled and built farms, at elevations between 6,000 and 8,000 feet. The farmers here were very concerned their farms would be taken away from them when Kenya gained its independence from Britain in the near future, and their properties would be divided up and distributed to local Africans. The farms were devoted to dairy cattle, coffee, peaches, tea, and a daisy-like plant, *Chrysanthemum cineraiifolium*, used in the production of pyrethrum (a chemical utilized in the manufacture of safe, natural insecticides). In the 1950's and 1960's the Belgian Congo and East Africa were the world's principal growers of pyrethrum. Roadside birds included 3 new trip additions: Brown-backed Woodpecker* (very like a Downy Woodpecker), Yellow-throated Greenbul* ("Leaflove"), and Gray-backed Fiscal* (a *Lanius* shrike). Other birds seen (in part) were 2 Hadada Ibises, 2 Long-crested Eagles, 3 Augur Buzzards, 4 Brown Parrots, 2 Sulphur-breasted Bushshrikes, 2 Copper Sunbirds, and 2 Golden-backed Weavers. That night I found a roadside rest house about 30 miles east of the town of Kisumu, at an elevation of about 8,000 feet.

I arrived at the Kakamega Forest the next day in the early afternoon (Mar. 2) and checked in at the Forest Rest House, where I presented my letter of introduction from John Williams. I was assigned to a small room with a bed, table and chair, at no cost to me. A bathroom was down the hall. Shortly thereafter, I took my binoculars, bird books, and field notebook and walked excitedly along a main trail into the forest. The Kakamega Forest, fluctuating between 5,500 & 6,000 ft. in elevation, was the easternmost, last remaining remnant of the once extensive Guinean-Congolese Rainforest of central Africa. Much of the forest was still pristine, having never been logged. Open glades and clearings, marshes, and small streams were found throughout. Rainfall averaged about 60" per year. The relatively small part of the overall forest where I would search for birds covered an area of about 100 square miles. I would spend four days birdwatching here (Mar. 2, 3, 4, & 5). There were an abundance of footpaths and forest roads for me to follow.

The forest trees were characterized by Elgon teak, Red Stinkwood, and several varieties of Croton trees. There were many ferns and a great diversity of orchids. Snails and arthropods were particularly abundant, including a myriad of colorful butterflies, beetles, grasshoppers, ants, spiders, and millipedes. In 1960, 325 bird species had been documented for all of the forest. Mammals included baboons, 4 species of monkeys (Red-tailed, De Brazza's, Blue, and Vervet), pottos, clawless otters, bats, mongooses, bush pigs, duikers, and bushbucks. Sadly, I had no references to help me identify the species of monkeys or other small mammals. When I wasn't birdwatching during the middle of hot afternoons, I devoted my time to chasing butterflies, which I caught and preserved in paper envelopes that I then placed in empty cracker tins.

My leisurely pursuit of birds in the Kakamega Forest was my first time in a tropical African rainforest, and was thus a particularly thrilling experience. During 39 hours of birdwatching over my 4-day period, I identified a total of 110 species, of which 54 (almost exactly half) were birds I had never seen before. Obviously, the number of new birds I saw on a particular day declined as the days passed. Out of curiosity, at the end of my time in the forest I drew a graph which plotted on the Y axis the number of new species I saw (per number of hours of observation on each of my 4 days) versus the day of observation (1, 2, 3, or 4) on the X axis. The data for my 4 points were: 4.8 birds per hr on Day 1, 1.3 birds per hr on Day 2, 0.87 birds per hr on Day 3, and 0.5 birds per hr on Day 4. The line that connected these four points formed an almost perfect inverse exponential curve! What this told me about my birdwatching in a new ecosystem was the predictable fact that I saw fewer and fewer new birds each day. However, I did not expect such a perfect relationship. What I concluded was that four days of birdwatching at a new location was long enough before I moved somewhere else. My total list for the four days was approximately one-third of all the birds known to inhabit the area. The families which contained the most species were bulbuls (10), cisticolid warblers (9), weavers (8), sylviid warblers (6), estrildid finches (6), bushshrikes (5), and sunbirds (5). Appendix B (for Mar.2, 3, 4, & 5, 1960) lists all the 54 new species I added to my trip list (none of which I had ever seen before). (One of these species, the Red-headed Malimbe in the ploceid genus *Malimbus*, I will have more to say about later.)

One of the birds I saw deserves special mention, the Gray-chested Illadopsis. I encountered a single individual on Mar. 4, after its low-pitched alarm notes directed my attention to it in some dense streamside vegetation. I cautiously approached it, by crouching down, to within 10 feet of the bird where I had a clear, mostly unobstructed view of it very near the ground, almost too close for me to focus on with my binoculars. I scrutinized it carefully for several minutes (mostly with my naked eye), memorizing its features. The bird was not particularly frightened by my very close approach. I did not recall having encountered it before, or having read about it or seen an illustration in my Praed & Grant during my many hours of evening study. Therefore, when the bird finally moved out of my sight I immediately took out my little field notebook and wrote as complete a description as I could of its features.

My written notes said: "Turdidae: alethe size & shape (somewhat smaller than white-chested?); above uniform bright russet, including tail, & extending to just below eye on the head; throat white; chest and flanks grayish; lower chest indistinctly streaked or mottled with gray; lower underparts whitish; scolding rather nervously from low down in dense streamside undergrowth with a low 'dirit, dirit'; observed very well (except posterior underparts) at close range for several minutes; eye dark; legs gray?; upper mandible gray; lower mandible light silverish-gray [I drew a very small sketch of the bill shape]; observed in shade at about 10 feet with 7x35 binoculars. The bird was in plain view."

I was unsuccessful at confidently matching these field notes to any species I could find in any of my bird books. Years later I sent a copy of my field notes to Dale Zimmerman (who was a friend of mine from our college days together at the University of Michigan in the early 1950's), knowing that he had spent numerous summers studying birds in Kenya and hoping that he might be able to identify the bird for me. (This was prior to the time he published his *Birds of Kenya*.) To my dismay, he responded that he could not be certain of the identification based solely on my field notes. So it was not until Jan. 24, 2000, when I was first perusing Dale's excellent *Birds of Kenya* (which he had published in 1996) that I came across his painting of the Gray-chested Illadopsis on Plate 90, and read his description of the bird on page 531. The match was perfect! Almost 40 years after I saw the bird in the Kakamega Forest, I was finally able to assign a name to it. Zimmerman (and subsequently Clements in his 2007 checklist) placed it in the family Timaliidae, in the monotypic genus *Kakamega*! (Recent studies of the DNA and other features of this illadopsis have indicated that its nearest relatives are the African sugarbirds in the genus *Promerops*, of which I will never be convinced.)

Since Noble was late in picking me up at the Kakamega Forest and I had spent as much time there as I wanted, I decided to hitchhike onward to Uganda where I had the address and phone number of a young birdwatching couple that I had met on the rim of the Ngorogoro Crater, who lived in Entebbe, a twin city of Kampala on the north shore of Lake Victoria. I left this information for Noble at the Kakamega Forest rest house (on one of our indispensable Jeep postcards) and set off on the morning of Mar. 6 for Entebbe.

The picturesque, green, rolling countryside in Uganda was dominated by many sugarcane and pineapple fields. Women wore colorful dresses with wide-flaring "butterfly" shoulders, and they carried bundles and baskets of all sizes on top of their head, as they had done for centuries. Influenced by Christian missionaries, about 20% of Ugandans had embraced Christianity, and small, plainly adorned churches with a cross on the top were situated along the roadside. Rural houses were very primitive, mostly made of mud, sticks, or wooden planks, with thatched roofs. Over-burdened bicycles were pushed along the highway with heavy loads of pineapples, reeds, and sugarcane. Small roadside markets and kiosks sold a great variety of fruits and vegetables, including bananas, pineapples, melons, squashes, potatoes, cabbages, and tomatoes. Rattan baskets, chairs, tables, and trays were manufactured by hand and displayed for sale in small shops everywhere.

My highway to Kampala paralleled the north shore of Lake Victoria, passing through the large, popular tourist city of Jinja on the lake's north shore at a point where water flowed south out of the lake as the "Victoria Nile" River. The Nile River is proclaimed to be the longest river in the world with a total length of about 4,250 miles. Its actual length is still slightly in dispute because of the difficulty in deciding which of two tributaries

Plate 51: I hitchhiked westward from Nairobi, stopping one day to witness the spectacle of flamingos at Lake Nakuru, then for three days of birdwatching in the remnant Congo rainforest in the Kakamega Forest, where I also caught butterflies.

is the longer. It is one or the other of two upper branches of the Kagera River, which flows into western Lake Victoria from Tanganyika. It is either the Nyabarongo River (with its headwaters in the Nyungwe Forest of Rwanda) or the Ruvyironza River (with its headwaters in SW Burundi). The Victoria Nile is the longest branch of the Nile.

I arrived in Kampala on the early evening of Mar. 6, checked into an inexpensive rest house, and then telephoned my friends in Entebbe (whom I had met on the rim of Ngorongoro Crater six weeks earlier). We made plans to go birdwatching together for the next two days. Kampala was the capital city of Uganda, located almost exactly on the equator at 3,800 ft. elevation, just above of its twin city of Entebbe on the north shore of Lake Victoria.

Lake Victoria was located on a plateau between the E & W branches of the Great Rift System. It was Africa's largest lake in surface area, covering 26,800 sq. miles, and was the world's largest lake anywhere inside the tropics. Owing to its relatively shallow depth (maximum of 270 ft. and average of 130 ft.) there were many very small islands in the lake. The largest of these were permanently inhabited by fishermen who regularly paddled back and forth to the shore in small canoes or flat bottomed boats, often through man-made canals in the emergent lake vegetation. Around the lake's shallow margins were extensive patches of papyrus, cattails, reeds, grasses, and lilac-colored water lilies that were home to birds and many other aquatic organisms and wildlife. There was a great diversity of native fish in the lake, more than 350 kinds of mostly endemic cichlids, many of which were very brightly colored. Members of the endemic cichlid genus *Tilapia* were the dominant commercial fishes in the 1950-60's, and the lake's shoreline was heavily populated with Bantu speaking people who had created Africa's largest inland fishing industry.

Ornithologically, the lake was renowned as one of the few sites in Africa where Shoebills could be found. These huge, ungainly, prehistoric-looking wading birds (also known as "Whale-headed Storks") stalked through the shallow, densely vegetated margins of the lake characteristically catching lungfishes, typically flapping their wings as they pounced on their prey. Their huge toes enabled them to walk on the top of floating vegetation. Other vertebrates, as well as invertebrates, were included in their diet. Shoebills possessed an enormous, broad, slightly hooked bill that reflected sunlight like a mirror, enabling observers to spot them through telescopes at great distances, which is exactly what I did on the morning of Mar. 7. Shoebills have always been a taxonomic puzzle, but recent DNA studies indicate their closest relatives may be Pelecaniforms! Currently, most authors, including Clements, place them alone in the monotypic family Balaenicipitidae. Shoebills make a loud hollow sound by clapping their mandibles together, and surprisingly they soar gracefully on thermals, with their head and neck tucked in.

I spent Mar. 7 & 8 with my 2 friends, birdwatching a total of 23 hours in the Entebbe area, including Lake Victoria, the botanical gardens, and the airport. During this time, I recorded a total of 138 species, which included 27 trip additions (Appendix B). There were 9 species of sunbirds and 11 species of *Ploceus* weavers on my list. Two species were local endemics to this area, the Red-chested Sunbird* and the Northern Brown-throated Weaver*. Among my more memorable and enjoyable species were the Gray Parrot*, Woodland Kingfisher (only my second sighting of this widespread non-aquatic kingfisher), African Fish-Eagle (always a favorite with its nostalgic, far-carrying vocalizations), the majestic Saddle-billed Stork, Gray-capped Warbler, Orange Weaver* (dominating in its bright orange plumage), Ross' Turaco* (a striking, dark purplish turaco with a prominent red crest and bright yellow bill, frontal shield, and bare eye patch), Black-and-white-casqued Hornbill (one of several large hornbills with a wide white trailing edge to the wings in flight), Snowy-crowned Robin-Chat* (with a black face and upperparts, rufous-orange underparts, and snow-white crown), Superb Sunbird* (a large, dark sunbird with an iridescent violet throat & upper breast, dark maroon belly, emerald green back, and unusually long, decurved bill), and Gray-headed Negrofinch (with black underparts, pearly-Gray back, and white eyebrow). In addition, I was intrigued by the mixed waterside colonies of nesting weaver species, each species with its own distinctive nest. It was exciting to witness a Gray-backed Fiscal (a laniid shrike) capture and kill a migrant Yellow Wagtail, a bird two-thirds its own length. The genus *Lanius* is found in both the Old and New Worlds, and members of this genus are one of the few groups of passerine birds with a hooked bill and raptorial behavior.

Noble caught up with me in Kampala on Mar. 9, and we discussed what we would each like to see in western Uganda on our way to the Belgian Congo. Noble wanted to travel north to see Murchison Falls (later renamed Kabarega Falls) on the Victoria Nile, but I preferred to spend this time searching for birds in the Semliki River rainforest along the Congo border, west of Fort Portal in western Uganda. Fort Portal was situated at 5,000 ft. on a plateau at the base of the northern Ruwenzori Mountains. It would be easy for us to join up again there in four or five days, where we had the address of a guest house, so we amiably agreed upon this plan. I would continue to hitchhike and Noble would continue to drive the van. It would be a longer distance for him, to go north and then return south. So we set off separately once again, from Kampala on the morning of Mar. 10.

Fort Portal was approximately 250 miles west of Kampala on a main road, gravel but fairly well maintained. The countryside was mostly a green rolling plateau between 4,000 and 5,000 ft. elevation with cattle range-land, coffee and banana plantations, cultivated fields (cassava, beans, sweet potatoes, and millet), and scattered patches of bush, woodlands, and marsh. There were few vehicles on the road (mostly trucks, which came in all sizes, kinds, and shapes). Virtually everyone stopped to offer me a ride, still with the query of "Safari a whapi, bwana?" It took me nine hours to make my journey (8:00 am-5:00 pm). I recorded the birds I saw along the way and finished the day with a list of 52 species, of which the African Pied Hornbill* (genus *Tockus*) and the Speckle-breasted Woodpecker* (genus *Dendropicus*) were the only trip additions. Other species included a Bateleur, Long-crested Eagle, Ross' Turaco, Great Blue Turaco, 7 Eastern (Gray) Plantain-eaters, 5 Broad-billed Rollers, 60 Little Bee-eaters, 15 Speckled Mousebirds, 4 Double-toothed Barbets, 6 Sooty Chats, 25 Rueppell's Glossy-Starlings, 30 Viellot's Weavers, 5 Fan-tailed Widowbirds, and 4 Pin-tailed Whydahs. It was a very satisfactory list for a day of hitchhiking. I stayed in a guest house that night in Fort Portal for $2.00, with very cordial and friendly European owners.

The next morning (Mar. 11) I obtained the name of a rest house where I could stay for several days in the small town of Bundibugyo, situated about 2,500 ft. in the tropical Semliki rainforest 25 miles west of Fort Portal on the east side of the northward flowing Semliki River. This river was the boundary between Uganda and the Belgian Congo for most of its 90 mile length between Lake Edward (to the south) and Lake Albert (to the north), in the Albertine Trench of the western Great Rift Valley. The Semliki rainforest, with an average annual rainfall of 60", was an eastern extension of the Ituri Forest in the northeastern Belgian Congo. The dominant tree was the Uganda Ironwood (*Cynometra alexandri*). Birds, mammals, and other wildlife were characteristic of the vast Guinean-Congolese tropical rainforest extending all the way across West and Central Africa on both sides of the equator. The most ancient human inhabitants are pygmies, of which only a very small fraction survive today.

I spent four days leisurely chasing birds and butterflies in the Semliki forest (Mar. 11-14). The penetrating songs and calls of doves, tinkerbirds, hornbills, cuckoos, turacos, greenbuls, bushshrikes, orioles, flycatchers, warblers, drongos, and many others resounded throughout the forest, but tracking down the vocalizing birds to identify them visually was a challenging, almost impossible task. Of the 93 total species that I identified, 26 were trip additions (Appendix B). Included on my list were 8 species of estrildid finches, 8 species of warblers (combined sylviids and cisticolids), 7 species in the cuckoo family, 7 species of pycnonotids (bulbuls, greenbuls, nicators, leaf-loves, and bristlebills), and 5 species of hornbills (White-thighed, Black-and-white-casqued, Afr. Pied, Piping, and Black-casqued). This latter species is one of the two largest African hornbills (along with the Yellow-casqued), with a huge casque and extremely noisy swishing wing beats in flight. Other species of note were the Openbill, Long-crested Eagle, Palm-nut Vulture, Gray Parrot, Woodland Kingfisher, Great Blue Turaco, and Ross' Turaco.

Two species were endemic to just the Congo rainforest, the Gray-headed Sunbird* (an attractive olive green species with a contrasting Gray head) and the Jameson's Antpecker* (an unusual warbler-like estrildid finch of the forest understory, with a narrow bill for feeding mostly on ants). I wrote two long descriptions of it in my field notebook and then spent over an hour perusing my bird books before finally solving its identity. Another difficult identification task was separating the Icterine and Xavier's bulbuls from each other. The only distinction Bannerman made was size; the Xavier's being very slightly larger. Such a feature was virtually useless in the field unless both species were side by side. (These two species were lumped together as a single species by Praed & Grant.) I also had considerable difficulty when writing this story in converting the name "Rufous Flycatcher" in my field notes to any bird in Clements (2007). I had obtained this name from Praed & Grant, who gave it the Latin name of "*Stizorhina fraseri*" (a species they placed in the family Muscicapidae). However, I could not find either this English name or generic name indexed in Clements. So I went to the indexed specific name of "*fraseri*" in Clements. Here I found 4 choices of genera. I chose "*Neocossyphus*" and went to page 401. Success! There I found "Rufous Flycatcher-Thrush, *Neocossyphus fraseri*", placed in the family Turdidae. (Avian taxonomy was a continual challenge and a constant source of frustration in writing this story.)

I was particularly enthralled by the Crested Malimbe, a striking, red and black, forest-inhabiting ploceid weaver in the genus *Malimbus*. There are ten species in this distinctive genus, all of which are confined to West and Central African forests. Their hanging nests have an entrance tube and are usually suspended from a palm frond or tree branch at a considerable height above the ground. In spite of their relatively heavy bill, malimbes are primarily insect-eating, often traveling in mixed foraging flocks in the forest canopy. This genus was one of the features that separated the avifauna of West Africa from that of East Africa. (As I mentioned previously, I saw my first malimbe, the Red-headed Malimbe, in the Kakamega Forest of Kenya, which was a remnant of the W. African rainforest.)

Noble met up with me in Bundibugyo on Mar. 14. We took a few photos alongside the road of a Batwa pygmy man with his wife and baby, next to Noble and the Jeep. Sadly, as their only source of income these people came out of the forest and stood by the roadside to solicit money for taking their picture from the few tourists who came by. The average height of males was just 4'4", and of females only 4'0." The woman who we photographed was wearing only a piece of cloth tied around her waist, and was holding a 1-2-year-old, naked, male toddler in one arm, up to her chest. He was attempting to nurse but her long pendulous breasts were hanging so low, to below her waist, that it was necessary for him to grab the lower end of one with both of hands and pull it, in a "U" shape, up to his mouth in order to drink. The pygmies are an evolutionary phenomenon that is approaching extinction.

Noble had experienced vehicle problems on his trip to Murchison Falls when the right front wheel developed an inboard lean. At a service station in the town of Masindi (just south of the falls) the Jeep was hoisted onto a lift and when the right axle housing was removed it was discovered that there were two crushed wheel bearings, and the axle was slightly bent. The bearings were replaced with a couple of old spares we had with us (until new ones could be purchased when he arrived in Stanleyville). To solve the problem with the axle he telephoned the Jeep factory in Toledo, Ohio (as he had done on several other occasions during our adventure) and ordered a new axle to be air freighted to us in Stanleyville. (Noble's most indispensable role in our adventure was to keep us on the road, and thankfully he managed to accomplish this even in the most remote parts of the world far off the beaten path, at minimal cost.) Kudos to him.

We drove together into Fort Portal for the night, and once again we discussed what we would both like to see in the remainder of our time in Uganda. Noble wanted to look for Mountain Gorillas near Kisoro on a day tour in the Virunga Mountains north of Lake Kivu, in extreme SW Uganda. As much as I would have liked to encounter a Gorilla in its natural habitat, my desire to see some of the endemic birds in the Ruwenzori Mountains was even greater. (Noble also wanted to drive south the short distance to the northern tip of Lake Tanganyika, in Ruanda-Urundi, but this option was also not one of my top priorities.) It would be easy for the two of us to join up again in the city of Bukavu on the southwestern shore of Lake Kivu just inside the Belgian

Congo. This then, was our plan. We had the name of the Royal Resident Hotel in Bukavu for a meeting place, and we estimated it would take us about a week of travel to arrive there (which we usually underestimated). I would continue to hitchhike and Noble would continue to drive our injured Jeep.

The Ruwenzori Mountains, known popularly as the "Mountains of the Moon", were famous to explorers and naturalists from around the world for their scenic beauty, rugged slopes, glaciers, lakes, snow-capped peaks, waterfalls, and botanic diversity. They became both a National Park (in 1991) and a World Heritage Site (in 1994). Six permanently snow-capped peaks were over 15,000 ft. high, the tallest one being Mt. Stanley at 16,750 ft. above sea level. The Ruwenzoris were located on the Uganda-Belgian Congo border, just north of the equator, stretching for 75 miles between Lake Edward to the south and Lake Albert to the north. The mountains were part of the "Albertine Rift" (the western branch of the Great African Rift system) and were one of the most important headwaters for the Nile River. An average annual rainfall of up to 100" produced luxurious, dense forests with tall tree ferns and bamboo at middle and upper elevations, and spectacular giant heaths, senecios, and lobelias on the high elevation moorlands. Mosses, liverworts, and lichens were abundant, often covering tree branches at higher elevations. The mountains were a paradise for botanists, ornithologists, photographers, mountain climbers, and others. Two dozen species of birds were endemic to the Albertine Rift, some of which were found only in these mountains.

I spent only one day in the Ruwenzoris, SW of Fort Portal on Mar. 15, hiking from dawn to dusk for 11 hours on a good trail, going up from an elevation of 6,000 ft. to 10,000 ft., then back down again. The habitat was almost entirely lush, pristine montane forest. At higher elevations, there were dense bamboo patches and elfin forests, and mosses and lichens clung to the tree branches. At 10,000 ft., I reached the very lower limit of moorland, where trees gave way to grasslands with giant heaths, lobelias, and senecios. It was as high as I got. Like most days in the Ruwenzoris, temperatures were mild and the weather was misty and cloudy most of the time, but fortunately without rain. I stopped frequently for photography or to try and find singing birds in the dense forest. It was an exciting challenge. I couldn't have been happier in my solitary pursuit of birds, alone with nature. There were occasional monkeys, squirrels, or other small mammals, and a duiker or two, but I did not encounter any Chimpanzees or Gorillas, although both of these reclusive "higher primates" inhabited the mountains here. (Nor did I encounter any other person.) Twice I found elephant dung on the trail, but I didn't see or hear any individuals in the forest. Even without direct sunlight, there were a small variety of butterflies, and other kinds of insects.

I recorded a total of 48 bird species during the day, 12 of which were additions to my trip list (Appendix B). Nine species were Albertine Rift endemics: Ruwenzori Batis*, Blue-headed Sunbird*, Regal Sunbird*, Red-throated Alethe*, Red-faced Woodland-Warbler*, Ruwenzori (Collared) Apalis*, Archer's Robin-Chat*, Ruwenzori Turaco*, and Strange Weaver*. The large, colorful, fruit-eating Ruwenzori Turaco was certainly the most exciting bird of the day. I counted seven individuals. It was easily located by its distinctive loud ringing vocalizations, and different enough from all the other 22 species of turacos that taxonomists placed it alone in the genus *Ruwenzorornis*! The Strange Weaver was unique among the 52 African species in the genus *Ploceus* by its behavior of foraging secretly, low in forest undergrowth or midstory in a tit-like manner. It was a handsome bird with an all black head, olive back, and yellow underparts with a chestnut breast patch. I encountered only one pair of these weavers. I glimpsed two crimsonwings (estrildid finches) flying into dense, low, damp vegetation in the montane forest-heathland ecotone at 9,500 ft. elevation. I tried to convince myself they were the rare, endemic Shelley's Crimsonwing, but I eventually concluded from my inadequate field notes that they were, alas, much more likely to have been Abyssinian Crimsonwings (which I had seen earlier on the rim of the Ngorongoro Crater).

Although I was tempted to diverge from following Clement's taxonomy of the mountain greenbul complex, I did not do so. He recognized only two species in this group (Eastern Mountain-Greenbul and Western Mountain-Greenbul), whereas Sinclair & Ryan (*Birds of Africa south of the Sahara*, 2010) separated the group into seven isolated species, one of which was the "Olive-breasted Mountain Greenbul" which I saw for the first time that day. I was happy to once again observe two Black-faced Rufous-Warblers, one of my favorite African birds. I also enjoyed getting good looks at Black-billed Turacos which were common in the forest here, and which I had first recorded in the Semliki Forest only two days previously. The most numerous (or at least the most conspicuous) species in the forest that day were 30 African Yellow White-eyes, 20 Regal Sunbirds, and 14 Yellow-bellied Waxbills (E. Afr. Swees). Remarkably, I recorded 12 different species of sunbirds, a record one day total for me. There were also 10 kinds of "warblers" (cisticolids & sylviids combined) and 6 kinds of "flycatchers" (in the combined families of Muscicapidae, Monarchidae, and Platysteiridae). My one day in the Ruwenzori Mountains will be a lasting memory for me of birds, scenic beauty, peace, and calm.

I spent the next two days (Mar. 16 & 17) in the Kibale (Mpanga) Forest ten miles east of Fort Portal at 5,000 ft. elevation. This forest was recommended to me for birdwatching by John G. Williams when I visited with him at the Coryndon Museum in Nairobi. It was a protected forest best known for its diversity of primate mammals, including Chimpanzees and several species of monkeys, but I encountered only the Black-and-White Colobus (also called the "Guereza"). It was a relatively large monkey about four feet long (head & tail) which traveled in small groups through the forest midstory and canopy. These monkeys were typically shy and did not allow me very good views. (Note: the taxonomy of African monkeys in both the genus *Colobus* and the genus *Cercopithecus* was hopelessly confusing, with no two authors agreeing on the assignment of populations to species or subspecies.)

My total list of 92 bird species from Kibale included 17 trip additions (Appendix B). These included the Jameson's Wattle-eye (a striking, very small, almost tailless black & white platysteirid flycatcher with a chestnut cheek patch and a very wide pale blue wattle encircling the eye). The most numerous avian species in the Kibale Forest were White-headed Woodhoopes, Grosbeak Weavers, African Yellow White-eyes, Splendid Glossy-Starlings, and White-breasted Negrofinches. A White-collared Oliveback, an endemic estrildid finch to this restricted region, was pleasantly colored in olive-green and light Gray, with a completely black head, bordered below by a conspicuous white collar. My birdwatching in the Kibale Forest necessitated that I constantly glance at the ground to avoid walking or stopping in the columns of ants. As usual, I did not see any snakes. It was a fun time.

I departed the Kibale Forest on Mar. 18 and hitchhiked south toward the town of Kasese. There were few vehicles on the road, and those that stopped to pick me up were usually not going very far, so it was not until evening that I arrived in this town, which was situated just outside the northern boundary of Queen Elizabeth National Park, at an elevation of 3,000 feet. This large park was partially located between Lake George (to the NE) and Lake Edward (to the SW), and was crossed by the equator in the far north. Established in 1954, it encompassed 764 sq. mi. of savanna, humid forests, lakes, wetlands, rolling green hills, and volcanic craters, and was was home to a great variety of birds, mammals, and other wildlife. Noteworthy mammals included three species of ungulates: Uganda Kob, Topi, and the rare Sitatunga (an inhabitant of swamps and reeds). March was the beginning of the first of two wet seasons during the year. That night I slept in the town of Kasese.

The next morning (Mar. 19), I documented birds and searched for other wildlife in the park for just 2½ hours. Among the few mammals that I observed was a small group of Kobs, my first sighting of this fawn-colored, reedbuck-like antelope with a white ring around its eyes and a black stripe on the front of its forelegs. Its heavily ringed two ft. long horns (present only in males) were lyre-like and S-shaped. It was yet another ungulate for my African list.

I continued my hitchhiking and proceeded south in the park for 10-15 miles, crossing the equator. Then I took a right-hand fork in the main road and headed SW for 65 miles, leaving the park and arriving at the town of Ishasha River on the boundary with the Belgian Congo. Here I crossed the border and entered the Belgian Congo, leaving British East Africa behind me and entering French speaking Africa. I received a sudden culture shock when the young African guard on the Congo side of the border addressed me in French. For a moment I was speechless, then I said to him (with a horrible English accent) "Je ne parle pas francais" (I don't speak French), which was almost the only French I knew. So the boy spoke to me in Swahili, and I just shook my head negatively. Then he tried one or two local native tongues, to which I still just shook my head negatively. He looked at me as if to ask how dumb could I be, threw up his hands in disgust, and pointed to my passport which I was holding in my hand. I gave it to him and he immediately stamped it, then with a look of great disdain, he motioned for me to continue. I comforted my wounded pride by telling myself that since English was spoken in Uganda, all border officials should be able to speak English. I had temporarily forgotten how much French-speaking persons hated to do this. Although the Congo was a Belgian Protectorate, the official language was French (which would change a little when independence came in just three months).

I spent the remainder of the day hitchhiking south along a scenic route just inside the Belgian Congo in green rolling foothills of the Virunga Mountains, through the town of Rutshuru all the way to the city of Goma on the north shore of Lake Kivu, where I spent the night. This area was one of considerable volcanic activity, past and present. Lake Kivu was in the southern section of the Albertine Rift, with a maximum depth of 1,575 ft. and a surface area of just over 1,000 sq. mi., lying at an elevation of 4,800 feet. Politically, the lake was divided down the middle by Ruanda/Urundi (Rwanda) on the east and the Belgian Congo on the west. Lake Kivu had very poor fish diversity, and two species of Tilapia and one species of sardine had been imported from elsewhere in Africa to help the sparse fishing industry. Goma was located in the southern Virunga Mts., just across the border from the town of Kisenyi (Gisenyi) in Ruanda/Urundi. In spite of ethnic conflicts, this whole area was popular with tourists.

The next morning, Mar. 20, I hitchhiked across the border from Goma into the town of Kisenyi (in Ruanda/Urundi). Here I came across the large, prominently advertised "Central African Curio Shop" on the main city street, an icon in the tourist industry with its huge selection of African gifts and artifacts. I took a picture from outside the shop and then looked around inside for 20 or 30 minutes. (In his travels through this town in our Jeep Noble spent considerable time visiting and talking about entrepreneurship with the Belgian couple who owned the shop, Jean Pierre Hallet and his wife.)

Rwanda was the most densely populated and one of the economically poorest of all African colonies. In 1960, it was allied with Burundi to the south as a single Belgian colony, "Ruanda-Urundi." It suffered from ethnic strife between the minority Tutsi who were tall statured herdsmen and had achieved social, economic, and political ascendancy over the majority Hutu, who were agriculturists. Uprisings between the two groups had started a year earlier, in 1959, and both Noble and I were cautious in our brief time in these two areas. (Widespread genocide eventually broke out in 1994, which killed tens of thousands of people.) In 1960, Rwanda had monarchial rule and a native Bantu language was the most widely used for communication, although a few people could speak French, English, or Swahili. Religion was 40% Catholic, 25% Protestant, 20% other Christian sects, and less than 10% Muslim.

My road in Rwanda followed along or near the eastern shore of Lake Kivu, where there were many small, densely populated settlements and marketplaces. Colorfully dressed women carried goods of all kinds on top of their head, to and from overcrowded markets. In the countryside were small fields of beans, sweet potatoes, sorghum, and bananas, along with a few coffee plantations. Cattle were the principal livestock, but there were also many goats and chickens. Subsistence farming was widely practiced and many people owned a tiny piece of property.

It was approximately 140 miles from Kisenyi back to the Belgian Congo border, on a very sinuous road that passed through the principal towns of Kibuye and Cyangugu. My hilly, scenic road was never far from the eastern shore of Lake Kivu. A popular tourist destination along this route was Napolean Island in the lake where a colony of 20,000 fruit bats spent the day, only a short boat ride out of Kibuye, but I did not stop. Hitchhiking was slow and it took me all day to reach Cyangugu (on the border).

Plate 52: In Uganda I searched for birds at Lake Victoria, where I obtained a distant view of the enigmatic Shoebill. Many people lived along the papyrus-bordered shore-line. Lake Victoria is the largest lake in the world inside the tropics.

Plate 53: The culture of the people changed greatly in the tropical rainforest of the Semliki River, which formed the Uganda-Congo border. Among the tribes were a very few primitive Batwa pygmies, where women averaged barely four feet tall and men averaged only 4'4".

Once more, I crossed into the Belgian Congo and then traveled 10 miles to the principal city of Bukavu, which was situated on the southwest corner of Lake Kivu at an elevation of 5,000 feet. Although I took quite a few roadside and marketplace pictures in Rwanda, I did not keep a record of the birds I saw there since I wanted to keep my binoculars out of sight.

When I arrived in Bukavu I checked for messages from Noble at the Royal Resident Hotel, as we had arranged. There was a telegram from him saying that the right front wheel spindle of the Jeep had broken once again (as it had in Argentina, in Patagonia) and he would be delayed for at least three days in his arrival in Bukavu. (One of the causes for Noble's vehicle problems might have been that he had a little trouble adjusting from flying his Banshee fighter plane to driving our overweight camping van, and he tended to drive a little faster than I did.) I spent the night in Bukavu at an inexpensive guest house.

The next day (Mar. 21) I decided that while I was waiting for Noble I would hitchhike south for two days to the northern shore of Lake Tanganyika, following the Albertine Trench 75 miles to the large city of Usumbara (Bujumbura), inside Rwanda/Urundi once again. It would give me a chance to glimpse a view of this famous lake, and perhaps find a new bird for my list from this country (within which I had not recorded any birds).

Lake Tanganyika was another Albertine Rift lake, at 2,500 ft. elevation. With a length of 420 miles from north to south, it was the longest lake in the world, from one end to the other! It was also the second deepest lake in the world (behind Lake Baikal in Siberia) with a maximum depth of 4,820 feet, and was second in the world in total volume of water (once again behind Lake Baikal) with 4,500 cu. miles. It was one of Africa's three largest lakes in the Great Rift System (along with lakes Victoria and Nyasa), with a surface area of 12,700 sq. miles. In addition to these impressive dimensions, Lake Tanganyika was famous for its tremendous diversity of more than 250 species of cichlid fishes, the majority of which were endemic to the lake. It was truly a natural wonder. The Ruzizi River flowed southward out of Lake Kivu into the northern end of Lake Tanganyika, just west of Usumbara. The Lukuga River flowed westward out of Lake Tanganyika as one of the headwaters of the Congo River.

The next morning, Mar. 22, I searched for birds a very short while just west of the city, in a grassy area. Here I found a breeding-plumaged male Black-winged Bishop* (a ploceid weaver in the genus *Euplectes*). In its stunning red-and-black plumage with its glimmering red back feathers puffed out it was performing an elaborate aerial courtship display, flying on fluttering wings between low perches in the grassland, uttering a rather weak twittering song as it did so. It was the only bird I added to my trip list from Rwanda/Urundi. I then walked into the outskirts of the city and ate a hot breakfast of steak and eggs at a small, home-owned restaurant, after which I returned to the highway to hitchhike back to Bukavu. I was still questioned by curious by-passers, "Safari a whapi, bwana?"

The Belgian Congo (soon to be called the Democratic Republic of the Congo, or DRC) was under Belgian rule from 1908 until June, 1960 (just after our departure). Lying astride the equator in the middle of the African continent, it was the third largest nation in Africa and was originally covered almost entirely by tropical rainforest (except for the Albertine Rift Mountains in the northeast and the widespread mesic woodlands in the southeast). Noble and I spent three weeks in the Congo, travelling from Bukavu to Stanleyville to Bangassou (across the northern Congo boundary in French Equatorial Africa).

My reference books for African birds now changed from the two volumes of Praed & Grant to the two volumes of Bannerman, *The Birds of West and Equatorial Africa*. Volume 1 (with 795 pages) dealt with non-passerines, and volume 2 (with 729 pages) was devoted to passerines. In spite of Bannerman's comment that these "handbooks" were made in a portable form to provide a guide for use in the field, they certainly didn't fit into my backpack very well. There were no range maps (only a written description of the range) and the plates of illustrations depicted only a very small fraction of the total avifauna. (In volume 1 there were 13 plates illustrating a total of 37 species in color, and 24 plates in black-and-white; in volume 2 there were 16 plates with colored paintings illustrating a total of 133 species, plus scattered black & white drawings.) Species names were adequately indexed, with English and scientific names together in the same index. These 2 volumes were my only bird books for central and west sub-Saharan Africa. (*Birds of the Sudan*, published by Cave & Macdonald in 1955, provided me some help for the Sahara). Birdwatching continued to be a challenge.

There was no new message from Noble in Bukavu the next morning (Mar. 23) so I decided to hitchhike north about 25 miles to a scientific research office of IRSAC (the French abbreviation for Scientific Research Institute of Central Africa) in the town of Lwiro, situated in the Itombwe Mountains on the west side of Lake Kivu. These were part of the Albertine Rift. Here I was fortunate to meet a very likeable young Belgian count, Michel de Meviens, and his wife. Michel spent most of his time photographing birds, travelling up and down the mountains in his Jeep. I was received as a visiting scientist, and invited to stay two nights with the Meviens, allowing me to spend all of one day (Mar. 24) with Michel in the mountains, between 6,000 & 8,000 ft., traveling in his old Jeep.

It was a memorable day. About mid-morning, at 7,000 ft. elevation, our narrow little dirt road through the montane forest was completely blocked by a very large tree that had quite recently fallen down across the road. Cutting a path with Michel's little chain saw through or around the tree was not a possibility. Michel said that there was a less traveled "bypass road" back down the mountain (from where we had come) which would also eventually take us to our destination, but it would require an extra hour or more of travel. This bypass road paralleled the one we were on, at 1000 ft. lower elevation. Michel told me that if I wanted to spend more time birdwatching I could simply walk straight down the mountain from where we were, through the forest, and I would eventually encounter this bypass road, at some point before he arrived there. He said it would be easy for me to follow some of the numerous elephant trails down the mountain. I pondered this plan with considerable skepticism and apprehension. I had no map and thought how easy it would be to get lost, but Michel assured me I could not get lost. I also considered the possibility that for whatever reason Michel, himself, might not be able to reach this undesignated meeting point, or that I might encounter an unfriendly elephant in the forest on my way down. However, my adventuresome nature prevailed and I said OK, that's what I would do!

Although my elephant trail was not difficult to follow at first, it soon became much less obvious and there were numerous side branches. I consoled myself by thinking that as long as I kept walking downward I would eventually have to encounter the bypass road, and anyway, we did not have a specific point on this road where we would join back up. Fearful of arriving too late, after Michel might have already driven past, I walked as rapidly I could, with very few stops to look at a bird. In my haste I stumbled over a hidden tree root and fell down, head forward, on the steep trail, barely managing to get my hands under me in time not to land on my face. However, my chest landed on top of a small, very sturdy upright projection from another root, which produced considerable pain in the middle of my sternum. I had no choice but to continue walking down the mountain. Finally, after about one hour of walking I reached the bypass road. Happily, Michel came along 30 minutes later, as he said he would. The pain in my sternum was easing slightly, but it was many weeks before it went away completely. (I suspect there could have been a tiny fracture.) I saw very few birds in this hour and a half in the forest.

We reached Michel's planned destination, a high elevation marsh where there was an old dugout canoe that he had used on other occasions for paddling along a tiny stream than ran through the marsh. The problem was that the canoe leaked rather badly, so that it was necessary to stop and empty out the water every 10 minutes or so. Then a very sudden, heavy rain shower soaked us completely. By the time we reached the other side of the marsh (from where we could walk back to the Jeep) we were not only very wet but also shivering from the cold because we were at 8,000 ft. elevation. Conditions were not favorable for birdwatching.

My day's list of birds managed to total 45 species, which included 9 trip additions (Appendix B). The Olive Ibis* was a thrilling sighting, at 6,500 ft. elevation, of a pair of birds flying just above the forest treetops at daybreak, calling loudly. This was a uniformly dark-plumaged, uncommon, forest inhabiting ibis that fed in open glades on the forest floor. The Handsome Francolin* was an Albertine Rift endemic, as were the Grauer's

Scrub-Warbler* (*Bradypterus graueri*), Purple-breasted Sunbird*, Krandt's Waxbill* (the highland form of the Black-headed Waxbill), and Dusky Crimsonwing*. Other species of particular interest were 3 Ruwenzori Turacos (my last sightings of this species), 1 African Hill Babbler (which I had first seen in the N. Pare Mts. of Tanganyika), 1 Cameroon Scrub (Evergreen Forest) Warbler (another *Bradypterus*), and Chestnut-throated Apalis. The three most numerous species, in order of their abundance, were the Krandt's Waxbill, Chubb's Cisticola, and Regal Sunbird.

I returned to Bukavu early the next morning (Mar. 25) and there was still no further news from Noble. Michel had told me of a small, primitive, IRSAC research station in the rainforest on the road to Stanleyville, at the 110 km. marker from Bukavu, and he gave me a letter of introduction to stay there at no cost for a few days if I wanted. Since this was on the road Noble and I planned to follow to Stanleyville, I decided that I could just hitchhike on ahead, once again, and wait for him to pick me up there, as he came through. So I left a message for him at the Royal Resident Hotel telling him of my decision, and I set off yet another time early that afternoon (Mar. 25) hitchhiking with my binoculars, bird books, sleeping bag, and a very limited supply of money, food, and water. It would become more of an adventure than I anticipated.

The Congo River drainage basin covered more than 1½ million square miles, 13% of the entire African continent. The river itself was just over 2,700 miles long, from the source of its most distant tributary (the Chambesi River) to its mouth on the eastern shore of the Atlantic Ocean. It was the tenth longest river in the world, and was one of the deepest with a maximum depth of 720 feet. It had the second greatest volume of annual flow in the world, behind only the Amazon. The Congo River had many stretches of shallow falls and cascades, but it was navigable for a stretch of over 1,000 miles between the cities of Leopoldville (now Kinshasa) and Stanleyville (now Kisangani), between the Livingstone Falls and the Stanley Falls.

Following the route of the river on a map for its entire length was an extremely arduous task, with a great many side branches and wide changes of direction (and confusing name changes as the colonies became independent). Its most distant beginning was the source of the Chambesi River at 5,800 ft. elevation in the Eastern Rift highland area at the lower end of Lake Tanganyika (in extreme N. Rhodesia) at latitude 9°S, longitude 31°E). From here, the river flowed southward through extensive marshes into Lake Bangweulu, which it left as the Luapula River and continued a short distance farther south all the way to 12½°S latitude. Here it turned westward and then northward and flowed into Lake Mweru. It then continued northwestward out of this lake, still as the Luapula (or Luvua) River, and some 220 miles later it joined the northward flowing upper Congo River (which in 1960 was considered to be a continuation of the Luapula River). To make the situation even more confusing, the name "Congo" River was eventually changed to "Zaire" River. (I finally gave up trying to unravel all the names and subsequent changes.) From Stanleyville the Congo River traveled in a very large arc northward and westward, all the way to 2° north of the equator before it turned southwest and flowed more than 1,000 miles into the Atlantic Ocean, at its mouth 6° south of the equator. The river's complex drainage pattern and its tributary streams required many years to decipher, name, and map.

The Congo rainforest was the 2nd largest tropical rainforest in the world, behind only the Amazon rainforest. It covered 700,000 sq. miles, which was ¼ of all the rainforests in the world. The daily temperature fluctuated around 80°F, and rainfall averaged 75-80" a year. Like all tropical rainforests it was home to a great diversity of wildlife, much of it shared with the adjacent Guinean rainforest in West Africa. Special mammals included the Bongo (a Nyala-like antelope), Giant Forest Squirrel (5 ft. long from head to tail), Golden Cat (like a large, buffy house cat), Giant Otter Shrew, diminutive-sized forest elephants, and Dwarf Galago (weighing only 90 gms, or about 0.2 lbs.) Mammals which were endemic to just the Congo rainforest included the Pygmy Chimpanzee (Bonobo) and the rare Okapi (a relative of the Giraffe). In addition, the lowland populations of gorillas which lived here were often considered specifically distinct from the mountain populations elsewhere. A wide variety of other primates also inhabited the forest, such as mandrills, mangabeys, guenons, colobus monkeys, and the marmoset-sized Talapoin (Africa's smallest monkey). They filled much the same niches as did the monkeys in

Neotropical rainforests, except they did not have prehensile tails and thus did not use their tail to swing from tree to tree (contrary to some Hollywood movies), as do many Neotropical monkeys.

As expected, there are Congo rainforest mammals which have ecological counterparts in the Amazon rainforest, inhabiting similar niches and very alike in size, overall appearance, and behavior. These include the following (with the African species named first): Leopard and Jaguar, duikers and Gray Brocket Deer, Aardvark and Giant Anteater, Chevrotain (Mouse Deer) and Paca, Giant Forest Hog and Peccary, Potto and 2-toed Sloth, pangolins and armadillos, cane rats and the Agouti, Large-spotted Genet and Margay, and the remarkably similar appearing Ratel (Honey Badger) and Grison (both of which are members of the family Mustelidae). In addition to these species, many of the bats, flying squirrels, and river otters replace one another in the rainforests of the two continents. Such ecological counterparts are evidence that similar environments around the world favor similar faunal adaptations.

Characteristic birds of the Congo rainforest included pigeons and doves, swifts and swallows, woodpeckers, hawks and eagles, coucals & other cuckoos, kingfishers, and flycatchers. Hornbills were ecological counterparts of New World toucans, bee-eaters were counterparts of jacamars, the Great Blue Turaco filled the guan niche, greenbuls were counterparts of tanagers, sunbirds filled the hummingbird niche, weavers & malimbes were counterparts of orioles & caciques, the Green-breasted Pitta filled the antpitta niche, and the White-headed Lapwing was the river sandbank counterpart of the Pied Lapwing in S. America. One of the most unique birds of the Congo rainforest was the Lyre-tailed Honeyguide, but perhaps the most iconic species was the rare, endemic Congo Peacock (a very large, strikingly-colored phasianid in which the male was deep purple and green with a bright red throat and a very long, pointed black & white plume on top of its head). During my brief time in the Ethiopian rainforest I was able to encounter only a small fraction of these birds (see Appendix B). As was the situation in the Amazon, birding in the Congo required patience and determination. (Adequate references would have greatly simplified the task.)

The original human inhabitants of the Congo rainforest were Mbuti and Efe pygmies, who first lived here at least 40,000 years ago. They were greatly reduced in numbers today. Modern day females averaged only 4'1" in height and males averaged just 4'10." Traditionally, these people were hunter-gatherers living on fruits, nuts, leaves, roots, fish, snails, crabs, ants, insect larvae, honey, lizards, snakes, birds, and such game animals as duikers, wild hogs, monkeys, and pangolins, which were obtained by spears, nets, traps, holes in the ground, and bows with poisoned arrows (from plant or insect poisons). There was no cultivation of any kind of food. They had no written language, wore only loin cloths made of softened bark tied around their waist with a vine (in both sexes), and lived in crude temporary circular huts made of sticks and leaves, in small family groups. They were semi-nomadic within the forest and regularly moved from one location to another. The forest was worshipped and considered sacred. Dancing and music were part of special occasions and ceremonies were typically carried out on moonlit nights, with sound provided by drums which were made from hollow logs, and by wooden or bamboo "trumpets." Women sometimes wore beaded necklaces, and both men and women filed their teeth to a point for decoration. Fire was used for cooking. The most common form of marriage was for families to trade sisters. Infants were often nursed by their mother for up to 3 or 4 years of age. The Congo pygmies were a rapidly declining race of people today, unable or unwilling to compete with other African tribal groups or with European colonists.

Stanleyville was approximately 550 miles through the rainforest from Bukavu, over a very primitive dirt and muddy road that was extremely poorly maintained for some of this distance, particularly in the eastern sections. On the afternoon of Friday, Mar. 25, I caught a ride with a truck all the way from Bukavu to the IRSAC research station at the km 110 marker, 70 miles from Bukavu. This location was on the boundary between the North Kivu and South Kivu provinces of the Belgian Congo, and it formed a triangle on a map with the Kivu lakeshore cities of Goma to the NE and Bukavu to the SE (both on the eastern boundary of the country). My truck driver wished me "bon voyage" as he let me out at the research station and drove on up the road, late in the afternoon.

It was my 1ˢᵗ night at the station.

I was greeted at the research station by a friendly, middle-aged African caretaker who spoke French and Swahili, but not English, so our only communication was by sign language. He was the only person present at the station (other than me). I gave him my letter of introduction from Michel de Meviens, and he showed me to a room in the guest house for visiting scientists. It had a bed (with a mattress, sheet, and pillow), a table and chair, a gas lantern, eating utensils, and a pitcher of rainwater for drinking. There were no glass panes on any of the windows, but fortunately there were virtually no mosquitoes, although there were an abundance of ants and spiders to keep me company. At the end of a short hallway was a bathroom with a shower, wash basin, and toilet. It was not much different from most of the other lodgings where I had been staying. The few research scientists which had been working there had all recently left the station, prior to the imminent chaos looming ahead when independence was soon to be granted. The IRSAC station was situated at 2,500 ft. elevation in the local "Irangi Forest", which was a southern subsection of the extensive Ituri Forest of the northeastern Congo. It was part of the Guineo-Congolian Biome. The area set aside for research encompassed only about 6 sq. miles of forest on the north side of the Bukavu-Walikale road, plus a smaller area of private forest on the south side of the road to which scientists had access, in the Mukowa primary forest along the Luhoho River.

The dominant canopy tree in the IRSAC forest was the Uganda Ironwood (*Cynometra alexandri*), an evergreen legume which grew to a height of up to 150 ft. and a diameter of 6½ ft., sometimes with hollow boles and buttressed roots. Surprisingly, it often occurred in pure stands (unlike most rainforest trees). This tree was used by Chimpanzees for nesting platforms, and its long pods produced seeds which were eaten by a wide variety of wildlife, including elephants, wild pigs, duikers, gorillas, and forest rodents. As in most tropical rainforests there were an abundance of fallen logs and hanging vines (lianas).

Mammalian research in the forest had documented the presence of duikers, tree hyraxes, 6-8 species of bats, giant rats (*Cricetomys*), Red-tailed (White-nosed) Monkeys, and the Giant Otter Shrew (along the Luhoho River). This latter species was a highly unusual, semi-aquatic mammal (family Potamogalidae, Order Insectivora) which lived along rivers or lakes in forested equatorial Africa, where it fed predominantly on crabs. It caught these and carried them to shore, where it meticulously extracted and ate the soft parts. It also ate fish, mollusks, frogs, and insects in its diet. It swam with lateral undulations of its 2 ft. long streamlined body and its laterally flattened tail. The Otter Shrew was primarily nocturnal with its peak activity at dusk, and it moved about on land with a humping otter-like gait. Avian research in the forest included an extensive ecological study of the 7 species of hornbills which were resident in the forest. The Congo Peacock was a rare inhabitant of the area, as was the colorful, widespread Forest Robin (*Stiphrornis erythrothorax*), but unfortunately I did not encounter either of these two species during my brief time there.

I spent 1 one night (Mar. 25) and 4 full days and nights (Mar. 26-29) at the IRSAC research station, departing on the morning of Mar. 30. This was more time than I needed. The days always started out calm, warm, and foggy, making birdwatching difficult for the first hour or two. Daily rain showers of varying intensities were frequent, mostly in the afternoons and during the nights. Birds were generally rather scarce and difficult to locate, contrary to my hopes and expectations. During the middle of sunny afternoons I devoted my time to catching butterflies or to watching the relatively numerous and conspicuous Red-tailed Monkeys. There were also squirrels, lizards, termites, and ants (for which I was constantly on the lookout). As usual, I failed to see a single snake. The "jungle" was often silent and serene, but none-the-less inspiring. However, it was less exciting and ecologically diverse, with fewer birds, than was the Amazon rainforest.

My total list of birds for the 4 days I spent in the forest was only 76 species, which included just 15 trip additions (Appendix B: Mar. 26, 27, & 28, 1960). (However, this total was very similar to the number of 82 species documented in this area during a two week hornbill study a few years later.) The birds which gave me the greatest thrills were: (1) White-throated Blue Swallow* (a swallow of forested rivers, filling the same niche as the White-banded Swallow in the Amazon rainforest), (2) Chocolate-backed Kingfisher* (a handsome forest kingfisher with dark chocolate brown upperparts, white underparts, mostly pale blue tail, pale blue wing patches, and bright red bill; although often found near water it was entirely insectivorous in its diet), (3) Chestnut Wattle-eye (a diminutive, almost tail-less, sexually dimorphic, platysteirid flycatcher with a prominent bare lilac colored wattle surrounding its eyes; the male was black & white but the female was chestnut colored with a Gray crown and white belly; I had first encountered this species in the Kakamega Forest); (4) Red-crowned Malimbe* (an endemic to the Congo rainforest and one of the least common of the several malimbes in the forest, slightly smaller than the others and with only a small red crown patch), (5) Cassin's Spinetail* (a broad-winged, very short-tailed swift seen regularly flying low above the forest, bat-like in overall shape and flight behavior, and for me an icon of the Congo rainforest), and (6) White-crested Hornbill* (see comments below).

The most numerous families of birds which I encountered in the IRSAC Forest were 8 species of the Pycnonotidae (bulbuls, greenbuls, & nicators), 6 species of the Cuculidae (cuckoos & coucals), 6 species of "fly-catchers" (in 3 families), 6 species of estrildids, 5 species of Hornbills (Red-billed Dwarf*, White-crested, Black-and-white-casqued, African Pied, and Black-casqued), 3 species of malimbes, 3 species of swallows, and 3 species of swifts. Only 2 species of raptors were seen, the African Cuckoo-Hawk and the Palm-nut Vulture. Two species on my list were endemic to just the Congo drainage basin, the Bate's Paradise-Flycatcher* (which I carefully distinguished from the several other possible *Terpsiphone* species) and the Red-crowned Malimbe (as previously noted). I was happy to see again Great Blue Turacos, Gray Parrots, and to obtain my first sightings of the Blue Cuckoo-shrike* and the colorful little Orange-cheeked Waxbill* (at the forest edge). I struggled to distinguish between Icterine and Xavier's greenbuls, depending almost entirely on the unreliable character of size. On several occasions I heard an African Broadbill but I could never track one down visually. I wrote in my notes that the song of the Rufous Flycatcher-Thrush (which I first saw along the Semliki River) was a single long mournful note followed by 3 upwardly inflected notes increasing in pitch.

The White-crested Hornbill was a most unusual forest hornbill, often placed by taxonomists in the monotypic genus *Tropicanus* (though not by Clements, who considered it a species of *Tockus*). It was intermediate in size, slender in shape with a long, very graduated tail, and was entirely black in color except for its all white head with a bushy, grizzled Gray crest. It foraged on ants, termites, and other insects (which it sometimes caught on the wing), and infrequently also on birds and mice. White-crested Hornbills typically traveled in pairs through the forest midstory, where they characteristically followed monkeys or large squirrels in order to capture the insects (often on the wing) which these mammals disturbed. Bannerman described their flight as "buoyant and graceful", and their vocalizations were said to be "strange, squawking noises" or "plaintive wails", unlike any other hornbill. The geographic range of this species extended throughout the dense forests of west and central Africa, where it was not uncommon. I recorded White-crested Hornbills several times in the IRSAC forest, singly or in pairs, where they became a permanent memory for me of the pristine African rainforest.

By the end of 2 days at the IRSAC forest I was ready to leave. I had run out of food and the caretaker had disappeared. I hoped Noble might come by. I went out to the road on Monday, Mar. 28, and waited several hours for any vehicle to come by which was headed west toward Walikale and Stanleyville. About mid-morning an ancient large black limousine of a make with which I was not familiar came along, headed west. It was completely overloaded with a Belgian husband, wife, 2 children, and household goods of all kinds and descriptions jammed in everywhere possible inside and outside the car. They stopped and explained they were on their way out of the Belgian Congo, prior to independence. They apologized greatly for the fact that there was no room for me, which was a vast understatement. I wished them well and said "bon voyage." It was necessary for them to drive extremely slowly on the very narrow, deeply rutted, muddy road. No other vehicles came along that day, going in either direction. So I returned to the guest house that evening. Although I was out of food, rainwater was still available for drinking.

I went back out to the road the next morning (Tu., Mar. 29) and eventually a large truck came along, coming from Walikale (to the west) and going east to Bukavu. The driver stopped and I explained to him that I wanted to travel to Walikale. The African driver spoke virtually no English but he made it clear to me that the 72 mile stretch of muddy road connecting the IRSAC research station to Walikale was so narrow and in such a bad state of repair that traffic was limited to only one direction, east or west, on alternate days. Today, vehicles were permitted to drive only east on this section of road! This was not the direction I wanted to go. How unlucky could I be? Perhaps Noble would come along with some food and clean clothing. He knew where I had gone. For a 2nd day I had nothing to eat.

Early on Wednesday morning, Mar. 30, I once again gathered my backpack, binoculars, bird books, and other paraphernalia and walked back out to the road, hoping to catch a ride to Walikale. Sure enough, about mid-morning a flatbed, dual-wheeled 1940's truck of unknown origin came along heading west, with an African driver and 8-10 male and female passengers of all ages sitting on wooden benches in the back. A canvas canopy provided protection from the sun and rain. An elderly woman sat in the passenger seat up front, alongside the driver. They were happy to have me join them, as the only European on board. Our very narrow badly rutted muddy road curved back and forth and undulated up and down in a generally NW direction, through the mostly pristine, hilly rainforest, between 2,000 and 3,000 ft. elevations. Although our progress was quite slow, generally between 10 and 20 mph, it was almost impossible to identify any roadside birds because of the constant vibrations and motions of the truck. A raptor occasionally soared over the forest, and turacos and hornbills perched along the forest edge or flew from one side of the road to the other. I saw no forest elephants, but now and then a duiker darted across the road ahead of us. It was frustrating not being able to stop (as I would have done if I were driving our camping van). Hitchhiking was not an ideal method of birding. Never-the-less, travelling through the Congo rainforest was a thrilling, once-in-a-lifetime experience. Because of our slow, laborious travel we did not arrive in Walikale until about 4:00 pm, after travelling the 72 miles almost nonstop for over 6 hours. Our overweight, top-heavy camping van would have found this hazardous stretch of road extremely difficult, if not impossible.

Walikale was a town of perhaps 600-700 people, with a paved main street, a hotel, a postal & telegram office, 2 or 3 eating establishments, and a Pentecostal Mission run by a 60 year old, white-haired, one-eyed Swedish man and his wife. Since I had almost no money with me, I asked him permission to stay, cost-free, at the mission. He looked at me quite disapprovingly with little sympathy for my plight. He spoke almost no English, nor did his wife. Nevertheless, he seemed to feel that it was his duty to assist me, so he showed me to a very small, barren room with a bed and a wash basin. A toilet was elsewhere. With some reluctance, he then invited me to eat dinner with him and his wife. (There were no other "guests" at the mission). A local African woman prepared the meal and served us.

Not having eaten in three days, I gulped down the whole meal in a matter of minutes, to the obvious disapproval of the pastor. I then sat there politely while the other two finished their meal in a normal amount of time. I'm sure I gave the appearance of wanting more to eat, but I didn't dare ask. However, to my surprise, the Swedish pastor asked me (in very broken English with a strong accent) if I would like more to eat. No sooner were the words out of his mouth than I instantly replied "yes - - please." He watched me as I quickly devoured my second large plate full of meat and vegetables, but when I finished there was not an offer of more food. I thanked the pastor and his wife and retired to my room for the night.

The next morning (Mar. 31), a Belgian policeman found me drinking coffee at the one hotel in town and to my great amazement he came up to me and said, mostly in English, that a telegram had come for me at the telegraph office, several days earlier! It was from Noble, of course, saying that when he arrived in Bukavu with a partially broken Jeep, and learned of the very poor condition of the road which I was now following through the rainforest between Bukavu and Stanleyville, he cancelled the plan we had made for him to drive this road. He chose instead a much longer, better maintained route around the eastern and northern edges of the rainforest. He said he would meet me in Stanleyville, and he provided me with the name and address there of an acquaintance he had recently met. I immediately sent a telegram back to him (at the Stanleyville address), telling him where I was and that it would probably take me about 2-3 more days before I would arrive in Stanleyville (maybe Apr. 2 or 3). Then I sat on the front porch of the hotel on main street and watched for any vehicle to come along which was headed northwest toward Stanleyville, over 400 miles away through mostly undisturbed rainforest. I didn't dare venture into the neighboring forest to birdwatch for fear of missing a ride. I sipped coffee and wrote a letter home as the day passed. No vehicles of any kind came along. The town of Lubutu was the next village on my route, about 175 miles from Walikale. Two heavy, 30 minute rain showers passed over in the afternoon. As the sun went down I returned again to the mission, where the pastor was obviously still not happy to see me, but once more he felt obliged to offer me a meal and a bed, which of course I gratefully accepted. I didn't have enough money to stay at the hotel, inexpensive as it was.

The next morning (Fri., Apr. 1) I ate a small very inexpensive breakfast of toast and eggs at the hotel, and was told by the owner that tomorrow (Apr. 2) a large commercial truck would arrive from Stanleyville and would then return to Stanleyville the next day (Sun., Apr. 3). This was very welcome news indeed, even if it meant one more night at the mission. Walikale was an extremely isolated place in the middle of the jungle, where it rained much of the time. Disappointingly, none of the birds which I was able to identify in the vicinity of the town were ones I had not already encountered. The adjacent countryside was mostly cultivated, although the Luhoho River was nearby, on its way to the Congo River.

A large, 15-year-old Mercedes-Benz truck with dual rear wheels arrived from Stanleyville Saturday afternoon as scheduled, bringing goods and supplies to the town. Early the next morning, Apr. 3, I put my sparse belongings in my backpack at the mission, said goodbye to the pastor, thanked him for his hospitality, gave him half of my few remaining coins, and walked down the street to the Mercedes truck which was being loaded for its scheduled 2 day return journey to Stanleyville, approximately 400 miles away through the rainforest. The African driver spoke no English, but he motioned for me to climb into the cab and sit next to him in the front! A few local passengers who weren't traveling very far, including a young teenage boy with a bicycle, found spaces to sit on top of the cargo. The morning mist had not yet lifted as we started out. The first of two towns on our route was Lubutu, isolated in the rainforest 175 miles distant.

A few birds flitted back & forth across the road, but it was impossible to recognize any except a very few because of the constant jostling of the truck. Our narrow, very sparsely-used track through the jungle was rutted and muddy, and we proceeded at speeds mostly between 10 and 20 mph. There were no other vehicles of any kind on the road. Having driven over such roads myself in our overweight, top heavy Jeep camping van, I appreciated the expertise and skill with which our experienced, middle-aged driver navigated the twists and turns, ups and downs, and side to side sways. It was necessary to constantly shift from one gear to another. The large dual rear wheels on the Mercedes were a great advantage on such terrain over the narrow 7x16" single rear wheels on our camping van. Noble's decision not to travel this route was a most wise choice. Being in the passenger seat was a luxury for me. Roadside birds included hornbills, African Gray Parrots, bulbuls, sunbirds, flycatchers, and mannikins. Monkeys scampered in the forest edge, and butterflies frequented the rain puddles in the middle of the road. My travel in a truck on a muddy road through the Congo rainforest was a once in a lifetime experience. However, the roadside was no longer entirely forest There were clearings where trees had been logged, or cleared for mining operations, and scattered small openings with cultivation and stick huts gradually appeared.

After several hours of travel one of the right rear dual tires on our truck went flat. There were 2 spare tires on board, but for an unexplained reason the vehicle jack was missing from its holder. (Theft was my guess.) So there we were, in the middle of the jungle with a flat tire and no way to jack up the truck. To continue with our heavy load on the very uneven road with one flat tire was not an option. The chances of another vehicle coming along were one in a hundred. There was no radio to ask anyone for assistance. The driver could think of only a single possible solution, as unlikely as it seemed to me. He could send the teenage boy on his bicycle up the road to look for any house, or logging or mining operation, where there might be a vehicle jack. So in mid-morning the boy willingly set off up the road. Luckily, there was no indication of impending rain. I was not at all optimistic. Our wait could be days. No one seemed to be particularly worried or in any hurry (which was undoubtedly an essential cultural trait in the rainforest). I decided to make the best of the situation and take the opportunity to look for whatever roadside birds I could encounter.

Remarkably, the boy returned on his bicycle just 3 hours later with a vehicle jack he had obtained from a truck driver who was involved in logging trees! What extremely good fortune (or divine intervention). I was delighted at being able to continue toward Stanleyville, although I had enjoyed my short chance to birdwatch, during which time I observed 2 birds new for my trip list: (1) the Little Green Sunbird* (a plainly colored, short-tailed, olive-green little bird, as was indicated by its name), and (2) the Magpie Mannikin* (a relatively large, forest edge estrildid finch with an oversized bill which was placed taxonomically in the African genus *Spermestes*, and which was black, brown, and white in color.) I also enjoyed watching a confiding, eloquent little Cassin's Flycatcher which was dappled Gray and whitish in color, with black wings and tail. It was a characteristic central African bird of river and water edge which I had first encountered at the IRSAC research station.

Without further stops or delays we arrived in the small town of Lubutu, shortly before dark. The last of our passengers got off here, leaving only the driver and myself. The driver indicated to me that we would spend the night there. Since there was no guest house in the town he invited me to stay with him at the house of a friend of his. I gladly accepted, and slept in my sleeping bag on a straw mat on the dirt floor. His friend's wife prepared dinner for us, a steaming hot broth with a wide variety of meats and vegetables, most of which were quite unfamiliar to me (perhaps for the better). It was a tasty, filling, and welcome meal.

We got underway at daybreak the next morning (Apr. 4) after a breakfast of bananas, bread, and some kind of dried meat (perhaps duiker). Lubutu was approximately 220-240 miles SE of Stanleyville. We acquired a few new passengers as we departed, but I continued to ride in the front. Our road gradually improved, and the forest had a few more logged areas and scattered small clearings with crude houses and patches of cultivation. The driver attempted to communicate with me, both in French and Swahili, but we had little success in understanding one another, although I could sometimes get the gist of what he was trying to tell me. Our road veered somewhat to the west, and there were occasional side roads going off to one side or the other of what appeared to be our main road, but without a map I never knew exactly where we were. There were widely spaced kilometer markers at the side of our road but I didn't know from where the numbering had begun. Forest continued to comprise 75% or more of the roadside edge. A heavy rain shower occurred shortly before noon, and the road became very muddy and slippery. Our rate of travel slowed to 5 or 10 mph, and I was very glad I was not driving. I had complete confidence in our driver.

Eventually, late in the afternoon, we came to a 200-250-yard-wide river flowing northward, from our left to right. I had no idea what river it was. Very unfortunately, because of recent heavy rains upstream, the river was flooding widely outside its banks on both sides of the river. A small, flat- bottomed motorized ferry was tied to a post out in the river, and was the only way for vehicles to travel back and forth across the river. However, the flood waters made it necessary for vehicles to drive through 1 or 2 ft. of rapidly moving water in order to reach the ferry. Our driver was very apprehensive, apparently not having done this previously, and he did not want to attempt such a risky maneuver. (I volunteered to do the driving, confident from my successful fording of the Rio Terraba in Costa Rica, but the driver adamantly refused.) He indicated that he would just sit there and wait until the river receded to its normal level, with the belief that it would not be longer than a day or two. Since I was already several days late in meeting Noble, I was not happy about any further delay. There being no other option I slept on the river bank that night, sitting up inside the cab out of the almost continuous rain. I was dry but not very comfortable. It was a long night. (The driver also sat up in the cab, and the 2 other passengers slept under a canvas covering on the back of the truck.)

The next morning (Tu., Apr. 5), the rain had stopped but the river, rather than receding, had risen even further. I envisioned a very long delay before we might be able to ferry across the river. Swimming across with my backpack, and the crocodiles, was not an option. A miraculous event then occurred. From the other side of the river, now swollen to 300 or 400 yards wide, came a long, narrow native dugout canoe with 2 native Africans, a man paddling in the back and another man riding as a passenger in the front. After the passenger had disembarked, the owner of the canoe turned around to go back across the river. Here was my chance. I grabbed my backpack, ran down to the river, and waved to the canoeist that I would like to ride back across the river with him. I held out my hand with some of the few remaining coins I had, and he willingly agreed. I then gave an appreciative hug to my truck driver, thanked him for his assistance, handed him the last of my coins, wished him "bon voyage", and cautiously climbed into the front of the canoe. My journey to Stanleyville would continue.

Paddling the very unstable dugout across the fast flowing river demanded great skill from my boatman, and it required all my effort not to lean left or right and overturn the canoe. However, my experienced helmsman took only 20 or 30 minutes to paddle from one side of the river to the other. He kept the bow of the canoe turned upstream at a considerable angle so that with the current pushing us downstream (to our right) we never-the-less followed a direct path straight across the river. When we arrived on the opposite bank of the river the muddy dirt road continued through the forest, following along the river on my right, heading north, downstream. I said goodbye to my boatman and set off at a moderate walking pace along the road, not knowing how far I might have to walk before I came to any kind of house or settlement. There were no people or habitations anywhere in sight. I paused only very briefly from time to time to look at a bird. Fortunately, it had not yet started to rain. It was still early in the morning.

Good fortune continued to be with me. In a little over an hour of walking, perhaps no more than 3 or 4 miles up the road, the forest suddenly opened out and to my great astonishment I came to a sizeable town, approximating that of Walikale, situated on the river I had just crossed. Without a map, I had no idea where I was or what the name of this town might be. I encountered a Belgian man who spoke a little English, and he told me the name of the town was Ponthierville (later changed to Ubundu, after independence), and the river was called the Luapula (later incorporated into the Congo, or Zaire River). The town was located just above the uppermost of a series of 7 widely separated cascades collectively referred to in 1960 (in English) as the "Stanley Falls", which were spaced along the river all the way from Ponthierville to Stanleyville, about 95 miles downriver (north). The river I had been transported across in a dugout canoe was none other than the Congo River! Ponthierville was an important historic city, connected to Stanleyville by a vastly improved road. My adventure had suddenly and dramatically changed. I was almost in civilization again.

My day of good fortune continued soon after I walked into the town. I very quickly found a small pickup truck which was preparing to depart for Stanleyville, about mid-morning. The African driver happily said I could ride with him, for the slightly less than 100 miles of remaining distance. I climbed into the front seat next to him and thought to myself that it couldn't get any better than this. However, the most astounding event of the morning was yet to come. When we had driven only 25 miles out of Ponthierville, at the km. 112 (70 mi.) marker from Stanleyville, the pickup in which I was riding encountered an approaching vehicle coming from the other direction. It was my first encounter with any approaching vehicle since I departed the IRSAC research station 400 miles ago. Even more astounding and unbelievable was the fact that the oncoming vehicle was none other than Noble in our Jeep camping van!! Of course Noble had no idea that I was the passenger in the approaching pickup truck so I shouted at my driver to stop, and as Noble approached I jumped out and waved wildly. His astonishment was overwhelming as he quickly braked to a halt. We were both equally amazed and delighted at this sudden, unexpected encounter.

We rushed up to each other and embraced warmly. It was a very happy moment. Noble said that I looked dirty and long-haired, and had lost some weight. He took a picture of me standing next to the Jeep. It had been exactly 3 weeks since we parted company in Fort Portal, Uganda (on Mar. 15). Noble had received my telegram from Walikale 5 days ago, and having just learned of the flooded condition of the river at Ponthierville he decided to come searching for me. I was already several days late from my estimated time of arrival. In spite of our

Plate 54: I hitchhiked for 12 days and 400 miles through the Congo rainforest. Birds were difficult to observe. The wide Congo River was out of its banks from recent rains. Noble (driving our Jeep) and I encountered each other 60 miles south of Stanleyville, in the jungle on the equator!

many differences we were good companions. Neither of us could have accomplished our adventure without the help of the other. Noble turned the Jeep around and headed back toward Stanleyville, just over 65 miles away. We almost immediately arrived at the equator, where we stopped for a picture of the two of us standing next to our Jeep van, as a permanent memory of our getting back together again. As further good fortune, our road at this point became hard surfaced ("bitumen") for the remainder of our travel into Stanleyville! with a bottle of wine in the van. Life was good.

Stanleyville was a city of about 70,000 people in 1960. The city was named by the Scottish explorer Henry Morton Stanley, for himself, when he first arrived there in 1883 and established the city as a trading post for King Leopold of Belgium. Stanleyville (now called Kisangani) was located on the Congo River almost at the exact geographical middle of the African continent, just above the equator approximately equidistant between Capetown & Cairo, and half way between the Atlantic & Indian oceans. This location was the meeting place of the Swahili language from the east and the Lingala language from the west. Stanleyville was an important hub for rail, road, and boat transportation. It was situated 1,300 miles upstream from the river's mouth on the Atlantic Ocean, at the upper end of a 1,000-mile navigable section of the river between Livingstone Falls at Leopoldville (now called Kinshasa) and Stanley Falls (Boyoma Falls) at Stanleyville (Kinsangani). After Noble had reached Stanleyville he took our van to a Jeep agency there and was able to have a few of our problems corrected while he waited for my arrival. However, the non-functional 4-wheel drive could not be repaired, so we still had no 4-wheel drive capability for our greatly anticipated crossing of the Sahara Desert. We hoped to commence this adventure in about 3 week's time and there was nowhere to correct this problem prior to starting out across the desert.

Noble and I departed Stanleyville the next morning, Apr. 6, once more with me behind the wheel and Noble in the passenger seat. It was great being back on the road together again. Our next major destination was Kano, Nigeria, almost 2,000 miles away and the staging point for our travel across the Sahara. On our way we would travel northwestward through the central African sub-Saharan countries of French Equatorial Africa, Chad, and the French/British Cameroons. However, we first had to traverse another 450 miles of the Belgian Congo, northward through rainforest and woodland on our way from Stanleyville to French Equatorial Africa. It was a part of the continent very little traveled or inhabited by Europeans. That first day we ferried across the Aruwimi River at the town of Banalia, where there were palm trees with nesting colonies of weavers, and dugout canoes carrying passengers and fishermen up and down the river. We continued northward through the diminishing tropical rainforest, and camped that night just south of the principal town of Buta, on the Itimbiri River at 1,500 ft. elevation. The puffy cumulus clouds had a silver lining. Of course it rained that night.

The next day (Apr. 7) we drove into the town of Buta where Noble searched out an Englishman, Mr. James, whose name he had been given in Stanleyville. Mr. James and his wife were the only persons in the town who spoke English. In an hour's conversation with him, over morning tea, he provided us with valuable road information for our forthcoming travel to Kano, on a route with which he was familiar. After finishing our tea we thanked him and continued on our journey. It was almost noon and the sky appeared as if it might rain again as we departed the town on a poorly maintained, rough gravel road. Our progress was slow and we traversed just over 100 miles the rest of the day, to a campsite 10 miles south of the town of Bondo, in a dense clump of large bamboos. Near our campsite I encountered 2 species of birds which were my first trip additions in several days: (1) the Black Bee-eater* (a lovely, small forest bee-eater with a red throat and pale azure blue and black plumage; it was my 12th species of bee-eater from Africa), and (2) the Lesser (or Yellow-eyed) Bristlebill*, which was a large member of the bulbul family, Pycnonotidae.

Early the next morning, Apr. 8, we crossed the wide, east to west flowing Uelle River on a motorized ferry, into the town of Bondo. We were now only 125 miles from the border between the Congo and French Equatorial Africa. Our poor gravel road zigzagged in a generally northwest direction from Bondo through gradually drier and drier forests, between the elevations of 1,500 & 2,000 feet. We slowly left the rainforest behind. There were scattered little thatched houses with pointed roofs. I stopped occasionally to identify the few roadside birds. My total for the day was just 26 species, of which only 2, the Egyptian Plover* (see below) and the Rufous-bellied Helmetshrike* (*Prionops fufiventris*) were additions to my trip list. Other species on my daily list included the African Harrier-Hawk, Long-crested Eagle, Lizard Buzzard, Wattled Lapwing, White-headed Lapwing, Great Blue Turaco (my last record of this species), African Palm-Swift, Cassin's Spinetail, African Pied, White-thighed, and Piping hornbills, Gray-throated Barbet, Velvet-mantled Drongo, and Veillot's (Black) Weaver. The Egyptian Plover and the lapwings were characteristic birds of the wide river banks.

The Egyptian Plover (*Pluvianus aegyptius*, popularly known as the "Crocodile-bird") was one of the most iconic African birds. Taxonomically it was most often positioned with the coursers in the family Glareolidae, but it had recently been placed by some authors in a family of its own. It was a confiding, rather plump, plover-like bird handsomely patterned in black, Gray, and sandy-buff colors, with a narrow black breast band, black crown, black eye stripe, and very conspicuous black & white patterned wings in flight. The Crocodile-bird frequented open sandbanks and sandy shores of lakes and rivers where it flew gracefully on widespread wings then ran short distances after landing. Books of lore said that it picked food particles from the open mouths of crocodiles (hence its unusual name). Its unique nesting behavior involved slightly burying the 2 or 3 eggs on a sandy shore and leaving them to hatch mostly unattended, except to cool them periodically by water carried on belly feathers of the parents. The young could run and swim almost immediately upon hatching.

We camped that night on the south side of the border with French Equatorial Africa, along the Mbomou River near the small town of Ndu. I finished the Belgian Congo with 30 trip additions to my bird list (Appendix B), 21 of them from within the rainforest. Although this was not a significantly large number, my 12 days and 550 miles of hitchhiking through the Congo rainforest was a never-to-be-repeated adventure, far off the beaten track. Savannas, primitive lifestyles, and the Sahara Desert lay ahead.

Chapter Sixteen – Central Africa

On the morning of Apr. 9 we ferried across the east to west flowing Mbomu River, from the small town of Ndu in the northern Belgian Congo to the city of Bangassou in French Equatorial Africa, which would be renamed the "Central African Republic" (CAR) in just 4 more months when the country gained its independence from France. Along the river were Reed Cormorants, a Common Sandpiper (from the Palearctic), a Banded Harrier-Hawk, Pied & Pygmy kingfishers, and a Tawny-flanked Prinia.

From Bangassou we would travel almost 1,500 miles in 9 days to the city of Kano in northern Nigeria, our staging point for driving across the Sahara. Our route would pass close to the southern shore of Lake Chad and take us through 5 countries: (1) French Equatorial Africa, (2) Chad, (3) French & (4) British Cameroons, and (5) Nigeria. The principal towns and cities along the way (with a number designating the country, and the distances between towns), were: Bangassou (1) 225 mi. to Bambari (1); 165 mi. to Fort Crampel (1); 175 mi. to Fort Archambault (2); 120 mi. to Lai (2); 90 mi. to Bongor (2); 150 mi. to Fort Lamy (2); 155 mi. to Maiduguri (5); which was 375 mi. from Kano (5). This entire route lay within a vast E-W belt of grassland and woodland savanna which stretched from Senegal and Guinea on the Atlantic Ocean eastward for 3,000 miles to the Sudan.

French Equatorial Africa had changed its boundaries several times during the past century, but in 1960 the country encompassed approximately 240,000 sq. miles (about twice the size of New Mexico). It was situated in the remote middle of the African continent between the Shari River on its northern boundary and the Mbomou/Ubangui rivers on its southern boundary. The land was a hot, relatively dry region lying on a rolling plateau between 1,000 & 1,500 ft. above sea level. Small streams and rivers were widespread, but most villagers obtained their water from deep wells. It was one of the poorest, least developed, and remote countries in all of Africa (or anywhere else in the world). Income came primarily from the sale of cotton, timber, and fish. Agricultural products included cassava (a starchy tuber also known also as yucca or manioc), plantains, coffee, tobacco, maize, sesame, peanuts, sorghum, and millet. Livestock were principally goats, pigs, donkeys, cows, and chickens.

Many different ethnic groups inhabited the country, each with their own language. European colonists and businessmen (mostly French) were very sparse, and there were virtually no tourists. Rural villagers waved at us happily as we drove along the dry, dusty roads, where we were very much an oddity. Children were often naked, and women and girls were regularly bare-breasted. Clothing for men was plain and simple, but women's dresses when worn were often colorful. Most rural people lived in mud and stick huts with pointed thatched roofs. Clay pots and jugs came in all sizes and shapes and were an indispensable household item, as were large, hollowed out, dry gourds, either entire or in halves. One of the most important functions of these was to carry water to one's house from the nearest well, a task that was always performed by women or girls who carried the water, cushioned by a thick head cloth, on top of their head. The lifestyle of most villagers had changed very little over the centuries. Tourists were non-existant.

Noble and I headed west from Bangassou on a fairly well-maintained dirt road which passed through scattered small villages with round stick houses which had tall, pointed, thatch or stick roofs. Eighty miles along the road we came to the Kotto River, where just above the bridge (on the north side of the road) was a deep gorge with a twin 40 ft. high waterfall. I stopped for a photograph. An estimated 100 Preuss (Cliff) Swallows* (genus *Petrochelidon*) were nesting in the gorge and under the bridge. This species was the only trip addition to my bird list for the day. Not far away, a group of 6 or 7 women were walking along the road carrying one or two wide, orange gourd bowls on top of their head. Several of the women were Hausa, and were wearing a long orange dress and scarf, a tight-fitting silver beaded necklace, and dangling silver earrings. Around her neck one of the other women wore a long, jeweled pendant with a hand-carved "Rose Cross", with 4 equal arms symbolizing resurrection, rebirth, the earth's axis, and the world of matter. Such a cross was a symbol of early Christianity. I was told that this woman was a Berber from much further north. As we continued, our road curved to the northwest, toward the principal town of Bambari. Sitting on the ground in front of his house was an almost naked, very dark-skinned man whittling on a long stick, perhaps shaping it into a spear. Shortly before sunset I picked out a campsite at the edge of a tall, green, grassy savanna 50 miles SSE of Bambari, at latitude 5°N of the equator. We had traveled 175 miles for the day. The two most conspicuous roadside birds during the day were aerial feeding African Palm-Swifts and Rufous-chested (Red-breasted) Swallows.

Early the next morning before our departure (Apr. 10) a bare-breasted woman came walking past our camping van, carefully supporting with both hands a large white porcelain tub of cotton on top of her head. Soon after leaving our campsite we came to a small village where, once again, many of the women were not wearing any garment above their waist. (In the very hot climate of equatorial Africa this custom was the most reasonable for keeping cool.) As we started out again along the road I was delighted to encounter a small, laterally compressed, 6" long, green and white, vertically striped chameleon which was slowly walking across the road in front of us. I stopped to pick it up for a closer inspection, and to take a photograph. Curiously, each eye was on the end of a short stalk sticking out to the side of its head, giving the animal a somewhat grotesque appearance, unlike any I had ever seen. I was greatly amused by the fact that each eye operated independently of the other, so that while one eye was stationary and staring directly at me, the other eye was wandering about looking in first one direction and then in another as it surveyed its environment. I laughed. This unlikely looking terrestrial vertebrate seemed to be an imaginary creature from a fairy tale, and my unexpected encounter with it in French Equatorial Africa became another unforgettable memory of our travels on this continent.

South of Bambari we drove through a small village where a large group of people - - men, women, and children - - were all gathered at a funeral service for an infant that had died. People were dancing to the sound of bells and drums, which were hollow logs beaten with sticks by teenage boys. When Noble and I stopped to watch and take a few pictures we became more of an attraction than the funeral service itself. However, people were only curious and not hostile, and simply stared at us in our safari shorts and long beards, apparently having never previously encountered persons quite like us. We were certainly off the beaten path. French Equatorial Africa did not attract tourists. Further along our road, we soon came to the town of Bambari. We stopped only long enough to fill up the 12-gallon water tank inside our van, with water which we obtained from a small hotel. As always, whenever we were in doubt about the purity of the water we simply emptied one of our small bottles of 50 halazone tablets into the water tank, which we used for drinking and washing dishes or our hands. (We left California with a huge supply of these bottles.)

From Bambari we headed northwest toward the town of Fort Crampel (subsequently renamed Kaga Bandoro). Our slow, narrow, dusty, poorly maintained dirt track took us through savanna woodlands at an elevation of 1,500 feet above sea level. We encountered no other vehicles. Fortunately it was the dry season, as this road would surely have become an impassable muddy quagmire for us if it rained. We thought that perhaps we made an error when we left Fort Bambari, and this track was not the main road to Fort Crampel. As usual there were no road signs anywhere, and none of the few people along the road could speak English. Our road map was almost useless, but we had ample gasoline in our fuel tank. The good news was that our compass told us we were traveling in the right direction. We were certainly in the middle of nowhere, somewhere near the center of the African continent. It was not an adventure for the timid - - only the naïve!

We passed by several clusters of primitive stick houses, but saw very few people anywhere, and no vehicles of

Plate 55: From Stanleyville we traveled northward to Bangassou, French Equatorial Africa, on the northern bank of the wide Mbomou River (the international boundary). Surprisingly, we encountered a small group of Hausa women, who were more characteristic of regions to the NW.

any kind. One group of houses had a nesting colony of Village (Black-headed) Weavers, which with its conspicuous long, hanging nests was one of the most characteristic, widespread, and best known of all African birds, an icon. Several groups of unidentified parrots flew by (either Meyer's Parrot, Senegal Parrot, or perhaps the uncommon, endemic Niam-Niam Parrot, which I never positively identified). I saw both Abyssinian* and Blue-bellied* rollers for the first time, gorgeous blue birds which sallied for insects from conspicuous high perches. That night we camped about 50 miles SE of Fort Crampel, after having driven 165 miles for the day, almost matching yesterday's travel.

The next morning, Apr. 11, I pursued birds around our campsite for 2 hours (0600-0800) before we began our day's travel, as I often did while Noble slept a little longer, wrote in his journal, or tidied up inside the van. There was considerable avian activity in the open woodlands and I documented 40 species of birds in this rather brief time, 7 of which were trip additions (see Appendix B, the first seven species for this date). Most memorable among my total of forty were the following: 1 Lesser Honeyguide (looking very sparrow-like with its short, stubby bill), 2 Black Scimitar-bills (in the woodhoopoe family), 1 Gray Woodpecker, 2 White-crested Turacos* (one of the most gorgeous of all birds with its green body, purple wings, conspicuous red primaries, all white crested head, black mask, and bright yellow bill), and 2 Black-bellied Firefinches* (an estrildid finch in which the male was vividly two-toned in plumage coloration, bright red and black).

We departed our campsite and drove first westward and then northward on a narrow, primitive, unmarked dirt road into the town of Fort Crampel, situated at 1,500 ft. elevation on the left bank of the small, northward flowing Gribingui River. Because of the primitive, mostly unused road over which we traveled it was necessary to proceed slowly, and we did not arrive in Fort Crampel until early afternoon. We stopped at an insignificant little café for a pastry and cup of coffee.

Leaving Fort Crampel, we followed a somewhat improved road NW out of town toward the distant border of Chad, less than 120 miles away, beyond which was the French governed city of Fort Archambault. The countryside was gradually becoming drier and warmer, and the savanna woodland was more open with trees of lesser height. Raptors along the roadside during the next 30 miles were 25 Black Kites, 1 Grasshopper Buzzard, 1 Gray Kestrel (only my 2nd sighting of this species, the 1st being in Tanganyika), and 1 Lizard Buzzard. There were also 2 Hammerkops, 1 Black-headed Heron, 2 Blue-headed Coucals, 1 Abyssinian Roller, and my first sightings of both the Purple Glossy-Starling* and Little Weaver*. In addition, I identified a White-browed Robin-Chat (Cossypha hueglini) about 30 miles NW of Fort Crampel. This location was approximately 100 miles outside (south) of the range of this species as described by Bannerman, but he remarked that the species "is quite likely to extend west of the Shari, although specimens have not yet been obtained." (The very similar appearing Blue-shouldered Robin-Chat was a rainforest species with a more southerly distribution.)

As dusk approached I was attracted to a small group of people gathered together a short distance off to the side of the road in an open woodland. They seemed to be engaged in some sort of activity involving a tent-like structure made of palm fronds and leafy tree branches. My curiosity was aroused and I stopped to investigate. I walked up to the group and observed that two people were making drum-like staccato rattling sounds by beating with sticks on the bottom of dried gourd bowls turned upside down on the ground, in the shade underneath the tent of palm fronds. I discovered there was a big colony of flying ants in the ground here, and in the shade provided by the palm fronds, the staccato drumming sounds caused these large insects to emerge from the ground. They were then gathered up by hand, placed in gourd bowls, and carried back to a village where they would be eaten as a delicacy! I later learned this was a common practice throughout much of Africa.

There was a clear sky with a full moon that night and I decided to drive for an additional 2 hours after dark to see what nocturnal birds I might encounter along the wooded roadside. My tally for this little excursion was 2 Bronze-winged Coursers, 5 probable Black-headed Lapwings (vocalizing loudly but never seen), 3 Long-tailed Nightjars* (apparently my 1st record of this species), 3 Black-shouldered Nightjars* (heard only; a taxonomic split by Clements and others from the Fiery-necked Nightjar), 5 probable Pennant-winged Nightjars (none with elongated wing feathers), and 3 calling African Scops-Owls. It was a delightful two hours. That night, we camped in a savanna woodland about 50 miles NW of Fort Crampel, having driven just 100 miles for the day.

I devoted the first hour of the day on Apr. 12 to looking for birds around our campsite, in woodland savanna at 1,500 ft. elevation. During this brief period of time I succeeded in finding 44 species, with 7 trip additions: Sun Lark*, Woodchat Shrike* (a Palearctic visitor), Vinaceous Dove*, Black-rumped Waxbill*, Bush Petronia*, Brown Babbler* (a *Turdoides*), and Black-faced Firefinch*. Other species which I was particularly happy to observe were 6 White-crested Turacos, 3 Gray Woodpeckers, 5 Meyer's (Brown) Parrots, 1 Double-toothed Barbet, 4 African Gray Hornbills, and 1 Black-bellied Firefinch. Most of these were species confined to the savanna belt of northern Africa.

As we started out on our day's travel we were still inside French Equatorial Africa. It was approximately 60 miles north to the border of Chad, at the Grande Sido River crossing. We soon entered a village (perhaps Batangafo) with stick houses and thatched roofs. Women wearing only a wide, short grass skirt were pounding grain into flour in large wooden "mortars" with heavy wooden "pestles." It was a tedious, energy demanding process. Tubs of cotton were being sold in a central square, as were ceramic pots and jugs. On the outskirts of the village were 2 young girls, the younger one appearing to be an early teenager and the other only very slightly older. They were both very elaborately and colorfully adorned and decorated in a manner which I had never previously seen. (Not far away was a small boy wearing nothing at all.) I inquired from a nearby, ordinary clothed man, using my best sign language, what the occasion was for the bright attire of the girls. He gesticulated and said to me in French (if I interpreted him correctly) that the younger one was getting married, and the other was the equivalent of her "maid of honor."

Of course I wanted one or two photographs of the girls, so I approached them and pointed at my camera, and then at the two of them. To my delight, they had no objection to being photographed. As always, I had to guess what shutter opening and speed to set for the amount of available light. (You will remember that my camera was entirely manually operated.) Happily, the color slides turned out very well when I was finally able to view them 2 years later, and the resultant picture of the young bride-to-be is one of my permanent treasures. She had black skin and a pensive, pretty face with prominent, narrow vertical scars on her forehead, obviously cut deliberately with a sharp knife for beautification. I suspect she was no more than 13 or 14 years old.

Her simple outfit and the elaborate beaded decorations she was wearing (as best they can be described from my picture) are as follows. She was barefooted, and the only pieces of clothing covering her anywhere were two long, narrow, dark brown loin cloths tied around her waist with a wide, red and white beaded band. One cloth hung straight down loosely in front of her to just below her knees, and the other hung straight down loosely in back of her to the same level. Attached at the top of her front loin cloth were some dangling, loosely tied, red and gold bell-like ornaments which hung down in front of the loin cloth to thigh level. Around her left wrist (but not her right) were 5 narrow, plain gold bracelets which when placed together formed a wide gold wristband. Around each upper arm was a rather narrow gold and red band. She wore no earrings. Around her neck were a close-fitting, red and white necklace of small beads and several narrow, longer, beaded necklaces. Also around her neck was a multi-stranded band of beads which hung straight down between her small, firm breasts. Tied to the bottom of this band of beads were 4 or 5 round gold medallions. Finally, her most conspicuous ornamentation was a wide multi-stranded band of red, white, and blue beads which was arranged in a double sash-like fashion, starting around the back of her neck and coming forward over both her shoulders then angling downward and crossing over the front of her, between her breasts. The bands then went down and around the sides of her waist and were tied together around her lower back. These two sash-like bands of brightly colored beads covered much of her front side above the waist. As I studied her attractive, serious, unsmiling face I pondered just what kind of life was ahead for her, and I hoped for the best. Her appearance and demeanor are a permanent memory for me of central Africa. (Noble persuaded the 2 girls to stand in front of our Jeep while he took a picture with his little box camera.)

From this village of uncertain name Noble and I continued our adventure northward toward the nearby border of Chad. Roadside birds included an African Harrier-Hawk, 8 Laughing Doves, 3 Black Scimitar-bills, 5 Bateleurs (which were always exciting to watch), 2 Abyssinian Rollers, 4 species of bee-eaters (8 Green, 7 Little,

Plate 56: French Equatorial Africa was isolated far off the beaten path, in the hot, dry savanna of middle Africa, with primitive villages, dusty roads (if it didn't rain), and no tourists. The young girl was decorated to be married!

15 N. Carmine, & 2 Black), 5 Beautiful Sunbirds (only my 2nd sighting of this tiny, emerald green sunbird with a broad red breast band and narrow, greatly elongated central tail feathers), 12 Pied Crows, and a stunning African Paradise-Flycatcher with long fluttering tail plumes.

We crossed the Grande Sido River from French Equatorial Africa into Chad at an elevation of 1,200 ft. and latitude of 8°N, between the towns of Sido (in Fr. Eq. Afr.) and Maro (in southern Chad). With an area of 496,000 sq. mi., Chad was 85% the size of Alaska, and was the 5th largest country in Africa. It extended N-S for over 1,000 miles, from the Shari River and Lake Chad in the south far northward into the southern Sahara Desert. Noble and I would travel across only the southwest corner of the country.

The major city of Fort Archambault (soon to be called Sahr) was only 65 miles away from our port of entry. The countryside remained savanna grasslands and dry open woodlands, with occasional low-lying wetlands and marshes. Some of the most unusual or satisfying birds which I added to my daily list were Dark Chanting-Goshawk, African Openbill, Purple Heron, African Sacred Ibis, White-faced Whistling-Duck, Grasshopper Buzzard, African Fish-Eagle, Gray Kestrel (my 3rd and last record of this local African endemic), Senegal Thick-knee, Gray Pratincole, Egyptian Plover ("Crocodile-bird", my 2nd sighting of this most unusual bird), Black-billed Wood-Dove, Bruce's Green-Pigeon* (genus *Treron*), 3 additional species of bee-eaters for the day (Blue-breasted, White-throated, and Red-throated), Wire-tailed Swallow, Gray-headed Batis (*Batis orientalis chadensis*, which was very challenging to identify), and Rufous Cisticola (which Bannerman called "Fraser's Rufous Grass-Warbler"). For the day, I recorded a very good total of 102 species, 16 of which were trip additions (Appendix B, in which all of them are listed from French Equatorial Africa since my field notes inadvertently did not record exactly where we crossed the border into Chad). A greatly satisfying personal achievement was to document seven species of bee-eaters today! Only in Africa could one attempt such a one day total, and certainly not by chance along a 125 mile stretch of un-scouted roadside. How lucky could I have been?

Fort Archambault was located on the west side of the wide, northwest- flowing Shari River, which supported a vital fishing industry. This shallow, slow-moving river was more than ½ mile wide. Scantily clad men and women with large seines waded and fished along the river shoreline, with 2 persons per seine, one on each end. Other fishermen used very long, narrow, hollowed out log boats to seine fish from the middle of the river. Such boats were also used to transport people back and forth across the river. There were no bridges, and vehicles had to be transported across the river on small, flat bottomed, diesel powered ferries. The Shari River was one of only 2 major rivers, along with the parallel flowing Logone River to the west, which supplied water to Lake Chad 350 miles to the north.

That night in Fort Archambault we were invited to dinner with a missionary family in their home. Noble remembers that the temperature was so hot in their non-air-conditioned house that hard butter taken out of the refrigerator and placed on the table all melted within an hour. After our meal we drove a short distance outside town and found a campsite along the road which we would follow tomorrow morning west and north toward the cities of Lai, Bongor, and Fort Lamy. We were at an elevation of 1,000 feet. Because of the very hot night, I slept on top of the van, under a mosquito net. The average rainfall here was 40", and the rainy season would begin in just one more month. Rainfall, of course, would decline steadily as we traveled north toward the desert.

We were underway shortly after sunrise the next morning, Apr. 13, heading west toward Lai on a very primitive, narrow, seasonally impassable sandy track through sparsely inhabited savanna woodland with scattered low-lying marshes and wetlands, between 1,000 and 1,500 feet. Kapok and ebony were once important commercial tree species here. The positive feature of this slow, remote road, from my viewpoint, was that in spite of the hot daytime temperature (up to 100°) the roadside birds were surprisingly active throughout much of the day, all except the hottest parts of the morning and afternoon, and I stopped at regular intervals for 10 or 15 minutes to document the avifauna. By the end of the day we had traveled 115 miles, and I chose a campsite just 5 miles before Lai, still in the savanna woodlands.

My bird list for the day totaled 67 species, of which 13 were trip additions (Appendix B). Many of the species I saw today were confined to this sub-Saharan savanna/woodland belt. Some of these were the Clapperton's

Francolin*, Rose-ringed Parakeet*, Abyssinian Ground-Hornbill*, Bearded Barbet*, Fine-spotted Woodpecker*, Hueglin's Wheatear*, Yellow-billed Shrike* (genus *Corvinella*, an unusual looking brown and white laniid with a long tail, yellow bill, and black mask, unlike any other in the family), Piapiac* (a unique, rather starling-like, entirely black, long-legged, sociable corvid with a long graduated tail and relatively heavy bill, in the monotypic genus *Ptilostomus*), and Brown-rumped Bunting* (a typical stripe-headed, golden-breasted *Emberiza* bunting separated from similar species by geographic range and the lack of white wing bars).

The next day (Apr. 14) we drove north from Lai to Bongor, traveling along the eastern (right hand) side of the wide Logone River, on a primitive road in one of the most remote areas of the world. It was a morning to remember. The dominant people here were the Sara (Kameni), a tall, very dark-skinned, physically powerful tribe whose ancestors had come from the Nile valley many centuries earlier. There were no other vehicles, but small villages were scattered along the hot, dry roadside. Peoples of several different ancestries lived in the region, but neither Islam nor Christianity had been established here. I was behind the steering wheel. There were virtually no visitors to this part of the African continent.

Early in the morning a tall, handsome, physically well-proportioned and muscled man, perhaps in his 30's, came striding nonchalantly toward us down the middle of the road ahead of us, carrying a coil of rope over one shoulder. Noble and I were astounded to notice that he was entirely naked! I slowed, grabbed my camera from the shelf behind my seat, and quickly snapped his picture through our wide wrap-around windshield as he approached. I almost came to a complete stop. Without altering his long stride he walked close past my open window, quickly glanced at me with complete indifference, and casually continued on his way. Noble and I were equally amazed. Our adventure had provided us with yet another first. As it turned out, this was only the beginning of a 40-50 mile stretch of road where we encountered small numbers of men, women, and children of all ages who, like the man, were completely naked. These persons (perhaps 10% of the total population here) were living compatibly with the other villagers, adults of whom wore at least a loin cloth tied around their waist. There was great variation in how much clothing was worn, and while many women and girls were bare-breasted, others were completely covered in a dress. Noble and I were intrigued by the lack of a uniform dress code, apparently resulting from the great mixture of ethnic groups here in the Logone valley, all of which seemed to be living together compatibly.

As we approached Bongor, the habitat of the Logone valley became much more open, with fewer trees and woody vegetation. Men and boys carrying shields and spears herded small groups of cattle along the roadside. The environment became hotter and drier and there were more wells for supplying drinking water, gathered in a bucket tied to the end of a long rope. Women carried well water back to their house in gourd containers balanced on top of their head with one hand. The overall appearance of villages changed as houses became round in shape and were now made of mud with a thatched bee-hive like domed roof. Dried fish from the nearby river hung on racks or on the branches of small trees. Many round, thatched dome-like structures served as storage for food and other items. Gray-colored pots and jugs of all sizes and shapes lay on the ground outside houses. Most adults wore very little clothing, and children were naked. It was an environment and lifestyle unlike any Noble and I had yet experienced, far off the beaten path in the middle of Africa. It was almost like we were in a fantasy world dating far back into history, having been transported there by a time machine.

Our day's travel in the Logone valley took us a total of 175 miles, all the way through Bongor to a campsite 75 miles beyond, half way to Fort Lamy. This was unspoiled Africa where life had changed very slowly and had resisted change from the outside world. My daily bird observations totaled 69 species, about the same number as yesterday's tally, but with half of the species today different from those of yesterday, so that my combined 2 day total was just over 100 species. Four species today were trip additions: (1) Black Crowned-Crane* (very similar in appearance to the allopatric Gray Crowned-Crane, from which it was distinguished by its black, not pale Gray, neck), (2) Spur-winged Plover* (a lapwing in the genus *Vanellus*), (3) Black Scrub-Robin* (an almost entirely black member of the genus *Cercotrichas*, confined to the Sahel belt), and (4) Chestnut-bellied Starling* (another endemic of the sub-Saharan savanna, closely resembling other members of the genus *Lamprortornis*). A single

Helmeted Guineafowl and 2 Abyssinian Ground-Hornbills were also recorded.

On the morning of Apr. 15 we drove northward the remaining distance into Fort Lamy, following in the narrow dry floodplain situated between the paralleling, northward flowing Logone and Shari rivers (to our left and right, respectably), at 1,500 ft. elevation. For the first 4 hours and 30 miles I drove slowly and stopped regularly to document roadside birds. The weather was partly cloudy and hot, with occasional gusts of wind, and the sandy road was dry, like the neighboring savanna grasslands, scrub, and scattered trees. Over this stretch of road I tallied 56 species of birds, including 4 trip additions: Western (Gray) Plaintain-eater*, Viellot's Barbet*, White-rumped Seedeater* (also called Gray Canary, in the genus *Serinus*), and Chestnut-crowned Sparrow-Weaver* (in the endemic African genus Plocepasser). During the last 3 days (Apr. 13, 14, & 15) Noble and I had driven approximately 360 miles through the savanna belt of northern Africa, during which travel I documented a total of 110 species of birds, with 21 trip additions (Appendix B).

We arrived at noon in Fort Lamy (now called "Ndjamena"), the capital and largest city of Chad, with a population in 1960 somewhere between 30,000 & 50,000. Uncharacteristically, we spent only about 2 hours in the city before continuing onward toward our primary destination of Kano. Fort Lamy was situated at an elevation of 1,000 feet at the confluence of the northward flowing Shari and Logone rivers, on the southern edge of a 60-mile-wide delta below Lake Chad. I stopped and got out several miles before we reached the city so I could walk and document birds on my way in, while Noble went on ahead to purchase gasoline and a few other supplies. As I was walking along the little traveled main street on my way into the city to join Noble, a French businessman passed me, driving a modern limousine. He stared at me as he drove past, then stopped, turned around, and came back. He asked me, first in French and then in English, if I needed any assistance, and he immediately reached into his wallet and offered me some money. I am certain that my bedraggled, dirty appearance convinced him I was destitute. No doubt very few Caucasians (except businessmen) traveled to Fort Lamy, on foot or any other way. I did not accept the money, but thanked him and explained my situation to him, after which he wished me "bon voyage" and drove on up the street. Yes, there were congenial, charitable, Frenchmen.

Lake Chad was Africa's 4th largest freshwater lake in surface area (at its maximum size), but it was extremely shallow in depth and its shoreline fluctuated greatly from season to season depending on rainfall and amount of human usage. The average depth was only 4½ feet. It was an endorheic lake with no outlet (a unique feature for a freshwater lake), and only two rivers supplied it with water, the Shari and Logone. It possessed many low-lying islands in the middle, covered and surrounded by reedbeds (including papyrus), marshes, perennial grasses, and algae, as was the lake's shoreline. Over 120 kinds of fish inhabited the lake, including the ancient lungfish and 40 commercially valuable species. Lake Chad was a very important year around refuge for many kinds of wildlife, and equally valuable for migrant and wintering waterfowl and shorebirds from the Palearctic. Elephants, hippos, rhinos, lions, and leopards were all residents or regular visitors to the lake, as were crocodiles, rock pythons, and spitting cobras. Breeding birds included Black Crowned-Cranes, storks, herons, ibises, spoonbills, pelicans, ducks & geese, crakes, and lapwings. Large numbers of Palearctic waterfowl and wading birds of all sizes utilized Lake Chad in migration or for "winter" residence. In addition to the wildlife, Lake Chad was extremely important to the surrounding human population as a water source for irrigation, livestock, and human consumption. A rapidly expanding human population severely threatened the future of the lake.

Noble and I departed Fort Lamy early that afternoon and ferried across the Shari River, with its fishermen and long hollowed log boats, to the town of Kousseri in the very northern tip of French Cameroon (which in 1960 was a separate colony from British Cameroon). Our route angled northwest for 75 miles across both of the Cameroons to the town of Gambaru on the border of Nigeria. This road followed along the southern edge of the wide, flat delta of Lake Chad, only a few miles north of us. It was not clear to me when we left French Cameroon and entered British Cameroon, so I have treated these as a single country ("Cameroons") in Appendices B & C. The afternoon was hot and dusty with a scattering of high, thin clouds. We were now almost on the southern border of the "Sahel", a narrow transitional belt of semi-arid grassland, open thorn scrub, and acacia trees which separated the savanna

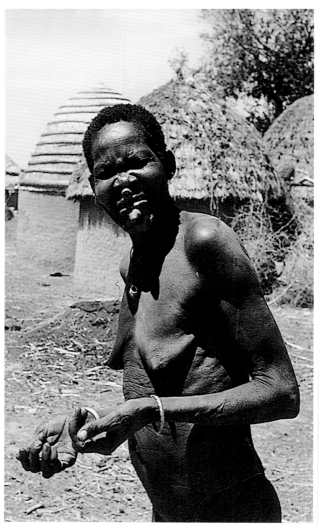

Plate 57: The Shari River in southern Chad flows northward into Lake Chad. It is an important source for fish. Very little clothing was worn in this hot, dry climate, and surprisingly, both men and women were sometimes entirely naked.

woodland belt on its south from the Sahara Desert on its north. The Sahel extended all the way from Mauritania on the Atlantic Ocean eastward in a very long, narrow band across northern Africa for more than 3,000 miles, to northeastern Sudan.

The Hausa people, of Islamic faith and noted for their horsemanship, were the dominant ethnic group in this region of the continent. They were our first encounter with a predominantly Muslim culture. Two Hausa horsemen trotted along the roadside next to us, sitting on bright red blankets, and bare-footed cloak-covered herdsmen walked behind burden-carrying bullocks. Colorfully dressed women carried babies on their back, along with all kinds of goods and items on top of their head (earthen and leather jugs, gourd bowls, straw mats, etc). In the fields, groups of workers with sticks beat the kernels from kaffir corn cobs, which were then stacked in piles at the edge of the fields. A slender young boy dressed in a long white gown carried a small mandolin in his hand. Music was an important part of Hausa life. I stopped to take pictures of all these cultural features. People were friendly and curious, and did not object to being photographed.

We camped that night in the far northeast corner of Nigeria just before the town of Gambaru, at an elevation of 1,000 feet. During our 3 hours of afternoon travel I added 2 additional species to my trip list of birds, the Ethiopian Swallow* (in the worldwide genus *Hirundo*) and the Black-headed Lapwing* (a handsome, distinctive *Vanellus* characteristic of open, semi-arid areas, with a striking, sandy-brown, black, and white color pattern, red legs, red facial wattle, red base of bill, and thin wispy black crest; I had heard this species vocalizing several night ago). I was also happy to carefully observe a Rufous-tailed Scrub-Robin ("Rufous Warbler" of Bannerman) and to obtain photographs of several N. Carmine Bee-eaters, which were perched in some low acacia trees near our campsite. It was a great finish to another day of adventure.

We left our campsite early the next morning (Apr. 16) and drove from Gambaru SW through the towns of Ngala and Dikwa, which as far as I could tell by our road map were just inside the border of Nigeria, though the customs post for Nigeria was not until Maiduguri. Nigeria was a British colony in April 1960, but would gain its independence in just 6 months. Although Nigeria stretched south to the Gulf of Guinea, Noble and I traveled only in the far north of this country. The climate was dry and the vegetation was interspersed between savanna grassland/woodland and semi-arid scrub ("Sahel").

It was approximately 60 miles from Dikwa to Maiduguri. The day was hot, dry, and windy, with occasional whirlwinds ("dust devils"). The countryside was mostly open grasslands interspersed with millet, sorghum, and wheat fields. The local Hausa people lived in stick houses and tended cattle, donkeys, and an occasional camel. In the fields, men sifted and winnowed grain by hand, pouring it from a gourd bowl held above their head to the ground, and from one bowl to another. The winnowed grain was then sacked and carried by donkeys or camels to the nearest marketplace. In this part of Africa the Hausa had assimilated with the partially nomadic Fulani with which they shared a Berber ancestry. Men were dressed in Muslim skull caps and white or indigo-blue gowns, and women wore colorful wrap-around, single piece "dresses." One woman with red stained teeth had apparently been chewing "betel nut", the sliced seed of the areca palm (*Areca catechu*) which was wrapped in a betel leaf to which lime had been added. This entire mixture was then chewed as a mild stimulant and to induce euphoria. Chewing produced a copious amount of red saliva, which was constantly spit onto the ground (and everywhere else), resulting in disgusting stains. The saliva also stained one's teeth and lips red. In spite of the many adverse health habits, chewing betel nut was widespread in tropical Asia but it was only a very infrequent habit in Africa, and this was our only encounter on this continent.

We arrived in Maiduguri in the middle of the afternoon and checked into the Nigerian customs and immigration offices, where we were issued temporary 7 day visas since we had not managed to obtain longer visas anywhere prior to our arrival. As we headed for Kano, we were astounded to encounter a newly asphalted ("bitumen" or "macadam") smooth surfaced, dust free highway. What a splendid surprise! We would discover that it went 375 miles, all the way to Kano. That night we camped by the roadside about 20 miles west of Maiduguri, at an elevation of 1,000 feet.

We had driven 130 miles for the day, during which time I documented 42 species of birds. Three of these were

Plate 58: Birds along our route in Chad included Black Crowned-Cranes and Northern Carmine Bee-eaters. During the hot middle of the day men gathered in non-working groups in the shade.

trip additions: African Collared-Dove* (*Streptopelia roseogrisea*, also known as the Pink-headed or African Gray dove), Pygmy Sunbird* (a very small plainly colored sunbird confined to the arid savanna belt, with a green head & upperparts, yellow underparts, and greatly elongated central tail feathers), and Beaudouin's Snake-Eagle* (virtually indistinguishable in the field from the Palearctic breeding Short-toed Eagle, which may or may not come to this part of Africa in the winter; Bannerman called this species the Beaudouin's Harrier Eagle). The most numerous species were African Mourning Doves (*Streptopelia decipens*), Namaqua Doves, Ruffs (a Palearctic visitor), Collared Pratincoles, Ethiopian Swallows, and unidentified *Ploceus* weavers in non-breeding plumage. I was delighted to obtain photos of two Abyssinian Rollers with my 300mm lens, as they were perched and flying. (I took very few pictures of birds with my 300 mm. lens because it was too time consuming and cumbersome to retrieve from my shoulder bag and exchange with the wide angle 35mm lens normally on my camera.)

On Apr. 17 (Easter Sunday) we drove west for 305 miles on our new asphalt highway, all the way from our campsite last night to our campsite for tonight, 50 miles E. of Kano. It is probable that this was the most miles we drove in single a day anywhere in our African travels. We could actually travel safely at speeds up to 50 mph (which we never exceeded). The countryside was hot, dry and mostly flat, with grasslands, scrub, and scattered small trees, ecologically situated between the savanna belt and the Sahel belt. Somewhat surprisingly there was a good variety of roadside birds and I kept busy jotting down names in my field notebook (names which would eventually have to be tediously converted to the nomenclature of Clements for the purpose of writing this story). My list included, in part, the following species (none of which were additions to my trip list): Dark Chanting Goshawk, Clapperton's Francolin, White-bellied Bustard, Mottled Spinetail, Abyssinian Roller, White-throated, Green, and Little bee-eaters, Gray and Red-billed hornbills, Viellot's Barbet, Yellow-fronted Tinkerbird, Flappet Lark (in the widespread genus *Mirafra*, 1 of 72 species of larks in Africa), Chestnut-backed Sparrow-Lark, Hueglin's Wheatear, Common (Yellow-crowned) Gonolek (a bushshrike in the genus *Laniarius*), Chestnut-bellied and Long-tailed Glossy starlings (both very common), Yellow-billed Oxpecker, Gray-headed Sparrow, Chestnut-crowned Sparrow-Weaver, Cut-throat (an estrildid finch), Red-billed Firefinch, Black-rumped Waxbill, and Red-cheeked Cordonbleu (another estrildid finch). Not surprisingly, there were many small seed-eating finches in the grasslands and scrub, most of which were common and widespread species throughout Africa.

In the late morning we approached the sizeable town of Potiskum, where there was a large colorful, Sunday market (somewhat reminiscent of those in the Andes of Peru). People (mostly Muslims) were walking into the town carrying goods on their head or following burden-laden donkeys. Among the most common items being transported were gourd bowls, bundles of reed thatch for rooftops, and a wide variety of ceramic pottery. Inside the bustling market people were sitting on the ground with their items for sale, or displaying them in crude wooden stalls. Pancakes were being prepared in a skillet over a small gas stove, and a man was being shaved by a barber with a long, straight-edge razor. The many items being sold or bartered included cotton, yarn, cloth, trousers, dresses, Muslim gowns, robes, & caps, sandals, straw mats, and a wide variety of fruits, grains, meats, and vegetables. A strangely clad non-Muslim woman of unknown ethnicity was carrying a baby on the top of her hips, wrapped in a large piece of plain Gray cloth which was tied around her waist. The woman was barefooted and not wearing any kind of garment above her waist. She was somehow balancing three large orange gourd bowls stacked vertically on top of her head and cushioned by a black cloth. Other people were paying no attention to her and my inquisitiveness as to her tribal affinities went unresolved, although I have a snapshot which perhaps one day will allow me to arrive at an answer.

We continued onward toward Kano on our asphalt highway. The afternoon was sunny and hot but I never-the-less regularly stopped briefly to identify roadside birds in the small marshes and elsewhere. As the sun was settling low over the western horizon ahead of us I found a suitable wide, uninhabited area of scrub for a campsite 50 miles from Kano, still at 1,000 ft. elevation. Some of the birds which I recorded during the afternoon were Black Kite, Great Spotted Cuckoo, Abyssinian Ground-Hornbill (an African icon which was always thrilling to see), Rufous-crowned Roller (a less frequently seen roller), Bateleur (with its amusing, teetering, gliding flight), Senegal

Parrot, Yellow-billed Shrike, Long-crested Eagle, Marabou Stork, White-faced Whistling-Duck (which breeds in both Africa and S. Am.), Green (Red-billed) Woodhoopoe, and African Jacana. My day's total of 82 species was quite satisfying. Only 4 of these were trip additions: Senegal Parrot*, Kordofan Lark*, Sennar Penduline-Tit*, and Common Gonolek*. Today was my last full day of birdwatching in Africa south of the Sahel.

As we neared Kano the next morning, Apr. 18, we drove past donkeys transporting goods and people into the city, heading for the main Kurmi market in the center of the city. Firewood was one of the most important commodities being transported, and as everywhere women carried jugs of water on top of their head.

Kano was the oldest city on the southern edge of the desert in West Africa, having first been settled at least 1,000 years ago. A wall was built around the city in the 12th century, and Kano became a major crossroad in trans-Saharan trade routes, including the slave trade. The Islamic Fulani people (an offshoot of Berbers) conquered the city at the beginning of the 1800's, bringing many slaves with them. In the early 1900's, long after the Atlantic slave trade had ceased, almost half the residents of the city was still slaves who lived around the edges of the city in "slave villages." Kano came under British rule in 1903. After 57 years of British domination, partial independence for Nigeria was scheduled to take place in just 6 more months (in October 1960). Muslims now constituted more than 90% of the city's population, and I photographed a beautiful, shiny new mosque.

Kano was not a major tourist attraction. Goats, sheep, donkeys, chickens, and naked children roamed through the narrow, dirty, and unsanitary streets, among the numerous carts which were being hand pushed everywhere. There were commercial trucks of all sizes and shapes, but very few privately-owned cars. The air was often filled with a disagreeable odor, and vultures everywhere fed on the street side garbage. Two young boys sat on a curb plucking feathers from a guineafowl, and a man was roasting the head of a sheep over a fire pit. Ragged children followed us through the streets, laughing and begging for money. Mud houses were in a constant state of being repaired. The many little shops were closed in the mid-day heat but stayed open far into the evening, operating by oil lamps. The huge Kurmi Market was known for its crafts and textiles from all over the world. A 500 year old dye pit produced the popular indigo-blue cloth worn by the Tuareg people (and many others). Leaves from the *Indigofera tintoria* shrub were soaked in water and fermented, then mixed with lye to produce a liquid violet-blue dye.

Noble and I had several important tasks to accomplish in Kano, the foremost being to repair our 4-wheel drive so we could drive across the Sahara. We discovered that our new front axle which Noble had ordered for us in Stanleyville from the Jeep home office in Toledo, Ohio, was not coming to Kano (where there was no Jeep agency) but was being sent to us instead in Casablanca, Morocco! So Noble set about seeing if there were any repair facilities in Kano which could get our 4-wheel drive operable. I took a day (Apr. 20) to drive 160 miles SW of Kano to the large city of Kaduna, where there was an American Consulate. After almost 15 months of travel through 34 countries, all of the original 15 pages in our passports were completely filled with visas and entry, exit, and other stamps. There was no more room! Fortunately, the problem could be easily solved at any American diplomatic mission by taking our passports there and having more pages added. Twelve more blank pages were added in Kaduna, folded up like an accordion and firmly sealed to one of the original pages. (Before we were completed with our 3-year adventure it was necessary to have pages added on five more occasions, and our passports each finished with 81 pages filled with visas and all kinds of other stamps and permits, from more than 80 different countries and kingdoms!)

In Kano, I mailed a box of butterflies to John G. Williams at the Coryndon Museum in Nairobi, as I had promised him I would do. Sadly, I never heard from him whether he received these or not. Very unfortunately, Noble was unsuccessful at finding any facility in Kano where our 4-wheel drive could be made functional. We both agreed it would be fool hardy to attempt a crossing of the Sahara by ourselves without 4-wheel drive. To do so would have been tempting fate, and beyond all rational reason. Although defeat was not on our agenda, we would have to cancel our dream of driving across the Sahara. There was no viable alternative. With great reluctance and extreme disappointment we resolved ourselves to bypassing the Sahara and driving from Kano west to Dakar, in Senegal on the Atlantic Ocean, and from there catching a freighter to Casablanca where a new axle was being sent. Our spirits descended to an all time low.

Chapter Seventeen – Across the Sahara

On Apr. 22, Noble and I set about finishing our tasks in Kano. One of these was to refuel our 40 gallon gas tank and the two spare 5 gallon jerry cans for our departure for Senegal. We asked directions to a Mobil station, where we would receive free gasoline for advertising their product with the flying red horse logo on the side of our van (placed there in Nairobi). When we arrived at the station we experienced a genuine miracle, far beyond the realm of all probability. There, right in front of us, was a huge, outdated British Bedford army truck outfitted for desert travel being driven by two young British boys, Tony Fleming and Collyn Rivers, who were about our same age. They were preparing to leave tomorrow to drive north across the Sahara to Algiers - - on the same route Noble and I had planned!! (This route, in fact, was the only road across the desert for almost 1,000 miles in any direction). Our joy and surprise could not have been greater. Here was a vehicle which could travel with us and assist us in getting out of soft sand, thus eliminating the necessity for our 4-wheel drive. This 11th hour encounter was beyond all imagination. What were the odds?

This was a return trip for Tony and Collyn. Four months earlier they had traveled this same route from north to south on the way to spending a holiday in Kenya. Now they were on their way back home to Britain. They were both as delighted as we were to have another vehicle with which to travel. Because of the greatly decreased risk of getting lost or stuck in the sand, it was not now necessary for them (or us) to purchase a $1,000 bond from the French government, as was required of all vehicles traveling alone, to pay for search & rescue if needed. All travelers were still required to have written, stamped permits from French authorities in Kano to drive this route. Neither of our two vehicles was equipped with a 2-way radio, and this was long before GPS was available. Their 7 ton, 4-wheel drive van was much heavier and larger in overall dimensions than was our Jeep van, and it was equipped with huge, 11x20 tires which were a big asset in driving across soft sand. It was much better suited for desert travel than was our Jeep. If Noble and I got stuck in sand we could simply attach our winch cable to their Bedford van (if it was within 200 ft. of us) and winch ourselves out, using the power from our own engine. What a godsend they were. Our four-wheel drive was no longer an absolute need for our Sahara adventure. We all celebrated our good fortune with a glass of red wine (from the supply Noble and I always carried with us)!

French colonization in the Sahara began during the early 19th century (1800's). In 1960 the French administered virtually the whole of northwestern Africa (Morocco, Algeria, Tunisia, Mauritania, Mali, Niger, & Chad), all of which were almost entirely in the Sahara Desert. Only a small coastal strip from Morocco to Tunisia was green, separated from the desert by the Atlas Mountains. The Sahara was by far the greatest hot desert in the world, encompassing 4,200,000 sq. mi. of area in which the average annual rainfall was 10" or less. It extended all the way from the Atlantic Ocean 3,400 miles east to the Red Sea, and 1,120 mi. north-south from the Atlas Mountains (in the western half) and Mediterranean Sea (in the eastern half) south to approximately 15°N latitude (in the upper Niger River valley, Lake Chad, and central Sudan). The overall desert covered about 1/3 of the African continent (which was slightly greater than the entire United States)! The sparse human population was largely semi-nomadic, comprised mostly of Berbers in the north and Tuaregs (an offshoot of the Berbers) in the south. Smaller tribal groups were descended from the Berbers, and they all shared languages related to the Amazigh language of the original Berbers. Although Arabs brought Islam to the Sahara in the 6th and 7th centuries, they did not succeed in assimilating these earlier Berber cultures.

The Sahara Desert can be defined by many features. Its surface, contrary to popular thinking, is not mostly sand and dunes (referred to as an "erg" desert) but instead is largely gravel (referred to as a "reg" desert). Barren rocky plateaus (referred to as "hamadas") and scattered rocky mountain ranges with elevations up to 11,270 ft. above sea level comprise much of the Sahara. Stream beds, referred to as "wadis" (or "oueds" in French) are dry for most of the year. The Sahara is overall the hottest desert in the world, with daytime shade temperatures regularly up to 115°F, and it is the 2nd driest desert in the world (to the Atacama Desert in Chile), with some areas receiving virtually no rainfall at all. The only two significant rivers in the entire area are the Niger River along its SW border and the Nile River along its far eastern edge. Only in the higher mountains in winter do temperatures occasionally fall just below freezing. Daily temperatures sometimes fluctuate as much as 70°F between the low at daybreak and the high in the afternoon. The northern and southern fringes of the Sahara, along with the higher elevations, receive most of the rainfall, and it is in these areas where most of the vegetation grows and wildlife survives, as well as in the wadis and around oases where water is near or on the surface of the ground. Elsewhere, there are vast areas without water, vegetation, or wildlife. Wind is an important feature of the desert and sandstorms occur regularly, sometimes in great intensity (particularly in Libya, which is also the hottest region). Much of the rocky landscape has been sculptured by the wind. Although sand constantly blows across the surface of the ground, surprisingly it often does not always obscure such ground features as tire tracks, merely blowing across the top but not filling in the depressions. The highest sand dunes are 500-600 ft. tall.

Despite the sparseness of suitable habitat, a variety of small to large mammals are resident in the Sahara, either locally or widespread. A partial list of such mammals includes the following (with an asterisk indicating that the species is endemic to the Sahara or its immediate edge): Topi, Addax*, Dama Gazelle*, Loder's Gazelle*, Red-fronted Gazelle*, Dorcas Gazelle*, Barbary Sheep*, Common Jackal, Fennec Fox* (see later comments), Small-spotted Genet, Sand Cat, and various rodent species - - gerbils, jerboas, spiny mice, sand rats, & others. (In our journey we would encounter only one of these mammals.) The Dromedary (or one-humped) Camel is not native to Africa, having been brought to the continent from Arabia about the time of Christ, or before. They were used for transport and trade by the Berbers. Camels are probably better adapted for life in a hot, dry desert than any other large mammal. Their hump stores fat which can be broken down into carbon dioxide and "metabolic water", aiding them to maintain water balance without drinking, and they can extract water from the dry thorns they eat and can thus go for active periods of more than 2 weeks without drinking water and eating only thorns. Their "digitigrade" foot posture and specialized toes with broad cutaneous pads provide them with greater support on soft sand than in other ungulate mammals. Camels are thus far superior for desert travel than were the horses previously used by the earliest human inhabitants of the Sahara. Their introduction from Asia enabled the nomadic lifestyle to flourish in north Africa.

Numerous resident African birds, including 18 endemic species, are confined to the Sahara Desert or its vicinity. Many of these also extend eastwards throughout the deserts of Arabia, the Middle East, and beyond. Other vertebrates found in the desert are such reptiles as agamid lizards, varanid lizards, and venomous horned vipers. Among many invertebrate animals are scorpions, spiders, ants, termites, locusts, and scarab beetles (considered sacred in Egypt). Characteristic plants are drought resistant grasses, saltbushes, artemisias, small thorny shrubs, and acacia trees. The Sahara was much wetter and greener 8,000-10,000 years ago during the last ice age than it is today, with rivers, lakes, and an abundance of vegetation and wildlife. Ancient cave paintings dated at 6,000BC in the Hoggar Mountains of Algeria depict rhinos, elephants, giraffes, jackals, and crocodiles.

Our south to north vehicle track across the middle of the Sahara between Kano and Algiers went through the following towns and French Foreign Legion outposts in the following order: Kano, Zinder, Agadez, In Abanyarit, In Guezzam, Tamanrasset, Arak, In Salah, El Golea, Ghardaia, Laghouat, Djelfa, and Algiers. We had a permit issued to us from the French foreign office in Kano which allowed us to travel this route, and we were obligated to check in with every French Foreign Legion post we came to en route. We told the commandant at each post how long we predicted our travel would take to reach the next post, and if we did not arrive within

Plate 59: Kano was the staging point for our 2,300 mile, 19 day adventure northward across the middle of the West Sahara Desert to Algiers. We joined two British boys driving a behemoth, desert-equipped, army surplus, Bedford van.

48 hours of our anticipated time the French government was obligated to begin a search for us. At each post we were able to obtain water and purchase gasoline. These posts were approximately 300 miles apart, and with a total fuel capacity of 50 U.S. gallons we were not concerned about running out of gas (even at 10 mpg or less). Both Noble and I had long dreamed about such an adventure, and we could hardly have been more excited when we started out. Without our 4-wheel drive we would not have undertaken such a folly by ourselves. I did almost all the driving. The desert looming ahead would be one of the greatest driving challenges of our entire 3-year adventure.

Tony and Collyn had a *Trans-African Highways* book which had a short chapter describing the route we were going to be following, including driving recommendations. The book stated that a 200-250 mile stretch of sandy desert between Agadez and Tamanrasset was without a well-defined track and would be the most difficult section of the entire route. A route across this sand was marked every kilometer by either a stone cairn 4 or 5 feet high or by a narrow 6 feet high steel fence post. However, the safest way between these route markers was always changing because of the shifting wind-blown sand, so that a driver had to choose his own way between and around the markers to avoid the softer, non-navigable patches of soft sand, driving around them as far as necessary.

So what advice did our highway guide give the driver? It stated: "Agadez – Tamanrasset: soft sand is encountered in the southern half of this sector, which should be negotiated in the early morning when the sand is firmer than at other times. Surfaces are either soft or gravelly sand, with several transverse channels, for which a careful watch is imperative. This portion is probably the most difficult of the whole desert crossing. The route is constantly changing, owing to sandstorms; the best method is to follow (but do not get into) a fresh set of lorry tracks regardless of any detour, which at times may seem strange; it will be found that the track keeps touching the line of cairns at intervals but avoids the worst patches of soft sand (do not follow the line of cairns if there are no fresh tracks that way). Sand mats and shovels must be carried; reconnoiter freely ahead and rush the soft stretches."

Our trans-Sahara adventure began on the morning of Apr. 23, 1960, when the four of us mid-20-year-old boys climbed into the two cabs of our very different camping vans and headed north from Kano to the nearby border between British ruled Nigeria and French ruled Niger, 75 miles away. I felt a surge of anticipation and excitement. Noble and I had looked forward to this moment ever since our arrival in Capetown just over 7 months ago. The day had finally arrived. The weather was hot and sunny but the beginning of the brief rainy season was only 1 or 2 weeks away. Our gravel road varied from moderately smooth to quite rough, and we drove at speeds between 15 and 25 mph. We were in no hurry and it was imperative to try and avoid any further mechanical breakdowns of our Jeep. Donkeys loaded with firewood were herded along the road, and a woman balancing four large ceramic jugs vertically on top of her head walked in an upright position toward the nearest little marketing stall. (I never ceased to be amazed at the ability of native women to accomplish this feat, which was obviously learned at a very early age.)

Formalities at the Nigerian-Niger border were friendly and efficient. We were now in French-speaking Africa once again, as we would be for all our remaining time in Africa. Niger was Africa's 6th largest country, almost twice the size of Texas, but among the poorest and least populated (except for the capital city of Niamey on the Niger River in the far southwest of the country). Eighty-five to 90% of the total area was in the Sahara Desert, all but the narrow Sahel belt along the southern edge of the country through which we were now traveling. We passed close by the tiny village of Kantche which was surrounded by a thatched fence, inside of which were small thatched houses. Further along we drove through the town of Zinder without stopping, and then onward for another 20 miles to a campsite for our 1st night, along the road in a dry, open, sparsely grassed area in the Sahel at an elevation of 1,500 feet. During the day I recorded a total of 24 species of birds. These included 8 Abdim's (White-bellied) Storks, 75 Hooded Vultures, 20 Pied Crows, 30 Cattle Egrets, 10 Black Kites, 12 Speckle-fronted Weavers (a small, unusual-looking ploceid), 2 Green Bee-eaters, 2 Black Scimitar-bills, 4 Black-crowned

(White-fronted) Sparrow-Larks*, 2 Northern Crombecs (a tiny, almost tail-less sylviid warbler), 4 Beautiful Sunbirds, 2 Black-crowned Tchagras (in the bushshrike family, Malaconotidae), and I Senegal Batis* (a small, sexually dimorphic species of flycatcher in the family Platysteiridae).

We ate a quick breakfast the next morning (Apr. 24) and were underway just after sunrise, at 7:00 am, toward the town of Agadez. Soon we came to a group of 2 or 3 dozen very dark-skinned, native men, women, and children, all gathered with their livestock around a water well in the open countryside. It was the first of several such wells we would encounter during the morning. The French government had built these as a water source for the local inhabitants. The wells were a circular metal or concrete lined tube which went down 10-40 ft. deep, and were in much use. The process of getting water was slow and tedious, with ropes, pulleys, and goatskin water bags which were typically pulled up by a bullock straddled by a young boy. The bottom end of a rope was tied to a water bag and the top end of the rope came up out of the well and over a wooden pulley, and was tied around a large bullock which was straddled by a small boy, who then led the bullock directly away from the well, pulling the water bag up as it did so. Four water bags could be pulled up at about the same time, with the 4 bullocks all walking away from the well like spokes on a wheel. We were fascinated as we watched this process. The well was surrounded by dozens of people along with all their livestock - - cattle, donkeys, camels, and goats. Two people carried a full water bag away from the well and tipped it into very large half gourd containers or ceramic pots and jugs, from which the people and animals all drank. Two donkeys sometimes drank at the same time from a half gourd container! This seemed to me to be a very primitive, inefficient way of getting water, and I thought that surely the method could be improved. Everyone was covered in dust, and when the wind blew they turned their heads away and shut their eyes. It was an all day process for almost everyone. The spectacle made for many photographs.

At this first well of the day for us the people were predominantly Hausa, with a few Fulani. Some of the children ran around naked. The semi-nomadic Fulani men were conspicuous by their peculiar dress with sandals, a tall, pointed straw hat adorned with a silver spire on top, a loose-fitting sleeveless black shirt, and a crude, rather stiff animal hide wrapped around their waist. The saddles for their camels were remarkable by having a tall "saddle horn" shaped like a Christian cross. We four travelers all filled up our own canvas water bags, not knowing when our next opportunity might be. These were hung over the sides and back of our vans where the moving air could help keep the water cool. The desert was just ahead. Life was an adventure!

Not long after we left the well our road angled to the northwest at the tiny village of Tanout. Shortly thereafter we encountered our first small group of nomadic Tuaregs, moving slowly across the desert on 5 or 6 camels carrying goods for trade, as they had done for centuries. These were all men with their faces covered by a turquoise blue veil, as was their tradition. (Contrary to most Muslim customs, women did not wear a veil.) In the southern regions of the Sahara, particularly in Niger and Algeria, the Tuareg people were the dominant cultural trading group, known as the "blue people" because of the blue coloring of the men's clothing, including the head covering and veil which left only their eyes exposed. (Not everywhere did all Tuaregs wear exactly these same traditional garments and colors.)

Ethnically, Tuaregs were closely related to Berbers, with a very similar language, and were not related to Arabs even though they had incorporated Islam into their culture. They were the dominant camel caravan traders in the southern parts of the Sahara, although their trade routes covered almost all of North Africa from the Mediterranean southward. Not only were they traders but they were once also well known and feared as pirates and marauders, widely attacking and robbing villages and other camel caravans, of their goods and people. They traded in such items as gold, salt, leather, kola nuts, cotton, animal hides, textiles, jewelry, henna, and slaves from sub-Saharan Africa. Henna was a reddish-brown dye used by both men and women for staining their hair, beards, fingernails, toes, and hands, as well for dyeing fabric, silk, wool, and leather. Kola nuts were a very important trading item from tropical West African rainforests, a caffeine containing seed chewed as a stimulant, and in more modern times as an ingredient in coca cola! During the slave trade period from the 10th until the

Plate 60: Our road initially was straight and well defined. At a desert well, leather waterbags were lowered by rope and pulley, then when full pulled up by a bullock which was led (or ridden) by a small boy.

Plate 61: Water was poured into clay pots (for people) or half gourd bowls (for livestock) to drink from. Large numbers of people, including a few Fulani men with peaked hats and camels, waited long hours for water.

end of the 19th century over 9 million slaves were transported from the south into and across the Sahara, as many as 10,000 in a year. I was very excited by our first encounter with the Tuaregs, icons of the desert. This small group did not alter their pace across the semi-arid countryside as I stopped to photograph them, and appeared not to object to my camera.

That night (Apr. 24 and day 2 of our Sahara adventure) I chose a campsite several hundred yards off the road among some small acacia trees at a point 110 miles south of Agadez and an elevation of 1,500 ft. above sea level. As the lead vehicle in our 2 vehicle caravan, I was allowed to choose our campsites. We were at the northern edge of the Sahel, at the gradual boundary with the Sahara Desert. Among 48 species of birds which I recorded for the day were 5 trip additions: Singing Bush-Lark*, Quail-plover* (a sandy-brown "buttonquail" with strikingly patterned black & white wings in flight, an uncommon & nomadic bird confined to the Sahel, in the monotypic genus *Ortyxelos*), Scissor-tailed Kite* (reminiscent of a Black-shouldered Kite with pale Gray coloration and black carpals but with a deeply forked tail; endemic to the Sahel), Rueppell's Griffon* (Vulture), and Cricket Longtail* ("Scaly-fronted Warbler"; a cisticolid warbler, rufous-brown above with a streaked black & white crown and a long white-tipped black tail; another Sahel endemic). Other birds of interest were the Chestnut-bellied Sandgrouse, Chestnut-backed Sparrow-Lark, White-billed Buffalo-weaver, Lappet-faced Vulture, African Silverbill (an estrildid finch), Helmeted Guineafowl, Black-bellied Bustard, Spotted Thick-knee, and Rose-ringed Parakeet (with a disjunct distribution between N. Africa & Asia).

I arose at daybreak the next morning (on Apr. 25, day 3 of our Saharan adventure) and while the others were still not up I went walking by myself in the open acacia bush. Soon after I started out I was greatly surprised to discover a Small-spotted (Common) Genet (*Genetta genetta*, family Viverridae) sleeping in the crotch of an acacia tree about 6 ft. above the ground. It was very reminiscent in overall size, appearance, and behavior, of the Ringtail (*Bassariscus astutus*, family Procyonidae) which lives in the SW deserts of the United States. Both species have a tail with conspicuous, wide, black & white bands. These 2 species are an excellent example of the evolutionary phenomenon referred to by ecologists as "convergence", where 2 geographically isolated, taxonomically unrelated species living in the same kind of habitat share similar morphological appearances and behavior, such as food habits. Such species are described as "ecological equivalents." I was enthralled by the find.

On the spur of the moment I decided I would like to collect this animal and save its pelt as a souvenir of my Saharan adventure. So I walked back to the Jeep and retrieved the .22 caliber pistol I had "borrowed" from Jean Roumanteau when we parted company in Bolivia. Returning to the tree where the genet was still sleeping I "collected" it (not without a little remorse). Having heard the shot, Noble arrived on the scene and took a picture of me holding the genet by its tail in one hand and the pistol in my other hand. I had prepared mammal specimens for the Museum of Zoology at the University of Michigan as an undergraduate student, and it required only a very few minutes for me to remove the skin, in one piece. Once again, at this very moment, a slender native tribesman appeared from out of nowhere and with sign language asked if he could have the carcass to take home and eat (just as happened with the zebra I shot in Tanganyika). Of course I said yes, as I had no desire to eat it myself. I wondered how such people always seemed to be present in what appeared to me to be the middle of uninhabited bush, waiting for someone to shoot anything they could take home and eat! I salted the pelt, stretched it out flat, and then nailed it to the flat rooftop of our van, where it dried out in only a few days. The pelt was about 2½ ft. long. I took another picture. (Eventually I had the skin tanned and then hung it on the wall of my den for many years before, inevitably, vermin finally destroyed it, as they also did with my zebra skin.) The event was another memory for me of the African bush.

For breakfast, Collyn and Tony prepared porridge for all of us to eat, as they did every morning. They had several cases with them of Settler brand, "U-like-me" porridge! We were then off again, heading north for Agadez - - with the dreaded sand not far beyond. The landscape was dry, flat, sparse grassland with occasional bushes and small trees, at 16°N latitude in the ecotone between the Sahel and the "true" desert (where the average annual rainfall was less than 10"). The road was narrow and rough clay or gravel but posed no problem for navigation, allowing me to drive between 15 and 25 mph. We passed another well, again surrounded by a large group of semi-nomads watering their livestock. It was the last such well we would encounter. We had now entered the real Sahara, which stretched almost 1,500 road miles ahead of us. It was an exciting challenge. What was our motive for such a risky undertaking? My father's remark once again came to mind - - "you must be crazy."

In the middle of the afternoon we were surprised to encounter a 250 ft. high cliff stretching across our highway from west to east. It was the "Falaise de Tiguidit", the southern face of the plateau on which Agadez was situated, 50 miles beyond. Our road ascended the cliff and then proceeded on the top of the plateau. We drove for 30 miles before stopping for the night, at an elevation of 1,750 feet. During our day's travel, I documented 53 species of birds, including 9 trip additions, 5 of which were larks (Appendix B). To help me identify the larks I sketched their tail patterns in my field notebook. Often, they went unidentified. Remarkably, during our last 2 days of travel along a 200 mile stretch of road between Zinder and Agadez, I identified nine different species of larks. These were the Singing Bushlark, Chestnut-backed Sparrow-Lark, Kordofan Lark, Black-crowned Sparrow-Lark, Dunn's Lark*, Desert Lark*, Rusty Lark*, Bar-tailed Lark*, and Greater Hoopoe-Lark* (a delightful bird, sandy in color with a conspicuous black & white wing pattern in flight, with long legs and a relatively long curved bill, which ran along the ground like a courser). My primary aid for lark identification was *Birds of Sudan* by Cave & Macdonald, 1955.

Other birds which I saw today included 4 Black-winged Lapwings (only my 2nd record of this species, and my last), 1 Quail-plover (also my 2nd and last record), 14 Green Bee-eaters, 5 Green (Red-billed) Woodhoopoes, 1 Scissor-tailed Kite, and 30 Helmeted Guineafowls. Olivaceous Warblers (a Palearctic migrant) were singing continuously with a not very pleasant squeaky warble. My field notes for this date made the following general comments about desert birds. "Desert birds are generally active only in the early morning, during the first 2 hours after sunrise, and again in the late afternoon one hour before sunset. At other times they sit quietly in the shade, concealed in whatever vegetation is available and can be very difficult to locate. One can walk within a few feet of a tree containing several dozen birds and not see a single one. To keep cool, birds held their wings slightly out from their body and panted with their mouth wide open."

I got up early the next morning (at dawn on Apr. 26, day 4) and pursued birds around our campsite for 2 hours, before others arose. The most numerous birds were 200 Pied Crows, followed by 25 individuals each of Black-crowned Sparrow-Larks and Red-billed Firefinches. Six species were trip additions: Egyptian Vulture* (a distinctive, small, black & white vulture occurring also in Europe and Asia), Arabian Bustard*, Barn Owl (a species of worldwide distribution but my first sighting on our adventure), Brown-necked Raven* (a desert bird ranging eastward across southern Asia), Fulvous Chatterer* (genus *Turdoides*), and White-tailed Wheatear* (*Oenanthe leucopyga* of Clements. This latter species was also called White-rumped Black Wheatear, Black Wheatear, White-rumped Black Chat, and White-rumped Wheatear by various other authors, each with a different scientific name, all of which caused me considerable confusion and frustration in my attempts to identify this species from my field notes, and assign a name to it from Clements)

Our 2 vans arrived in Agadez about 9:00 o'clock that morning. This well established city was located in central Niger at the southern end of the Air Mountains at an elevation of 1,750 feet. It was a provincial capital and an important trade and market center, a crossroad on camel caravan trade routes. Its principal mosque, the "Grand Mosque", was first built in the 1500's and later rebuilt in the 1800's. Hausa, Fulani, and Tuareg ethnic groups were all represented, and there were still a few dark-skinned slaves from south of the desert. Along the unpaved city streets women constantly carried water in jugs on their head, back and forth from the city wells to and from their clay houses. Both donkeys and camels carried goods and people in and out of the city and to and from the bustling open marketplace, where vendors sat on the ground or inside open street side stalls and shops. A myriad of items of all kinds were available for sale or bartering - - silver, jewelry, hats, sandals, clothing, camel saddles and whips, glass bottles of all sizes and shapes, pottery, leather goods, and a great variety of fruits, meats, and vegetables. There was an open air meat market where goats were being slaughtered and chopped into

Plate 62: On our second night we camped in the open scrub of the Sahel. The next morning I shot a Small-spotted Genet and saved its pelt as a souvenir. Nearby, a man was making thatched roofs and walls for a house.

Plate 63: Early in our travel we encountered our first Tuareg nomads on their camels. These traditional trading people of the Sahara adhered to their ancient customs, including pirating villages. Soon we arrived at the small town of Agadez.

pieces with machetes, and then the various parts were hung in the open air to be sold to those passing residents who could ignore the flies. We four "tourists" took advantage of the many photo opportunities. The scenes were like a step backward in time. There was a small adobe mosque which had been newly painted on the outside in white and henna colored designs. A few of the resident men were dressed in traditional white Muslim caps and long white gowns. Life changed very slowly in this isolated desert community. It was our last town of any size in the southern Sahara. It wasn't until late afternoon that we finally departed, and continued northward toward our next destination, the foreign legion outpost of In Abanyarit, 150 miles away. We drove only an hour before stopping for the night.

We were still in Niger all the next day (Apr 27), day 5 of our Sahara transit, as we traversed the desert all the way to In Abanyarit. The countryside was almost entirely flat, barren brown sand and rock with sparse grass and scattered thorny trees, between 1,000 and 1,500 ft. elevation. Areas of dark Gray dried mud were an indication that a little rain sometimes fell. The road surface was often rough and our two vans traveled at speeds between 15 and 25 mph most of the time, with Noble and me in the lead. Soft sand was not yet a problem. We encountered no other vehicles traveling either north or south. To the far east of our road we could see the distant Air Mountains, with peaks up to 6,500 ft. above sea level. These mountains, confined to the Niger, were one of the more significant mountain ranges in the Sahara, and one of the few regions with enough water and vegetation to support a good variety of wildlife. Our trans-Saharan road did not pass through them.

In Abanyarit was a French Foreign Legion site for gasoline and water, with a few resident houses and people. The French word "In" (also sometimes spelled "Ain"), indicated a water source - - a well or an oasis. We arrived late in the afternoon and filled up with both gasoline and water. We were about to continue on our way when a group of 6 Tuareg men approached us and asked if they could ride with us, all the way to Tamanrasset 400 miles up the road, and at least 3 days of travel. Each of the men was carrying a rawhide camel saddle and whip, and they wanted to buy a camel there. Collyn was able to converse with one of them in French. They said they would sleep on the ground outside our vehicles, on straw mats, and they had their own food, water, and firewood with them. After much discussion between the four of us we decided we could offer 2 of them a ride. There was room in the big Bedford van for all their gear, and one each could ride inside our two vans (or on top if they preferred). The group was reasonably satisfied with this decision, and chose which two would go with us. One of these spoke French, having been in the French Foreign Legion at one time, and he proudly showed us a tattered card proving this fact. They wore a very long scarf which covered the top of their head and the bottom of their face, and wrapped around their neck. It left them only a narrow slit to look through. A long, one piece, loose-fitting, half-sleeved, blue or white "gown" hung from their shoulders almost all the way to the ground. They wore a short or long-sleeved, thin cotton shirt under their gown. Open, wide leather sandals were worn as footwear. Each had two prominent religious articles hanging around their neck in front of them, on narrow leather straps to below their waist. One of these was a leather bound Moslem prayer booklet and the other was a decorative cloth of sacred or ritual significance.

The sun was setting as we drove out of In Abanyarit, and we soon found a campsite for the night. As they had said they would, the 2 hitchhikers spread out two straw mats next to each other on the ground outside our vehicles, one for each of them to sleep on. They made a small fire with some of their firewood so they could boil water for tea, and soon went to bed, having already said their evening prayers to Allah at sunset, kneeling on the ground, bowing, and facing to the east toward Mecca. They each had several blankets with them to protect from the cold nighttime temperatures, which could be down into the 40's before morning on a cloudless night. During the day I identified 17 species of birds, but only the Spotted Sandgrouse* was a trip addition. It was one of the slender species with a long pointed tail. Only females had black spots (on both her upperparts and breast). Sandgrouse were one of the most characteristic birds of the Sahara, where they were almost always seen as they flew to and from water early in the morning or late in the afternoon. Other birds recorded today were 200 Pied Crows (a ubiquitous species which was found widely throughout most of Africa, and was at the very northern limit of its range here) and the astounding number of 750 Black-crowned (White-fronted) Sparrow-Larks, which occurred in flocks everywhere in our travel today. It had been a long, eventful day, and the four of us "tourists" were happy to crawl into our warm sleeping bags inside our two vans.

April 28 (day 6) was a day of memories. We finally reached the beginning of the 200-250 mi. stretch of our trans-Saharan route that was entirely sand, with no discernible road for most of the way. This large area was located between In Abanyarit (on the south side) and the Hoggar Plateau (on the north side). As usual, Tony and Collyn made us porridge for breakfast before we started out on our day's adventure. Our route across the sand was marked every kilometer with either a 5 feet high stone cairn or a 7 foot high narrow iron post (neither of which was visible very far away). However, it was necessary for us to find our own way around the soft sand patches as best we could, which sometimes required long meandering detours of several miles or more, during which we had no markers in sight at all. A major difficulty was that because of the constantly shifting wind-blown sand the non-navigable patches of soft sand were always changing. Our guide book warned us that travelers who got lost in the desert (or got permanently stuck in the sand) did so because they could not remember which side they were on from the line of markers. (Remember, this was before the days of GPS!) It was by far the greatest and most dangerous challenge we would face in our entire desert crossing. Noble and I would be the lead van in our two-vehicle convoy. Noble would help me look for soft patches of sand ahead of us while I tried to steer around them. Our two Tuareg hitchhikers who were riding on the top front of our vans provided us no assistance of any kind.

I tried to remember all the traveling advice we had read or been told. Drive in the cool of the morning when the sand was more firm. Deflate your tires to about 15 or 20 lbs. of air pressure (to flatten them out and provide your tires with more surface contact area with the sand). Keep constant vigilance ahead of you for sand which was slightly paler in color, which indicated softer patches of sand. Speed up as you enter a soft patch of sand if you must travel across it. Make all turns in the sand as wide as possible. DO NOT STOP OR SLOW DOWN IN SOFT SAND. So off we went, with our Jeep in the lead, traveling about 20 mph in 2nd gear (necessary because of our deflated tires). I depended on Noble to help me see soft sand patches ahead of us before we got so close that we could no longer avoid them. This day and tomorrow would be the most adventuresome and hazardous part of our drive across the Sahara. (Without our 4-wheel drive this section of road would have been prohibitive if we were traveling alone, without the Bedford truck to provide us with an anchor for our winch cable if it were needed.) Luckily, it was a mostly overcast day and somewhat cooler than normal. We were diligent in remembering which side of the marker line we were on, although the markers were often out of our view for extended periods of time, and we never became misplaced or confused about where we were, relative to the markers. It was challenging and almost fun!

In 3 hours of driving I got stuck in small sand patches (200-400 yards across) on only 4 or 5 occasions. Usually I was half way across, or more, before coming to a halt. The British Bedford was following us at a distance far enough behind so they could see us if we got stuck, in time for them to find a way around the soft patch and position their van ahead of us on firm sand, within reach of our 200 foot cable. It thus required no more than 15 or 20 minutes to winch us forward and out of the soft patch. (Our winch was operated by the engine of our Jeep.) Only on one occasion could the Bedford not get within reach of our winch cable, thus necessitating that we use sand mats and shovels for almost 2 hours to slowly move forward, bit by bit, until we were within 200 ft. of the Bedford. Our two hitchhikers did not offer us any assistance.

By noon we reached the international boundary between Algeria (to the north) and Niger (to the south), at a latitude just above 19°N and a longitude of 5°E. We were at an elevation of 1,500 feet, heading north. In Guez-zam was only 10 miles north of us, inside Algeria. The desert here was almost devoid of vegetation and I saw only 2 birds all morning, a single Black-crowned Sparrow-Lark and a single Greater Hoopoe-Lark. I stopped to take a picture of the border sign and our 2 hitchhikers climbed down to say their noonday prayers, facing Mecca as always. Algeria was the largest country in all of Africa (almost 20% bigger than Alaska) and extended south from the Mediterranean Sea more than 1,250 miles into the Sahara Desert. Our point of entry was at the extreme southern tip of the country.

Very soon after entering Algeria we were greatly amazed to encounter another "tourist" vehicle along our route. It was a small, air-cooled French Citroen van with 2 young Frenchmen (maybe in their early 30's). They were heading north, as we were, and were stuck in a small patch of sand. They were cursing and sweating as they shoveled slowly out of the sand, placing their 10 ft. long aluminum sand tracks in front of their wheels as they went. It was a very slow process. Of course they were as surprised to see us as we were to see them. Neither of us had seen any other vehicle in our travel across the desert. The back of their van was just within the reach of our winch cable so it was no trouble for us to attach our cable to the back of their van and pull them out backward to where we were, on firm sand. Typically of some Frenchmen they barely managed to say "merci", but they did ask if they could then travel with us as a 3rd member of our convoy. Of course we said yes. However, they were very impatient and soon decided we were traveling too slowly, so they speeded up and before very long disappeared over the horizon ahead of us, and we never saw them again. It was the only non-commercial vehicle which the four of us encountered anywhere in the entire 2½ weeks of our Sahara Desert crossing.

Tony and Collyn decided they would like to be the lead vehicle for awhile. However, this turned out to be a mistake because within a few miles they failed to see a patch of soft sand ahead of them and became bogged in the middle of it half way across. Their enormous van was too heavy for our little winch to pull, so the four of us once again had to spend 2 hours shoveling and sweating and sand-matting to get their van out of the sand. As previously, neither of our 2 hitchhikers offered to help. (Perhaps such work violated some aspect of their culture.)

About mid-afternoon we reached the French Foreign Legion post at the small desert settlement of In Guezzam. Here we had our passports stamped for entry into Algeria, and we purchased gasoline at a price of 85 US cents a gallon, the highest price we would pay anywhere in the world during our 3 year adventure! (After all, it had to be trucked in over a thousand miles on a very primitive desert road.) The army sergeant in charge of the post said he had been there almost 10 years. When I asked him how many tourist vehicles came through every year, going either north or south, he responded (in French and interpreted by Collyn) "there are lots more than there used to be; last year there were 15!" The sergeant told us if we wanted to stay until evening we could have a steak and fries dinner for the equivalent of $2.50 U.S. each. We rousingly said yes, and brought out some of our own red wine to go with it. After all, when would we be dining in the middle of the Sahara Desert again, having traveled there by road?

That gave me 2 hours to search for birds in the trees and shrubs around the settlement, which were watered from a well. It was spring migration time and a variety of Palearctic breeding birds, mostly passerines, were on their way back north to Europe from their wintering locations in sub-Saharan Africa. They congregated at any green oasis they passed over, such as the one where we were at In Guezzam. I recorded 15 species and counted the following numbers of individuals: 60 Eurasian Golden Orioles, 2 Barn Swallows, 2 Spotted Flycatchers, 1 kestrel (not identified to species), 1 Rufous-tailed Scrub-Robin (*Cercotrichas galactotes*, "Rufous Warbler" in my field notes of 1960), 5 Greater Whitethroats, 8 Garden Warblers, 1 Black-eared Wheatear*, 1 Olivaceous Warbler (not identified as Western or Eastern, which were not separated in 1960), 3 Melodious Warblers, 2 Willow Warblers, 2 European Bee-eaters, 1 Western Bonelli's Warbler* (identified from the Eastern only by its range), 2 Brown-necked Ravens, and 1 Thrush Nightingale.

After our wonderful steak dinner we departed In Guezzam, just before sunset. There were camel tracks and gazelle tracks in the sand, and a very dried camel carcass which probably had been lying in the desert for many years, where there were few organisms of decay or predators. The camel's skeleton was intact and was entirely covered by its skin, which was now hard and leathery. We stopped for the night a short distance up the road. Once again there was a colorful orange band on the western horizon. The night was still and cloudless. We were alone on the planet. It was a wonderful feeling.

We got underway at daybreak early the next morning, Apr. 29 (day 7 of our 19 day Sahara adventure). We wanted to take advantage of the cooler morning temperatures when the sand was more firm for the continuation of our travel across the sandy desert, where our route was still marked only by stone cairns or narrow iron posts. We were at an elevation of 1,250 ft. above sea level. Noble and I continued to be in the lead, with Noble diligently looking ahead for treacherous, soft patches of sand. Our tires were under-inflated as we drove 15-20 mph, undulating and weaving around the soft sand, keeping in mind at all times where the cairns or posts were which marked our route. There was a thrill of excitement in the air. Tamanrasset was 250 miles ahead of us. There were no birds in sight anywhere, not even any sandgrouse flying to water.

Vehicle tracks were widespread in the sand, giving the false conclusion that many vehicles were crossing the desert. However, one of the most inexplicable features of the desert was the fact that vehicle tracks sometimes remained visible in the sand for a great many years after their origin, the desert sand simply blowing over them but not filling them in with sand. I have not yet found an explanation for this phenomenon. Every now and again along our track there was the shiny skeleton of a car which had perished en route, and had become polished by many years of wind-blown sand. One such carcass which we stopped to examine had a nameplate on it indicating that it was a 1925 Rolls Royce! (The fate of the driver can be left up to one's imagination.) Several times again I got stuck in a patch of soft sand, but the behemoth Bedford truck was always available to help us out, as an anchor for our winch.

Inevitably, however, I once became stuck in the middle of a much larger patch of sand, 400 or 500 hundred yards from the far side of the patch, well out of reach of our winch cable. The Bedford was too far away to initially be of any assistance. (We could not winch ourselves out backwards because the winch was on the front end of our Jeep.) It was necessary, once again, to get out the shovels and sand mats. It then required over two hours of sweating and toiling by all four of us, in the rapidly increasing heat of the morning, before our cable would finally reach the Bedford. (We could drive forward only very short distances at a time, the length of a sand mat between our front and back wheels.) For the moment we were not having any fun! When our Jeep was once again on firm sand we all four drank 1-2 quarts of water each. As before, our two hitchhikers still did not offer us any assistance. The experience was a permanent reminder of the risks involved with trans-Saharan travel in 1960. Without Tony and Collyn, and without 4-wheel drive or a very strong anchor for our winch cable, Noble and I would have had no chance at all to conquer the desert on our own.

By early afternoon we had navigated across almost all the 200-250 mile wide region of roadless sandy desert. We were back on a firm surfaced well defined track, or road. The worst of our Sahara travel was now behind us. Our attention could once again return to enjoying the vastness, beauty, and loneliness of the desert. The flat landscape was broken only by small pinnacles of black granite or lava rocks. Granite or basalt cliffs could be seen in the distance. At one point where we all stopped for a drink of water I took out my camera to photograph the scenery, after which I laid it on the back bumper of our Jeep van while I drank. When we had finished drinking, all of us climbed back into the vans and continued on our way, northward toward Tamanrasset. An hour or so later, 15 or 20 miles up the road, I stopped for another photograph of the desert and was surprised not to find my camera at its customary spot on the shelf behind my driver's seat. A horrifying thought suddenly came to my mind - - I had laid it on top of the back bumper at our last water stop. What were the odds that it would still be there after we had bounced and jostled up and down for the past 20 miles? I stopped and got out to look, fearing the worst. There it was, right where I had laid it down! It was yet another life in the nine lives of my camera. How many more would there be?

We stopped for the night just before sunset, about 100 miles south of Tamanrasset in an area of vast, open desert at 2,000 ft., a slightly higher elevation than this morning. Once again the sun sank below the horizon in a brilliant narrow band of orange. I had not seen a single bird all day long, in 150 miles of travel! There was no wind that evening and the absolute stillness was an indescribable feeling, soothing to one's soul. I was enjoying the moment when far in the distance, up the road, I heard the faint sound of what I interpreted to be a slowly

Plate 64: At the tiny French Foreign Legion outpost of In Abanyarit, our convoy of two vans picked up 2 Tuareg hitchhikers for 5 days of travel to Tamanrasset (where they wanted to buy camels). Our way across the featureless desert was sometimes marked by a narrow, tall steel post every kilometer.

Plate 65: The border between Niger and Algeria was 1/3 of the distance from Kano to Algiers. We ate dinner at the French Foreign Legion post of In Guezzam, and then camped in the desert.

approaching vehicle. It gradually became louder and I strained my eyes to see any headlights, but in spite of the openness of the desert I saw nothing. Perhaps 15 minutes after I first heard the sound there appeared on the far horizon ahead of us a faint glimmer of light, from the same direction the sound was coming. It, too, gradually became more apparent. Eventually, perhaps 30 or more minutes after I first detected the sound, a commercial truck drove past us, about ½ mile off to one side of our campsite, and without stopping it continued slowly southward, toward In Guezzam. We had no lights on anywhere at our campsite, and the truck driver almost certainly was unaware of our presence. I was fascinated by this event and attempted to understand why I heard the vehicle long before I saw it. My best conclusion, in the absence of any solid information, was that perhaps sound followed around the curvature of the earth, whereas light did not. Regardless of the explanation, it was another permanent memory for me of our Sahara adventure. The truck was the 2nd vehicle we had encountered in the desert (and our last).

At day break the next morning (Apr. 30, the 8th day of our travel) the temperature was a chilly 45°. The 2 hitchhikers outside our vans said their morning prayers, as they always did, then built a small fire over which to warm their hands and boil water for their morning tea. Noble and I ate our usual hot porridge with Tony and Collyn inside their Bedford truck. Tamanrasset was 100 miles northward, across a mostly featureless landscape. Never-the-less, rock structures were at times fascinating and picturesque, and I occasionally stopped for a picture or two. For a 2nd straight day I saw no birds at all. I was intrigued by this fact since in my navy days, even in the middle of the vast Pacific Ocean if I stood on deck and looked for birds as long as 2 hours, I always saw one or more albatrosses, shearwaters, tropicbirds, or other birds. I concluded that in the ocean there were always food organisms for birds available at the water's surface or just below, but in the dry sands of the Sahara there were large areas without any attainable food or water for most birds. We drove only 65 miles on this date and stopped to camp early, when we were still 35 miles from Tamanrasset. We were at an elevation of 3,000 feet, having gained 1,000 ft. during the day. The British boys wanted to work on a minor mechanical problem with their van. Noble wrote in his journal and I went walking in the hopes of finding a bird, which I did not. Our 2 hitchhikers said their prayers, drank tea, and loafed on their mats. This would be their last day with us. Although Collyn could speak French with one of them, they were reluctant to engage us in conversation, and they always ate their own food rather than any of ours, although we often offered it to them.

May 1st (our 9th day of desert travel) was an exciting day. We finally arrived at Tamanrasset! During the last 15 miles our sandy and rocky road gained 1,500 feet in elevation, up to 4,500 ft. on top of the Hoggar Plateau where the city was situated. In the dry wadis there were sparse grasses and bushes where I found a variety of birds: Brown-necked Raven, White-tailed Wheatear, Barn Swallow, Bank Swallow, Crowned Sandgrouse*, Yellow Wagtail (genus *Motacilla*, in the pipit family), Rock Martin, S. Gray Shrike, and Desert Lark. Some of these were breeding residents and others were Palearctic migrants. To the north and east of the city were the barren, rocky, moonscape-like Hoggar (Ahoggar) Mountains with peaks up to 9,500 feet. We were now in the Palearctic zoogeographic region of the world (see Brown & Gibson, 1983, *Biogeography*, p. 8).

On our way into the city we encountered a long Tuareg camel train of 15 or 16 camels, with a group of 10 or 12 people. Some were walking and others were riding. Both men and women wore a knee length black gown which hung from their shoulders. Underneath they wore long trousers, dark colored in men and more colorful in women. Everyone had a white head covering, and in the men this also covered their face. Open flat sandals protected their feet from rocks and the hot desert sand. As we reached the outskirts of the city there was an attractive new diamond-shaped road sign on the top of an iron post which announced "Tamanrasset." We had arrived! Tamanrasset was the chief cultural center of Algerian Tuaregs. This ancient oasis city and trading center, a crossroad of camel caravan routes, was situated in almost the exact middle of the Sahara Desert, virtually on the Tropic of Cancer and the 5°E meridian. It was one of the most remote human settlements anywhere in the world. In 1960, the only way in and out was on foot, on a camel, or in a vehicle over an inhospitable road which began more than 1,000 miles away from either the north or the south. (There was no airport at this time.) It was not a tourist destination except for the very most adventuresome. In our last 1,000 miles of travel (since beyond Zinder) we had encountered only one tourist vehicle (the French citroen which was stuck in the sand) and one commercial truck (two nights ago).

Tamanrasset, like virtually all Saharan communities, was built around an oasis where date palms were widely planted, as were citrus fruits, figs, almonds, corn, and cereals. Daytime temperatures were very hot, up to 117°F, throughout much of the year. Streets were unpaved and houses were built of clay (adobe). As always, there was a conspicuous mosque and a large central marketplace. Our 2 hitchhikers directed us to a location where camels were being sold, and they disembarked there with their camel whips and saddles and other belongings, thanking us warmly. Noble and I each wanted one of their handmade, rawhide camel whips as a souvenir, and they readily agreed (perhaps feeling they owed us a little something for their 3½ days of transportation). Noble gave each of them a Polaroid picture of themselves, a gift which never ceased to please and amaze most persons.

I walked around the town and surrounding cultivated fields from noon until 3:00pm, taking pictures and sleuthing for birds. I succeeded in documenting 19 species, which included 5 trip additions: Western Orphean Warbler* (a Palearctic migrant and recent taxonomic split, which I identified on the basis of its non-overlapping range with the E. Orphean Warbler), European Pied Flycatcher (a Palearctic migrant), Common Redstart (a Palearctic migrant), House Bunting* (a breeding resident), and Trumpeter Finch* (a breeding resident characteristic of the northern Sahara; a very attractive, small fringillid finch with a short, stubby red bill, grayish head, reddish-brown upperparts, and pinkish underparts). Other species which I listed included 2 Egyptian Vultures, 3 Woodchat Shrikes, 4 Northern Wheatears, 3 White-tailed Wheatears, and 2 Laughing Doves (somewhat out of place from their usual more southerly distribution, perhaps released cage birds here). That night we camped just outside the town under a date palm tree. With the departure of our 2 hitchhikers there were now just the four of us again.

During the next two days (May 2 & 3, days 10 & 11) we 4 Saharan adventurers left our main N-S trans-Saharan route and drove on a narrow, winding, sightseeing route that took us through, over, and around the Hoggar Mountains east and north of Tamanrasset. We devoted our time to taking pictures, hiking, and searching for birds or other wildlife. It was a welcome change from our usual travel. These spectacular, naked lava (basalt) and granite mountains were of both igneous and volcanic origin, with peaks and pinnacles rising to over 9,000 feet, creating a surreal lunar landscape appearance. Our road took us as high as 8,725 ft. above sea level, which was the highest point we traveled in our Jeep anywhere on the whole of Africa! Noble and I were greatly surprised at such a rugged spectacle in the middle of the Sahara Desert, a geologic feature of which neither of us was previously aware. (Tony and Collyn were knowledgeable because they had traveled north to south on this route four months earlier.) Sadly, there was little vegetation except in the difficult to ascend canyons and ridges, or along wadis which were not accessible to us, and we saw no wild mammals and only a few birds. Native flora consisted of xeric-adapted grasses & shrubs, and trees such as myrtles, wild olives, cypresses, wild figs, and acacias. Noteworthy mammal species known to inhabit the mountains were Barbary Sheep, Dorcas Gazelles, and Cheetahs. Ancient cave paintings (not yet open to tourists) depicted flora and fauna from 10,000 years ago which are characteristic of much wetter conditions than currently present. The Hoggar Mountains were included by ecologists today in the "West Saharan Montane Xeric Woodland Ecosystem."

My complete bird list for our 2 days in the mountains contains only 15 species: 3 Rock Martins, 7 Rock Pigeons* (my 1st sighting of wild birds in their original, non-introduced range), 15 Brown-necked Ravens, 1 Yellow (or possibly Gray) Wagtail, 15 White-tailed Wheatears, 8 Desert Larks, 6 House Buntings, 2 Lanner Falcons, 2 Crowned Sandgrouse, 10 Barn Swallows, I Long-legged Buzzard* (*Buteo rufinus*, my first sighting of this widespread SW Palearctic species), 2 Bank Swallows, 1 Common Redstart, 5 Willow Warblers, and 2 Eurasian Golden Orioles. These birds were seen mostly in sandy wadis where there were small pools of water with trees and other vegetation. As previously, these species were a mixture of Palearctic spring migrants and resident breeding birds. We camped the night of May 3rd alongside a wadi at 6,500 feet elevation, near a site named "Hirhafok" approximately 75 road miles NNE of Tamanrasset, and 50 road miles east of our main trans-Saharan route.

Plate 66: Our Jeep, with its heavy weight and narrow tires, got stuck in the sand on several occasions, and the four of us had to winch and sometimes also shovel to get us out. Noble needed a drink of water after taking his turn at shoveling.

Plate 66a. We finally arrived at Tamanrasset in the very middle of the Sahara Desert, reached only by camel tracks and a single, unmaintained vehicle road. Our 2 hitchhikers left us here to buy camels.

On May 4 (day 12) we descended westward down out of the mountains on a scenic road to an elevation of 3,500 ft., where we rejoined our main trans-Saharan road which was headed northwest toward the town of Arak (and beyond to the city of In Salah). A short distance up this road, just before the tiny Tuareg settlement of In Amguel, we came to a 50-75 yard wide sandy wadi, the "Tekouiat Wadi." Our "road" as it crossed this wadi was nothing more than scattered individual vehicle tracks in the sand. The wadi had not been modified in any way to aid vehicles in crossing. As the lead vehicle, I stopped to inspect the situation very carefully, and then said to the others there was no chance at all that our Jeep could cross the wadi without almost immediately getting stuck. However, there was no other option if we wanted to continue our journey. (I was reminded of fording the wide, fast flowing Rio Terraba in Costa Rica.) It was decided that Noble and I should go first, and see what happened. So I got as much of a running start as I could on our slightly deflated tires. I floor-boarded the accelerator in 2nd gear, and at a speed of 20 mph I entered the wadi. Predictably, I traveled no more than 15 yards before sinking into the soft sand and coming to a halt, with the sand half way up to my wheel hubs. Now it was Tony & Collyn's turn in the Bedford, with tires almost twice as wide as ours, plus 4-wheel drive. Sure enough, they made it all the way to the other side! There was great jubilation. Now all we had to do, once again, was to fasten our winch cable to their van and slowly winch us across. Our cable was just long enough to reach across the wadi. We would not have to resort to the dreaded sand mats! After we were successfully across I took a photograph showing the very deep tire tracks our Jeep left behind in the sand.

We did not stop in the oasis village of In Amguel but continued onward in a NNW direction, at an elevation of 3,500 feet. Our road was narrow and generally straight, with a sandy to gravelly, rough to smooth surface. We proceeded at our customary speed between 15 & 25 mph. The desert once again was mostly featureless, but 40 miles up the road we passed on our left an isolated, 6,770 ft. tall granite massif called "Taourit Tan Afella." During the 1960's (very shortly after our travel) the French conducted underground nuclear tests inside this mountain. Along our road at this site was a large, sturdy white road sign which gave distances to major settlements along our route, both north and south. These were in km but after we converted them to miles the sign said it was 115 miles back (south) to Tamanrasset and 675 miles back to Agadez, which we had passed through 9 days ago. Heading north it was 325 mi. to In Salah, 575 mi. to El Golea, 1,130 mi. to Algiers, and 2,200 mi. to Paris! This meant that after 13 days on the road we were only exactly half way in distance between Kano and Algiers. That night we camped by the roadside 110 miles from Arak, at an elevation of 3,500 feet. I recorded 19 bird species for the day but only 1 species, the Common Chiffchaff, was a trip addition (a Palearctic breeding sylviid warbler which I had seen in Europe on my midshipmen cruises).

On May 5 (day 13) or two vans set out across the desert heading northwest toward the small village of Arak. Our road was slowly dropping in elevation as we traveled through a scenic region of sand, rocky outcrops, and gorges on the Immidir Plateau in the "Mountains of Mouydir." The region was vegetated with sparse grass, and in the sandy wadis of the gorges there were scattered trees and waterholes. Wildlife living in this region included several species of gazelles, Barbary Sheep, Leopards (rare), jackals, and Fennec Foxes. In some of the gorges I found footprints of gazelles and burrows of foxes.

The tiny, endemic Fennec Fox (*Vulpes cerda*) was one of the most endearing, common, and widespread mammals of the Sahara. It was the smallest canid in the world, standing just 8" high at the shoulder and weighing only 1½-3½ pounds, with a total length of just 2 ft. (including an 8" long bushy tail). Its wide, tall pointed ears were unusually large for the size of the animal. The pale sandy coloration of its upperparts and the whitish colored underparts made it very difficult to see in the daytime, but it was almost strictly nocturnal in its time of activity. The Fennec Fox lived in small family groups in widespread, connected underground sandy burrows with many entrances. Its diet consisted of rodents, lizards, small birds, lizards, eggs, snails, insects, berries, and other fruit. Its large ears served to dissipate heat in the daytime and to detect the slightest sounds made by prey animals at night. The thick coat of fur (even on the soles of its feet) protected it from daytime heat and also kept it warm at night. This little fox's main predator was the Pharaoh Eagle-Owl (*Bubo ascalaphus*), and to avoid it the fox was capable of suddenly making both vertical jumps 2 ft. upward and horizontal leaps 4 ft. forward. The Fennec Fox had an intelligent, cute little face, and it was sometimes trapped and kept as a pet, both in Africa and elsewhere. It was the national animal of Algeria, and also served as the nickname for the Algerian national soccer team, "Les Fennecs"!

Our road passed two tall, isolated buttes to our west: (1) "Tintejert" at 4,800 ft. elevation and (2) "Moulay Hassan" at 5,700 ft. elevation. In between these buttes was an area of "queer Gray stone" (as stated in Tony & Collyn's little guide book). These were small basalt boulders 4-5 ft. tall scattered across the open desert and sculptured into many weird shapes and patterns by the persistent wind-blown sand. I photographed more than a half dozen of them for a pictorial record. That evening we camped near Arak, on the northeast side of the road at an elevation of 1,750 ft., at the mouth of a large, scenic gorge which had been carved over centuries by an intermittently wet and dry wadi flowing down through the gorge from the elevated Immidir Plateau above it. It was one of our more picturesque campsites. There were trees, bushes, and more footprints of gazelles. The Sahara, in spite of extensive barren and featureless regions, had many scenic areas and habitats for wildlife. For the day, I documented just 7 species of birds, all of which I had seen commonly during the past several days.

We were underway again early the next morning (May 6, day 14), headed NNW for the principal desert town of In Salah, 175 miles distant. Before very long we passed through the tiny village of Tadjemout, at the edge of a wadi. It was a typical Tuareg settlement with dirt streets, a small market center, few people, clay houses, a water well, and date palms. We did not stop. As we continued onward, our narrow, sandy route was remarkably straight for long distances, with high sandy edges to the road as if it had once been graded. In the distance to the east of us (our right) were occasional buttes, flat-topped mesas, and low undulating ridges of mountains. We were averaging our usual rate of speed, 20 mph. Driving was more relaxing for me and I could take time to enjoy the magnificence of the desert. At one point, a road sign announced the presence of a 6 meter deep well, "In Takoula", which was 2 km off the road to our left, but we did not investigate this. We had an ample supply of water with us. Further along, a track led off to our right, to a "guelta" with "eau bonne" (good water). A guelta was a deep hole in a rocky boulder which held rainwater for long periods.

With only a few stops during the day we were nearing In Salah by late afternoon, about 4:00 o'clock, when suddenly the main leaf of our right front wheel spring broke. We were in a picturesque area of extensive sand dunes, so I pulled off the road several hundred yards and parked at the base of a 200-300 foot high sand dune, with a beautiful rippled surface (like a photograph from a travel book). Here, the 4 of us began replacing the broken spring with a spare leaf Noble and I had in our tool compartment, having brought it all the way from California with us for just such an emergency. We quickly jacked up the front of the Jeep, and in the hot sun with the temperature over 100° we began replacing the broken spring. Noble and Collyn did most of the manual labor, being more robust and mechanically knowledgeable than either Tony or I were. (Fortunately, Noble was much better at mechanical repairs than I was.) It required about 2 hours of toil and sweat before the new leaf had been installed. Since it was now nearing sundown we decided to camp at this site for the night. We all enjoyed some cold coca cola from our refrigerator. We were now only 5 miles from In Salah. In more than 8 hours and 170 miles of travel during the day I saw just 3 individual birds: 2 Brown-necked Ravens and 1 White-tailed Wheatear. As the sun went down the temperature rapidly began to drop for the much colder night.

Before our departure the next morning (May 7, day 15), I walked up to the top of the tall sand dune next to us and took some photographs, including our two trans-Saharan camping vans parked at the bottom, side by side, looking very much like David and Goliath next to each other in the sand. From our campsite it was less than an hour's drive to In Salah, which was situated around an oasis at an elevation of only 900 ft. above sea level, the lowest point of our Saharan travel. The city was at the almost exact geographical center of Algeria, at 2½° E longitude and 27°N latitude, 250 mi. north of the Tropic of Cancer (at the same latitude as Corpus Christi, Texas)! In Salah was one of the hottest spots in the Sahara, or anywhere else on earth, with an average daytime high temperature in July of 115°F. A major concern, and constant task, was to protect the city from creeping sand dunes blowing from the west.

Plate 67: The spectacular Hoggar Mountains rose to above 9,000 feet elevation and looked like a lunar landscape. Borders of oases were sparsely populated and cultivated. It was 3,518 km (2,200 mi) from the middle of the Sahara Desert to Paris!

Plate 68: Lovely canyons were bordered by trees near Arak. The ancient, picturesque adobe city of In Salah (latitude 27° N) featured a 25 foot high arched adobe entrance gate, which was guarded by a dark-skinned sentry with a rifle.

A 15 ft. high brick and clay wall surrounded the city and protected it from marauding pirates and other enemies. On each side of the city was an attractive, 20-30 ft. high arched adobe entrance gate where a black-skinned guard stood at attention with a carbine rifle, and inspected our entry permits. The guard at our gate was dressed with a white turban & scarf, elbow length white shirt, long ankle length white Muslim gown, and a sleeveless henna-colored (orange-red) vest. Over one shoulder was a bandolier for carrying ammunition. At one time these dark-skinned sentries were African slaves from south of the Sahara, but whether this was still true or not in 1960 I did not know. The city's importance as a trading center for gold, ivory, and slaves was now declining.

A large well in the center of the city was encircled by a high adobe wall and provided water for the city's inhabitants. Above ground adobe aqueducts, supported by arches up to 15 or 20 ft. high, carried water from the well to all corners of the city. There were no stone or paved streets, only dirt, and few people were moving about. An occasional, heavily burdened donkey was herded along the main street, but there were no motorized vehicles. In Salah was sub-divided into 4 red or violet brick-walled villages, each with its own citadel and watch tower. Outside the city were irrigated areas for growing dates, fruits, and vegetables, which were protected from the encroaching desert by hedge fences. Water came from artesian wells. The 4 of us enjoyed several hours of meandering around the city and taking pictures, finally departing about mid-day. In Salah provided me with the memory of a desert community which had been isolated from the rest of the world for many centuries but which was now changing, sadly, as it gradually came into contact with more modern civilization.

Tony, Collyn, Noble, and I continued our travel northward from In Salah toward El Golea, our next principal town and oasis stop, 250 miles away. Although there was still a long distance to go, we were approaching the end of our trans-Saharan adventure. The landscape was more irregular, with isolated buttes in the distance, and our flat desert road began gaining slowly in elevation. Fifty miles later we reached the southern side of the extensive Tademait Plateau and ascended to the top of this rocky plateau, at 2,600 ft. elevation. The surface was broken with cliffs, pinnacles, and ravines. We stopped for 45 minutes so that we could all climb and hike around on the surface, and take photos of the higher ridges in the far distance. The view was spectacular, with the road we had just traversed far below us. Thereafter, we drove for another hour before stopping for the day, in time to take pictures of another gorgeous desert sunset.

During the afternoon of travel I recorded a total of only 5 individual birds: 1 Brown-necked Raven, 1 Peregrine Falcon (which flew up from the roadside in a flat, barren area), and 3 Lanner Falcons. Two of the Lanner Falcons were chasing each other in the far distance, and the other was generally very white in overall color, much paler than birds of this species farther south on the continent (which caused me difficulty in arriving at an identification). I later discovered that the desert populations of Lanner Falcons in northwestern Africa were known to be paler in color than other populations elsewhere on the continent, and this species thus conformed to "Gloger's rule", which states that populations in hot, dry environments were very often paler in color than other populations of the same species living in wetter, cooler environments. (The most generally accepted explanation for this phenomenon is that it allows organisms to better match the overall color of their environment, and thus make them less conspicuous to predators, or prey, but this answer has been questioned, and other factors may be involved.) Whatever the explanation might be, Lanner Falcons here in the desert seemed to conform to this widespread rule. That night we camped approximately 60 miles north of In Salah, at 2,600 ft. elevation on the Tademait Plateau.

The next day (May 8, day 16) we traveled north the remaining 190 miles to El Golea, across the top of the Tademait Plateau for the first 75 miles. Noble and I continued to be the lead vehicle. As we approached El Golea there were picturesque sand dunes with rippled wind-blown surfaces extending to the horizons in both directions. These were the eastern edge of the vast "Grand Erg Occidental." They presented a scene which for most persons characterized the entire Sahara Desert. (In fact, only about 10% of the Sahara was erg desert.) Luckily, our narrow road surface followed between the dunes, across sand firm enough to easily support both of our vans. There was no need to break out either the winch or the sand mats.

This was a road that on much later maps was designated with a green line, indicating a scenic road. I was thrilled at the sight of these dunes, and of course we all stopped for photographs. Indeed, this was the "real" Sahara in almost everyone's mind. The only bird I saw was a single Greater Hoopoe-Lark, which displayed its conspicuous black & white wing pattern in flight. What bird could possibly have better fitted in this scene? It was a thrilling moment for me.

El Golea, at 1,500 ft. elevation, was a small, typical oasis town with a surrounding protective wall, unpaved streets, and a tall Muslim prayer tower ("ksar"). An old fort was on a hill top not far away, and date palms grew at the water's edge just outside the town. The town and its surrounding oasis with date palms appeared to have come directly out of a travel book. It was almost too perfect to be true, and provided all of us with an unforgettable memory. We had taken a very long time to arrive here! That night we camped just outside the town.

The next morning (May 9, day 17) I searched for birds around the oasis for an hour before we departed. My list of 15 species was: 1 Purple Heron, 1 Squacco Heron, 5 Little Stints (a sandpiper), 4 Common Sandpipers, 1 Green Sandpiper (a trip addition from the Palearctic), 1 Black Kite, 1 Western Marsh-Harrier* (a much later taxonomic split from the Marsh Harrier), 3 European Bee-eaters, 5 Common Swifts, 15 House Martins, 10 Barn Swallows, 1 Bank Swallow, 35 Brown-necked Ravens, 50 House Sparrows, and 1 female Common Redstart (or possibly a female Black Redstart). The only resident, breeding birds on this list were the raven and House Sparrow (and maybe the Black Kite). The others were all Palearctic migrants on their way north. Having entered the Palearctic Zoogeographic Region at Tamanrasset, I could now use *A Field Guide to the Birds of Britain and Europe*, 1954, by Peterson, Mountfort, and Hollom as my primary field guide and reference for bird identification. What a pleasant change! (I had brought this guide with me from California.) It was the only modern, easy-to-use, comprehensive field guide I had with me for my entire adventure. There were colored illustrations and range maps for almost all species! It would not any longer be necessary for me to write descriptions of the birds I encountered, and I could concentrate more on finding birds rather than on writing down what they looked like. My lists would be more representative of the total avifauna in an area. It was a change I enthusiastically welcomed!

As we left El Golea at mid-morning to continue our journey northward, our primitive sand and gravel desert road suddenly became a one lane asphalt highway! For Noble and me our surprise could not have been greater. Neither Tony nor Collyn had given us advance knowledge of this. (Perhaps this section had been completed just during the last 4 months, after they had first traveled here on their way south.) For us, it was entirely unexpected. The French were exploiting for oil in this region of the desert and needed such a hard-surfaced road for transporting their heavy machinery and equipment. What it meant for us was that our driving speeds could now be doubled, up to 40 or even 50 mph. However, I felt saddened by the thought that tourism would soon surely follow. The desert would become less of an adventure and less pristine, and the unique human cultures which had evolved here in isolation from the rest of the world would become assimilated into the modern world. Camel trains for trading would become obsolete. I felt fortunate to have come at the time we did. I can cherish my pictures and memories.

It was only 150 miles to Ghardaia, across a stony desert varying between 1,500 and 2,000 ft. elevation. I drove at speeds between 30 and 40 mph, still in the lead of our 2 vans, stopping regularly to identify roadside birds or to take pictures of goats, camels, and camel trains alongside our highway. We camped that night only a few miles prior to Ghardaia, at 2,000 ft. elevation. My final tally of birds for the day was 8 White-tailed Wheatears, 5 Desert Larks, 4 Barn Swallows, 1 Yellow Wagtail, 1 Woodchat Shrike, 1 S. Gray Shrike, 5 Brown-necked Ravens, 1 Collared Pratincole, and 6 small shorebirds.

The next morning (May 10, day 18) we drove into Ghardaia, which was the largest city through which we traveled in the Sahara, with a population of perhaps 40,000-50,000 in 1960. It was a historical oasis city situated at 2,000 ft. elevation, which was founded by the Mozabite Berbers in 1048, on a hilltop in the Mzab Valley on the south side of the Wadi Mzab. (The Mzab Valley, including the city of Ghardaia and 4 smaller nearby towns, all

Plate 69: Vast, scenic, magnificent sand dunes and picturesque oases are iconic scenes of the Sahara. Berbers were the earliest inhabitants of the region, and are today still one of the most numerous ethnic groups.

became a UNESCO World Heritage Site in 1982.) In 1960, one-third of the city's population was Berbers, of the Ibadi sect of Islam. Arabs made up a majority of the other ethnic groups. (Arabs overcame much of North Africa in the early days of Islam in the 6th and 7th centuries, and although they converted Berbers into Islam, they never succeeded in assimilating them into their cultural or language group.) The two ethnic groups have fought with each other almost continuously for over a thousand years. In Algeria (and later in Morocco), I found it difficult to separate most Arabs from most Berbers by their traditional dress or their general outward appearance. Both men and women typically wore a head covering and a long, loose, garment with full sleeves (a robe or tunic, or "kaftan"). Women wore decorated dresses. Footwear was a soft leather slipper with no heel. Their language always distinguished them, as did their customs and lifestyle. Today, more than 2,000 years after their first appearance in Africa, Berbers were the dominant nomadic group in the northern Sahara of Algeria and Morocco, and they were the most numerous ethnic group in the Atlas Mountains.

Ghardaia was a miniature citadel surrounded by a high wall, with a large central marketplace and an arcaded square with a pyramid style mosque. A tall minaret served for calling out prayers as well as a watchtower. Distinctive white, pink, and red houses were built of sand, clay, and gypsum, and were situated on terraces in a circular pattern, in a labyrinth of alleyways. A unique system of underground tunnels harvested rainfall and distributed it throughout the city, and surrounding irrigated fields. Additional water was obtained from numerous wells, some as deep as 350 feet. Dates were a principal economic product, as were rugs and carpets (made from both goat and camel hair). Remarkably, thieves were said to be almost non-existent! Climatically, the city received only 4½" of rainfall annually. The four hottest summer months had an average high temperature of 101°-108°F, and the four coldest winter months had an average low temperature from 41°-49°F. Tony, Collyn, Noble, and I thoroughly enjoyed strolling through the marketplace and meandering through the alleyways for most of the morning, taking pictures and assimilating memories. This city, like others, would change with the new highway and influx of tourists (and thieves). There would be more conflict between the Arab and Berber communities. Civilization had its consequences.

An hour or so before noon, the 4 of us climbed back into our camping vans and commenced the final 400 miles of our Kano-Algiers adventure. We were still in a stony, sparsely vegetated desert, traveling on our new, gradually widening asphalt highway. Laghouat was 132 miles away, and then it was another 70 miles to Djelfa. We would climb in elevation from 2,000 ft. at Ghardaia to 4,000 ft. at Djelfa, which was situated on a high plateau on the north side of the "Saharan Atlas Mountains" (a low-lying, southern, isolated range of the Atlas Mountains).

With only a few stops we were able to reach Djelfa before sunset, and we camped nearby there for the night. My list of birds for the day totaled 18 species, of which 6 were trip additions (Appendix B). I was happy to add another wheatear to my list, the rather plainly colored Desert Wheatear* (distinguished from the black-throated form of the Black-eared Wheatear by its all black tail). My favorite bird of the day, however (which became one of my all time favorites), was the Cream-colored Courser*, which I considered a desert icon. Its pale sandy coloration, entirely black underwings (conspicuous in flight), and distinctly patterned head (black & white stripes through and above the eye) were particularly appealing to me, as were its long legs, slender, upright posture, and habit of running short distances rapidly in quick spurts across the desert sands.

The next day (May 11) was the 19th and last day of our 2,300 mile trans-Saharan adventure. From Djelfa to Algiers was 185 miles, crossing the Atlas Mountains en route. Not long after we started out we overtook a long camel train of Berbers moving slowly northward, in the same direction we were going, close to and paralleling the right (east) side of our highway. Surprisingly, the other side of our highway, to the west, was paralleled closely by a railroad track. I wanted to get in a position with the sun behind me where I could take some pictures of the camel train as it proceeded past us. Therefore, I drove ahead a half mile or so and parked just off the highway, from where I walked east across the desert a short distance into the morning sun, to a small mound where I could face west with the morning sun at my back, and take photographs of the camel train as it proceeded just below me. Unbelievably, at the exact same time as I was photographing the camel train, along came a several car passenger train traveling quite rapidly north on the other side of the highway. I instantly snapped a picture. Thus, one of my photographs shows a rapidly moving passenger railroad train, a paved highway, and a slowly moving camel train all side by side next to one another in the same frame! (One could wait a lifetime in an attempt to duplicate, unrehearsed, such a scene which portrays one ancient and two modern means of travel all together at the exact same time - - at the edge of the desert!). As if this combining of the old with the new were not enough, there was also a modern, tall electrical power line on telephone poles running between the railroad track and the highway. The northern Sahara Desert under French colonization was succumbing to modern civilization. Such changes, of course, were inevitable almost everywhere in the world.

The camel train contained both older camels with heavy burdens (including firewood) and younger, unburdened camels. All the people - - men, women, and children - - were walking alongside the camels, with no one riding. They wore white and henna-colored garments which covered them from head to foot, but their faces were uncovered. All wore sandals. Several small dogs trotted along playfully with them. The group of nomads looked at me curiously as they walked close past, but no one stopped, approached me, or said anything to me. A slender, short-statured European-looking young man with a long black beard, blond moustache, and a camera, must have been an unusual sight for them.

Berbers (also known by their ancient name of "Amazighs") were one of the earliest groups of people to appear in northern Africa, having come from Europe, the Middle East, or the Nile (but not from Arabia). They were either "pagans" or a few converted Christians, after the time of Christ. Over the centuries they split into many subgroups, including the Tuaregs, Moabites, Kabyles, Almoravids, and Almohads. All of these people were converted to Islam after northern Africa was overrun by the Arabs in the 6th & 7th centuries, but they did not give up their traditional languages. In the beginning they were all mostly farmers, but as time went on many became nomadic pastoralists or camel traders, a way of life they still followed today. In 1960, Berbers outnumbered Arabs in North Africa by perhaps 2:1. Berber women were well known for their singing and musical ability, and for their manufacture of jewelry, ceramics, and rugs. They wore a great variety of necklaces, bracelets, brooches, earrings, and pendants. (I am reminded of the picture I described in my last chapter, of the ornamented woman near the Kotto River west of Bangassou in French Equatorial Africa, who I was told was a Berber woman, in spite of the locality.)

The Atlas Mountains of North Africa extend in a slightly upward curving ENE arc for 1,600 miles in several long, narrow ranges, all the way from the Atlantic Ocean in SW Morocco eastward across northern Algeria to northern Tunisia, paralleling the Mediterranean coast. A narrow, fertile, green strip of land is situated along the Mediterranean coast between the mountains and the sea. The southern edge of the Atlas Mountains forms the northern border of the Western Sahara Desert. The tallest peaks are all in the "High Atlas" of Morocco, with two peaks above 13,000 feet. The highest mountain top in Algeria was only just above 7,600 ft., and there were only a few peaks as high as 5,000 feet.

When one traveled south to north over the Atlas Mountains (as we did between Djelfa and Algiers) he started in a desert ecosystem and then passed upward through the "N. Sahara steppe and woodland ecosystem" on the south side of the Atlas, over the top, and then down through the "Mediterranean forest, woodlands, and scrub ecosystem" on the north side of the Atlas. These woodland and forest ecosystems had a high floral and faunal diversity, with reptiles being unusually diverse. Noteworthy mammals (most of which were declining in abundance) were the Barbary Macaque, Leopard, Barbary Stag (a form of the Red Deer), and Barbary Sheep. Among birds, the Algerian Nuthatch (first discovered in 1975) was endemic to mountain coniferous forests in NE Algeria. Trees included such conifers as cedars (*Cedrus*, growing up to 120 ft. tall), firs (*Abies*), pines (*Pinus*), and junipers (*Juniperus*), and such broad-leafed species as oaks (*Quercus*, including commercially valuable cork oaks), maples (*Acer*), and native olive trees. The "Mediterranean climate" biome (which I have discussed previously from Chile and South Africa) was locally called the "maquis."

Plate 70: The 4 of us arrived in Algiers after 19 days and more than 2,000 miles of travel together in the Sahara Desert. The beaches and luxurious green seacoast were a pleasant change. We parted company over two bottles of wine, leaving our addresses with each other for future correspondence.

307

This distinct biome occurred where the climate was hot and dry in the summer and mild and wet in the winter, on the west or south side of continents in both hemispheres between 30° & 40° latitude. Its characteristic vegetation was stiff, twiggy shrubs with sclerophyllous (leathery) evergreen leaves (as in the "chaparral" of the southern California coast). Geologically, the Atlas Mountains were rocky with steep narrow gorges and numerous waterfalls. The entire geographical region of NW Africa, west of Libya, was known as the "Maghreb", throughout which Berbers were usually the dominant ethnic group. This region included the northwestern Sahara, the Atlas Mountains, and the western Mediterranean coastal region west of Libya.

When Tony, Collyn, Noble, and I drove across the Atlas Mountains on May 11, 1960, there was a fierce guerilla war going on between French government forces and a determined group of Algerian independence fighters known as the FLN (National Liberation Front), called "fellaghas" by the French. The French had not only colonized Algeria in the early 1800's, but they proclaimed that Algeria was an integral part of France! Now, at long last, the indigenous people were fighting to regain their independence (which they finally succeeded in doing in 1962, after many years of warfare). In 1960 we were caught in the middle of this war and were advised not to camp along the roadside anywhere in the mountains. All major mountain roads and highways were officially closed to traffic after dark. Every bridge, large or small, was guarded by a concrete pillbox with one or more machine guns. It was not a relaxing way for us to end to our Sahara travel.

During this last day of travel, I stopped only occasionally and very briefly in the mountains to investigate birds. Never-the-less, I was able to document 30 species, including 9 trip additions: Calandra Lark*, Thekla Lark*, House Sparrow (my first wild, non-introduced birds), Black Tern, Black Wheatear*, Corn Bunting*, Short-toed Treecreeper*, European Stonechat*, and European Goldfinch. The Black Wheatear (*Oenanthe leucura*) was a bird strictly of rocky ravines and cliffs on both the northern and southern sides of the western Mediterranean. Other birds on my list were 4 Egyptian Vultures, 3 Black Kites, 15 Greater Flamingos, and 2 European Bee-eaters.

That evening, May 11, we arrived in Algiers just before dark, found a campsite in a large park, and broke out 2 bottles of red wine to celebrate our achievement. It was a joyous moment. The four of us had challenged the greatest desert in the world and won! For Noble and me, we could not have accomplished this feat without some assistance from Tony and Collyn, whom we miraculously encountered in Kano at the 11th hour. Now we were on the Mediterranean Sea after a 19 day, 2,300 mile journey over a mostly unmaintained, remote desert road, one that had been traveled by very few. It was time for jubilation. Noble and I toasted to the remaining 3 continents of our adventure yet to come. After 15½ months on the road we were not yet at the half way mark of our global circumnavigation. In Africa, Morocco was still ahead.

On May 19, Noble and I departed Algiers and drove west along the Mediterranean Sea on a good, paved, scenic highway that wound just inland through mountain foothills up to 2,000 ft. elevation, all the way to the large coastal city of Oran. Burden-laden donkeys were herded or ridden along the highway by both men and women, who were all mostly bundled up from head to foot because of the chilly sea breeze. Larks and warblers sang from green olive groves, vineyards, and maquis scrublands. Peasant farmers dressed in baggy pants and long or short-sleeved shirts cut wheat from yellow fields with long, hand-held sickles, and then tightly bound the long stems in tall stacks, which were subsequently carried away on donkeys. Orange, purple, and yellow wildflowers covered the hillsides. White Storks were abundant almost everywhere, and several roadside nests had 1 or 2 storks standing on top of them. We arrived in Oran late in the afternoon. On a hill overlooking the city was a church with a tall bell tower, on the top of which was a statue of Christ. We found a campsite overlooking the harbor on the outskirts of the city, where we hoped we would be safe from the fellaghas. Above the harbor was a distant, beautiful, orange-red sunset.

The next morning (May 20) we proceeded leisurely SW along the Algerian coast on a scenic highway that curved around on top of the rocky Mediterranean seashore, then turned south, inland toward the city of Tlemcen at the base of the Atlas Mountains. Again, there were vineyards, olive groves, hillsides of colorful wildflowers, and burden-laden donkeys. Our highway skirted around Tlemcen and then headed west toward the international border with Morocco. We crossed a deep ravine on a high bridge in the hills west of Tlemcen, where there was the usual machine gun pillbox protecting the bridge. Soon thereafter (at 3:00 pm) we reached the border of Morocco, at an elevation of about 2,000 feet. We each had a visa in our passport so there was no delay. In Morocco there were no more fellaghas to worry about because Morocco had gained its independence from France 4 years earlier, in 1956.

During the day so far, in Algeria, I had tallied 35 species of birds, with 3 trip additions: Wood Warbler, Dartford Warbler* (genus *Sylvia*), and Spanish Sparrow* (genus *Passer*, closely resembling a House Sparrow). The Dartford Warbler was seen on a hillside of dense maquis scrub, its preferred habitat. A common, closely related species, the Sardinian Warbler, was feeding full grown young just out of the nest. Other species documented included the Corn Bunting, which was an unusual looking, heavily streaked *Emberiza*, the largest species in the genus.

Morocco, situated on the northwest corner of Africa, was the last country Noble and I would visit on the continent. The majority of the population was either Berber (in the Sahara Desert and the Atlas Mountains) or Arab (mostly in the coastal regions). These two ethnic groups were separated by language and cultural differences, with virtually no intermarriage between them. They had a long history of hostility toward one another. Noble and I would travel across only a small northern fraction of the country.

Once inside Morocco we soon came to the city of Oujda, which we drove through without stopping, and then continued westward for another 40 miles before finally pulling over to a suitable wide area for the night's campsite, in a stony, arid riverbed at the low elevation of only 1,000 ft. above sea level. Birds from Morocco which I added to my day's list (in part) were Short-toed Eagle* (very closely allied to and perhaps conspecific with the Beaudouin's Snake-Eagle which I recorded from Nigeria), Little Owl (similar in appearance and closely related to the New World Burrowing Owl), European Roller, Calandra Lark, and Greater Short-toed Lark.

Prior to breakfast the next morning (May 21) I pursued birds for 2½ hours in the vicinity of our campsite, along the dry, stony riverbed and in the neighboring bushes, trees, and fields. Among the 21 species which I encountered were 3 trip additions: (1) a pair of Demoiselle Cranes* (a truly exciting and unexpected little crane with light Gray plumage, white post-ocular plume, black face & foreneck, and black trailing edge of the wings in flight), (2) a pair of Black-bellied Sandgrouse* which flew past me on their way to or from water (one of the plump, short-tailed sandgrouse species), and (3) Spectacled Warbler* (one of 24 species in the genus *Sylvia*, very similar in appearance to the Whitethroat). Some of the other birds present were 1 European Roller, 4 European Bee-eaters, 1 Egyptian Vulture, 6 Cream-colored Coursers (what a thrilling bird), 2 Desert Larks (singing and displaying), 15 Crested Larks (many of which were singing), 5 Rufous-tailed Scrub-Robins, 6 Black-eared Wheatears, 1 Olivaceous Warbler, 4 Sardinian Warblers, 1 Woodchat Shrike, and 3 Trumpeter Bullfinches. These birds and the still, cool, sunny weather produced one of the most pleasant and memorable mornings of our adventure.

We departed our campsite at 0930 and began our day's travel toward the cities of Fes and Meknes. There were once again extensive fields of yellow wheat which were being harvested by hand with long sickles, by peasant workers dressed in baggy trousers, a turban, and sandals. Along the roadside was a picturesque, white-bearded, friendly goat herder who wore open, flat sandals, a white turban, a long white shirt, knee-length white baggy pants, and a wide blue sash over one shoulder. He looked at me quizzically as I stopped to ask him if I could take his photograph. The subsequent picture became one of my all time favorites. Not much farther along the highway was a very stately-looking Moroccan nobleman with a neatly trimmed black beard, sitting upright on a henna-colored blanket on the top of a magnificent whitish-Gray Arabian stallion. He was wearing a white turban and a long white robe. It was like a scene from the "Arabian nights." Again, I stopped to ask for a picture, and he willingly obliged me as he sat motionless on his horse, statue-like. This picture, too, became a permanent treasure. The day was off to a very good start with both my bird encounters and my picture-taking. Life was great.

Plate 71: The Mediterranean coast of Morocco was a strip of green valleys and foothills. I was delighted to photograph an iconic Arab horseman, and nesting White Storks.

Our highway continued through stony, semi-arid scrublands, plains, and open fields to the wide valley of the Moulouya River, which we crossed at 1,000 ft. elevation, just before the town of Guercif. As we climbed up into the hills on the other side of the river, a young girl sitting side-saddle on a donkey came slowly plodding down a dry gulley, with a large jerry can of water strapped to the side of the donkey. There was obviously a water source further up the gulley. Donkeys were indispensible in the life of the rural people. There were many roadside nests of White Storks during our 5 hour drive between Guercif and Fes.

We entered the walled city of Fes in mid-afternoon, through a tall, ancient entrance gate with 3 arches. However, we did not stop to sightsee in this historic city but continued through the narrow streets to an exit gate on the other side. On the outskirts of the city was an impoverished wine peddler walking along the road with a goatskin wine flask over his shoulder, carrying a tall brass wine container for drinking, and a long-handled ladle. He wore a white turban, dark Gray pullover shirt, dark knee-length pants, and well-worn, open sandals. He had no customers. As we proceeded onward toward Meknes, in stony hill country, we encountered a young girl walking behind an over-laden donkey. She was colorfully and ornately dressed with long, dangling earrings and a tight-fitting white head covering, attached to which was a long dangling pale pink "sash" which hung far down over her back. Her long-sleeved shirt was pale pink patterned with dark flowers, as was a piece of cloth tied around her waist which covered the top of her ankle length baggy trousers. Amazingly, in spite of the very stony roadside terrain over which she was walking, she was bare-footed!

From Fes, it was only 40 miles to Meknes, an ancient city 2,000 ft. above sea level, founded by Berbers between the 7th and 9th centuries. A 25 mile wall encircled the city, within which was a fortress completed in the 11th century. Meknes was renowned for its many mosques, more than 100 minarets, monumental gates, gardens, palaces, and mausoleums. In 1996 the city became a UNESCO World Heritage Site. We drove through the city and continued onward toward Rabat before stopping for the night alongside the highway 25 miles west of Meknes, at an elevation of only 500 ft. above sea level. We were in a cultivated river valley bordered by juniper and scrub-covered hillsides.

The next morning (May 22) I got up at daybreak, grabbed my notebook, binoculars, and Peterson Field Guide, and headed off into the patchy maquis hillsides for 4½ hours of ferreting out birds. It would be my last morning of birdwatching in Africa. My efforts produced just 26 species, only one of which was a trip addition, the Cirl Bunting, an *Emberiza* which was vaguely reminiscent of the N. Am. Dickcissel. A few of the other species were 9 Sardinian Warblers, 10 Rufous-tailed Scrub-Robins, 4 Black Kites, 3 Woodchat Shrikes, 5 Corn Buntings, 13 Crested Larks, 2 Willow Warblers, 1 Lesser Kestrel, 1 Eurasian Buzzard, 1 Barbary Partridge, (which flushed with loud, harsh, squealing vocalizations), 3 Egyptian Vultures (1 with a nest), 2 Orphean Warblers, 1 probable sub-adult Marsh Harrier (in a plumage with which I was not familiar), 7 Cattle Egrets, 4 White Storks (with 2 nests), and 1 Common (Garden) Bulbul (ubiquitous over all of Africa).

Noble and I packed up our campsite at 1030 and headed west toward the Moroccan capital city of Rabat, 100 miles away on the Atlantic coast. En route we stopped for an hour in a cork oak woodlands at 1,000 ft. elevation, to allow me a chance to investigate birds in a different habitat. Additional species to my early morning list were a pair of Eurasian Golden Orioles building a nest, 4 Great Tits, 5 Blue Tits (which were patterned with a distinct vertical black stripe down the middle of their breast), 2 Common Ravens, and 4 Eurasian Serins. When we arrived in Rabat 2 hours later there were three species of swifts circling over the city: 1 Little Swift, 4 Common Swifts, and 50 Pallid Swifts. It was the best chance I had experienced for critically comparing Common and Pallid Swifts under good lighting conditions, and I concluded that some (maybe even most) of the swifts I had been documenting recently as Common Swifts may actually have been Pallid Swifts. In my notebook today I described Common Swifts as being "sooty black" in color, and Pallid Swifts as being "dark brown" (and not very pallid at all to my eye).

A scenic lighthouse, which I photographed, was situated on the coast just outside the city of Rabat. Like other Moroccan cities, Rabat was enclosed by a very high wall with a huge entrance gate. Inside the city was the usual "clop, clop" sound of donkeys on the streets. Stalls of oranges lined the streets, and ceramic jugs of all sizes and shapes were displayed in shops everywhere. There were many modern areas within the city but the older parts with narrow streets and ancient architecture were a more interesting step backward into time. Noble and I found a scenic site on the coast just outside the city where we camped for the night, where we could listen to the constant sound of the waves.

The next day (May 23) Noble and I drove north from Rabat along the coast 25 miles to a small town which in 1960 was called Port Lyautey (but was later renamed "Kenitra"). Surprisingly, there was an active U.S. Naval Air Station here, having been established in 1951 six years after the end of WW2. This was significant in our lives because Noble and I were each in the inactive naval reserve and we each had received a letter from the U.S, Navy several months after we departed from California in January, 1959, ordering us to report to "the nearest U.S. naval facility" for a physical exam required for promotion to full lieutenant. Hah! At the time we received this letter we had just started our travel in South America, and from there of course we traveled to South Africa. Nowhere in our journey was there a U.S. naval facility within a thousand miles. We sent a letter to the navy department explaining our dilemma, but received no response. So here we were in Morocco almost a year later, and sure enough there was a U.S. naval facility!

On Monday morning, May 23, 1960, we took our orders to the base hospital at the U.S Naval Air Station in Port Lyautey, and dutifully reported for our physical exams. We said to the hospital corpsman at the reception desk that we were there to carry out our orders. He looked at our orders, and then at the date they were issued, and said they were very long out of date. Did it really matter? We argued that we were correctly carrying out our orders and that Port Lyautey was indeed the first U.S. naval facility we had encountered since receiving our letters. The corpsman was a good natured intelligent young man and saw the humor in the situation, so he willingly agreed to give us each an exam requiring about 30 minutes per person. He said he would send the results back to the USA. His only comment to me about my exam was that my weight of 128 pounds was a little too low (even for my short stature), and he asked if I had been sick. I assured him that in 16 months of travel I had not been sick for even one day, only constantly active.

After our exams were completed it occurred to me that here was a good opportunity for us to have all our international health immunizations renewed, so I asked the corpsman if he could do this for us, and he said yes. He inquired how many we each needed and I responded "six." He then asked how long we were going to be there. When I answered "only today", he pondered this for a moment and then queried "well, how do you want them"? After I realized what he meant, I gave this question a moment's thought and then replied "three in one arm and three in the other." So that's what he gave both of us, in a total time span of about 10 minutes each. These were for protection against typhus, yellow fever, cholera, tetanus, and typhoid (if I remember correctly). Neither of us suffered any dire effects afterwards, although Noble complained for several days that his arms hurt. That afternoon Noble and I spent a couple of hours telling the story of our travels to some of the children in the American school at the air station.

After 8½ months and 20,000 miles on the roads of Africa our Jeep very badly needed servicing, some new parts, and a great many repairs. Fortunately there was an excellent, modern Jeep agency in Casablanca back down the coast not too many miles away. Noble had already contacted them, and the Jeep head office in Toledo had air freighted a new front axle for us there, so our 4-wheel drive could at long last become functional again. Therefore, on May 24, we drove from Port Lyautey southwest along the good coastal highway 105 miles to the large, renowned, port city of Casablanca, at a latitude just below 34°N on the east side of the Atlantic Ocean (the same latitude as Myrtle Beach, S. Carolina). To the far south of the city, almost 200 miles away, one could barely see the snow-capped top of Mount Toukbal, which at 13,670 ft. above sea level was the tallest peak in the entire Atlas Mountains.

Casablanca was famous around the world for its casbah, the historic inner city with its many bazaars, as well for as its modern parks, gardens, monuments, and office buildings in the newer parts of the city. It was a tourist paradise and Noble in particular relished some time there as an aspiring entrepreneur. We checked our Jeep van into the agency that afternoon, and they were expecting our arrival. Noble, as always, took charge of making all the arrangements, both mechanical and financial. He enjoyed doing so, and nothing gave him more pleasure. It wasn't long before he had met numerous new acquaintances and friends, and we were soon invited into restaurants with them as their guests, or even into their homes. Noble enjoyed the people to people interactions.

After two full days of repair at the Jeep agency (May 25 & 26), it was apparent that all of our mandated repair would require considerably longer than we had originally anticipated, likely 2-3 weeks or longer. Noble was patient and said that he would wait as long as it took, and told me that during this time I could go anywhere else I wished, as he did not need any assistance from me. I gave the matter careful thought and decided I would like to hitchhike on ahead to Spain, where I could walk, take pictures, pursue birds, and enjoy the beach while I waited for Noble. It would be very inexpensive for me to do so, and being able to speak a little of the language would be a big asset for me. Therefore it was an easy decision for me to make, and I began packing my rucksack for yet another hitchhiking adventure. Before I left, Noble and I both decided to have our long, scraggly beards shortened and neatly trimmed at a local barber, but not entirely removed. Afterwards I still possessed a short black beard, and a pale blond mustache.

I left the next morning (May 27) and spent the day catching one ride after another, eventually arriving in Tangiers just before dark, at which time I found a very inexpensive rest house (50 cents) for the night. I had a total of less than $100 with me. It was a carefree life.

On Saturday morning at 9:00 am on May 28, I boarded a ferry for the short 2½ hour ride from Tangiers across the Straits of Gibraltar to Algeciras, Spain. I was leaving one continent and arriving on another. The day was partly cloudy and the sea was a little choppy. En route I was able to identify 1 Audouin's Gull*, my only record of this uncommon, local gull. It was my 1st trip list species from Spain.

During our 8½ months in Africa I recorded 1,090 bird species for my trip list, which when added to the 1,101 species from Central & South America and the 10 species from the South Atlantic brought my complete trip list at this time to a total of 2,201 species for the first 16 months of our 3-year adventure. Since leaving California I had filled over 25,000 lines with bird notes in eight 4x6 in. spiral field notebooks. (It required many years to identify the notes, sketches, and descriptions in my field notebooks, and the above totals were not determined until long after our adventure ended.) My father was right, I must be crazy.

Plate 72: The ancient city of Rabat was situated on the Atlantic coast. Nearby was a friendly, kind-faced goat herder. I smoked my pipe while waiting for a ferry in Tangiers to take me across the Straits of Gibraltar, from Africa to Europe.

Chapter Eighteen – Europe

Europe was an entirely different adventure from Africa and South America. We would be in modern "civilized" countries with cultures much like our own, and would be traveling on mostly paved roads, with road signs! My story will no longer be a day by day account, but abbreviated, selected excerpts. During our 8 months in Europe Noble and I frequently did not travel together, as we wanted to visit different places and see different attractions. On these occasions Noble drove the Jeep and I either hitchhiked or traveled by train (which was very inexpensive). We both meandered back & forth across the continent. My birdwatching was limited but I enjoyed taking pictures, visiting my twin sister and her husband (in both Holland and Austria), and stopping by in Denmark to say hello to Ruth (from the sisal estate in Tanganyika). The Palearctic avifauna (Europe and much of Asia) was closely related to that of the Nearctic (Canada and the USA), and was not as diverse as in either the Neotropical (C. & S. America) or Ethiopian (Africa) regions. I had visited Europe previously, briefly in the summers of 1954 & 1955 when I was a midshipman in the NROTC, and was thus familiar with the more common and widespread species. Since I had just encountered many of the Palearctic breeding birds which wintered in Africa, my trip list of birds did not increase very much when Noble and I were in Europe. Appendices B & C summarize our travels, by date and country, and the birds which were added to my list.

Spain

I arrived on the southern coast of Spain, in Algeciras, on May 28th. It wasn't until 5 weeks later, on July 2nd, that Noble finally joined me, in Granada. In the meantime, I spent my days wandering along the country roads or in the small villages taking pictures, looking at birds, and trying to communicate with the local residents. My initial plan was to sleep out on the ground at night in my sleeping bag under a tree, saving expense for lodging. The first night, in a grove of cork oak trees, I was kept awake by a group of goats with very loud bells hanging around their necks, which clanged and banged around me all night long. The second night, in an olive grove, it rained all night and since I had no waterproof protection I got very soaked and wet. After these two initial disasters, I happily discovered that there were small, cheap "pensions" with clean beds almost everywhere which cost the equivalent of only 20 cents per night, and I always slept in such places from then on. My normal food cost was 15 cents a day - - for coffee, a loaf of bread, and a banana (occasionally more).

I was stared at by most of the persons I encountered because the authoritative dictator of Spain, Franco, had only in the last two years opened the borders of his country to foreign tourists, who were closely watched by the local "Guardia Civil." Most people had not seen a bearded backpacker with unruly hair and dressed in a short-sleeved T-shirt, shorts, and hiking shoes, carrying binoculars and a camera. I was walking through a small seaside town one morning and noticed that 15-20 men and boys (and a very few women or girls) were closely following me through the streets. When I turned around and asked the man just behind me (in Spanish) why everyone was following me through the streets, he responded (in Spanish) that people had never seen an "extranjero" (foreigner) before! On another occasion, I was walking in the countryside and came across two very colorfully dressed peasant girls hoeing a new field of cotton. I thought they would make a nice photographic memory for me so I walked out toward them to get a little closer for my picture. One of the two girls immediately fled as fast as she could, but the other girl stood her ground as I approached and raised her hoe over her shoulder, making it very apparent that if I got close enough she would hit me with her hoe. So I stopped and asked her in Spanish why she was afraid of me. Her answer, in Spanish, was that she had "never seen a man like you before." Another memory is of an evening in a small town where I took a young girl to a movie, after which she asked me if I would marry her!

I spent 3 days (June 4, 5, & 6) hiking, taking pictures, and recording birds in the Sierra Nevada Mountains just east of the historic city of Granada. The cooler atmosphere between 2,000 & 8,000 ft. elevations was a refreshing change from the stifling summer heat of lower elevations. The mountains were characterized by cliffs, streams, hillsides with scrub and open woodlands, and at the higher elevations by sparse vegetation and patches of snow. I tallied a total of 51 species in my 3 days there during about 20 hours of birdwatching. These included 12 trip additions, 6 of which I had never previously seen (Appendix B). Particularly noteworthy were several Rock Petronias (family Passeridae), 3 or 4 Red-legged Partridges, 1 unidentified *Aquila* eagle, 2 Alpine Accentors (at 8,250 ft. elevation; a species I had first seen on Mt. Fuji 3 years earlier in Japan), 140 Red-billed Choughs (also at 8,250 ft. elevation, which was as high as I walked), 1 Great Spotted Cuckoo, 2 Eurasian Griffons (vultures), 1 Subalpine Warbler (my only trip sighting of this species), and 3 Hoopoes. At the family level my notes included 5 sylviids, 5 muscicapids (3 species of which were wheatears), 5 motacillids (3 pipits & 2 wagtails), 5 corvids, 4 swallows, 4 fringillids, 3 emberizids (all in the genus *Emberiza*), 2 turdids, 2 larks, and 2 tits (Paridae). (After Noble arrived we would return together to this site in the Jeep, at his request.)

When I was in Malaga, a large popular tourist city on the SE Spanish coast, I happened to encounter a young American boy about my same age by the name of Frederick Griscom, who by a remarkable coincidence was a twice removed cousin of the famed American ornithologist Ludlow Griscom! When I told him that I was planning to visit the "marismas" for 2 days of birdwatching at the mouth of the Guadalquivir River, he said he would like to accompany me. On our 2nd day of hitchhiking we were unable to obtain a ride and finished the day by walking more than 20 miles on hard, paved, hot highways, with our backpacks, which produced blisters on our feet for both of us. It was not a good beginning. The marismas were the wide flat grassy and marshy delta of the river at its mouth on the southern coast of Spain, on the Gulf of Cadiz near the city of Sanlucar de Barrameda, and were well known as a breeding and migration site for great numbers of all kinds of water birds - - "waders", herons, egrets, spoonbills, ibises, rails, gulls, terns, ducks, and others. It was very hot and there were swarms of mosquitoes which were such a nuisance that we retreated to the open breezy beach to try and sleep for the night (which only partially solved the problem). On the beach, we met a Guardia Civil policeman who made us pay him $1 each for an entry permit, since he told us we were in a protected area for wildlife.

On 2 days there (June 20 & 21) I documented 65 species of birds, including 8 trip additions (Appendix B). Some of the birds I listed were 40 Black Kites, 30 Azure-winged Magpies (with the questionable taxonomic decision of being conspecific with the very far separated, similar looking magpies in eastern & central Asia), 1 Short-toed Eagle, 45 White Storks, 25 Zitting Cisticolas (the most widespread member of the genus *Cisticola*), 2 Eurasian Buzzards (genus *Buteo*), 250 Collared Pratincoles, 400 Northern Lapwings, 1,200 Black-tailed Godwits, 350 Whiskered Terns, 2,500 Black-winged Stilts, 25 Little Terns, 10 Pied Avocets, 75 Great Crested Grebes (with many young), 5 Purple Herons, 1 Little Bittern, 3 Greater Flamingoes, 75 Marbled Ducks, 1,000 Common Redshanks, 2 Eurasian Thick-knees, and 2 Western Marsh-Harriers. During my two days of observation the only 3 species I had never seen previously were 8 Eurasian Spoonbills*, 1 Red Kite*, and 3 Green Woodpeckers*.

Noble finally joined me in Granada on July 2nd, 5 weeks after we had parted company in Casablanca. He wanted me to drive him up to the highest road elevation in the nearby Sierra Nevada Mts. (where I had already been), so I did. The end of the road was near the top of Veleta Peak at 11,000 ft. above sea level (which turned out to be our highest elevation in Europe, for either us or our Jeep). In the barren rocks at the top of the road I saw Common Ravens, Red-billed Choughs, Alpine Accentors, Northern Wheatears, Black Redstarts, and Eurasian Linnets.

We then drove together in our camping van to Madrid, and after several days there we drove 250 miles to the north to the city of Pamplona to participate in the annual weeklong "Festival of San Fermines" (from July 6-14). This very colorful spectacle features the running of bulls through city streets and attracts throngs of tourists from all over Spain & Europe, as well as such faraway places as N. America and Australia. For an entire week almost all work in the town comes to a halt, and all day long people dance and sing and drink wine in the streets. Small glasses of wine are sold for 15 cents each by street vendors standing every 20 meters along most of the streets. Every morning more than a dozen bulls run down the middle of 4 narrow city streets for a distance of 1,000 yards, chasing all those tourists and others foolish enough to try and outrun them, at a speed of about 15 mph. Noble participated one morning but I refrained from doing so. In the afternoon, there was a traditional bullfight in a huge arena with many thousands of spectators. One time was enough for me.

Noble and I were willing participants in the day long wine drinking and dancing in the streets, with as many different college age girls as possible. One of the girls I met was Margaret Rankin from Australia, and we became temporary friends for the week, drinking wine and dancing together. (She was traveling with 3 other girls from Australia.) One afternoon Margaret and I were sitting on a park bench together after both of us had consumed a considerable amount of wine. On the spur of the moment, I put one of my arms around her and kissed her on the mouth. Almost instantly I felt a very heavy hand on my shoulder and when I turned around here was a stern-faced Guardia Civil policeman who admonished me severely, proclaiming that kissing in public was prohibited by law! Our week ended too soon and Margaret and I parted company and went our separate ways, but not before exchanging names, addresses, and telephone numbers, as young people always did in such circumstances. Margaret lived in Perth, Western Australia.

Paris to Russia

I will fast forward to Paris, on Aug. 18. Noble and I wanted to travel in our Jeep through communist controlled East Europe to Russia, and then northward through Finland to the tip of the European continent in northern Norway. This would require obtaining visas for us and a permit for our Jeep to visit Russia, at the height of the Cold War in 1960. What were our chances?

We made an appointment with the Russian embassy in Paris, where we were directed to the office of "Intourist", the official government travel agency for Russia. Here we explained what we wanted to do. To our very pleasant surprise, we were told that such travel would be permitted under certain rules and restrictions: (1) we could travel only on designated highways, (2) we could not camp in our Jeep but would have to stay in government designated hotels, in designated cities, and when we departed from any hotel we had to inform the Intourist agent there what our next hotel would be, and (3) all of our 3 daily meals and each of our night's lodging had to be paid for ahead of time with a voucher from a booklet which we were required to purchase prior to our travel, at the time our visa was issued to us. Furthermore, this booklet had to be purchased from Intourist and paid for with Russian rubles! What a monopoly, but of course Russia was a communist country. When we inquired where we could purchase rubles to pay for our booklets you can guess the answer to this question - - from Intourist of course. Our next question was how many rubles we would be given for each dollar. The answer was 4, take it or leave it. (Keep this exchange rate in mind as I proceed with my story.) So that's what we did, and sure enough we were issued visas for ourselves and a permit for our Jeep. We were quite excited at the thought of driving into Russia! We paid for 13 days and 12 nights in the country. Noble determined it was costing each of us about $11.25 a day for food and lodging (which of course was far more than what we normally spent for those two essentials). It was also necessary for us to purchase as many rubles as we anticipated spending in Russia (for gasoline, souvenirs, entertainment, and other items), since we were allowed to take with us into the country only $100 each in cash for possible exchange at any government bank.

We outlined an itinerary to enter the country from Warsaw, in Poland, on Sept. 17, and then spend our 12 nights in the following Russian cities: Minsk (1 night), Smolensk (1 night), Moscow (5 nights), Kalinin (1 night),

Plate 73: I hitchhiked for five weeks in southern Spain while the Jeep was being repaired in Casablanca. The countryside reminded me of Latin America. After Noble arrived we attended a traditional bullfight in Pamplona.

Novgorod (1 night), and Leningrad (2 nights). We would then exit Russia from Leningrad, and drive to Helsinki, Finland, on Sept. 29. We were both very thrilled.

On Sept. 7, Noble and I departed Paris to begin our travel to Russia. We drove ENE through Luxembourg to W. Germany, then SE from Frankfort to Nurnberg and E to the border with Czechoslovakia, where we arrived on Sept. 10. Here there was a 12 ft. high barb-wired fence, police dogs, and guards armed with machine guns. It was the first country with communist rule which we had entered. We spent 2 nights in the capital city of Prague then continued eastward to the scenic Tatra Mts. along the polish border in the eastern part of the country, where we arrived on Sept. 14. I birded for several hours in the lovely, luxurious conifer-deciduous mixed forest between 5,000 & 6,000 ft. elevations. Here I saw my first ever Black Woodpecker* (a close relative of the Pileated Woodpecker), Eurasian (Spotted) Nutcracker* (a counterpart of the N. Am. Clark's Nutcracker), and Ring Ouzel* (an overall black *Turdus* similar to the Eurasian Blackbird but with a wide white collar). Other species in the forest here were the following trip additions: Golden Eagle (a Holarctic species), Eurasian Nuthatch, Song Thrush, Hooded Crow, and Eurasian Siskin.

We ate dinner that night in a small rustic hotel in the mountains. In the dining room with us in one corner were 4 elderly men playing checkers and smoking long-stemmed, curved pipes. After dinner, I lighted up my own pipe, and shortly thereafter one of the men playing checkers walked over to our table. He very much liked the smell of my pipe tobacco (which I had recently purchased in Germany because of its curious name of "Kansas"). He gestured that he would like a little of my tobacco for his pipe. Since there was not very much left I handed him the whole pouch, which he received very gratefully. He then walked back to his table and immediately returned with a gift for me, a small oil painting of the mountains which he himself had painted that day. It was a very well done little painting and I thanked him sincerely. Today it is one of my most fond travel memories, and is displayed in our family room on the wall near our fireplace. It was such pleasant, unexpected people to people interactions that were among the most rewarding of all our travel experiences.

Noble and I drove north across the border into Poland the next day (Sept. 15), to the city of Krakow and then on to the capital city of Warsaw. What a distressing sight the country was, having been rebuilt very little since the end of the war. The hard working, proud, patriotic Polish people were under a brutal communist regime which had done virtually nothing for the people. Everyone fervently hated the Russians (and the Germans). We were asked repeatedly if we could help them come to the USA (which of course we couldn't). Offering them our sympathy wasn't any assistance. We spent only 2 nights in the country. In a field along our highway was a large flock of 500 migrating Northern Lapwings. It was a relatively short drive eastward from Warsaw to the border of Russia.

We went through immigration and customs at the Russian border city of Brest, on the morning of Sept. 17th. It required 2 hours for the Russians to inspect our van thoroughly, inside and out, including putting our van up on a hoist to examine the under carriage. Luckily, our little locked safe with all our (illegal) dollars was never discovered from its secret hiding place. The officials carefully looked at our books and documents, noting that we possessed only one Bible (Noble's), which was all we were permitted to bring into the country with us. We were questioned about our correspondence with the U.S. Navy but our answers were satisfactory. We had removed all firearms and weapons of any kind, leaving them in custody outside Russia with friends or relatives. Surprisingly, in 2 hours of searching the Russians found nothing at all to confiscate.

As we got underway for Minsk the countryside was mostly farmland, with scattered patches of birch, willow, and fir woodlands, interspersed with lakes and canals. It was autumn and many of the trees were a colorful yellow. Our highway was pavement in relatively good condition. Sugar beets were the most frequent crop, and some sort of grain was spread out drying on the shoulders of the highway. Over 40% of the population in Russia worked on farms, where the majority of workers were middle aged women. Small girls tended goats in the fields. There were virtually no vehicles on the road but occasionally we passed horse drawn carts or wagons. Billboards cluttered the landscape, with government "propaganda" extolling the benefits of communism and the economic progress of the country. The usual advertisements for commercial products were nonexistent. Instead there were statues of ballet dancers, portraits of Lenin, and replicas of wildlife, such as elk. I stopped once and got out of our Jeep to take a picture of a local farmer who was chopping firewood in front of his house.

Almost immediately from out of nowhere a uniformed policeman appeared and stationed himself between me and the farmer, in a rigid, military "at rest" position with one arm behind his back. No words were exchanged between anyone, but the farmer hastily retreated inside his house. On another occasion at a road junction I mistakenly took the wrong highway, but within 2 miles there was a police checkpoint and I was motioned, not unpleasantly, to turn around and go back. Gasoline stations were very few and far between but fortunately we always had plenty with our 50 gallon capacity. Communism had a long way to go economically to catch up with capitalism.

Between Brest and Minsk on Sept. 17, I counted 1 European Roller and approximately 750 migrating Barn Swallows, and the next day (Sept. 18) between Minsk and Smolensk, I documented the following birds: 2 Eurasian Sparrowhawks (an *Accipiter*), 50 Hooded Crows, 1,500 Rooks, 800 Jackdaws, 100 Eurasian Jays, 60 Eurasian Magpies, 3,000 European Starlings, 2 Red-backed Shrikes, 10 White Wagtails, and 15 Eurasian Tree Sparrows (genus *Passer*, a close relative of the House Sparrow). A majority of these were autumn migrants heading south.

We arrived in Moscow on the afternoon of Sept. 19 and checked into the Metropol Hotel, a clean, relatively new hotel popular with foreign businessmen, and of course approved by the government for tourists. It was necessary to park our camping van on the street in front of the hotel because the hotel had no parking lot, but there was plenty of space on the street because so few people owned cars, and there were virtually no other tourists. The wide city streets were almost completely void of privately owned vehicles, with the traffic being only buses, taxis, and delivery trucks. As on the highways, gasoline stations in the city were very sparse and hard to find.

We had a university girl with us for the entire time we were in Moscow, as a translator and travel guide provided to us by Intourist at no cost to us. Galina spoke good English but she was quite stiff in her personality and not very talkative. Never-the-less, she was very much an asset to have with us most of the time, although on one occasion she was not helpful at all. I had turned the wrong way on a one-way street, and a sidewalk policeman blew his whistle at me to stop. In an irate manner, he began to write out a ticket for me. I asked Galina to explain to him that I was a tourist and couldn't read the one-way sign in Russian, but she only said to me "I can't do that", and would say nothing to the policeman at all. Eventually the policeman gave up shouting at me and in an exasperated manner motioned for me to turn around and go back.

On our 2nd morning in Moscow as Noble and I were leaving the hotel a young American businessman also staying at the hotel (the only other American patron) said to us "I suppose you boys know you are being followed when you leave the hotel." No, we didn't. (It didn't seem necessary since we had a Russian guide with us.) However, as I drove off with Noble and Galina that morning I looked in my rear vision mirror and sure enough a big black limousine immediately began following us - - and continued to do so throughout the entire day, everywhere we went. I thought to myself that surely the Russian security police (the KGB?) had something better to do with their time and money. Our van was very dirty and badly in need of a wash, which allowed an unknown passerby who was knowledgeable in English to take his (or her) finger and in the dust on the back of our van write the following message: "We know you are good boys and we like the American people but we don't like the American government." It was not the first time in our travels that such sentiments had been expressed to us.

Being an aspiring entrepreneur Noble was very interested to see if he could exchange $20 for some rubles at any government owned bank (where the rates were all the same, being set by the government). Galina agreed to accompany us to a bank, and to act as an interpreter for Noble. We had been told by Intourist at the Russian embassy in Paris that we could each exchange up to a maximum of $100 each during our stay in Russia. You will remember that at the embassy we were sold rubles at the rate of 4:1. So what was the official exchange rate at a government bank in Moscow? To our astonishment, it was 10 rubles for 1 dollar, 2½ times more than in Paris even though in both cases the seller was the same - - the Russian government! This was certainly a first experience for both of us and a particularly strong memory for Noble, to inscribe in his daily journal.

There is yet more to this story. That same night after we had eaten our evening meal at the Metropol (each paid for with one of our coupons), Noble and I took a stroll along the street outside. It was already dark. Soon we encountered a college-aged boy walking toward us on the street. He recognized us as tourists and in very broken English asked if we had any money which we would like to exchange. Of course, both Noble and I knew that to do so was against the law, and a person could be sent to jail if caught. (For all we knew the boy was an undercover agent for the government.) Noble did not respond to the boy's question, so illogically and uncharacteristically (without sensibly considering the possible consequences) I took the initiative and asked him what kind of money he wanted. He replied either American dollars or British pounds. I pondered this briefly and then continued the conversation by asking him how many rubles he would give me for one dollar. His answer was 40! I thought his conversion rate must be faulty so I said to him "that means if I give you 20 dollars you will give me 800 rubles?" He replied, "yes." I was so astonished that I immediately reached into my pocket to get a 20 dollar bill. (Noble still said nothing.) "No, not here" the boy said, "come with me." Now was my chance to terminate this risky undertaking, but as dangerous as this suggestion was I followed the boy around a corner and into a dark vacant entranceway, where he stopped and I handed him a 20 dollar bill. He quickly gave me a huge wad of bills and then hurriedly disappeared down the street. Noble and I returned to our hotel room where I sat down and carefully counted the money I had been given. It totaled exactly 800 rubles! What a bargain. (I would later discover that all the money was authentic, not counterfeit, and I was never arrested by the security police.) Now I had to find something to buy inside Russia with all my blackmarket currency, since rubles were worthless outside Russia.

Of course, we visited all the popular tourist sites in Moscow, most notably the Kremlin with its famous Red Square, St. Basil's Cathedral (built in 1679), palaces, and historic gold-domed orthodox church of Byzantine architecture (now a national museum). We also toured the 36-storied national university, the Lenin Library, and the world famous, ornately decorated Moscow subway, constructed in 1935 with marble walls, murals, and chandeliers, of which Russians were justly proud. We attended the Bolshoi Theater on two different nights, one time to see an excellent ballet performance and the other time to sit through a very poor rendition of the opera Carmen. We watched a movie on another occasion where there was a short cartoon showing a Russian missile shooting down Gary Power's U-2 spy plane (an event that had occurred just 4 months earlier and greatly alienated and angered many Russians toward the USA). Never-the-less, a majority of the Russians we met were curious and friendly, although only a very few spoke any English at all. Most Russians had been convinced by their government's constant barrage of propaganda that communism was superior to free enterprise, and would eventually provide everyone with a better life, even though current productivity and wages were very low, consumer goods were scarce, and people stood in long lines waiting to get inside shops just to purchase basic commodities. A pair of shoes cost one month's wage. Communism replaced the church for providing hope to the people that their lives would get better if they just had faith. In 1960, life was very bleak for most Russians.

Noble and I departed Moscow on Sept. 24 in our Jeep and headed NW toward Leningrad, 500 miles distant. We arrived there 2 days later, having spent one night in Kalinin and one night in Novgorod on the way. The countryside was low (500 ft. above sea level), flat, and mostly very wet with lakes, marshes, streams, rivers, and canals. Much of the time we were in the Volga River watershed. There were many fields, and patches of woodland with mixed fir, aspen, birch, willow, and alders, with colorful yellow leaves. Log production was a major industry. Roadside birds which I documented included Eurasian Buzzard, Eurasian Kestrel, Mew (Common)

Plate 74: Russian peasant women worked long hours daily harvesting sugar beets. The Kremlin was an important sightseeing stop. Between Moscow & Leningrad, as everywhere in Russia, were roadside billboards extolling the virtues and successes of communism.

Gull, Black-headed Gull, Eurasian Skylark, Barn Swallow, Rook, Eurasian Magpie, Eurasian Jay, Red-throated Pipit, White Wagtail, Northern Shrike, Brambling, Chaffinch, Yellowhammer, and House Sparrow. The 3 most numerous species, in order of abundance, were Eurasian Jackdaws, European Starlings, and Hooded Crows.

Leningrad (later changed to St. Petersburg) was known as the "Venice of the north" because of its many canals. It was a very picturesque city, even though skies were Gray and overcast for the entire duration of our two full days there. The most popular tourist attractions were the world-renowned Hermitage art museum, the statue of Peter the Great on horseback, the Smolny Cathedral, the Winter Palace, the St. Petersburg Mosque (the largest mosque in Europe outside of Turkey), and of course the great many canals with their boat rides.

On the first of our 2 days (Sept. 27) I was walking by myself and chanced to meet a pleasant university graduate student who, along with his wife, was studying chemistry. He spoke very good English and we stopped on the street to chat with one another. I had seen a rather expensive 2 volume set of books on Russian birds (written in Russian) in the Moscow University, and I asked the chemistry student if he might be able to find and purchase these books for me (which I could pay for with my blackmarket rubles). He agreed to try, and then he invited both Noble and me to have dinner with him and his wife that evening at 6:00 pm in their small upstairs student apartment. He wrote down their address and directions how to get there on a piece of paper which he gave to me. I thanked him and said we would be very happy to eat with them. Noble and I arrived exactly on time and were walking up the outside wooden back stairs to his very modest apartment when he opened his door and came down the stairs toward us. He passed us on the steps without stopping and quickly said to us in a sincere low voice as he passed "I am sorry but I cannot eat with you." He continued walking off, down the street at the bottom of the stairs, without glancing back at us. I can only assume that a plain clothed security agent had seen him talking with me earlier in the day, on the street, and told him that he was not to talk with me again. Life in Russia was not easy for the people. They were constantly spied upon, without knowing who was watching them, or when.

Now I had to find something else to buy with my 800 rubles before I left Russia, as they would be worthless outside the country. Ordinary tourist souvenirs were almost nonexistent in any shops in 1960 (as were tourists themselves). I happened to find a shop selling a wide variety of goods and household items, and I walked inside to look around. After 10 or 15 minutes when I was about to leave I noticed on a back shelf almost out of sight a dust covered black, solid cast iron statue. It depicted a Cossack warrior with a traditional cap, coat, and boots sitting on his horse with a cased rifle over his shoulder and leaning over to hold a girl up to him as he kissed her farewell. I knew immediately that I had to have this statue. It was a perfect souvenir for me of Russia. When I asked a salesgirl (through hand gesturing) if I could have a closer look she was quite amazed. Apparently, the statue had been sitting on the shelf a very long time without anyone even wanting to look at it, as evidenced by its covering of dust. (Very few Russians could have afforded such a non-essential, luxury item.) I was surprised when I picked it up how heavy it was (20 lbs. I discovered later), and it stood 16" high, 13" long, and 6½" wide. It was not something one could easily carry or transport. The price of 600 rubles was indeed quite expensive for almost any working person, but with my extraordinary black market exchange rate it came to the equivalent of $15 for me. I was happy to have found something I liked on which to spend my money (before it became useless outside Russia). Today this statue sits in front of my living room fireplace. With my remaining $5 worth of rubles I purchased more than a dozen old time vinyl, long-playing 33 rpm phonograph records of Tchaikovsky songs, which the government had underpriced because of their Russian cultural significance.

Noble and I left Leningrad on Sept. 29 and drove 90 miles NW to the city of Vyborg, on the NE end of the long, narrow Gulf of Finland at the eastern end of the Baltic Sea. We were still in Russia but it was only 50 mi. west to the border of Finland, on the way to Helsinki. We had experienced 13 memorable days in Russia. Just before reaching the border at 2:00pm I stopped in a mixed forest and field habitat for 15 minutes to pursue birds in Russia for one last time. Here I added 3 species to my trip (and life) lists: 2 Black Grouse*, 20 Fieldfares* (a

Blackbird sized *Turdus* thrush with a Gray head, chestnut back, and black spotting or streaking on its breast and flanks), and 5 Redwings* (yet another *Turdus* thrush, smaller in size with heavy black spots over most of its underparts and reddish flanks and under wing coverts). The border crossing into Finland that afternoon was without incident, and we both breathed a sigh of relief when we left Russia and communism behind us, not withstanding our memorable adventures there.

Finland and Norway

As we drove west in Finland toward Helsinki along the northern shore of the Gulf of Finland on that same afternoon I observed 5 species of ducks which were further additions to my trip list on this date: Eurasian Wigeon, Common Goldeneye, Northern Pintail, Tufted Duck, and Common Pochard. We camped that night along the Gulf of Finland west of the town of Kokta. The next morning (Sept. 30) we drove to Helsinki, arriving at noon. Along the way, I documented 2 more trip birds, the Common Crane* (filling the same niche in Eurasia as the Sandhill Crane does in N. America) and the Rough-legged Hawk (another species of Holarctic - - circumpolar - - distribution).

Having driven to the southern tip of continental South America, as far as one could by road, Noble and I wanted to drive to the northern tip of the European continent, as far as one could by road. (In both cases one could then ferry to a nearby island and continue somewhat farther, south on the large island of Tierra del Fuego or north on the small island of Mageroy.) Therefore, we planned a route to drive north in Scandinavia, traveling across the Arctic Circle in Finland and into "Lapland" in northern Norway, until we reached the northern end of the road. Before starting out on this adventure we relaxed for one day in Helsinki (Oct. 1) where we experienced a traditional hot sauna, followed afterward by sitting outside in the cold for a short time. (There was no snow present yet.) We also enjoyed a 4-course dinner with smoked dried fish, and strolled along the picturesque waterfront to take photos of the great many fishing boats in all sizes and shapes. Fishing was one of the most important industries of Finland. People everywhere were very warm and friendly and a surprising number could speak at least a little English. (The Finnish language was quite distinct from other Scandinavian tongues.)

On Oct. 2nd, we headed north from Helsinki for a 6-day journey on well-maintained gravel roads for 600 miles to the quaint little town of Honningsvog on the small island of Mageroy, just off the northern tip of the European continent. Our route would take us through the following towns and villages, in order: FINLAND - - Helsinki to Tampere to Jyvaskyla to Oulu to Kemi to Rovaniemi (on the Arctic Circle at 66½°N) to Ivalo (on Lake Inari), to NORWAY - - Karasjok to Lakselv to Russenes (next to Kistrand on Porsangerfiord) to Honningsvog (by ferry), where we would arrive on Oct. 7. Honningsvog was situated almost exactly on the same latitude, 71°N, as the town of Barrow on the northern coast of Alaska (where it was the northernmost town on the N. Am. Continent).

The countryside through which we traveled was remarkably like that of Minnesota, with 60,000 lakes (rather than 10,000)! There were swamps, tamarack bogs, mosses, lichens, birches, alders, willows, pines, firs, spruces, scattered open fields, red barns (for cattle or sheep) and occasional small villages or towns. Squirrels scampered through the forests, birds flitted across the fields or between the trees, and yellow Birch and Aspen leaves added lovely autumn color. There was a gorgeous full moon over Lake Inari at our campsite on Oct. 5. As we traveled farther north the forest became predominantly coniferous, a circumpolar "taiga" forest which formed a wide band around the world south of the arctic tundra.

Some of the birds which I documented were Great Crested Grebe, Northern Goshawk, Rough-legged Hawk, Eurasian Capercaille* (a very large grouse of coniferous forests), Great & Lesser Spotted* woodpeckers, Black Grouse (a bird of swampy heath lands in which the male was entirely black with a conspicuous lyre-shaped tail, white wing bar, and small red bare area above the eye, which I first saw on our last day in Russia), Northern Hawk-Owl* (which I watched catch and eat a small rodent in open birch scrub), Black Woodpecker, Parrot Crossbill*, Gray-headed Chickadee* (Siberian Tit), Common & Hoary* redpolls, Siberian Jay*, and Bohemian

Waxwing. The northern forests could be either very still and quiet or active with birds. It was an awe inspiring environment and one which brought back fond memories of my university years in Michigan. An unusually early snow storm on the night of Oct. 5 covered our road and the whole countryside with several inches of snow, northward from the town of Karasjok at 69½° latitude in far northern Norway. The wind created drifts between 1 & 2 ft. high along the roadside and across the road, which greatly complicated our travel northward and our efforts to reach North Cape (Nordkapp).

The far north of the Scandinavian Peninsula and the adjacent Kola Peninsula of Russia collectively comprised an ethnic region known as "Lapland" (Lappland, or Sapmi). This region was bordered by the Barents Sea on the north and east, the Norwegian Sea on the west, and the Arctic Circle on the south. The indigenous European people in this region were most often referred to as "Lapps", but they preferred to be called the "Sami" people. They always dressed in their colorful, traditional reindeer fur & hide caps, shawls, mittens, jackets (men) or dresses (women), with high reindeer leather fur-lined boots. Reindeer (called Caribou in N. Am.) were an important part of their life, providing them with food, clothing, and transportation. Their bright red, blue, green, and white clothing was decorated with beads, embroidery, and copper or silver jewelry. About 10% of the total population was semi-nomadic, following their reindeer herds throughout the year. We first encountered a few Lapps in Karasjok, and thereafter sparsely along the road and in small villages all the way to Honningsvog and the island of Mageroy. Unfortunately, they tried to avoid my camera and I succeeded in getting only a few pictures of them.

On the morning of Oct. 7, Noble and I arrived in far northern Norway at the tiny port town of Russenes, at 70.5° latitude on the west side of the southern end of the long north to south fiord known as Porsangerfiord, one of the longest fiords in all of Norway. From the northern mouth of this fiord it was only 5-10 miles across the Mageroysund Strait to the town of Honningsvog on the small island of Mageroy. The northern tip of this island was known as Nordkapp (North Cape in English) and was claimed as the northernmost tip of Europe (even though it was not actually situated on the continent). A 20 mile road connected Honnigsvog to North Cape, at 71.2°N latitude. This was the destination which Noble and I hoped to reach.

In Russenes the ground was covered with a few inches of snow and there were a few Lapps shopping in the village. Noble and I booked passage for us and our Jeep on a small ferry leaving at 10:00am for Honningsvog (Oct. 7). Our 9,000 lb. camping van was almost too heavy to be hoisted aboard the ferry, and the hoist creaked and swayed while I took pictures and hoped it wouldn't collapse. Although Honningsvog was only 40 miles away (mostly in the fiord) it took our little ferry boat 4 hours to make the journey since its maximum rate of speed was only 10 knots. While we were traveling northward down the fiord toward its mouth I counted 1 Yellow-billed Loon* (very similar in appearance to the Common Loon), 60 Long-tailed Ducks (which I still prefer to call Old-squaws), 5 European Shags, 200 Common Eiders, 3 jaegars (most likely Parasitic), 8 Black-backed Gulls, 350 Herring Gulls, 5 Mew Gulls, 500 Black-legged Kittiwakes, 10 Arctic Terns, 2 Razorbills (an alcid), 75 Black Guillemots, 1 Common Murre, 2 Atlantic Puffins, 15 Common Ravens, 5 Hooded Crows, and 2 Eurasian Magpies.

When we finally arrived in Honningsvog at 2:30 pm, on Mageroy Island, we were discouraged to find that the snowstorm of several days ago had deposited considerably more snow on the island than it had on the mainland. The island's one snowplow had been very busy trying to clear the city streets of snow, and the 6-10 inches of snow covering the 20 mile road to Nordkapp. We would have to wait until tomorrow morning to see how well the plow succeeded. At this latitude (almost exactly 71°N) in early October there were 10 hours between sunrise and sunset (0700-1700) with long periods of pink twilight at dawn and dusk, and the sun very low above the horizon at noon. At night there was a lovely whitish aurora borealis. The harbor was open with no accumulation of ice on the surface. As we were the only tourists in town we were invited by the owner of the one restaurant to have dinner with him, where we drank local spirits and participated in a lively long conversation (in English) well into the night, where Noble took the lead in talking entrepreneurship, social injustices, and the local economy.

The next morning (Oct. 8) Noble and I set off to see how far we could get across the island toward Nordkapp, only 20 road miles NE and a gain of just 0.2° in latitude. The answer was not very far. The snow plow had pushed a path in the middle of the road, piling up snow as high as 2 or 3 ft. along the sides of the road as it did so. The snow on the road gradually deepened, and after just 7½ miles the snow plow could make no more headway. The road ahead was blocked by almost 2 feet of snow. Of course, this was as far as Noble and I got in our Jeep, with the snow on the road in front of us at a level up to our headlights, blocking any further progress. We recorded this situation - - the end of the road for us - - with pictures of us, our Jeep, and a nearby sign which said "71° Nord." This point was our greatest distance from the equator during our 3 years of travel.

The next day (Oct. 9) we boarded a ferry boat to take us back to the mainland via a different route, to the prominent town of Hammerfest on the west coast of Norway, in a fiord situated on the opposite side of the peninsula from Russenes (at the same latitude of 70.5°). From here we began our long drive back down the Scandinavian Peninsula along the west coast of Norway, curving alongside and around its many fiords. The last 2 trip additions to my bird list in far northern Norway were, fittingly enough, the White-tailed Eagle* (a close relative and ecological counterpart of the Bald Eagle) and the Snow Bunting!

We now began a long, irregular, wandering route of travel for almost 4 months, back and forth south, east, and west across Europe, terminating finally at Istanbul, Turkey, our last destination on the continent. Driving on paved highways, crowded, narrow city streets, or wide boulevards with bumper to bumper traffic in our bulky camping van with very limited rear view vision was a new challenge for both of us. Our adventure had taken an 8-month detour from the primitive roads of Central America, South America, and Africa. (They would return in Asia.) My trip list of birds grew by just 162 species (Appendix B) during our European "holiday." Two continents remained.

Plate 75: Heavy snow already covered the ground on October 7 on the island of Mageroy at 71° N latitude in Norway. (Note our tire chains.) This was the heart of "Lapland." Half way across the island we reached the end of our snow-plowed road.

Chapter Nineteen -- Turkey, the Middle East, & Afghanistan

On Feb. 6, 1961, Noble and I crossed the very narrow Bosporus Strait at Istanbul, Turkey, on a 20 minute ferry boat ride, leaving Europe behind us and entering Asian Turkey. As we headed east along the southern shore of the Black Sea there were more than 1,000 Hooded Crows along a 50 mile stretch of highway. I also counted 75 Yellow-legged Gulls, 40 Black-headed Gulls, 100 Eurasian Magpies, 20 Rooks, 150 Eurasian Jackdaws, 25 Mistle Thrushes, 250 Fieldfares, 200 European Goldfinches, and 175 Chaffinches. My assumption was they were mostly breeding birds from farther north, wintering here in the southern Palearctic. The avifauna of Turkey was not much different from that of Europe. All of Asia Minor, the Near East, and the Middle East were included in the Palearctic zoogeographic region, as far east as Afghanistan and northwestern Pakistan. In addition to Peterson's European field guide I had 2 other ornithological references to help me identify birds in this part of the world: (1) Vaurie's "The Birds of the Palearctic Fauna, Order Passeriformes", published in 1959 (with no illustrations of any kind), and (2) the now discredited "Birds of Arabia" by Meinhertzhagen, published in 1954 with very few illustrations. Both of these books were brought with me from California.

As we drove across the central plateau of Turkey south of Ankara on Feb. 8, I enjoyed seeing my first Ruddy Shelducks*, a truly handsome duck. The next day our route across the Anatolian Plateau took us along the western edge of the vast "Tuz Golu" salt flat, at an elevation of 3,000 ft. between the cities of Ankara and Konya. Here we stopped for more than an hour to photograph a large group of workers shoveling salt into 100 lb. bags and then loading two such bags on a kneeling camel, one on each side of their back. All of the labor was very heavy, manually intensive. When 4 or 5 camels had been loaded they were led off together across the desert to an unknown destination. In this same area a large group of 20 Black-bellied Sandgrouse flew past our Jeep, and later another pair also flew by. I always wondered just how far these birds flew in their twice daily round trips to and from water. Further along the plateau I was happy to closely observe and identify a Saker Falcon, since my only other record of this species (from the Serengeti Plain in Tanganyika on Jan. 20, 1960), was a little less certain.

On the morning of Feb.10 we drove across a mountain pass in southern Turkey north of the city of Mut, at an elevation of 6,000 feet. Here we encountered a heavy snowstorm which produced several inches of snow on our highway, making it almost impossible for us to ascend to the pass. With considerable difficulty we plowed through the snow in 4-wheel drive and our lowest gear, finally making it to the top. We took pictures of ourselves all covered with snow, both inside and outside our van. As we descended from the southern side of the pass we were in a very rugged area of rocky cliffs and gorges with pines, junipers, and smaller scrubs. Two Long-legged Buzzards were very strikingly patterned in rufous and white, with prominent black carpals. There were also kestrels, Crested Larks, a single Crag Martin, many Hooded Crows, magpies, a pair of Blue Tits, 7 Rock Nuthatches, a Winter Wren (the only one of 80 species in the family Troglodytidae which lived in the Old World), Fieldfares, Eurasian Blackbirds, stonechats, 3 Black Redstarts, a Blue Rock-Thrush, 5 Sardinian Warblers, goldfinches, chaffinches, serins, 2 Dunnocks (an accentor), House Sparrows, and a single female Cirl Bunting. We drove between Adana and Antakya (Antioch) on

Feb. 11 and arrived at the Turkey/Syrian border that afternoon, west of the Syrian city of Aleppo. A White-spectacled Bulbul* (*Pycnonotus xanthopygus*) was a trip addition to my bird list prior to crossing the border.

On the outside of the door to our van we had painted a list of the names of the countries we intended to visit. The name "Israel" was included on our list. However, at the Syrian border the officials there made us take out a pocket knife and scratch out completely this name. Tensions were very high between all the eastern Mediterranean countries and Israel, which had been designated a Jewish nation by the United Nations only a dozen years earlier, in 1948, when a decision initiated by Britain and the United States gave Palestine to the Jews, since it was their ancient homeland which they say God gave to them, and where Christianity arose. Most of the Palestinians then fled to the West Bank of Jordan, and Israel became the only nation in the entire region where a majority of the population was non-Muslim.

Noble and I drove from Homs in Syria to Beirut, Lebanon, and after 3 days in Lebanon we drove back into Syria again, to Damascus. From there we drove south into Jordan, crossing the border at Dara (in Syria) on Feb. 19, and then continuing south to the Jordanian capital city of Amman, not far NE of the Dead Sea. At the Iraq embassy in Amman, Noble and I obtained visas in our passports for the two of us to visit Iraq, but we were told it might take 1 or more weeks before a permit from Baghdad could be obtained for our camping van. While we were waiting for this permit to arrive we drove to the nearby Dead Sea, center of biblical history and site of the "Dead Sea scrolls." It was a popular tourist destination. We also visited eastern Jerusalem one day, where we walked down some of the ancient narrow streets, took pictures of the Garden of Gethsemane, and drank tea at a small sidewalk café. (Eastern Jerusalem belonged to Jordan, not Israel, as mandated in 1948.)

The Dead Sea was the earth's lowest land elevation, with its surface being 1,400 ft. below sea level. It was an endorheic lake (with no outlet) and was fed only by the Jordan River, coming from the north. The lake was 30 miles long and 10 miles wide, with a salinity of 34% (more than 10 times that of sea water) and a maximum depth of 1,000 feet (which was quite deep for a saline lake). No fishes lived in the lake and the only kinds of life were bacteria, algae, and fungi - - hence the name "dead" sea. Some of the birdlife which I found in the vicinity of the sea were the Sand Partridge* (also called See-see), Brown-necked and Fan-tailed* ravens, Desert Lark, Tristram's Starling* (or Grackle, genus *Onychognathus*), Mourning*, Desert, & White-rumped wheatears (many singing), Blackstart* (*Cercomela melanura*), Blue Rock-Thrush, European Stonechat, Chiffchaff, Graceful Prinia*, Zitting Cisticola, Spectacled & Sardinian warblers, Eurasian Linnet - - and 9, singing, Dead Sea Sparrows* (in the genus *Passer*)! It was already springtime here in this part of the Middle East.

After waiting a week for our vehicle permit to arrive, we decided to drive south in Jordan for a few days to witness the famous archaeological ruins of Petra. We would then continue further south to the city of Aqaba, the southernmost Jordanian city at the northern end of the Red Sea, at the tip of the Gulf of Aqaba where 4 countries come together - - Jordan, Israel, Egypt, and Saudi Arabia. It was a spur of the moment plan.

When we arrived at Petra on Feb. 23 we were required to hire a young girl guide to show us around. This ancient city carved by hand out of towering red sandstone cliffs several hundred years before the time of Christ is one of the most spectacular of archaeological sites anywhere. It is one of the world's man-made wonders. Scientists marvel at the intricate, perfectly designed water conduits, dams, and cisterns which brought water from distant mountains and stored it for irrigation, bathing, drinking, and cooking. Petra was a major center on camel trade routes for 700 or 800 years before being destroyed beyond repair by two separate earthquakes, one in the 3rd century and one in the 5th century. It became a world heritage site in 1985. Birds which I saw in the vicinity of Petra included Isabelline Wheatear (singing), Streaked Scrub-Warbler* (in the monotypic genus *Scotocerca*), and Pale (Sinai) Rosefinch* (genus *Carpodacus*, my only sighting of this species).

It was approximately 75 miles from Petra south to Aqaba. The surrounding countryside was barren desert with scattered grasses, bushes, and acacias at an elevation of just 200 ft. above sea level. In the distance were rocky cliffs and ravines. There were no farms or villages or people and virtually no traffic on the road. As we

Plate 76: Istanbul is renowned for its famous "Blue Mosque" with six minarets. 100 lb. bags of salt were being loaded on camels at the "Tuz Golu" salt flat on the Anatolian Plateau between the cities of Ankara and Konya in central Turkey.

Plate 77: Rocky cropland was being plowed by bullocks and horses in northwest Syria. Colorfully dressed women and girls worked in the fields. Men prayed daily at mosques everywhere, as pictured here in Damascus, Syria's capital city.

approached to within 10 or 12 miles of the town in the middle of the morning on Feb. 24, I decided I would like to spend some time looking for birds in the area. I parked off to the side of the road and said to Noble he should drive on into the town and I would hitchhike there to join him in about 2 or 3 hours. He proposed that I keep the Jeep with me and he would hitchhike into town, and in a few hours we could then join up at the most prominent hotel there. (We practically never had any difficulty finding each other because no other persons looked like we did.) We were both happy with this plan, so I grabbed my camera, binoculars, compass, canteen, and hat, and set off across the hot, dry desert. Noble put a water bottle and a few items in a shoulder bag and stood by the roadside, where 10 minutes later a car came along and took him into Aqaba.

The desert was mostly pristine and gave me a thrill of anticipation as I began slowly meandering back and forth, scrutinizing the bushes for birds. As always, they were shy and difficult to observe as they flew from the middle of one shrub to the middle of the next, except on those occasions when they sang from the top of a thorny scrub or acacia tree. It was a challenging task. I soon encountered a Bedouin man in traditional dress walking across the desert with his camel, and with his permission I took his picture. By mid morning the temperature had rapidly warmed up and the avian activity had mostly ceased, so I decided to walk back to the Jeep and drive into town. In my 2 hours of desert birding I had observed only one trip addition for my bird list, the appropriately named Red Sea Warbler* (*Sylvia leucomelaena*). It had a black crown, gray back, and white underparts (very like an Orphean Warbler, which was only a migrant at this location and should not have been here yet at this early time of year).

When I arrived at the outskirts of the town there was the customary military check post. As I stopped a friendly looking young lieutenant who spoke almost perfect English walked up to my driver's side open window and said calmly "you are under arrest for spying." I was momentarily speechless, but then managed to say that I had not been spying but merely birdwatching. He remarked that several people had been watching my morning activities from a tall airport tower (which I had not noticed), and they all said I was spying. He got in the cab with me and told me to follow the car in front of us, which took us to the military headquarters. Here I was taken to an office where there were chairs, a table, a filing cabinet, a couch, a small bathroom, and not much else. I was told that I was to remain there until further notice, and my camera was taken from me. The young, not un-friendly, lieutenant then left the room, locking the door behind him and leaving me alone. So I sat and waited. Several hours later the lieutenant returned with Noble, and in my presence he asked him if I were spying. Of course he said no, only birdwatching. Noble was then allowed to leave.

In the early evening the lieutenant unlocked my door and came in again with a scribe, and asked me to tell him in my own words what it was I did in the desert that morning, and what I took pictures of. I tried to remember as best I could, and he translated this to the scribe, who wrote it all down, in Arabic. (Of course, I didn't really know what he told the scribe to write since he spoke to him only in Arabic.) I was then told that my statement would be compared with what the "witnesses" said who had watched me. A committee of 3 judges would then evaluate all of the written statements and decide whether I was spying or not. I would have to wait for them to make a decision. I concede that the process seemed to be as fair as possible given the circumstances, assuming that everyone told the truth.

In another hour, now about 8:00 pm, the young lieutenant once again unlocked my door and came into my room, holding a very long 8x15 inch piece of paper completely filled out in Arabic. He said to me "we have decided that you were not spying and if you will just sign this paper you may go." Of course I was elated at what he told me, but how did I really know what the paper said, since it was written entirely in Arabic, which looked like scribbling to me. So I asked him "what does this paper say?" He answered that it was the statement I gave to him telling him what I did in the desert that morning. I pondered this for a moment and then said to him "I don't want to sign anything I can't read." He looked at me straight in the face and in a calm, quiet tone he said to me "either you sign this or you will spend the rest of your life in jail here"! I thought to myself that if ever there were a lose/lose situation, this was it. Only one option seemed absolutely certain, and that was if I didn't sign it I would go to jail forever. Therefore, with great apprehension I signed the document. The young lieutenant then smiled at me, returned my camera (with all the film still inside), and apologized to me for the inconvenience he had caused, explaining that this location (Aqaba) was very strategically located and security was an utmost necessity. He was correct, of course. Life was an adventure!

During our continued wait for a vehicle permit we spent 3 days (Feb. 26-28) at the Azraq desert oasis located 65 miles east of Amman, where I could track down birds and Noble could write in his journal while he relaxed under the shade of date palm trees. This isolated wetlands at 1,500 ft. elevation encompassed mudflats, marshes, and a shallow lake where there were a great many waterfowl, sandpipers & plovers, herons, cranes, egrets, crakes, harriers, falcons, eagles and a wide variety of passerines. I had my shotgun with me and collected a few birds, both for identification and for eating. Only 4 birds were added to my trip list - - Greater Spotted Eagle*, Little Crake* (coll.), Temminck's Stint*, and Pin-tailed Sandgrouse*. One of the birds I collected was a Water Rail which was my 1st sighting of this species (although it was on my list as having been heard). I also shot 2 ducks and a dozen snipe which we fried up and ate for dinner.

Noble finally became impatient waiting to receive our vehicle permit at the Iraq embassy in Amman, so he told me that he would hitchhike to Baghdad and talk to the customs people in person, to get a permit issued. (He could do this with the visa in his passport, allowing him to enter.) His 600 mile route would follow along the Kirkuk-Tripoli oil pipeline for much of the distance. He said that after he had received permission he would telegraph the permit for our Jeep to me at the American embassy in Amman, and I could then drive it to Baghdad to pick him up there. We had overcome many different obstacles in our travels and this one seemed quite accomplishable. It would be the first time that Noble had hitchhiked for more than one day and left the Jeep with me. So off he went, on Mar. 5.

The day after he left I was walking along an isolated country road by myself with my camera and binoculars when I encountered a local rural farmer walking toward me from up the road. As he approached me he studied my appearance and then immediately pulled a dagger out from under his coat, raised his arm, held the dagger against my chest, and blurted out "Jew?" I responded in English (which he probably did not understand) that I was an American tourist and not a Jew. (Of course I could have been both.) He was not convinced and appeared about ready to stab me when by very good fortune a private car came along the road just at that exact moment, driven by a well-dressed man who spoke English. Upon seeing the man holding a dagger against my chest, he stopped and asked if I needed some assistance (which must have been obvious). Of course I said yes and explained the situation to him. The driver then turned to the farmer and spoke very harshly to him, at great length, in Arabic. The farmer immediately turned around and proceeded rapidly back in the direction from which he had come. My life had been spared one more time. This incident is evidence of the great dislike toward Israel, and Jews, throughout almost all of the Middle East.

On Mar. 8, I received a telegram from Noble saying a permit had been approved and would be sent to the Jordanian/Iraq border, where I could pick it up on my arrival there. The border was 300 miles away, half way between Amman and Baghdad. I left the next day, traveling on a relatively smooth very straight paved road which paralleled the pipeline across a largely uninhabited, scrubby, sandy and rocky desert. The only people along the way were scattered Bedouins with their camels and tents. I was not in a hurry and stopped regularly to identify the few roadside birds. By late afternoon I was nearing the border. It was half an hour before sunset and sandgrouse, both Black-bellied and Pin-tailed, were making their evening flight to and from a waterhole.

On the spur of the moment I drove off to the side of the road and across the desert for a short distance toward an apparent water source which was attracting sandgrouse. I stopped and retrieved my shotgun from the tool box under the back of the van where I always kept it, locked and out of sight. During the next 15 minutes I shot 4 or 5 Black-bellied Sandgrouse as they flew past, after which I cleaned them, put them in boiling water,

Plate 78: As we drove south in northern Jordan our road descended to below sea level. We camped one night at the Dead Sea, the world's lowest land elevation, 1,400 ft. below sea level. Squatters lived along the shore.

Plate 79: The magnificient palaces carved out of red sandstone cliffs at Petra in South Jordan are one of the archaeological wonders of the world. On my way across the desert from Jordan to Iraq, I shot Black-bellied Sandgrouse near the border at dusk.

and cooked them for dinner! The sun went down and I went to bed early, shoving my shotgun behind all the clothes hanging up in our small inside clothes closet. (This was also where I kept my navy ceremonial sword, which for a short while at the beginning of our adventure we hung from the roof of our cab for decoration, until we tired of retrieving it from the floor every time we hit a bump and it fell from its overhead hook.) I was camped at 3,000 ft. elevation in the solitude of the still, open desert, enjoying this vast biome, and the sunset, for yet another time.

A few of the 2 dozen birds which I recorded during the day (in addition to the sandgrouse) were Hoopoe, Egyptian Vulture, Cream-colored Courser, kestrels (which I couldn't distinguish to species), perhaps a Merlin, several kinds of unidentified larks, and SIX SPECIES OF WHEATEARS (Northern, Isabelline, Desert, Mourning, Finsch's, and White-tailed). As I have previously commented, the genus *Oenanthe* was one of my very favorite. Species in this genus were pleasantly tan, black, and white in color patterns, and were often conspicuous in their aerial courtship displays and singing. Wheatears were widespread and frequently common throughout much of the Palearctic Region, where some species occurred in the most barren and inhospitable deserts. They were distributed more sparsely throughout much of Africa, and one species, the Northern Wheatear, bred in the New World, in arctic N. America (as well as in the Palearctic). A total of 20 species were recognized in this genus by Clements, although taxonomists were in considerable disagreement on species and subspecies boundaries in the genus owing to the many disjunct populations and color morphs. (I would encounter 16 of the 20 species of wheatears during my 3 year adventure.)

The next morning (Mar. 10) I arrived at the border at 8:00 am, just as it was officially opening for the day. Several military personnel commanded the post, a captain and 2 subordinates, a corporal and a sergeant. The not unpleasant but authoritative Iraqi captain spoke quite good English, but the other two spoke none. When I identified myself, the captain said immediately that he remembered my name and having received a telegram permitting to enter, but when he looked for it in his office he couldn't find where he had put it. After almost 30 minutes of searching unsuccessfully, he finally said that he would just write me out another permit which would be all that I needed for my short time of travel in the country. That was OK with me. He then directed the beady-eyed, mustached, dark haired little corporal to inspect the inside of my van. Such border inspections were routine whenever Noble and I entered a new country. I suddenly remembered that I hadn't put my shotgun back inside its usual inconspicuous tool box, but I thought to myself that very few inspectors ever looked inside our clothes closet, and even if they did they simply opened the door, looked at all the clothes hanging up, and shut the door again. Not this time! After opening the clothes closet door the corporal then pushed all the clothes to one side and looked behind them. Sure enough, there was my shotgun, and my sword. Both were "prohibited weapons." Somewhat gleefully, the corporal took them inside and gave them to the captain.

The captain told me that I needed a permit to bring "weapons" into the country. When I asked where I could obtain such a permit, he responded "from the army." I said to him that he was a member of the army and asked if he couldn't issue me such a temporary permit for the very few days I was going to be in Iraq. The answer, as you might suspect, was no, he couldn't do that. It would need to be someone from the army headquarters in Baghdad, and until I had my permit I was under arrest for having illegal weapons in my possession! (This was the 2nd time in the past 2½ weeks that I was officially under arrest.) This matter was quickly getting out of hand. I told him I would hand over the weapons to him and he could transport them to Baghdad for me, where I could retrieve them after a permit was issued. Again, the answer was no, since the weapons were mine they had to remain in my possession until I handed them over personally to the military in Baghdad. I really found it hard to believe what I was hearing. Talk about ridiculous legalities! There was more to come. Since I was under arrest while I drove from the border to the military headquarters in Baghdad, because I had non-permitted weapons with me, I had to be under the supervision of a military "guard." You guessed it, none other than the beady-eyed little corporal. This beady-eyed little man traveled with me for 2 whole days between the border and Baghdad, following me around everywhere I went, always carrying his ancient carbine with him as if I were going to escape and run away somewhere - - in the middle of the desert! It was impossible to communicate in any way with him, and he refused to eat any of my food which I offered to him.

We crossed the Euphrates River at Ramadi on Mar. 11, and later that same day my guard and I arrived at the military headquarters in Baghdad. Here there was considerable discussion among all those present as to exactly whom the weapons should be delivered. There didn't seem to be any precedents for this situation. I was told that I could not personally obtain a permit from anyone at the headquarters, but would have to go through the American embassy. So I talked to the Chief of Mission, the U.S. Army Attaché, and even the Ambassador himself, none of whom had ever experienced such a situation previously. Finally, I was given the name of an Iraqi army colonel, a personal acquaintance of the attaché, who I was told would issue me a permit if I came to his office early the next morning. When I walked into his office at the assigned time he immediately said to me "I am sorry but I cannot issue you a permit." That was the end of our conversation.

Since my Iraqi visa was good for only one week, the time available for me to retrieve my shotgun and sword was quickly expiring. A young American vice consul at the embassy, Robert Maule, came up with a suggestion. He told me that if I "sold" the weapons to him, and they therefore were no longer my property but his, that because of his diplomatic immunity he could retrieve them from the military. We signed a hand written "bill of sale", for the value of one dollar. He told me that when he got them he would mail them to me wherever I wanted, so I suggested the American embassy in New Delhi, where I anticipated arriving in about 6 week's time or so. (When I eventually arrived in New Delhi, there was a letter for me from Robert saying he had successfully retrieved my weapons from the military, but since they now belonged to him they could leave the country only with him! The final, happy, ending to this story is that in August, 1962, 17 months later when Robert was visiting his parents in Arcadia, California, only a few blocks from where I was staying with my parents in Long Beach, he called me on the telephone, to my great surprise, and said that if I came by his parent's house he would return my shotgun and sword to me, which he did!)

Noble did not meet me in Baghdad as we had planned, since as usual he had become acquainted with an entrepreneur, an American oil administrator who offered to take him to Kuwait and explain to him some of the oil operations there. There was nothing that Noble enjoyed more, so of course he accepted the offer and left me a message at the American embassy in Baghdad saying that he would join me in Isfahan, Iran, in about 10 days time. He would leave me a contact address at the American consulate there. So I stayed by myself in Baghdad for several days, wandering through the bazaars and searching for birds along the Tigris River. On Mar. 12, I located 4 new trip species: Black Francolin* (calling frequently), Red-wattled Lapwing*, Bluethroat* (in the genus *Luscinia*, along with nightingales), and Common Babbler* (another *Turdoides*). In addition, I documented 12 White-eared Bulbuls. On Mar. 15 there were 2 Pallid Harriers, 8 White Storks, 6 Red-wattled and 3 White-tailed lapwings, 3 Black-winged Stilts, 8 kinds of migrating scolopacids (sandpipers, etc.), 2 large flocks of 75 & 50 individuals each of Pin-tailed Sandgrouse, 2 White-breasted (White-throated) Kingfishers, and 30 Spanish Sparrows.

On the morning of Mar. 16 I ferried east across the southward flowing Tigris River at Baghdad and headed upriver toward the international border with Iran, 100 miles away on the eastern edge of the wide Tigris River watershed, at an elevation of 1,000 ft. above sea level. Willow trees and date palms were abundant. (Iraq supplied 80% of the world's dates.) There were many nesting White Storks on the tops of telephone poles and the taller trees in villages. The border crossing was quick and without incident (since I no longer had any weapons with me) and I camped not far on the other side, in Iran about 100 miles west of the major city of Kermanshaw. The next morning (Mar. 17) as my gravel road began climbing eastward and upward into the Zagros Mountains I was thrilled to see a Great Bustard* (*Otis tarda*) fly across the road not far in front of me, from right to left, with its wide white wing band clearly displayed in flight. As indicated by its name this species was one of the largest flying birds in the world, with males weighing up to 25 lbs.

Plate 80: Plowing was done with horses (once again) in the Tigris-Euphrates river valleys of Iraq, west of Baghdad. Women did much of the work in the fields and villages.

Plate 81: Baghdad was bustling with people working at myriads of tasks to earn enough for their families. (I spent considerable time trying to get my shotgun and sword back from the military, without success).

The Zagros Mountains were a very long range of mountains extending in a mostly NW to SE direction along the entire western border of Iran for a distance of almost 1,000 miles. The highest peaks were 10,000-11,000 ft. above sea level. My 500 mile gravel route from the Iraq border to Tehran took me through the towns and cities of Kermanshaw, Hamadan, Avej, Qazvin, and Karaj. This scenic route across the rugged mountains took me up and down between valleys at 5,000 ft. elevation and snow-covered passes as high as 8,000 feet. There were dry scrub-covered slopes and higher, mostly barren rocky plateaus with frequent cliffs, gorges, and waterfalls, as well as scattered fields and small villages with mud houses. I stopped in the ancient mud village of Avej to take a few pictures. A bolt in the van's accelerator linkage broke early in my travel across the mountains, and the spare replacement bolts we had with us were not strong enough to hold very long and they soon also broke, one by one. Replacing the bolt every 20 or 30 miles in the cold weather by crawling under the front end of the van became a very exasperating and unpleasant task. I camped between the towns of Kermanshaw and Hamadan on Mar. 17, in a valley at 5,000 ft. elevation. It was very cold outside. During the day I had observed many large soaring raptors, including 2 Lammergeiers (one of which I photographed, in the distance), 4 Griffons, 1 Imperial Eagle, 1 Spotted Eagle, 1 Osprey, 1 Long-legged Buzzard, 1 unidentified harrier (probably Pallid), and 5 unidentified *Aquila* eagles. I also documented numerous kestrels (without being able to separate Eurasian from Lesser), 7 Chukars, 4 Ruffs, 2 Hoopoes, 250 Rooks, 50 Hooded Crows, 25 Common Ravens, 5 Eurasian Magpies, 30 Red-billed Choughs, 600 Skylarks, 250 Calandra Larks, and 500 European Starlings.

On Mar. 19, I arrived in Tehran, a very metropolitan city and the capital of Iran, situated at 5,000 ft. elevation in the north of the country at the southern base of the high Elburz Mountains. Here I checked at the American embassy for a message from Noble. Success! A telegram there for me said that he would arrive in Isfahan by airplane on an afternoon flight from Kuwait on Mar. 21. (We will have been separated for 16 days.) The only important item of business for me in Tehran was to get a sturdier bolt for our accelerator linkage, which I did. Unlike Noble, I spent as little time as possible in the big cities, as our personalities were very different in this respect.

Much of Iran was a desert, on high plateaus between 3,000 & 5,000 ft. elevation, with no major rivers at all. There was just a single species of bird endemic only to Iran, the Iranian (Pleske's) Ground-Jay (*Podoces pleskei*), one of 4 closely related, allopatric species in the genus *Podoces*, all of which were found only in central Asian deserts. I did not encounter this species during my time in Iran, nor in fact was I even aware of its existence then.

I drove south from Tehran on Mar. 20 and soon came to the small mud village of Aliabad, encircled by a mud wall with a large central courtyard inside which was surrounded by shops and houses. I stopped to take some pictures. People sat on the ground on "Persian rugs" drinking tea from copper pots, smoking water pipes ("hookahs"), or eating a picnic lunch. Women wore a brightly colored "chador" (a single large piece of cloth which served as a scarf, veil, and coat all in one), and men wore a high or close-fitting Muslim cap and a white cloak. Small boys inside the courtyard tended goats, sheep, donkeys, or occasionally even a cow. Outside the wall was a dirty stream where women gathered to wash clothes and dirty dishes, and to gather water which was carried in jugs on their head back to their house, where it would be boiled for drinking. Life in Aliabad had changed very little over time.

Not far south of Aliabad was the sacred city of Qom (sometimes spelled Gom) with numerous very sacred, century old mosques, where people came to worship from many miles around, often by walking or on donkeys. Non-Muslims ("infidels") were not allowed inside. However, I very much wanted to look inside one of the largest and most ornate of the mosques, so I parked the Jeep a few blocks away, put my camera in a small brown bag, and walked up to one of the mosques, wearing casual long brown trousers and a short-sleeved T-shirt. With my brown hair, black beard, and blond mustache no one could possibly have mistaken me for a Muslim.

I stopped to take some photos from outside the mosque, and then I walked up to the main entrance at the front, hoping to go inside and take a few pictures. A holy man guarding the door vigorously motioned for me to go away. So I walked around to the other 3 sides of the mosque, one at a time, each of which had an entrance. The first two of these were also guarded by a holy man who waved me away, but on the 4th and last side there was, surprisingly, no guard at the entrance, so I cautiously walked inside. Here was a very large room of unbelievable grandeur and splendor, the like of which I had never seen, including mirrored walls, gold and silver chandeliers, small hand woven prayer carpets on the floor, and many pools for washing one's hands and feet. Only a small number of holy men (with henna colored beards) and worshippers were present, and amazingly no one noticed me at first, allowing me to quickly snap some pictures, as inconspicuously as possible. After several minutes I sneaked up close to a holy man with a henna colored beard who was sitting on a carpet and reading out loud from the Koran to two worshippers sitting next to him. At the same instance that I snapped his picture I felt a heavy hand on one of my shoulders, and I was very roughly dragged out the front entrance and shoved into the street. Not long after this incident I related it to a European tourist I met in Isfahan. He told me that I had been very lucky, and that a European young man ventured into this same mosque several years earlier by himself, also with a camera, and was never seen again! (I cannot vouch for the authenticity of this story).

Camels and donkeys loaded with goods of all kinds were being herded, singly, across the desert as I approached the outskirts of Isfahan at 5,000 ft. elevation. A horse-drawn cart with oversized wooden wheels was transporting a high stack of large burlap bags into the city. Isfahan was an historic city situated on the Zayandeh River, with its roots going back 3,000 years (1,000 BC). With its many palaces, mosques, gardens, fountains, and ancient arched stone bridges of the Safavid era, the city would one day become a world cultural heritage site. It was a photographer's paradise. I picked up Noble at the airport on the afternoon of Mar. 21, and we went sightseeing together in the city all day on Mar. 22.

A young boy was selling enormous sized round, flat loaves of bread which he carried on top of his head from one shop to another. An elderly man with a kind face and a white beard, and wearing a long-sleeved patched coat, loose-fitting trousers, and a white knit turban, stood on a street corner calmly smoking a long-stemmed pipe. He willingly allowed me to take his picture, which became an all time favorite of mine. In the huge plaza bazaar skilled craftsmen were hammering out copper trays and pans, carpet makers were tediously weaving rugs and carpets, silk textile workers were embroidering scarves, dresses, and ladies coats, and jewelry makers were creating gold and silver brooches, bracelets, necklaces, earrings, and watch bands. The bazaar was teeming with people and activity. A barber on the sidewalk with a long straight edge razor was shaving all the hair from the top of a man's head. A few of the women were dressed in a black or a white gown, covering them from head to foot. Noble and I enjoyed relaxing and drinking tea for 30 minutes in the traditional tea shops. Isfahan was one of our most memorable cities.

Prior to Noble's arrival, I spent the morning of Mar. 21 looking for birds in the countryside outside the city, in an area of irrigated fields, orchards, scattered trees, shallow ponds, water filled ditches, and salt flats. I was able to locate 32 species of birds, but none were new additions for my trip list. The most noteworthy were 5 European Kestrels (finally distinguished to my satisfaction from Lesser Kestrels), 2 Common Cranes, 250 Northern Lapwings, 2 Black-tailed Godwits, 2 Common Redshanks, 40 Ruffs, 8 Black-winged Stilts, 75 Black-bellied Sandgrouse, 3 Syrian Woodpeckers, 15 Calandra and 20 Crested larks (both engaging in aerial courtship singing and displaying), and 2 Red-billed Choughs. Many of these were spring migrants on their way north.

Noble and I drove together to Tehran on Mar. 23 (which was my second time there). We found a wooded estate where we were given permission to camp. Throughout the night, I could hear a European Scops-Owl calling repeatedly just outside our van, but as always, I was unable to locate it visually with my flashlight as it sat hidden among the dense foliage (a situation which continuously frustrated me with this group of small owls throughout our entire adventure). The next morning (Mar. 24) Noble caught up on some business matters in Tehran, while I took the Jeep and drove up into the nearby Elburz Mountains just north of the city. Here I tallied birds for 3 hours in rocky gorges and on scrub covered slopes with patches of snow, at elevations between

Plate 82: Still without Noble, I drove across an 8,000 ft. high pass in the Zagros Mountains of western Iran on my way from Baghdad to Tehran. Women gathered drinking water, and washed clothes and dishes in a small muddy stream. An ancient stone bridge crossed the Zayandeh River in Isfahan.

Plate 83: Isfahan, Iran, is a World Heritage Site and is famous for its 600 year old architecture, such as the Jame Mosque in the central square. Long-stemmed pipes and hand-made copper trays and pots were cultural traditions.

6,000 & 8,000 feet. I observed and wrote descriptions of 2 new exciting birds, the Fire-fronted Serin* and the Crimson-winged Finch*.

Before we left Tehran it was necessary for us to decide whether to drive to Pakistan via Afghanistan or not, as we had a choice of two routes, one traversing Afghanistan and the other circling around it to the south. We asked advice from two different groups of European tourists which we encountered, mostly British. Almost everyone said there was nothing of interest to see in Afghanistan, and the drive would be difficult. That made our decision easy - - we would travel via Afghanistan! We had learned that what most tourists considered of interest was not usually the same as ours.

Three days later, on Mar. 27, we left Tehran to travel via the Caspian Sea across NE Iran to western Afghanistan, to the city of Herat. Our highway north to the Caspian Sea traversed the Elburz Mountains, over a pass at 8,500 feet. It was cold and very foggy at the pass but I wanted to search for birds there briefly as we went over the pass. Noble was content to sit in the Jeep and catch up on his journal while I hiked between 8,000 & 9,000 ft on the grassy, thinly brushed mountainsides and stream sides, looking for birds between the patches of fog. Soon I encountered 2 ptarmigan like birds walking slowly on the barren ground over and between patches of snow, allowing me to approach them very closely, to within 15 or 20 feet through the mist. I studied them carefully then drew a sketch of one and wrote a lengthy description of it in my field notebook. Quite some time afterwards I was exuberant to discover they were Caspian Snowcocks*! This chance encounter became one of my all time favorite bird sightings, with memories of the foggy high Iranian mountains above the Caspian Sea. This species was one of 7 mostly allopatric species in the genus *Tetraogallus*, all of them Palearctic birds of high mountains in western and central Asia. Other birds which I recorded here at the pass were 250 Horned Larks, 2 Hooded Crows, 8 Eurasian Magpies, 4 Rock Nuthatches (not separated to species), 3 White-throated Dippers, 1 Eurasian Linnet, and 15 White-winged Snowfinches (1 of 7 species in the genus *Montifringilla*, and the only one I saw during my travels). This appealing genus was characteristic of higher elevations and open habitats of central Asian mountains and plateaus, and this species was the single one which extended westward into Europe, where I first saw it in the French Alps.

Our highway now descended to the southern shore of the Caspian Sea, the world's largest completely inland body of water, with a surface area of 143,000 sq. miles, lying 90 ft. below sea level. It had a salt content of 1.2% (about ⅓ that of sea water) so its brackish water was not considered to be a "lake" but rather a "sea." Therefore, although it was 4½ times larger than Lake Superior, the latter was claimed to be the largest "freshwater lake" in the world.

The Caspian Sea was fed by many rivers, the largest of which was the Volga River, but like other lakes Noble and I had encountered it was an endorheic lake, with no outlet. It had a maximum depth of 3,360 ft. (in the southern part of the lake) and a shoreline length of 4,300 miles. A surprising aquatic mammal that inhabited the lake was the endemic Caspian Seal (now an endangered species). Another noteworthy inhabitant was the world's largest non-marine fish, the Beluga Sturgeon, in which old males could weigh as much as 3,400 lbs., attain lengths of 16 ft., and live to over 100 years. This fish was very important commercially in the production of caviar (from the eggs of the female). Ornithologically, the cosmopolitan nesting Caspian Tern is named for this lake and breeds here, as well as elsewhere on lake shores and sea coasts around the world. (Other global nesting birds are the Great Egret, Peregrine Falcon, and Barn Owl.) I pondered some of these interesting facts as I took a few photographs of this unique body of water. (Unfortunately, the Caspian Sea has become very badly polluted today.)

Noble and I headed eastward and followed along or near the southern shore of the lake for approximately 75 miles, to the town of Babol just inland. During this stretch of road some of the birds which I documented were 40 Great Cormorants, 2 Gray Herons, 5 Little Egrets, 1 White-tailed Eagle, 20 Lesser Black-backed Gulls, 1 Little Gull, 2 Common Terns, 1 Pallid Harrier, 3 W. Marsh-Harriers, 2 Red Kites, 5 Black Kites, 5 Eurasian Kestrels, 1 Little Owl, and 6 Hoopoes. At Babol we continued eastward for 25 mi. to the town of Sari, and then proceeded another 20 miles east to a campsite for the night, at an elevation of sea level, between the Elburz Mountains to our south and the Caspian Sea to our north. It had been a very productive day with contrasting scenery, habitats, and birds.

During the next 4 days (Mar. 28-31) we traveled for 500 miles in a long, northward curving arc through the cities or villages of Gorgan, Qabus, Bujnurd, Quchan, Mashad (Iran's 2nd largest city), Torbat-e Gam, and Tay Yabet, all the way to the border of Afghanistan. This route started at sea level and fluctuated in hills and mountains as high as 5,000 ft. elevation, following in valleys between the Kuh-e Aladag mountains to the south of us and the Koppet Mountains to the north, where they formed the border with Turkmenistan. Our rate of travel was slow, undulating up and down and curving first in one direction and then in another. Forests covered some of the mountain slopes. The smaller villages, as elsewhere in Iran, were made of mud. We made very few stops, except rarely for me to search for birds in productive looking habitats. As always, I kept records of the roadside birds I was able to identify. The 5 additions to my trip list were Hawfinch, Bimaculated Lark*, Variable Wheatear*, Siberian Stonechat, and Pied Bushchat*.

We crossed the border between Iran and NW Afghanistan on the afternoon of Mar. 31, in the desert about 3,000 ft. elevation, approximately 75 miles west of the city of Herat. That night we camped in arid scrub along a wadi 10 miles east of the border, still at 3,000 ft. elevation. It rained during the night. The next morning (Apr. 1st) we continued toward Herat, and had to ford numerous wadis because of last night's rain. Fortunately, the ancient arched adobe bridge over the wide Hari River was still standing, since the water in the river was too deep for us to ford. I took a picture of our van on top of the bridge. Many such bridges over smaller streams had collapsed, necessitating that we ford the stream. At his request, I took a picture of Noble in our Jeep as he drove through 2 ft. of water in one of the deeper wadis.

Along the roadside this morning was an elderly, shabbily dressed goat herder with a white turban and white beard. He happily allowed me to take his photograph, which was very reminiscent of paintings representing biblical times. On the outskirts of Herat were two enormous, ancient, round, watch towers made of clay bricks which had been redecorated on the outside with modern lilac and green tiles. These towers were perhaps 100 ft. tall with a base diameter of an estimated 12-15 ft., and they completely dwarfed two women who were walking past them, where they appeared as ghosts under their completely concealing white burkas. Our Jeep looked like a miniature toy when we parked it next to one of the towers to take a photograph illustrating its gigantic size. Adobe circular stairs on the inside of each tower led to the top. Only in an extremely arid climate could these minarets have withstood their many centuries of age.

Herat was the largest and most important town in the remote western half of Afghanistan. What a culture shock it was for us! Isolated in space and time the city had remained virtually unchanged for over a thousand years. It was like stepping into a time machine and suddenly being transported back to biblical days. Camels and donkeys carried goods and people (men, women, and children) in and out of the town on dry, dusty roads. Merchants sat or reclined lazily just inside or outside their opened door shops on the unpaved main street. In one shop, wheat was pounded into flour with a mortar and pestle. The flour was made into dough in a nearby bakery by a young boy who treaded the dough with his bare feet, and then formed it with his hands into long flat oval loaves of bread which he placed in an earthen, charcoal fired oven. The baked loaves, when done, were then carried on top of his head and sold to customers along the street. It was the best tasting bread I had ever eaten! A coppersmith hammered out trays, pots, and pans in a small shop, and in another shop a cobbler sewed and repaired shoes by hand. Further along the street an ancient Singer sewing machine was used in making all kinds of garments. (I was amazed to find Singer sewing machines everywhere we went in our travels, even the most remote towns and villages.) In yet another shop, charcoal for cooking was sold by measuring quantities on a hand balance. Textiles for sale were hung over the top balcony of a two-level store where a young boy was playing a reed flute to himself, just for pleasure. In a lovely, quaint little tea house a boy played a mandolin, again simply for his own contentment. Noble and I relaxed there for almost an hour, leisurely drinking tea, and my photograph of the mandolin player is an all time favorite of mine. No one anywhere was in a hurry. We did not encounter anyone who could speak English (or any other European language). People were friendly and curious.

The few women who were present on the streets kept their distance, although young girls would sometimes giggle as we walked past them. All post-puberty girls and women were completely covered from head to foot in a "burka" which allowed them to see only through a narrow slit in a heavily netted veil. They came and went from the large central courtyard which served as the town's marketplace for vegetables, fruits, glassware, pottery, and many other items. On the outskirts of the town was an area where gray mud was shaped into bricks, which were then spread out on the ground to dry in the sun, after which they were used in constructing almost all the houses and shops. The only taxis in town were carriages drawn by colorfully decorated horses, but few people could afford such a luxury. Other than our Jeep there was only one other vehicle in the town, an ancient WW2 model Russian transport truck parked next to a vacant lot. To fill up with gasoline we had to be guided to the one gas pump on the outskirts of town. On one of the side streets, a group of men and boys watched and bet small sums of money on a quail fight which was taking place in the shade under the cover of a sheet. There was one large, beautifully designed and decorated mosque in the town, with intricate dark blue and pale aqua tiles. Herat was like nothing we had ever experienced. The opinion of those tourists who told us "there was nothing to see in Afghanistan" could not have been further from the truth. Men and boys constantly gathered around Noble and me to stare at us and our Jeep. We were as different to them as their culture was to us. We were far off the beaten path of tourists. Herat was the adventure of a lifetime.

Late on the afternoon of Apr. 1st we reluctantly departed Herat. As we drove out of town I saw my first Common Mynas* (a member of the starling family). This species was one of the most common, widespread, and best known city and countryside birds of southern Asia, from Iran eastward throughout SE Asia, and was widely introduced many other places around the world. Our planned route took us south from Herat to the town of Farah, approximately 175 miles distant over a very primitive, barely maintained dirt and gravel road. In 1961 this road was not for the timid. We began climbing upward in elevation from the Hari River valley in which Herat was located, and camped that night at 4,000 ft. elevation 5 mi. south of the city. We were in an area of desert where there were irrigated fields, willows, bushes, and even planted pine trees (*Pinus*) along the roadside. I birded in this area for an hour early the next morning (Apr. 2), before starting out. Here I recorded 20 kinds of birds, including 3 trip additions: Red-breasted Flycatcher*, Common Rosefinch* (in the widespread genus *Carpodacus*), and Desert Finch*. This latter species was a lovely bird with subdued brown plumage, pink and white in the wings, mostly white tail, and heavy black bill, in the monotypic genus *Rhodospiza* (placed by Clements in the Carduelinae of the family Fringillidae). Many of the birds on my list this morning were spring migrants on their way north. It was a pleasant, rain free beginning to the day.

Our road surface had not been graded for a very long time and was badly rutted, corrugated, and washed out. It was necessary that I drive extremely slowly and cautiously, at speeds between 10 & 15 mph. The situation worsened considerably in the middle of the morning when we suddenly encountered a prolonged, very heavy downpour of rain, which then continued on and off for the rest of the day, with occasional heavy winds. Not surprisingly, the rain was soon followed by severe flash flooding in all the numerous desert wadis. Our travel became much more difficult and risky. Even the camels which were transporting people and goods along the road had considerable difficulty wading across the usually dry but now rushing streams, which were sometimes 2 or 3 ft. deep. We were the only vehicle on the road in this very remote part of the world. Our road in 1961 deviated through the town of Sindand (now bypassed by a shorter road) and then climbed upward all the way to a mountain pass at 6,000 ft. elevation. Shortly thereafter our road turned southwest and descended toward Farah (a deviation that has also since been shortened, bypassing Farah). We camped for the night along this section of road, about 40 miles from Farah at an elevation of approximately 3,500 feet. In spite of the rain I documented 5

Plate 84: Having joined Noble again, we traveled across Afghanistan together, into the past. Life was changed little from 2,000 years ago. Although there was an ancient stone bridge over a river, the smaller streams had to be forded.

Plate 85: Semi-nomads lived in tents. It was springtime in Afghanistan and shepherds needed warm clothes for the cold nights. Four women near Herat were dwarfed by an ancient tower. They were required by local Islamic law to be completely covered.

Plate 86: Herat, in far western Afghanistan, was much as I imagined it must have been in biblical days. The pace of life was leisurely, and we were as much a curiosity for them as they were for us.

Plate 87: Women were always completely concealed everywhere they went. Men relaxed with a water pipe, and a young boy sought solace by playing a mandolin to himself in a tea parlor. Noble & I joined him. Nowhere was there another town like Herat.

species of wheatears on this date, and I saw 2 more Desert Finches, plus my first listing of the Persian Nuthatch* (which was almost identical in appearance and behavior to the Rock Nuthatch, a slightly smaller, more westerly occurring species). The rain finally stopped, and we slept very soundly that night.

The next morning, Apr. 3, there was no more rain but wadis were still in flash flood stage and we soon encountered so much water running over the road that we had to stop and wait until 1:00pm before the level dropped to under 3 ft. and we could proceed again, slowly. At mid-afternoon we finally entered the small quaint town of Farah, located with its mud houses and central marketplace at 2,500 ft. elevation on the Farah River in SW Afghanistan. We joyfully exclaimed "we made it to Farah, hurrah!" Because of our long morning delay, we did not stop to take pictures but immediately headed east out of town toward the distant city of Kandahar, more than 250 miles and 2 days away. Our route would skirt to the south of the extensive mountainous terrain of central Afghanistan. This region of Afghanistan would prove to be a particularly fascinating 2 days of travel. That night we camped 25 miles east of Farah in a barren rocky desert at 3,000 ft. elevation. During our shortened day of just 6 hours of travel I counted 5 Cream-colored Coursers, 20 Hoopoes, and both my first Rufous-tailed (Isabelline) Shrike* and Hume's Wheatear* (in a rocky ravine).

The countryside on the next morning (Apr. 4) was mostly sparsely grassed desert and rocky ravines between the elevations of 2,500 & 3,500 feet, with occasional small villages, trees, and green fields in river valleys where there was water for irrigation. It was springtime with new grass appearing and larks, wagtails, chats, and wheatears were all singing. The sun came back out and driving became more fun once again, although our poorly maintained gravel road remained hazardous much of the time. Camels were the only mode of travel we encountered, and nomads were moving north across the desert steppes for the coming summertime.

Much to our great surprise we encountered a group of 4 New Zealand and Australian boys on motorcycles, traveling west across the remote southern Afghan desert! Of course we all stopped for a chat. They were on their way to Europe and had been traveling for more than 6 months. When we told them of our intention to drive from India to SE Asia via Burma, they said we would not be given permission by the Burmese authorities to do this. They had tried every way possible to convince the government to issue them permits for travel on their motorcycles through the country, but were denied permission to do so. They said that other travelers with vehicles all told them the same story, that they, too, had been denied a permit to bring their car, camping van, or motorcycle into the country (ostensibly because the very authoritative, military Burmese government could not guarantee them complete safety from rebellious groups of people within the country). Noble and I said we would just have to wait and see what the situation was when we eventually arrived there, but of course Burma was a very important link in our proposed itinerary.

As we continued east across the desert we could see a long ridge of snow-capped mountains far to our north. Men and boys who were tending goats and sheep wore heavy white coats for warmth, with their arms underneath and sleeves dangling empty at their sides. They were curious and gathered around our Jeep whenever we stopped for a picture, and were amused to look through Noble's movie camera. Taking photos of a woman were, of course, strictly prohibited. On one occasion a man threatened to throw a rock at me when I pointed my camera in the direction of a far off woman. Women here did not hide completely under a burka as they did in Herat, and wore a scarf covering only the top of their head, which kept their face uncovered. The small villages through which we passed were like those in Iran, with dome roofed mud houses built next to each other in an arcade like fashion around a large, open central courtyard. Men smoked water pipes, and children of both genders sat around lazily next to the abundant camels and donkeys, apparently with little or nothing to do. In one of the villages there was, once again, an antiquated Russian built transport truck (without a hood over the engine) parked along the dirt main street. It was the first vehicle of any kind we had seen such leaving Herat. There was never a sign indicating the name of a town, or any signs to the next town up the road. We became very adept at guessing which road to follow (though often there was only one main road going through a town). We camped that night at an elevation of 3,500 ft. in the open desert, about half way between Farah and Kandahar. The temperature dropped rapidly after dark.

We got a very early start the next morning (Apr. 5) on our way to Kandahar, 135 miles away. The sun had just come up when we spotted far off the roadside a nomadic group of people very busily involved in packing up their tent and all their livestock, pets, and personal belongings in preparation for their day's journey. Nomadic pastoralists utilized camels and donkeys for transporting their tents, firewood, and all their personal belongings. Tents were made of heavy wool and animal hides, supported firmly by ropes and a few poles. Water was gathered prior to their departure by women from the nearest pool or drainage ditch, in goatskin water bags. Both men and boys were dressed in white turbans and ragged shirts, coats, and trousers. Women wore shabby dresses, sometimes entirely black, with a scarf over the top of their head, but with their faces exposed. Small children were simply dressed, often quite colorfully, but hid behind their mother when they saw us approaching with our cameras. Men were sometimes barefooted, and sat on the ground outside their tents smoking water pipes.

I walked a long distance across the desert to get close for some pictures of one small family group as they were packing up and preparing to depart. They paid little attention to me or my camera but kept continuously busy with their chores, with never a wasted motion. It was obvious they had done this on many previous occasions. There was a husband, a younger wife, and a very small baby, all of whom were heavily bundled up for warmth on the cold morning. For transportation, they had 3 camels and 2 donkeys, all of which were very heavily loaded, with no wasted space. I was amazed at the efficiency and speed with which they accomplished all their tasks, and I took a great many photos (with no objection from them). Just prior to their departure they sat for several minutes warming their hands over a small fire pit, and drinking one last cup of tea. They then tucked the baby and several small lambs securely on the backs of camels and set off on foot across the desert for their day's travel, the woman walking in front and leading the first camel with a rope. This lifestyle was centuries old, as pastoralists moved north in the spring and south in the autumn. I was thrilled at the opportunity to photograph it. The tourists in Tehran who told us there was nothing to see in Afghanistan, once again, could not have been more mistaken.

We arrived late that afternoon in the large, important regional city of Kandahar, where our road turned north-eastward toward Kabul. However, we stopped long enough only for gasoline and a cup of coffee, then drove a short distance out of town to a campsite for the night. My bird lists for Apr. 4 & 5 (between Farah & Kandahar) totaled just over 30 species, with 4 trip additions (Appendix B), of which the Hooded Wheatear* was the most exciting for me. It was strikingly handsome in its bold black & white plumage pattern, and the rocky habitat in which it lived epitomized the rugged environment of Afghanistan. The 2 males I saw were actively singing and displaying, and chasing a female from one rock to another.

It was 320 miles from Kandahar to Kabul, via the principal town of Ghazni. Our moderately well maintained gravel road ascended as high as 9,000 ft. in the arid, rocky, brush-covered mountains. The higher elevations in the distance were covered with snow. I was happy to find a pair of Red-tailed Wheatears* (*Oenanthe xanthoprymna*), which was the last of 16 species of wheatears which I listed during my 3-year adventure.

Kabul, the capital city of Afghanistan, was situated in the northeastern sector of the country on a high, arid plateau at 7,500 ft. elevation. We arrived on Apr. 8, and spent only one full day there before departing eastward for Pakistan on Apr. 10, via the city of Jalalabad and Khyber Pass, 140 mi. E of Kabul. We camped for the night east of Jalalabad at an elevation of 3,000 ft., about 10 miles before the border. Along this last stretch of birding in Afghanistan I added 6 species to my trip list (Appendix B), the two most noteworthy being the lovely White-capped Redstart* (often called a Water Redstart because of its preference for rocky streamside gorges) and the Eurasian Eagle-Owl* (*Bubo bubo*), a widespread Palearctic counterpart of the Great Horned Owl. It was seen at dusk on the top of a roadside telephone pole, and was my last trip bird from the Palearctic Region (which was poorly separated in this area from the Oriental Region). Afghanistan had been one of our most exciting adventures.

Plate 88: Early one morning at 3,500 ft. in the desert of southern Afghanistan there was a family of nomads busily taking down their tent and quickly packing all their belongings on camels and donkeys for their day's travel. I marveled at their expertise and efficiency.

Chapter Twenty -- Pakistan & India to SE Asia

Noble and I drove across the much publicized Khyber Pass between Afghanistan and Pakistan at 3,500 ft. elevation on the morning of Apr. 11, 1961. In spite of its historical significance in east-west trade, the pass was hardly noteworthy geographically, lying at a low elevation between unspectacular hills to either side. To the west, the road was much higher than at the pass.

We were now in the Oriental biogeographic region as outlined by Alfred Wallace in 1876 (see the map on p. 8 in *Biogeography*, by Brown & Gibson, 1983). This artificial boundary was not sudden but quite gradual. Never-the-less, the overall birdlife was different at the lower elevations. Noble and I would spend 5 months in the Oriental region, on the continent of Asia. Appendices B & C name the countries we visited, the dates of entry, the total number of days in each country, and the birds which were added to my trip list by country and date. The 7 bird books I had with me in 1961 for the Oriental region (all of which I had transported from California) were: (1) *Popular Handbook of Indian Birds* by Hugh Whistler, 1949 (with 560 pages, 5 color plates, and 19 black & white plates), (2) *The Book of Indian Birds* by Salim Ali, 1955 (a small book of common lowland birds with 144 pages, 56 color plates, and 18 black & white plates), (3) *Indian Hill Birds* by Salim Ali, 1949 (another small book with 188 pages, 64 color plates, and 8 black & white plates), (4) *The Birds of Burma* by B. E. Smythies, 1953 (with 668 pages, and 30 color plates), (5) *A Guide to the Birds of Ceylon* by G. M. Henry, 1955 (with 432 pages, 30 color plates, and numerous black & white drawings), (6) *The Birds of the Malay Peninsula, Singapore, & Penang* by A. G. Glenister, 1951 (with 282 pages, 16 mostly color plates, and a few black & white drawings), and (7)*Birds of Malaysia* by Jean Delacour, 1947 (with 382 pages and 83 black & white figures only).

I was in my customary driver's seat as Noble and I set off eastward on a wide, dusty, mostly smooth surfaced gravel road leading away from the Afghanistan border toward the Pakistan city of Peshawar. I was traveling at 50 mph, a speed I never exceeded. Very soon after we left the border a huge, heavily loaded truck came toward us driving at a high rate of speed - - on the same side of the road we were on! As we neared each other the driver flashed his lights and sounded his horn at me, but gave no indication of moving over. I moved as far to the right side of the road as I possibly could. At the very last minute he swerved toward the middle of the road just enough to avoid a serious collision. He shook his fist very angrily at me. I was not only frightened by this very close encounter but also quite puzzled. Then instantly, I arrived at the answer. Pakistan had been under British rule until very recently (1947) and still followed many British customs, one of which was to drive on the LEFT side of the road! This custom obviously had not changed. I thought to myself that when a driver entered Pakistan from Afghanistan there should be several very large signs announcing this fact, but there were none at all. (Noble agreed that he, too, had seen no such sign.) This brush with death was a mistake I never made again. Our 80,000 road miles by Jeep were constantly filled with risks, and attentive driving was one of the most critical factors essential for our success.

Our initial travel in Pakistan took us through Peshawar to the large city of Rawalpindi. During this travel, I added 2 widespread, conspicuous Oriental raptors to my trip list, the Oriental Honey-buzzard (*Pernis ptilo-rhynchus*, which I had first seen in Japan in 1956) and the White-eyed Buzzard* (*Butastur teesa*). In his usual outgoing manner, Noble met a European resident on a downtown street of Rawalpindi and engaged him in conversation. After 10 minutes of talking, Noble was invited to fly with him in a small airplane to a city in the far north of Pakistan, to spend a few days with him there. Noble immediately accepted the invitation and told me I could take the Jeep and go anywhere I wanted during this short time. There was a "hill station" at Murree, in the mountains at 7,000 ft. elevation NE of Rawalpindi not far away. (The British had established hill stations throughout India and Pakistan as retreats from the hot summer days of the low-lying plains.) Therefore, on Apr. 13, I drove to Murree, during heavy rain most of the way.

Here, in 2 full days of hiking (Apr. 14 & 15) in mixed pine-hardwood forests between 6,500 & 7,500 ft. elevation, I documented a total of 72 species of birds, with 42 trip additions (including 35 on Apr. 14ᵗʰ, one of my best days ever; see Appendix B). At this elevation many of the species were characteristic more of the Palearctic region than they were of the Oriental region. My total list included 9 muscicapids, 7 raptors, 6 sylviids, 5 woodpeckers, 4 tits (family Paridae), and 2 laughingthrushes (the Streaked* & Variegated*, 2 of the 52 species in the genus *Garrulax,* family Timaliidae). Members of this genus were noisy, gregarious, handsomely colored birds of forest edge and understory, and were one of the most characteristic groups of tropical and subtropical birds in Asia, Indonesia, and the SW Pacific islands. Though secretive in their behavior they called attention to themselves by loud chuckling vocalizations, and I found them particularly pleasing, among my most favorite birds. Other birds at Murree included Slaty-headed Parakeets* (in flocks flying over the treetops), Black Bulbuls*, Ashy Drongos*, Gold-billed Magpies*, Blue Whistling-Thrushes, Blue-capped* & Blue-fronted* redstarts, Gray-hooded Warblers* (singing), Rufous-bellied Niltavas*, Bar-tailed Treecreepers* (in the genus *Certhia,* like the Brown Creeper of N. Am.), White-browed Shrike-Babblers* (one of the many species in the family Timaliidae), and Rufous-breasted Accentors*. I also carefully observed, sketched, and wrote a long description of a soaring Steppe Eagle* (*Aquila nipalensis*) which was only a non-breeding visitor to this location. My two leisurely, fun-filled days at Murree were an exciting introduction to birds in the Oriental zoogeographical region.

Back on the road with Noble again on Apr. 16, we drove for 3 hours from Rawalpindi SE toward Lahore, on the low, hot plains of eastern Pakistan, between elevations of 1,200 & 1,500 feet. The most conspicuous and/or common roadside birds were White-cheeked Bulbuls, House Crows, Jungle* & Common babblers (both in the genus *Turdoides*), Bay-backed and Long-tailed shrikes, White Wagtails, Black Drongos. Common Mynas, Crested Larks, Purple Sunbirds*, Ashy Prinias*, Indian Rollers, White-throated (White-breasted) Kingfishers, Rose-ringed (also called Long-tailed, or Green) Parakeets, White-rumped (White-backed) Vultures, Egyptian (Neophron) Vultures, Black-shouldered Kites, Black Kites, Tawny Eagles, White-eyed Buzzards, Laughing (Little Brown) Doves, Eurasian Collared (Ring) Doves, and Hoopoes. (I had great difficulty with the different common names used by varying authors for the same species.)

On Apr. 18, we crossed the border between Pakistan and India, on the low plains between the cities of Lahore (Pakistan) and Amritsar (India). Driving continued to be on the left hand side of the road! I added Bank Myna*, Blue-tailed Bee-eater, Asian Paradise-Flycatcher*, Ashy-crowned Sparrow-Lark*, Indian Gray Hornbill*, River Tern*, and Indian Pond-Heron* to my rapidly growing trip list.

The next day, still on the low plains, we curved in a northward direction from Amritsar to the city of Jammu, where we entered the far SW corner of the province of Kashmir. This political region had a large majority Muslim population, but never-the-less was given to India by the British in 1947 when they arbitrarily drew international boundaries between India and Pakistan, to the immense anger of the many Muslims in Kashmir. (To maintain peace in this disputed region today requires the presence of tens of thousands of Indian military troops.) Politics aside, Kashmir was a lovely scenic area characterized by lakes, green pastures, yellow fields, and high snow-covered Himalayan Mountains, surely one of the most delightful regions anywhere on earth. On this date (Apr. 19) I recorded the following species for my first time: Rufous Treepie*, Yellow-eyed Babbler*, Small Minivet*, Common Woodshrike*, Savanna Nightjar* (seen), Common Hawk-Cuckoo*, and Alexandrine Parakeet*.

Noble and I both thoroughly enjoyed our 9 days in Kashmir, where I hiked, pursued birds, took pictures in the mountains, and relaxed in tea parlors. Once again, we drove as far north as the snow plow allowed us, which was at 9,500 feet, 5 miles east of the town of Sonnamarg. While hiking by myself in the snow at this site early on the morning of Apr. 24, I was thrilled to discover the very recent, indisputable tracks of a snow leopard! I searched everywhere in the surrounding environs with my binoculars but could not locate the leopard itself, which must have been quite close by. I was told later that same morning by the owner of a tea parlor in Sonnamarg that, indeed, a snow leopard lived at this site, and was regularly observed. Sadly, my photographs of the tracks don't do justice to the excitement of the moment.

A few of the most memorable birds I encountered in the mountains of Kashmir were Brown Dipper, White-capped Redstart, Kashmir Nuthatch* (a local endemic very similar in appearance to the allopatric Eurasian Nuthatch), Rufous-breasted Accentor, Snow Pigeon*, and a Himalayan Griffon* (a vulture which was almost indistinguishable from the sympatric breeding Eurasian Griffon).

Photographically, two of my all time favorite pictures were taken at Lake Wular, in Kashmir. One of these was a close-up of a smiling, turbaned, bearded plowman with a very kind, friendly face. The other one was a scenic view of the lake, with snow-topped mountains in the distance and a long, slender boat in the foreground, which was being poled across the lake by a solitary boatman at the back. The boat was silhouetted against the still waters of the lake. On another occasion, early one sunny morning, I photographed a large congregation of Muslim men in white gowns and turbans, all kneeling and bowing together, side by side on a green hillside as they faced west toward Mecca, praying in unison. These pictures are preserved in my memory and in my photo archives. Kashmir will be remembered as one of the most enjoyable places of all my travels.

Our route to Dharamsala through the foothills of northwestern India on Apr. 28 took us through Punjab district, in Himachal Pradesh. This was the home of bearded, turbaned Sikhs. A few of the new trip birds along the road were Plum-headed Parakeet*, Red-headed (Black or King) Vulture*, Pallas' Fish-Eagle* (genus *Haliaeetus*), and the large, stately Black-necked Stork* (a close relative of the Saddle-billed Stork in Africa). On Apr. 29 we continued traveling east in the Himalayan foothills between 3,000 & 5,000 feet. I tracked down 12 new trip additions during the day, including the Brown Fish-Owl*, Spotted Owlet* (genus *Glaucidium*), and the little Black-chinned Babbler* (*Stachyrus pyrrhops*, also called the Red-billed Babbler). This latter species gave me considerable difficulty in its identification and I not only wrote a lengthy description in my field notebook, but also drew a sketch of its facial pattern to eventually figure out what it was.

We camped that night 5 miles south of the town of Mandi, at 2,500 ft. elevation. Throughout the night, I could hear the loud, slow, resonant "chaunk, chaunk, chaunk" notes of Large-tailed Nightjars*, produced once every 3 or 4 seconds. These birds were not only heard but were also seen flying at dusk. The next morning (Apr. 30) I got up at daybreak and went walking in the forest by myself for 4 hours in a densely wooded, brushy riverside ravine. Of 13 birds which I added to my trip list (Appendix B) the Changeable Hawk-Eagle* and the Indian Peafowl* were the two most thrilling. Even though the 2 male peafowls were not displaying, they were among the most magnificent of all birds, with their elaborate, greatly lengthened upper tail coverts which they raised and spread in dazzling visual displays.

On the morning of May 2, I searched for birds 5 mi. N of Simla in a dense oak-pine-spruce forest at 7,500 ft. elevation. Here I found 42 species, 10 of which were trip additions (Appendix B). I listed 3 kinds of vultures (including a Lammergeier) and a handsome Kalij Pheasant* (with its long white crest and bare eye patch), but my greatest thrill came from sleuthing out 5 kinds of laughingthrushes (White-throated*, Striated*, Chestnut-crowned*, Variegated, and Streaked). We drove all the way up to 10,500 ft. in Himachal Pradesh east of Narkanda on May 3, and here I saw my 1st Black Eagles* (*Ictinaetus malayensis*) and 6 Himalayan Monals* (a pheasant; 5 males and 1 female). At this same site at dusk I was delighted to watch and listen to a male Eurasian Woodcock displaying over the forest, the first time I had witnessed this species displaying (having first seen a

Plate 89: During our nine days in Kashmir we drove as high as 9,500 ft. in the forested mountains just east of the town of Sonamarg, where one morning I saw tracks of a Snow Leopard. The birds and vegetation were characteristic of the Palearctic Region.

Plate 90: Lake Wular in Kashmir was situated at 5,500 ft. in a river valley surrounded by mountains. A kindly-faced plowman was happy for me to take his picture, and a large gathering of Moslem men on a hillside faced toward Mecca and prayed in unison.

wintering bird in Greece in January). It was a great finish to my day in this high elevation, stimulating forest environment, home of the Sikhs.

On May 4 we headed south from Narkanda out of the Himalayas to begin a 4 day drive to the twin cities of Delhi and New Delhi, the latter of which was the capital of India, an impoverished, overpopulated nation struggling with Hinduism, a caste system, and great inequalities. Our narrow, paved highways were hazardous driving, being overcrowded with buses, trucks, taxis, ox carts, bicycles, pedestrians, cows, and everything else imaginable. Traffic was as bad as it can get, with motorists ignoring whatever road rules and regulations there might have been, and constantly blowing their horns at each other. Driving was a nightmare, both on the highways and inside the towns and cities.

Traditional dress for women was a head scarf, a colorful sari (a single piece of cloth tied around their waist with one end draped across a shoulder, over an underlying blouse), lots of jewelry, and a painted red dot in the middle of their forehead. Men wore a "dhoti" (a loose skirt around their waist) and a short or long sleeved shirt. Cows meandered back and forth on the streets, and everywhere else, in the towns, cities, and countryside. They were followed by women who scooped up the wet manure as soon as it was deposited, patted it with their bare hands into flat round patties, and then stacked these in 3 or 4 feet high pyramids along the streets and sidewalks for drying. Later they would be collected and taken inside their houses, where they were used as fuel for cooking and boiling water. Cows, of course, were considered sacred (as were almost all animals), and they were allowed to meander everywhere and do whatever they pleased. I was particularly annoyed by the many beggars, mostly children, who followed us everywhere we went along the streets, constantly holding out their hands and asking for money. The lifestyle and culture of lowland India did not make my top 10 list!

Noble and I traveled south from Delhi on May 12 to the Taj Mahal in Agra, where like all tourists we took pictures of this justly renowned work of architecture. It was built between the years 1631 & 1648 by Shah Jahan, the 5th emperor of the Mughal Dynasty (of Muslim rule) as a tomb for his favorite wife. The Taj Mahal was constructed of white marble and situated on a 42 acre complex (along with the Red Fort and other buildings) on the south bank of the Yamuna River. Outside the mausoleum were extensive gardens, fountains, and a long marble pool of water which reflected the mausoleum. The entire complex became a UNESCO world heritage site in 1983.

While Noble walked around the walled town of Agra on May 13, I drove in the Jeep 50 miles west to the Ghana Bird Sanctuary, a large wetland area near the town of Bharatpur, at an elevation of just 800 ft. above sea level. Here I spent the morning with my binoculars and telescope counting the many kinds of birds, particularly those associated with an aquatic habitat. My complete list for the morning totaled exactly 75 species, of which only 7 were trip additions (Appendix B). A sampling of the birds I saw includes the following water associated species: 4 Clamorous Reed-Warblers*, 3 Baya Weavers, 20 Pied Kingfishers, 1 Brown Fish-Owl, 2 White-breasted Waterhens, 1 Purple Swamphen, 200 Sarus Cranes, 3 Bronze-winged Jacanas*, 15 Pheasant-tailed Jacanas, 12 Red-wattled Lapwings, 12 Black-winged Stilts, 16 Darters, 15 Black-headed Ibises, 3 Red-naped Ibises*, 1 Lesser Adjutant Stork*, 4 Black-necked Storks, 5 Painted Storks, 3 Asian Openbills, 7 Gray Herons, 6 Purple Herons, 10 Great Egrets, 8 Indian Pond-Herons, 10 Little Grebes, 8 Comb Ducks, 20 Spot-billed Ducks, 2 Ferruginous Pochards*, 1 Ruddy Shelduck, 8 Garganeys, and 10 Northern Pintails. (The last 4 ducks were non-breeding visitors from the Palearctic.) Other birds of note were the Rufous Treepie, Jungle Babbler, Bay-backed Shrike, Yellow-headed Wagtail, Mahratta Woodpecker, Indian Roller, Green & Blue-tailed bee-eaters, and Rose-ringed Parakeet. It was a productive, challenging morning.

From Agra we traveled SE to the city of Banaras, situated on the 1,000 mile long Ganges River, the 3rd largest river in the world in annual volume of water. It was the most sacred of all the rivers in India and was worshipped as the Goddess Ganga. The river's origin, at 13,500 ft. elevation, was at the bottom of the Gangotri Glacier in Garhwal (Uttarakhand) State in NW India, and the mouth of the river was in the small country of Bangladesh (to the southeast) at the northern end of the Bay of Bengal. The threatened Ganges River Dolphin lived in the heavily polluted river. According to Hindu teaching those who bathe in the river attain purity, and souls of dead bodies (or their ashes) which are placed in the river after death become one with God and go straight to heaven, becoming holy. Banaras was one of the most sacred cities in India and there were hundreds of bathers in the river. Many holy men ("sadhus", dressed in orange robes and characteristically carrying a trident) walked along the streets or sat outside the numerous temples. A sign along the river front said in English (for the benefit of tourists) "it is prohibited to take pictures of beggars, holy men, or dead bodies." I wondered about the last of these so I took out my binoculars and scanned the surface of the river. Sure enough, every 2 or 3 minutes a dead body came floating by! (I sneaked photos of the beggars and Sadhus but did not take any pictures of the dead bodies.) We spent as little time as possible in the cities on the central plains.

From Banaras, Noble and I headed east and north for approximately 500 miles, to the small town of Birganj on the Nepal side of the India/Nepal border, at 2,500 ft. elevation. Birds in the forest here on May 18 & 19 included raptors, swifts, treeswifts, doves, cuckoos, parakeets, kingfishers, bee-eaters, barbets, woodpeckers, minivets, cuckoo-shrikes, laughingthrushes, magpies, drongos, bulbuls, leafbirds, mynas, forktails, small timaliids, flycatchers, warblers, sunbirds, flowerpeckers, and white-eyes. My bird list for 10 hours of birding here (over a 2 day period) totaled 41 species, with 16 trip additions (Appendix B). I was particularly happy to see a Crested Serpent-Eagle*, a Stork-billed Kingfisher*, and Silver-backed Needletails* (large swifts in the genus *Hirundapus*, which flew very rapidly, high and low over the forest).

It was 100 miles from Birganj north to Katmandu, the capital city of Nepal at an elevation of 4,500 ft. in a mountain valley. We spent 2 delightful days here, taking pictures and visiting Bhuddist temples with their prayer wheels and ornamental lion entrance statues. The friendly people smiled at us and the overall atmosphere was entirely different from that of Delhi, in spite of widespread poverty. Both men and women carried goods to and from the market in baskets on their back, or on both ends of a pole carried across their shoulders. Firewood and flowers were popular items for sale. Persons of all ages were often barefoot. Clothes were washed by hand on the river bank. Noble and I enjoyed relaxing, as always, in the little tea shops. I found the Buddhist culture much more pleasant than the Hindu culture.

I sought birds in 4 or 5 different areas of forest in Nepal (including Birganj), at elevations ranging from 1,500 to 8,000 feet. My total list for these areas was somewhere between 125 & 150 species, including 74 trip additions (Appendix B: Mar. 18-25). Birds which gave me the greatest thrill were the Red-headed Trogon*, White-crested Laughingthrush* (an incredible bird which became an instant favorite with its chestnut brown plumage, distinctive crested white head, and black mask), Hoary-throated Barwing* (a timaliid), Hooded Pitta* (one of the very few pittas I was able to locate and identify anywhere in my world travels, since I had no tape recorder to call them to me), Gray-bellied Tesia* (a tiny, almost tailless sylviid warbler of the forest floor and undergrowth), and Purple Cochoa* (a totally unexpected, rare, stunning bird of uncertain taxonomic affinity but placed by Clements, surprisingly, between forktails and bushchats in the family Muscicapidae). This species was a gorgeous, sexually dimorphic, thrush-sized arboreal fruiteater in which the male was dark purplish-brown in color (almost black below) with a beautiful lilac-blue crown, which was also the color of a wing patch and much of the tail. Two males and 1 female sat quietly in the midstory of a fruiting tree while I studied them for quite some time. In addition to the cochoas, I wrote notebook descriptions for more than a dozen other birds which I couldn't recognize on first sight. Included among these was the Striped Tit-Babbler* (*Macronus gularis*), a common, widespread little timaliid which required extensive field notes and effort to finally identify.

After crossing the Nepal/India border once again at Birganj, on May 25, Noble and I proceeded on a shortcut route which would take us more quickly to Darjeeling, as shown on the inadequate road map we had with us. We camped that night "5 mi. E of Motihari" (as recorded in my field notebook). The next day (May 26) we drove to Darbhanga (probably via the town of Sitamarhi), arriving there in the late afternoon. We continued

Plate 91: Sacred cows rested in the middle of a city street in northern India. Women were draped in saris, those of lower castes being barefoot. The most famous architectural icon in all of India was the Taj Mahal in Agra, built in the 1600's as a tomb.

Plate 92: Young boys tended cattle in the fields. The outfit of the man on the left was at the bottom end of the dress code! The paint on the forehead of this Hindu man was mostly worn off.

Plate 93: Not only did I photograph this sign, but also beggars and bathers. A holy man washed alongside some bathers in the very polluted sacred waters of the Ganges River at Banaras. Scavenging White-rumped Vultures congregated on the river bank nearby.

Plate 94: Indian elephants were domesticated and used for work and transportation. Grain was winnowed by hand. Oxcarts were a common hazard on the narrow, overcrowded roads.

eastward for another 1½ hours to "4 mi. E of Narahia" (a name I cannot find on any map), and camped there for the night, not far from the Kosi River. A new bird for my trip list this day was the Brahminy Kite, a favorite raptor of mine which I had first seen in my Navy days, in the Philippiines, in 1956. It was splendidly rufous in color with a white head and breast. Another new trip bird today was the Cinnamon (Chestnut) Bittern, an oriental relative of the Least Bittern. Several Coppersmith Barbets were heard during the day, of which Hugh Whistler wrote in his *Popular Handbook of Indian Birds* that it was found in every type of open country, frequenting the top of large trees where its voice was more familiar than its form and that its loud, mellow "took" note sounded like the tap of a small hammer on metal, repeatedly indefinitely at regular intervals throughout the day, monotonously and exasperatingly. I found these sentiments to be also my own.

The next morning (May 27) we drove a short distance to the wide, shallow Kosi River where there was no bridge, only a crude raft for ferrying small vehicles, cows, and pedestrians across the river, by 4 men with long, hand held poles and 2 men with paddles. There was no engine of any kind. The raft was a platform of bamboo logs and boards tied together on the top of and between two long narrow boats (in a catamaran like fashion). It was barely big enough for our overweight camping van to fit, and I thought to myself "oh no, here we go again" (remembering our attempted crossing of the Rio Terraba in Costa Rica). However, the twin hulls gave it greater stability and carrying capacity than I imagined. It was necessary for me to drive up two rather narrow board ramps to get from the muddy river bank to the top of the raft. I was quite skeptical about our chance for success, but the only alternative to get to Darjeeling was a long detour to the south, where there was a highway bridge across the river.

With great trepidation, I very cautiously navigated up the sloping narrow ramps, with Noble's helpful directions. Success! Although the raft sank down more than a foot into the river, the water level remained well below the platform of boards and bamboo on which our Jeep was parked. It required all of their effort and more than 20 minutes for the 6 men to push and navigate the raft directly across the wide, rather slow moving river, to the landing on the other side. They were sweating heavily when we finally arrived. Noble and I gave them a few more coins than they asked us for (perhaps $2 worth). Our adventure could continue! Some of the birds which I saw in the vicinity of the river here were Indian Roller, Green Bee-eater, Rose-ringed Parakeet, White-eyed Buzzard, Brahminy Kite, Small Pratincole (only my 2nd record), Bronze-winged Jacana, Red-wattled & River lapwings, Little, River, & Black-bellied terns, Little Cormorant, Asian Openbill, 15 Adjutant Storks (not identified to species), and 5 species of herons and egrets.

We now headed east toward the city of Purnia, via the town of Saharsa, but stopping to camp for the night 25 miles before reaching Purnia, at the low elevation of only 500 ft. above sea level. The next morning (May 28) we drove through Purnia and then followed a gradually more northward direction toward the distant town of Suliguri (later changed to Shiliguri) on the northern edge of the plains, where we arrived in the early afternoon. Along our road were 3 Stork-billed Kingfishers, my first Cotton Pygmy-geese*, and many Black Kites and vultures. The most numerous species were House Crows, Common Mynas, and Cattle Egrets. As we continued north toward Darjeeling our road soon began climbing sharply upward in elevation and became much narrower and more scenic as it curved back and forth through forests and ravines, following a narrow-gauge railway which continuously wound back and forth across our paved highway. Several times we waved at the passengers on the train as we passed each other. The driving was slow and I stopped whenever I could to search for birds in the forest, or take pictures. I saw my first Ashy Woodswallow* (family Artamidae), and I wrote a very long description of a handsome small understory flycatcher which I couldn't immediately recognize, but eventually I identified as a Ferruginous Flycatcher* (*Muscicapa ferruginea*). Also in the forest were 3 species of attractively blue-colored niltavas (muscicapid flycatchers). We camped that night at 7,000 ft. elevation just below the popular tourist city and hill resort of Darjeeling, at the southern border of Sikkim Province.

We spent the next 8 days (May 29-June 5) hiking, sightseeing, and pursuing birds in the forested mountains of the Darjeeling area between the elevations of 7,000 & 12,000 feet. From the top of nearby "Tiger Hill" (at 8,500 ft.) we could look north through a telescope at the far distant snow-covered peak of Mt. Kachenjunga, one of the highest Himalayan peaks, rising incredibly to over 28,000 ft. above sea level! As I marveled at this natural wonder I realized just how insignificant on the planet I was. We encountered 2 boys from Australia & New Zealand, and joined them for a 4 day, 3 night backpacking trip, following a well-marked mountain trail all the way up to a youth hostel at 12,000 feet elevation, at timberline, and then back down on a different route.

Birds which I identified at 10,000 ft. or above, in the bamboo and rhododendron thickets and the alpine scrub, were the following (with a + denoting those which I recorded as high as 12,000 ft.): Olive-backed Pipit, Black-faced Laughingthrush*, Spotted Laughingthrush*+ (one of my most exciting birds), Fulvous* & Black-throated* parrotbills (both in the enigmatic genus *Paradoxornis*), Chestnut-tailed Minla+ (a timaliid), Fire-tailed Sunbird*+, Dark-breasted*, Dark-rumped*, & White-browed* rosefinches, Red-headed Bullfinch*+, Collared (Allied) Grosbeak, Gray-hooded Warbler+ (very common), Buff (Orange)-barred Warbler*, Greenish Warbler+, Flame-colored Minivet+, Orange-gorgeted Flycatcher+ (very common), Gray-sided (Rufous-capped) Bush-Warbler*+, White-collared Blackbird, Golden Bush-Robin*, White-capped Redstart+ (breeding at 12,000 ft.), Blue-fronted Redstart, Indian Blue Robin, Rusty-flanked Treecreeper*, White-browed Fulvetta+, Stripe-throated+ & Rufous-vented*+ yuhinas, Gray (Brown)-crested & Rufous-vented* tits, Eurasian Nutcracker, and Large-billed Crow+. Pursuing birds at a high elevation in the Himalayas was one of the most delightful experiences of my 3- year adventure, in the exhilarating atmosphere above tree line.

On the late rainy, overcast afternoon of June 4, I was walking by myself up a small paved road through mountain forest at 8,000 ft., on my way back to our Jeep campsite on Tiger Hill, and I happened upon a once in a lifetime event. As I came to a little bridge over a rather swift running, rocky, small mountain stream, I noticed a group of 4 or 5 men, and 1 boy, standing alongside the stream just above the bridge. In the middle of the stream was a tall stack of dried logs, crudely built as a funeral pyre. On top of the logs was the dead body of an elderly man wrapped in a thin white gauze-like cloth. The people were preparing to cremate the body so that his ashes would fall into the stream and be washed downstream and eventually into the Ganges River, as was the Hindu tradition. The preparation and the burning followed long held beliefs and rituals. I walked down to the stream and stopped a short distance from the group of men so as not to offend anyone, and prepared to watch, but not take any photos. There were no women present. I suspect that the majority (if not all) of the people present were close relatives of the deceased man, most of them probably his sons. One of the men, probably the oldest son, was in charge of the activities (and I shall refer to him as the "eldest"). I felt certain that the ritual and proceedings which occurred were traditional, and they took place in an orderly sequence following time honored custom. It was obvious to me that none of those present perceived the body in the same way as they did when it was a living being, but now viewed it only as a structure to be made into ash in order to preserve its departed soul for eternity. The corpse was lying on its back with the face uncovered, looking upward. Here are the following series of events as I witnessed them.

First the eldest took out a long, straight-edge razor and personally shaved all the hair off the top of his head. He then placed some rice in the mouth of the corpse, and afterwards placed a coin between the teeth of the corpse. A bag of coins was then placed on the chest of the corpse and his hands folded over the bag. A small piece of paraffin was placed on the corpse's mouth, lighted, and then blown out. At this point, the eldest gathered some moss from the streamside and sprinkled it on top of the corpse, after which each person present gathered some sand and sprinkled this over the corpse. The young boy then shaved bare the heads of all the other men present. More logs were gathered for the pyre, on top of which cut ferns were placed. A bundle of clothes and blankets, presumably those which had belonged to the dead man, were thrown into the stream, where they washed only a short distance before they became entangled on rocks.

Now it was time to light the fire. Unfortunately, the logs had become so damp from the day's rain that they wouldn't light, so the boy was sent up the road on his bicycle to the nearest house to bring back a can of kerosene.

Dusk was beginning to settle as he returned 15 minutes later, at which time the fire was lighted. Soon it was burning vigorously. The cremation proceeded rather slowly, and the body did not burn evenly throughout. After many minutes had passed a foot fell off the corpse and into the stream, where it was retrieved by the eldest and tossed back onto the fire, like a piece of firewood. The stomach of the corpse was punctured by the eldest with a pointed stick, to relieve expanding gas inside. After 40 or 50 minutes, most of the corpse had burned except for the skull, which burned slower than the rest of the body. To solve this problem the eldest vigorously crushed the skull with a log (an event I could barely manage to watch). By this time, it was getting dark and I determined that I had seen enough, and walked quickly up the hill to our campsite. This experience left me with one of my most unpleasant memories ever, and reinforced my dislike of the Indian culture.

The initial itinerary which Noble and I drew up in California before we started our adventure called for us to travel from India to SE Asia by road, through Burma, which was the only road route available to us since our passport was not valid for travel in China. However, it had recently become apparent to us through numerous inquiries and conversations with diplomatic officials and others that the Burmese government under no circumstances was going to give us permission to bring our Jeep into the country. Our original plan would have to be cancelled. After careful consideration of all the alternatives and their ramifications we finally settled on a new travel plan. We would send our Jeep via freighter directly from Calcutta to Perth, Western Australia, which was the cheapest freight transportation by ship available between India and Australia. Noble and I would then take our backpacks and together travel by bus, train, hitchhiking, or ship, from India to Ceylon, Malaya, Thailand, Laos (briefly), Cambodia, Vietnam, and Perth. This plan was cheaper than any alternative. We would have to do without the luxury of having our own home with us for this period of time, and I would not be able to camp where I wanted or stop to look at a bird when I wanted. Some of the time we would have to sit or stand on crowded and uncomfortable buses and trains. Our adventure would not be the same. We placed the militarily controlled Burmese government at the very bottom of our list.

Therefore, we departed Darjeeling to travel to Calcutta. On the way we stopped for a day (June 8) in a pristine forest about 10 miles north of Suliguri, at 1,000 ft. elevation. Here I added 14 species to my trip list (Appendix B). I was particularly pleased to identify 3 Oriental Pied Hornbills*, 10 Blue-bearded Bee-eaters* (in the genus *Nyctyornis*, along with the Red-bearded Bee-eater), 4 Rufous Woodpeckers* (placed in the New World genus *Celeus*, which seems very unlikely to me), 12 Greater Racket-tailed Drongos*, 15 White-rumped Shamas* (a marvelous songster), and 8 Asian Fairy-bluebirds* (1 of 2 species in the family Irenidae).

The next morning (June 9) we drove to a nearby site about 5 miles E of Sukna, in a lowland teak forest with scattered pools of water. Here I was happy to SEE my 1st wild Red Junglefowl, and to locate a single adjutant stork next to one of the pools, where I studied it very carefully through my binoculars and telescope. It possessed the distinctive long, hanging, pinkish to yellowish bare gular pouch of the Greater Adjutant*, and it thus became my first positive identification of this local, irregularly occurring stork (which was an Asian counterpart of the African Marabou Stork). With its bare head and neck it was one of the least handsome of all birds, but an icon of the Oriental region. Like its African relative it possessed an enormous wedge-shaped bill and stood 5 ft. or more in height. There was a narrow pale band visible across the black wings in flight.

Not only did I encounter the Marabou Stork this morning but I also came across a single Great Hornbill* in a tall forest tree, one of the most magnificent and the largest of all the world's hornbills (up to 5 ft. long). It possessed a very large, slightly down curved yellow bill with a prominent yellow casque. It, too, was an icon of the Oriental region. In flight the wings produced very loud, far-carrying "droning" sounds which were said by some authors to be the loudest of all avian produced sounds. Complementing the sounds in flight were almost equally loud vocalizations, variously described as cackles, roars, bellows, and rattles. The stork and hornbill were a thrilling finish to my northern Indian birding. That night we camped again at this site. It was the last time we slept in our Jeep in a forest environment on the Asian continent.

Plate 95: We crossed the Kosi River in Bihar Province of northeastern India on a crude catamaran raft. The raft was pushed across the wide, shallow river with poles. A narrow-gauged railroad track crisscrossed back and forth across our winding road up the mountains to Darjeeling.

Plate 96: From Darjeeling, there was a view of the far distant snow-covered high Himalayas to the northwest, with peaks just over 28,000 feet! A Buddhist woman was on her way to a nearby temple. Noble and I trekked for two days to a youth hostel at 12,000 ft. elevation.

Two days later (June 11) we arrived in Calcutta, and Noble took command of handling all the many arrangements and financial transactions necessary for the freighting of our camping van from Calcutta to Perth. After several days we sadly left our beloved Jeep in the hands of a shipping company, and on June 14 we booked 3rd class transportation on a train from Calcutta south along the Bay of Bengal to the city of Puri, 10 hours and 400 miles distant. There was standing room only after everyone crowded and pushed and shoved to get aboard. Noble cheerfully said it was costing us only $10 for 1,000 miles! I had to stand on top of my backpack much of the time. Most of the passengers were dirty, unkempt, and smelly. (For our next train ride we upgraded to 2nd class, which was a little more costly and only slightly better in comfort, as the wooden seats we obtained were hard and unpleasant for extended sitting.) I wondered if we had chosen the right plan to circumnavigate Burma.

We spent the night in a cheap youth hostel in Puri and then went sightseeing around this historic, sacred town the next morning, where we visited and photographed the outside of the magnificent Hindi Jagannath Temple. Non-Hindis were not allowed inside. (I recalled my incident in the Muslim mosque in Qom.) The temple, one of the most holy in India, was 200 ft. tall and had been built in the 12th century. Many thousands of religious pilgrims visited the temple each year on an annual pilgrimage. As I gazed at the temple I pondered the fact that the most magnificent and elaborate structures built by humans were for the purpose of worshipping supernatural beings.

It was necessary for Noble to make a return train trip to Calcutta for several days, to finalize the booking arrangements and payments for the shipment of our van. I chose to continue south on the Indian peninsula by train to the Nilgiri Hills, where I could document birds for several days in the forested hills, then rejoin Noble in the major city of Madras. From there we would proceed together to the island of Ceylon.

I arrived in the Nilgiri Hills and enjoyed 6 days of birding in the forests here (June 24-29) at elevations between 3,000 & 7,000 feet. During this period, I recorded 40 trip additions to my bird list (Appendix B), many of which were endemic to this separate ecological region of India. My most favorite species were the Malabar Trogon*, Rufous-breasted (Nilgiri) Laughingthrush*, Gray-headed Bulbul* (with its bubbling vocalizations), and Heart-spotted Woodpecker* (a peculiar-looking, small, distinctly patterned Gray, black, & white woodpecker with an unusually short tail and very pronounced long pointed crest). The Nilgiri Hills birds were my last additions from India, and I finished this country with a total of 250 new trip birds during our 10 weeks of travel here.

A memorable meal took place during my time in the Nilgiri Hills, on the evening of June 25. I had been hiking up and down on forest roads all day for over 20 miles, with nothing to eat and only a single canteen of water for drinking. By sundown I was hot, tired, thirsty, and hungry when I walked into the small foothill town of Mettupailayam. I walked into the first restaurant I came to and sat down on a wooden bench in front of a long wooden table. In this one room restaurant with open windows and doors there were only 4 other persons eating, a single group of men sitting at the far end of one of the several long tables. My table was devoid of any plates, glasses, eating utensils, or napkins. The only item on the whole length of the table was one large wooden bowl sitting in the middle of the table with some sort of sauce in it. There were no waiters, waitresses, or any working personnel in sight. The 4 men eating paid no attention to me at all, nor said anything to me.

Five or ten minutes after I sat down a young boy finally appeared in the doorway to the kitchen and looked around the room. When he saw me sitting he walked back into the kitchen and then quickly reappeared with a very long, wide, green banana leaf, walked up to my table and without saying a word of any kind he placed it on the table in front of me, then disappeared into the kitchen again. Before long he returned with a very large wooden bowl of steamed rice and a big wooden spoon. Still without saying anything he spooned out a huge pile of rice onto the banana leaf in front of me, and then once more left. I sat there waiting for something else to happen, but it never did. I glanced at the men at their end of the table to see what they were doing. It was very simple. They picked up a handful of sticky, steamed rice from the banana leaf in front of them with their bare hands, rolled it around into a mouth-sized ball, dipped it with one hand into the sauce bowl in front of them

Plate 97: In Calcutta we put our Jeep on a ship to Perth in Western Australia, then took a train to the Southern tip of India. Along the way we stopped to view the Hindu temple in Puri, where I photographed this worshipper. Very dark-skinned fishermen inhabited the southern tip of India.

(fingers and all), and then opened their mouth as wide as possible and tossed the sauce-covered ball of rice into their mouth. That was it.

I was very hungry and I thought to myself that when in India, do as the Indians do. So I picked up a handful of rice from the leaf in front of me, rolled it into a mouth-sized ball, dipped it into the sauce in front of me (dirty fingers and all) and tossed it into my mouth. WOW! It was extremely spicy hot, and burned all the way down my throat. I gasped for breath, looking around me anywhere for something to drink. My eyes instantly began to water, and I dabbed them with a tissue from my pocket. I glanced again at the others to see what they were doing and sure enough, they, too, were all dabbing their eyes with a white handkerchief. I wasn't alone in my reaction to the spicy sauce. Most fortunately the young boy reappeared again just then, and I motioned him to my table. With hand movements to my throat I made it very clear that I needed some water to drink. He understood and left, then quickly returned with a very large glass pitcher full of water, and a large empty drinking glass. At that point I lost all concern from where the water might have come. In no time at all I drank the entire pitcher of water. The boy watched me but still did not speak to me, or I to him. (It occurred to me later that perhaps the boy couldn't speak, for some reason or another.)

I ate all the huge pile of rice on my banana leaf, but never again dipped it into the sauce. The total price of my evening meal in Mettupulaiyam came to the equivalent of 12 cents. It was a meal which I will never forget - - or repeat. I said to myself afterward that one of the primary reasons for traveling abroad is to experience cultures different from our own, and that is exactly what I had just done. Luckily, I developed no negative re-percussions of any kind from either the food or the water. My digestive system, fortunately, had always been very good in this respect.

Noble and I joined up in Madras and then took the very short ferry boat ride from Dhanuskodi, India, to the nearby island nation of Ceylon (which name was later changed to Sri Lanka). However, we were delayed for one day while I forged the date and signature on an old navy immunization card which was in my backpack, to say that I had recently been immunized for cholera, because the health inspector on my first attempt to board the ferry had refused me permission, saying that the date on my new card which I showed him had expired (which it hadn't, but he couldn't read it correctly). .

During the 15 days we were in Ceylon (July 3-17) I increased my trip bird list by 24 species (Appendix B). This included 17 of the 23 species endemic to the island, the most exciting of which for me was the Ceylon Blue Magpie*.

After a 5-day voyage across the northern Indian Ocean by ship from Colombo, Noble and I arrived in Singapore on July 22. Singapore was not an independent country at this time, but was incorporated with Malaya. We had successfully bypassed Burma on our way to SE Asia! I spent most of my time on board the ship studying the two bird books I had with me for Malaya. The only pelagic birds I identified during our voyage, in my casual periods of looking in this avifaunal poor region of the world's oceans, were the Wilson's Storm-Petrel and the Wedge-tailed Shearwater (a light phased individual). Four shearwaters went unidentified.

Noble and I did not always spend our time together in SE Asia. As elsewhere, Noble preferred the cities and I preferred the forests. We traveled by hitchhiking, bus, and train. When we arrived in Singapore in 1961 this small island was politically a part of the "Federation of Malaya", at the southern end of the Malay Peninsula. This region had become independent from British rule just 4 years earlier, and in 1961 it did not include any part of Borneo. Continental "Malaya" was a peninsula approximately 500 miles long and 75-200 miles wide, lying in a NW-SE direction just north of the equator.

Noble and I left Singapore on July 23 and traveled by bus 250 miles NW to the Malayan capital city of Kuala Lumpur, in Selangor Province at an elevation of just 200 ft. above sea level. The city had many parks and gardens in which I could look for birds. I almost immediately contacted a friend of mine, H. Elliott McClure, whom I had first met in Japan just 3 years earlier at the end of my navy days. Elliott was 50 years old in 1961, and had a Ph.D. from Iowa State University. He worked for the U.S. Army Medical Research Unit which was collaborating with the Walter Reed Institute of Research in a study of avian borne diseases in Asia, a task he first became involved with in Japan, before being transferred to Malaya two years ago. He captured thousands of birds annually with mist nets, secured blood samples from them for viral studies, and then released the birds unharmed back into nature. Elliott was a very enthusiastic birdwatcher and he and I birded together in Japan on several occasions, and were compatible companions. Just before I left the navy in 1958, when Noble and I were initiating plans for our world adventure together, Elliott told me that he was going to be transferred to Malaya very soon, and perhaps we might meet up there in a year or two. Sure enough, here I was! What very good fortune it was for me.

Elliott invited me to stay with him at his house in Kuala Lumpur, and he took as much time off his work as possible in the next 2-1/2 weeks to go birdwatching with me in his car. Noble stayed awhile in Kuala Lumpur where he kept busy enjoying conversations with new acquaintances, but eventually he left to go on ahead to Bangkok, in Thailand, where he said I could join him in one or two weeks. We were each happy with our own activities. On those days when Elliott could not go with me in his car I traveled to nearby forests either by bus or hitchhiking. It was a fun time for me, birdwatching, taking pictures, and enjoying nature. The days went by quickly.

Almost all the forest areas within 20 or 30 miles of the city, north, east, or west, were tropical rainforests, from sea level up to 2,000 ft. elevation. Most of these had been logged at one time or another in the past, but a few accessible, sizeable areas were still pristine rainforest, and it was in these areas where I experienced my most exciting and memorable moments, and thrilling birdwatching.

The Malayan rainforests shared many features with the rainforests of Africa and South America, but with different species. In the Malay forests the tallest, dominant trees were dipterocarps, occasionally up to 250 ft. tall and often with the wide buttresses characteristic of rainforest trees everywhere. Other typical rainforest features were the great variety and abundance of orchids, insects, and fungi. Such giant insects as walking sticks (over 2 ft. long), rhinoceros beetles, empress cicadas, and atlas moths all lived in the Malayan rainforests. The largest bovid mammal in the world (by weight) was the Gaur, a huge "cow" in which large bulls sometimes weighed up to over 3,000 lbs. It lived sparsely in the forested foothill rainforests of southeastern Asia, and was the probable wild ancestor of all modern day cattle. Other mammals were the Sun Bear (or Honey Bear, the world's smallest bear), the Clouded Leopard, lorises (primitive sloth-like primates), pangolins (armadillo-like in appearance, found also in Africa), gibbons, and giant squirrels (both flying and non-flying). Elephants, tigers, and rhinos once lived in these forests but were now virtually exterminated. Among reptiles were flying lizards, large monitor lizards, one of the longest snakes in the world (the reticulated python), and the longest venomous snake in the world (the king cobra). Birds are discussed below.

The sounds of a tropical rainforest, or "jungle", were unlike anything else. At times, just before a rain, there could be ominous silence, but as the rain approached it could be heard coming through the tree tops from a very long way. During the first hour after sunrise, the sounds of birds, monkeys, insects, and frogs were almost deafening. By far the loudest and most thrilling of all the sounds in Malaya were those made by 2 species of gibbons, the larger Siamang and the smaller White-handed Gibbon. Words cannot do justice to their vocalizations, and my best attempt to describe them is to compare them to the singing of a female opera singer - - loud, up and down the scale, short or prolonged, reverberating, and carrying for long distances throughout the forest. They were the most spine-tingling of any sound produced by animals which I have ever heard. Parrots and parakeets flew overhead in screaming flocks, and huge Rhinoceros Hornbills flew through the canopy singly or in pairs, sounding like the horn of an old model A Ford. Barbets, pigeons, and doves called continuously all day long with hoots, coos, and various single or double notes, often given repetitiously for long periods of time. White-rumped Shamas and some of the bulbuls produced the most musical and pleasant songs. Monkeys, when alarmed by the approach of a person, crashed loudly through the tree tops. Squirrels of all sizes and colors chattered constantly at each intruder, as did babblers working through the forest understory. These were my most memorable sounds of the rainforest.

To see the animals of the forest was more difficult than to hear them. Nearly all stayed hidden within the forest, from the canopy to the ground, each species preferring a different vertical level. Among the most brilliantly colored birds were the pheasants, trogons, parakeets, bee-eaters, kingfishers, broadbills, pittas, leafbirds, minivets, and sunbirds. I found the cryptic, barred, and spotted plumage patterns of the babblers, laughingthrushes, and other timaliids to be particularly attractive and appealing, along with their colors and shades of ochre, chestnut, rufous, and brown. If a person stood quietly in one spot, monkeys and squirrels commenced moving about him, chattering and feeding in the trees. Long-tailed lizards glided with amazing maneuverability from branch to branch or trunk to trunk. It always came as a surprise to glimpse a streak going past my nose and upon looking in front of me to where it landed find a lizard clinging to a branch or tree trunk! Within almost every animal group there were rainforest species adapted to flying or gliding, which included not only the the birds, bats, and insects, but also squirrels, snakes, and lizards. To stand in a forest clearing at sundown and watch bats and nightjars replace swifts and swallows as aerial insectivores was a sight never to be forgotten (an observation I first witnessed, and described, from the Amazon rainforest).

One often hears or reads of the many possible annoying or life threatening dangers in a tropical rainforest, but the chances of encountering all but a very few of these are quite minimal. My two greatest nemeses in my travels were ants (in the Congo) and terrestrial leeches, which I experienced for the first time here in Malaya. I was walking on a forested hillside for several hours one afternoon (July 29) when I suddenly discovered that my trouser legs, shoes, and socks were all stiff with dried blood. I couldn't imagine what might have happened (except stabbing my legs with thorns) so I sat down on a log in the forest and pulled up my trouser legs to have a look. There, to my horror, below my knees on each leg were 10-12 very fat leeches, up to ¾" in length! I couldn't believe my eyes. They must have been feeding there for hours, completely unknown to me. I shuddered at the sight of them. Their blood- sucking, remarkably, was quite painless and unnoticeable, resulting from both an anticoagulant and an anesthetic in their oral secretions. As I sat on a log pulling the leeches off and throwing them as far away as I could, one by one, others were slowly crawling toward me on top of the log, in an undulating worm-like fashion. If I put a finger directly behind one it would immediately turn around and follow my finger, presumably by detecting the body heat in my finger. The worst part of this unpleasant experience was that the "bites" began to itch quite severely within a few hours, after I had thrown away the leeches. The itching continued for more than 2 weeks afterward, causing me to scratch my legs vigorously and uncontrollably, while people stared at me in wonderment. The leeches were my most unpleasant experience ever with any organism in a rainforest, (plant or animal). I prevented the problem from happening again by keeping my pant legs tucked securely in the tops of my boots.

I documented birds in the Malayan rainforests on 6 different days, at 4 different locations. All the sites were below 1000 ft. in elevation except for one at 2,000 ft. (on July 29, mentioned above and to be further discussed below). My complete list of birds totaled 180 species, of which 85 were trip additions (Appendix B for the dates of July 27, 28, 29, & 30, and Aug. 5 & 6). The avian groups or families with the greatest number of species were 26 babblers & relatives (Timaliidae), 21 bulbuls, 12 sunbirds & spiderhunters (Nectarinidae), 10 woodpeckers, 8 warblers (combined Sylviidae & Cisticolidae), 8 flycatchers (combined Muscicapidae, Rhipiduridae, and Monarchidae), 7 diurnal raptors (vultures, hawks, eagles,& falcons), 7 cuculids (5 malkohas & 2 cuckoos), 6 columbids (pigeons & doves), and 6 barbets. (I found it of interest that during my 6 days in the Malayan rainforests the total number of species I added to my trip list, 85, was just over half the total number of 162 birds I added to my trip list the entire 8 months I was in Europe!)

July 29 (previously mentioned) was a day to remember, the day I went with Elliott to the rainforest in the hills at 2,000 ft. elevation, 20 miles N of Kuala Lumpur. Here, Elliott had built a wooden, very simple tree platform 100 ft. above the ground in a tall emergent forest tree on a hillside, from which he could look out for birds in the forest canopy around him, and into the distance above the trees. The tree platform was a magnificent accomplishment, erected by just Elliott and a Malay boy as a helper. But it definitely was not for the timid - - such as me! The plain wooden platform was built on the very first horizontal branch of the tree, exactly 100 feet above the ground. It was 5'x5' square and had no railings or anything else at all to hang onto. It consisted only of flat wooden boards attached horizontally and firmly to the large tree branch, nothing more. In my entire life, I had never liked heights or getting my feet very far above the ground, so when Elliott scurried up the ladder and then shouted for me to come on up, I was quite hesitant. The narrow (8" wide) aluminum, 100 ft. ladder had been purchased from a Swedish catalog, and was fastened around the wide, smooth-barked tree trunk with an aluminum band every 6 feet. There were small aluminum horizontal bars spaced evenly 10" apart which served as steps - - and the only thing to hold onto! Nothing else was available for support, and there was no safety harness of any kind. You either held onto the rungs as you climbed 100 feet straight up, or you let go and fell. It was that simple. If you got tired on your way up you could "rest" by hanging your arms over the rung just above the one on which you were standing. (In reality the tree was not exactly straight up but had a very slight backward lean on the way up!)

Elliott could not understand what was taking me so long to get started. I didn't dare confess my fear of heights. So I started up. I was determined never to look down at the ground but stare only straight ahead, at the tree trunk just in front of my nose. I "rested" several times on the way up as my arms became more and more tired. When I finally arrived at the top I was horrified to find out that the top end of the ladder was no more than a few inches above the wooden platform, with nothing available for me to grab onto to help pull myself up onto the platform, a critical task for which I didn't have any strength left. Fortunately, Elliott saw my plight and pulled me up over the edge, where I immediately collapsed. Thereafter, I refused to stand up at all but only crawled around on top of the not very wide platform, without railings and barely enough room for two people. I couldn't understand Elliott's complete abandonment of any fear at all as he happily searched the trees and the sky for birds, standing up and moving around while holding his binoculars in both hands. After all, he was almost twice my age, and we both possessed much the same small body frame and height. I found it difficult to get any enthusiasm at all for looking at the canopy birds, but I did succeed in locating and identifying, without Elliott's help, a Long-billed Spiderhunter, which Elliott said was a canopy specialist and practically never observed from the ground. Was my death defying climb worth one new bird? Definitely not! My father was certainly right on this occasion, and I was crazy.

Once I made up my mind to start back down, the descent was much less tiresome and frightening than the ascent had been, but I still never looked down at the ground, not even once. It was an experience I vowed never to repeat - - and I haven't, and won't. Elliott had to return to Kuala Lumpur at mid-day but he said I could stay and look for birds in the forest if I wanted, and afterwards I could either hitchhike or take a local bus back into the city. I opted to stay. I needed to calm down from my very traumatic morning. However, such was not my fate. On that very same day, in the afternoon, I encountered the terrestrial leeches in the forest (which I have already discussed). How could two such emotionally and physically draining events possibly happen on the same day? Surprisingly, in spite of these two exhausting experiences I somehow managed to finish the day with 19 new trip additions to my bird list (Appendix B). It was a day in the rainforest that was etched in my memory forever.

My last 1½ days pursuing birds in the rainforest were Aug. 5 (all day) and the morning of Aug. 6, at the very low elevation of 100 ft. above sea level in a pristine forest on the floodplain of the Bernam River. I spent the night of Aug. 5 in a guest house, in the small nearby village of Slim River. During these 2 days I pursued birds, gibbons, monkeys, squirrels, lizards, and butterflies in the forest and its clearings and edges for a total of 17½ hours. Nothing could have been more exciting. My total bird list was 95 species, of which 13 were trip additions (Appendix B). Among those species which I enjoyed the most were the following: 7 species of wood-peckers (5 Gray & Buff, 3 Rufous, 2 Checker-throated, 1 Buff-necked, 1 Buff-rumped*, 1 Common Flameback, and 2 White-bellied Black), 2 Gray-headed Fish-Eagles (genus *Ichthyophaga*), 10 Black-thighed Falconets, 120 Long-tailed Parakeets* (flying back & forth over the forest), 2 Chestnut-breasted Malkohas (filling the Squir-rel Cuckoo niche), 4 Sunda Scops-Owls (calling outside my guest house all night long, but frustratingly never seen), 12 Malaysian (Eared-) Nightjars* (genus *Eurostopodus*, seen & heard as they circled over the forest clear-ings at dusk and dawn, calling loudly with their 3-syllabled "drink your beer" vocalizations, with accents on the first and last syllables, one of the most characteristic sounds of the rainforest at dusk), 30 Crested Tree-Swifts, 80 Silver-rumped Needletails (a swift in the genus *Rhaphidura* with a completely white tail, flying high and low over the forest, filling the same niche as the congeneric Sabine's Spinetail in the Congo rainforest), 3 Dollarbirds ("Broad-billed Rollers" in the genus *Eurystomus*), 4 Rhinoceros & 15 Bushy-crested hornbills, 10 Blue-eared (Little) & 10 Yellow-crowned barbets (calling throughout the day), 50 Blue-throated Bee-eaters, 1 Straw-head-ed Bulbul (a large, handsome bulbul with a golden yellow crown & ear coverts, a black line through the eye, a white throat, and a prominent black moustache stripe), 10 Fluffy-backed Tit-Babblers (*Macronus ptilopus*, a denizen of the forest undergrowth, like Amazon antbirds), 2 species of *Stachyrus* babblers (Chestnut-rumped & Black-throated, both very handsome and distinctly plumaged), 1 Purple-throated (Van Hasselt's) Sunbird* (a very pretty, small sunbird with a relatively short bill), and 50 White-headed Munias ("Pale-headed Manni-kins" in the genus *Lonchura*, family Estrildidae). My experiences in the Malayan rainforests ended with these 2 memorable days.

Between my days in the rainforests I traveled by bus from Kuala Lumpur to the Cameron Highlands for 2½ days of birding (Aug. 1-3), in the forested mountains between the elevations of 4,000 & 6,500 feet. It was a relief to be out of the hot lowlands for a few days, in the cool, refreshing mountain air. My 9 trip additions for these days are listed in Appendix B. The birds that provided me the most pleasure were the Greater Yellownape (a woodpecker), Chestnut-capped Laughingthrush (which I had first encountered on Fraser's Hill near Kuala Lumpur on July 25), Silver-eared Mesia (a timaliid), Streaked Wren-Babbler, Large Niltava (a vivid, dark vio-let-blue muscicapid flycatcher), Black-throated Sunbird (genus *Aethopyga*, with elongated central tail feathers), Streaked Spiderhunter, Fire-tufted Barbet, Red-headed Trogon, Golden Babbler, Black-eared Shrike-Babbler, Chestnut-crowned Laughingthrush, Blue Nuthatch* (my only record of this species), Sultan Tit* (the largest of all the traditional members of the family Paridae), Green Magpie, and Orange-bellied Leafbird.

On Aug. 10, I said goodbye to Elliott McClure, thanked him sincerely for his hospitality (excluding the tree platform), slung my backpack over my shoulders, and caught a bus to take me east across the peninsula to the city of Kuantan on the southern coast of the South China Sea, where I spent the night. The next morning (Aug. 11), I headed north along the coast on another bus to begin an 800-mile bus trip to Bangkok (Thailand), where I would join up with Noble once again.

Along the way, I stopped for a day and night (Aug. 11 & 12) to watch giant sea turtles come ashore and lay their eggs on the sandy beaches north of the city of Dungan. A 400-500 pound female was amazing in her ability to dig "post holes" several feet deep with just her hind flippers, in a matter of only about 30 minutes. It then required her just another 10 minutes to deposit up to 100 eggs in the hole, about one egg every 5 or 6 sec-onds. The female turtle then spent up to an hour meticulously covering the eggs with sand, obscuring all visible signs of the nest hole to hide it from the many predators which scavenged the beaches for the eggs. Sadly, these included humans, who collected the eggs to eat them. About 2 hours after first coming ashore a female would crawl laboriously across the beach again and back into the pounding surf, where she disappeared into the sea, not to return to shore again until next year's breeding season. For a brief period of time her distinct footprints were left in the sand. It was a marvelous spectacle, a cycle of events that had evolved over millions of years.

Plate 98: In Southeast Asia we were no longer traveling in our Jeep camping van. An ornate gateway welcomed visitors to Kuala Lumpur, Malaya. Goods were sold or bartered in open markets. Durians were a popular local fruit but their odor was repugnant.

Plate 99: Elliot McClure, an American ornithologist I knew from my Navy days in Japan, invited me to stay with him. Together we experienced the birds and thrilling vocalizations of the White-handed Gibbons in the tropical rainforests – along with a treetop viewing platform & terrestrial leeches!

Plate 100: Local residents collected sea turtle eggs on the Pacific coast of northeastern Malaya near the town of Dungan. Teenage boys enjoyed riding on the backs of the turtles. Fishermen cast their nets for marine fish.

On Aug. 13, I continued northward to Kota Bharu for the night, on the coast just below the border with Thailand. It was my last full day in Malaya, and I added 2 final trip birds to my list, the Collared Kingfisher (a very widespread coastal kingfisher in southern Asia and a great many islands in the southwest Pacific, to the coasts of northern Australia), and the German's Swiftlet* (one of 27 confusing little swifts in the genus *Aerodramus* which inhabited the Oriental, SW Pacific, & Australian regions). These 2 species brought my trip list for Malaya up to 149 species (one of the top 10 countries in my three year adventure, see Appendix C).

I entered Thailand on Aug. 14, and found Noble in Bangkok 2 days later. After several days of sightseeing in this overpopulated city where many people chewed "betel nut" (see chapter 16), Noble and I booked a 2-day train & bus ride to the hill country of northern Thailand, in and around the popular tourist city of Chiang Mai. The Thai people were warm and friendly almost everywhere.

From Chiang Mai we traveled to a small, nearby, hill country village where the Maew tribe people lived in thatch-roofed houses of boards and sticks placed on an elevated bamboo floor. Many of the inhabitants in the village, both men & women, daily wore their colorful traditional clothing, with silver hand-made jewelry, which included wide ring necklaces, bracelets, and frontal chest plates. Often, people were bare-footed. They were friendly people and happy for us to take their photographs (without asking for money). Only a few tourists came to visit this little village. Young boys had crude, hand-made wooden bows and arrows with which they hunted birds, hares, and other small game animals. I offered one of the boys a few coins for his bow and he willing obliged me. (It didn't appear to me that constructing another one would involve very much time or effort.) I managed to fit the bow in the top of my backpack (which will be a topic for later mention).

Birds in the mountains here at 3,500-5,000 ft. featured woodpeckers (7 kinds), bulbuls, timaliids, and small flycatchers. In a day of hiking in the Doi Suthep mountain area on Aug. 21, I documented 66 species. Fifteen of these were trip additions (Appendix B). I was particularly excited by 2 woodpeckers (the Maroon Woodpecker* and the Bay Woodpecker*) and by 2 timaliids (the Red-billed Scimitar-Babbler* and the Chestnut-fronted Shrike-Babbler*). We returned to Bangkok the next day.

Before leaving Thailand, Noble and I decided we would like to visit Laos for a couple of days, to add this nation to our list of countries which we visited. There was a railroad track going from Bangkok all the way north for roughly 400 miles, to the small town of Nong Khai on the Mekong River, opposite Vientiane, the large capital city of Laos. This adventure was somewhat off the beaten tourist path, and required a 12 hour, not very comfortable, 2nd class train ride. We obtained an entry visa good for one week, but we planned to spend only 2 nights and 1 full day in the country.

We arrived in Nong Khai and spent the night there (on the Thailand side of the border). The next morning (Aug.26) we got up early to catch the first ferry ride across the wide Mekong River. It was a large ferry with room for several hundred passengers (mostly standing), which went several times daily back & forth across the river, a one-way trip of only 20 minutes. We were the only "Europeans" on board, and Noble towered over almost everyone else, but people paid us very little attention. Until quite recently (1950), Laos had been part of "French Indochina."

Our time in Vientiane was mostly uneventful and without much excitement, except for one event. Noble and I knew that the city was known for its many "opium dens", so we decided to see for ourselves what the big attraction was of this drug. Smoking opium was supposed to relax a person and provide pleasant feelings of euphoria and ecstasy. It was a narcotic used in medicine to treat nervous disorders and depression, but it could become addictive, and made into heroin. Opium was produced from the fruiting capsules of the poppy *Papaver semniferum*, and the dried powder which was obtained from these capsules could be heated and smoked in special long, narrow opium pipes. Such pipes were traditionally made from bamboo, but also were manufactured artistically from silver, jade, horn, ivory, porcelain, or cloisonné. Pipe stems were 12-15 inches long, to which a door-knob shaped bowl was attached near the far end. Cultivation of poppies for opium dated all the way back in history to Sumerians, Babylonians, and Egyptians, 2,000 - 3,000 years BCE.

On our first night in Vientianne, we walked along a main street after dark looking for an opium den. Soon we found some narrow stairs leading down from the street to an unlighted wooden door at the bottom. (The setting was much as I had imagined.) When we walked down the stairs and knocked on the door, a thin, elderly, sly-looking Chinese man with a long, wispy, white beard opened the door, carefully scrutinized the two of us, and then without saying anything motioned us to come inside. His appearance was just like in Hollywood movies. He must have immediately realized that we were complete novices and knew nothing about smoking opium. He spoke no English but motioned for us to follow him down a long narrow hallway, on both sides of which were many doors, each leading separately into a small room with two wooden benches. He led us into one of the vacant rooms and we each sat down, separately, on a bench. We bargained with our hands, asking how much money he wanted us to pay him, and finally settled on a price equivalent to about $7 each (which I feel certain was much too high and certainly a lot of money for our limited budget). Never-the-less, we paid the old man and he left. It was an experience we might never have a chance to repeat.

Quite some time later, a much younger Asian man came into our room with a burning opium pipe for each of us, which he handed over and then left, without saying anything. On our separate benches Noble and I sat back and relaxed, and puffed on our pipes, very carefully at first. The hot vapor did not in any way change my feelings or emotions, so I puffed a little harder. Ten minutes later I still felt no different at all, so I asked Noble if he was experiencing any changes in his feelings. His answer was no. After 20-25 minutes our pipes burned up whatever it was in the bowl, and they stopped producing vapor, none of which had caused either Noble or me to detect any kind of change in our feelings or emotions. It was obvious that we had been purposefully deceived and cheated, and had not been given opium at all. As we walked down the hallway, out the door and back up the stairs to the street, we never saw either of the two men again, or anyone else. It was the first and last time either of us attempted to smoke opium, and one of our few poor decisions.

On our last night in Vientiane we went to a hotel for a beer. Here there was a French war correspondent who spoke English, and the 3 of us became involved in conversation. When Noble mentioned to him that we were planning to go back tomorrow morning on the ferry to Thailand, he asked us if we had obtained our EXIT visas yet, which he remarked were required of all non-Asian visitors. Our answer was no, that we were unaware of this government regulation. He then said to us that we could circumnavigate this necessity by hiding among the many passengers boarding the ferry, in the middle of all the Asians where we would not be spotted by the immigration officials as we walked on board. Of course this meant that Noble would have to stoop very low. However, this is exactly what we did the next morning, without being detected! Life was an adventure.

We traveled by train back to Bangkok, and from there we headed east on a bus to the border of Cambodia, and entered this country on Aug. 30. Cambodia had gained independence from France in 1953. From the border it was about 35 miles east to the town of Sisophon, and from there another 70 miles east to the city of Siem Reap, where we spent the night. During this bus ride I saw my first Rufous-winged Buzzard* (*Butastur liventer*). Siem Reap was only 4 miles south of the world-famous temple ruins of Angkor Wat, which Noble and I wanted to visit. We spent the next 3 days and 2 nights there.

Angkor Wat was the largest religious monument anywhere in the world. It was built of sandstone blocks in Kmer architecture in the early 12th century, originally as a Hindi temple but later transformed into a Buddhist temple. An outer rectangular wall was 15 ft. high, 3,360 ft. long, & 2,630 ft. wide. It enclosed 200 acres. In 1961 this outer wall was almost completely overgrown with huge tree roots, which gave the ruins a distinctive, ancient feeling of time gone by. Of course, I took many pictures. Surrounding the wall was a 600 ft. wide moat, protecting it from intruders and conquerors. There was a single main entrance to the temple complex, and several lesser entrances, all of which led inside to 3 main towers, each with galleries on 3 separate levels above the temple base. It was on one of these levels where I slept one memorable night, with bats of all sizes flying in and out over my head all night long! Inside the temple there were extensive stone carvings on the walls, and paintings depicting Hindi and Buddhist deities, spirits, and mythical beings. The top of the central tower was 210 ft. above ground level and 140 ft. above the elevated man-made base of the temple. Today, the temple is a UNESCO world heritage site. It was the most fascinating ancient ruin I have ever visited. Again, I pondered at the effort put forth by people everywhere to build enormous, elaborate, highly decorated and ornamented temples and places of worship to supernatural beings, all of them to answer their infinite questions about life and death, and provide them with comfort. The human mind was the greatest creation of all.

During the 3 days, we were at Angkor Wat (Aug. 31 and Sept. 1 & 2), I devoted many hours to searching for birds around the ruins. The countryside was open, dry scrub & jungle, and small villages. My total list for the 3 days came to 75 species, including 8 trip additions (Appendix B). The most numerous birds identified, in order of abundance, were 120 Red-breasted Parakeets*, 80 Greater Racket-tailed Drongos, 75 Black-collared Starlings, 50 White-crested Laughingthrushes (wow), 50 Jungle Mynas, 35 Lineated Barbets, 35 Striped Tit-Babblers, 30 Common Mynas, 25 Fork-tailed Swifts (a Palearctic fall migrant), 25 Barn Swallows (another Palearctic migrant), 25 Streak-eared Bulbuls*, 20 House Swifts, 20 Brown-backed Needletails (the largest swift in SE Asia), 20 Common Tailorbirds, 20 Abbott's Babblers, 20 Chestnut-tailed Starlings, and 20 Large-billed Crows I was particularly pleased to see my first Black Bazas* (a small, distinctly patterned hawk in the genus *Aviceda*, with a long prominent plume on top of its head) and Banded Broadbills* (5 seen or heard). Also recorded were 7 Brahminy Kites, 1 Collared Falconet, 1 Barred Buttonquail, 8 Oriental Pied Hornbills, 5 Cotton Pygmy-geese, 1 Spot-billed Pelican, 6 Lesser Whistling-Ducks, 2 Stork-billed Kingfishers, 10 Alexandrine Parakeets, 1 Barn Owl (!), and 8 Asian Barred Owlets (active diurnally).

On Sept. 3, we traveled SE by bus and hitchhiking from Siem Reap to Kompong Thom, where we spent the night on our way to Phnom Penh (the capital city of Cambodia). The countryside had many palm trees and rice paddies, where peasant workers wearing wide, pointed straw hats planted and harvested rice by hand, standing and bending over in the shallow water of the rice paddies all day long. Some of the birds along the roadside were herons, bitterns, whistling-ducks, storks, ibises, darters, 1 Rufous-winged Buzzard, and 6 Green Bee-eaters.

While Noble walked up and down the streets of Phnom Penh visiting with all those who could speak English, I took the bus on Sept. 6 & 7 to the towns of Kampot & Bokor in forested foothill mountains near the Gulf of Siam, about 100 mi. SW of Phnom Penh. Here I pursued birds for 1½ days. During this time I documented 51 species, 6 of which were trip additions (Appendix B). Particularly appealing were 7 Great Hornbills, 30 Wreathed Hornbills, 2 Blue-bearded Bee-eaters (genus *Nyctyornis*), 5 Moustached Barbets*, 2 Green-eared Barbets*, 2 Laced Woodpeckers, 20 Black-crested Bulbuls, 15 Fairy-bluebirds, 12 Blue-winged Leafbirds, 3 Velvet-fronted Nuthatches, and 6 Yellow-breasted (Indochinese or E. Green) Magpies*. Barbets vocalized throughout the day, reinforcing this permanent memory for me of the Oriental forests. During my bus ride back to Phnom Penh on Sept. 8, I observed dozens of Painted Storks soaring above the rice paddies. I also carefully studied 2 Indochinese Bushlarks* next to the roadside. This latter species was my last trip addition from Asia, bringing the number of birds I added from this "continent" up to 587 species.

From Cambodia, Noble and I wanted to travel to Saigon, in South Vietnam. Such travel was greatly complicated by the civil war going on between the North & South Vietnamese. Buses were no longer carrying passengers back & forth, and very few private vehicles were risking this short journey. We were thus very fortunate to encounter a French war correspondent at a hotel bar (once again), in downtown Phnom Penh, the evening I returned from Kampot. When we mentioned that we were looking for transportation to Saigon, he informed us that he was traveling there the next morning and would be very happy to take us with him, since he did not like traveling that route by himself! How lucky could we get?

Plate 101: Thailand was characterized by vast areas of rice paddies. Noble and I traveled together from Bangkok to Chiang Mai in the far north of the country, a two-day train and bus ride.

Plate 102: From Chiang Mai, we traveled to a "hill tribe" village of the Maew people, situated at 5,000 ft. in wet forests. The primitive houses were built on stilts and made of board planks, with thatched roofs.

Plate 103: People were friendly and did not object to being photographed (and they did not ask for money). They characteristically wore traditional dress. For a very few coins, I purchased a crude cross-bow from one of the boys.

385

Plate 104: On a spur-of-the-moment decision we decided to ferry across the Mekong River from northern Thailand into Laos, where we spent two days in the capital city of Vientiane, and attempted to smoke opium!

Plate 105: We traveled across Cambodia mostly by bus. Rice paddies, canals, shallow lakes, and Oil Palm trees dominated the landscape. At one eating stop there was a Green Peafowl, an iconic bird of Southeast Asia, tethered to the bamboo porch.

Plate 106: The 12th century ruins of Angkor Wat, a sandstone block temple in Cambodia, were overgrown with giant tree roots. The ruins provided me with one of my most thrilling archaeological moments, when I slept inside one night with bats flying in and out over my head.

The drive the next morning (Sept. 9) required less than 3 hours. We did not stop en route for fear of encountering hostile combatants. We were told by the Frenchman that the countryside was not safe. Our intention after arriving in Saigon was to spend no more than 2 or 3 days there before flying to Singapore, from where Noble had booked us cheap passage on a slow freighter to Perth, scheduled to depart in about one week. (It was coming to Singapore from somewhere else.) Our Jeep would be waiting for us in Perth when we arrived! However, as often happened, Noble almost immediately encountered a person on the streets of Saigon, an American military advisor, who after a short conversation with him, invited Noble to accompany him to the demilitarized zone near the border between South and North Vietnam, to learn what was going on there, and why. They could fly there together for a week. Nothing gave Noble greater pleasure than such excursions, and of course he immediately said yes. I gave this situation some careful thought, and decided I would rather wait for Noble in Singapore than in Saigon, which was fine with Noble. Therefore, on Sept. 12, Noble flew north from Saigon and I flew south to Singapore.

Events happened rapidly after that date. When I arrived in Singapore, I discovered that the freighter on which we had passenger accommodations was going to be at least 10-15 days late in arriving to Singapore. I was certain Noble would not want to wait that long, nor did I. Therefore, without consulting Noble, I took our travel into my own hands (which I had never previously done). I canceled our ship passage and booked an airline flight on Sept. 14 for me to fly directly to Perth from Singapore, which was a popular flight for businessmen. (I think I must have paid for my ticket with a refund from our ship cancellation.) I sent Noble a telegram to the Am. embassy in Saigon, telling him what I had done, and why, and saying that I would plan to remove our Jeep from customs and be waiting for him whenever he arrived in Perth. He could contact me through the Am. Consulate in Perth. Australia would be a new adventure!

Chapter Twenty-one --- Australia

T he country of Australia includes the whole of the Australian continent plus the island of Tasmania (off the southern coast). The country is subdivided politically into six states (five on the continent plus Tasmania) and two territories (the "Northern Territory" and the small "Australian Capital Territory" surrounding the capital city of Canberra). The continent is almost exactly the same size geographically as the contiguous 48 states in the USA, and although it is the smallest of all seven "continents" in the world, it is still 3 times larger than the largest "island" of Greenland. The single state of "Western Australia" comprises 1/3 of the continent and is 50% bigger than the largest U.S. state of Alaska. Australia is often referred to as "The Land Down Under" because the entire country is situated below the equator, and almost 2/3 is below the tropics. It is unique in being the hottest, the most uniform in topography, and the driest (outside of Antarctica) of all seven continents. The highest mountain (Mt. Kosciusko in the state of New South Wales) is only 7,350 ft. above sea level. Most of the interior receives less than 15" of rainfall annually, and 40% of the continent receives less than 10 inches annually.

The Australian flora and fauna are equally unique, and are placed by Wallace in the "Australian Biogeographic Region" along with New Guinea, New Caledonia, and New Zealand. Most notable are the facts that the only three species of egg-laying mammals in the world (two echidnas and one platypus) live only in this region, and it is the only region where marsupials are the dominant group of mammals, filling ecological niches that are occupied everywhere else in the world by placental mammals. There are 27 families of endemic birds in the Australian region, rivaling the endemism of the Neotropical region. The 174 species of honeyeaters (family Meliphagidae) are virtually all confined to this region (and neighboring islands in the southwest Pacific as far east as the Hawaiian Islands). Almost all cockatoos are endemic to this region, and the overall diversity of parrots is rivaled only in the Neotropics. More than 80% of all snakes are venomous, monitor lizards ("goannas") grow to 6' in length, and the 250 species of skinks (family Scincidae) show much adaptive radiation. The Myrtaceae and Proteaceae comprise a large proportion of the flora. The genera *Acacia* (with almost 1,000 species), *Eucalyptus* (with 700 species and a majority of all individual trees), and *Banksia* (with 180 species) are the dominant woody genera.

I arrived in Perth just after midnight on Thursday, September 14, 1961, on a Qantas flight from Singapore, a flight in which most of the two or three dozen passengers were business men dressed in coats and ties. I was wearing hiking shorts and a clean, neat, short-sleeved shirt (with no rips or holes). My beard was short and trimmed, my hair had been recently cut, and my only luggage was a backpack in which the crossbow from Thailand was sticking out of the top. I walked down the steps from the airplane onto the tarmac runway below, and walked with the other passengers into the terminal building 100 yards away. It took only ten minutes for me to clear immigration and customs. As I walked out of customs through the gate into the terminal lobby I was approached by a nice looking young man with a large camera who stopped me, introduced himself to me as Ian, and said he was a journalist working for "The West Australian", Perth's largest newspaper.

He asked me if I would be willing to provide him with some information about myself, and a photograph for an article which would come out in today's morning edition of the newspaper. Of course, I said yes. I presumed I was stopped because my appearance was radically different from almost all the other passengers. When the story and picture appeared on the newsstands later that morning it described me as being "bearded and roughly dressed, with a crossbow sticking out of my backpack" (which could clearly be seen in the photograph). I was perturbed by the "roughly dressed" description, thinking that the journalist should have seen the way I looked most of the time for the past two and a half years if he thought I looked roughly dressed that day. (It's true that I was not dressed in a coat and tie as were almost all the other male passengers.)

After the interview, Ian asked me where I was going to spend the remainder of the night, and I told him that I had no reservations anywhere. Therefore, he said I could come with him to his "flat" (apartment) where he had an extra bed in which I could sleep. I thanked him and readily accepted his kind invitation. He drove me to his flat and then went back to the press to write his article. Later that morning after I got up, I miraculously found the telephone number that Margaret Rankin had given me in Pamplona, Spain, 14 months ago. Neither of us had contacted the other since we parted company. Ian said I could use his telephone to call the number, which was in Perth. I was very surprised when Margaret answered the telephone at her mother's house where she was still living. I identified myself and asked her if she remembered meeting me in Pamplona. She emphatically answered "Of course I do." She then immediately informed me that she now had a steady boyfriend, but she would introduce me to one of her girlfriends and the 4 of us could all go on a picnic together on Sunday, two days from today, in the forested hills of the Darling Range just SE of the city. That sounded fine to me.

Later that same morning I contacted the Australian Customs Office in the port city of Fremantle, where our Jeep van had been shipped from Calcutta three months ago. I was anxious to remove it from customs as quickly as possible. However, I should have known there would be government regulations and red tape. I was told that the van could not be removed until it had been thoroughly fumigated, and this process could not be performed without written permission from the vehicle's owner (Noble or me). Furthermore, after the fumigation was completed there was a mandatory waiting period of 5 days before it could be removed. If I signed the papers today, which I did, then it would be next Tuesday (Sept. 19) before I could drive away in the Jeep. I was not at all happy about this situation but my protests were in vain. The Australian government was extra careful not to bring unwanted organisms of any level of complexity into the country from anywhere outside the country, particularly the places where Noble and I had driven our Jeep. Furthermore, in their careful inspection of all the articles inside our van they discovered my souvenir raw cowhide Masai shield from Tanganyika, and I was told that under no circumstances would it be allowed into the country without separate fumigation and a long waiting period. Therefore, I would have to package it up and mail it directly somewhere else. I obliged by putting it in the mail via slow boat to my parent's address in Long Beach, California. Life in the modern world was not easy. Perhaps Noble's persuasiveness would have produced better results, but I suspect not in these situations. The good news was that Margaret's mother said I was welcome to use a spare bedroom at her house until I had obtained our camping van.

I checked regularly at the American Consulate in Perth for any word from Noble, but so far there was nothing. This consulate was of interest because it was more distant from Washington, DC than any other American diplomatic post anywhere in the world. Not only that, it was also more isolated from the next nearest American post (which was Djarkata, in Indonesia). In other words, the American Consulate in Perth was as remote as it could be from the American government anywhere else.

During the waiting period for my Jeep I visited the West Australian Museum and introduced myself to the curator of birds, Glenn Storr, who permitted me to examine specimens in the bird collection. My only Australian bird book with any illustrations was Neville Cayley's, 1954, *What Bird Is That?* This book had 315 pages and 36 color plates with small paintings of almost all the native Australian birds. In addition to this guide (which I had brought with me from California) I also had a small paperback *Systematic List of the Birds of Western Australia* published in 1948 by H.M. Whittell & D.L. Serventy (with no illustrations or descriptions of the birds, only their names and geographic distribution).

Perth had a population of about 500,000 people in 1961, almost exactly half the total population of the entire state of Western Australia. Although I was not a city oriented person, I never the less found Perth to be a very pleasant and attractive city, clean and neat with appealing architecture, parks, gardens, and friendly people. City streets had cafes, arcades, modern department stores, sufficiently wide sidewalks, and streets with relatively little traffic noise or congestion. Adequate public transportation was supplied by many buses and commuter trains. The city was situated on the wide estuary of the Swan River, which was bordered on the north by a high hill where Kings Park was located, with magnificent scenic views overlooking the city and the estuary. The spacious park itself was a botanist's paradise with its gardens, flowers, and stately eucalypts, many of them with their names attached. A war memorial was an additional attraction, as were small kiosks and a large restaurant. For a first-time visitor to Australia such as I was, it was a great introduction to the flowers, trees, and birdlife. Noisy and colorful Port Lincoln ("Twenty-eight") Parrots* flew back and forth, Red Wattlebirds* and other honeyeaters chased each other from flower to flower, and Australasian (Western) Magpies* (more appropriately called "bellmagpies") sang with lovely flute-like and bell-like vocalizations. Perth was as close to being a utopia as any city I had ever visited. Adding to its appeal were many miles of clean, white sandy beaches stretching along the Indian Ocean to the west and north of the city, and south of the city a short distance were the seaside towns of Rockingham and Safety Bay, with more beaches.

Saturday morning, September 16, Margaret took me to meet her girlfriend, Laurie Weaver, who lived in a mid-sized, two-story flat with two other girls (her one and a half year younger sister, Robin, and Jenny, a recent arrival from England). Laurie came downstairs to greet me wearing a short-sleeved blouse and fashionable short shorts, and happily said hello to me with only a slight hint of an Australian accent. Laurie was 24 years old, and had a very pretty face and a cute little giggle. We all drank a cup of tea together while Margaret and Laurie made plans for tomorrow's picnic, in the nearby forested hills above the city to the southeast.

The next day at the picnic (Sept. 17) it was difficult for me to decide whether to spend my time talking with Laurie and the other two or walking by myself in the Jarrah forest, which was full of birds I had never seen before. The birds won! Trip additions to my bird list at the end of the day totaled 30 species (included under Sept. 16-18, in Appendix B). Some of these were endemic to just the SW corner of the Australian continent, the "Southwest." Among the more attractive smaller birds were Splendid Fairywrens*, Scarlet Robins*, Rufous Treecreepers*, Western Spinebills*, and Tawny-crowned Honeyeaters*. I enjoyed listening to the loud, pleasant songs of shrike-thrushes, whistlers, magpies, and butcherbirds. The far-carrying, almost obnoxious "kook, kook, kook, caw, caw, caw" cackles of the Laughing Kookaburra were one of the most characteristic sounds of the forest. This species was introduced into western forests from eastern Australia about 1900, and was often regarded as a pest because it ate both the eggs and young of other birds, although its diet also included a wide variety of vertebrates and invertebrates. It was a large terrestrial member of the kingfisher family. In addition to the birds that I encountered, I also observed my first Australian wild marsupial, the Western Brush Wallaby (*Macropus irma*) which was in the kangaroo family but only about half as large in dimensions as kangaroos, and weighing just 15-20 pounds. They bounded off through the forest when I startled them.

Botanically, my day was also a success. Along with the dominant Jarrah trees (*Eucalyptus marginata*) there were cycads (primitive gymnosperms) and blackboys (also called grass trees) in the understory. Blackboys were peculiar looking monocotyledon flowering shrubs in the endemic Australian genus *Xanthorrhea*. They grew 10 to 15 feet tall and were characterized by a bare, hollow, resinous, glabrous trunk (usually blackened by repeated bush fires), on top of which was a whorled clump of long, thin, stiff, grass-like leaves. White flowers (which attracted bees and butterflies) were produced once a year along a tall slender spike rising from the middle of the plant. They were slow growing, long lived (up to 600 years), and fire resistant, and were one of the most characteristic understory plants in the open Jarrah forest, particularly on rocky soils. Like forests everywhere in Western Australia, there were a great variety of wildflowers, many of them endemic to the Southwest. Those which I found that day (and identified from a booklet I had purchased in Kings Park) included orchids, kangaroo paws, begonias, cone flowers, and leschenaultias (some of which I photographed). My day in the forest was a pleasant introduction to the Australian flora and fauna, and I was happy with the choice I made on this occasion. Back in Perth that afternoon after the picnic I said to Laurie that I would like to see her again (having spent virtually no time at all with her at the picnic), so we arranged to go to a movie together on the coming Tuesday evening. (She later told me that never before had she been on a first date with any boy who ignored her so completely.)

The next morning (Mon., Sept 18) I took a bus to the Yanchep Nature Park about 10 miles north of Perth where there were open woodlands, coastal heath with colorful wildflowers, and scattered ponds with a variety of aquatic birds (see Appendix B). Black Swans* were unique in being almost entirely black in color rather than white. (The locally brewed "Swan Lager" beer used this species as its logo.) There was also a small zoo at Yanchep with a variety of native birds and marsupial mammals, including Koalas, which roamed freely throughout the park without being constrained in captivity. A prettily dressed girl walked up under one in a tree with her camera to get as close as she could for a picture and it promptly urinated on her, to her great consternation. Such behavior was a common defense mechanism I was told. My Australian bird list grew rapidly just within the environs of Perth, even without our Jeep for transportation.

On Tuesday, Sept. 19, I was finally able to retrieve our Jeep camping van from the Australian customs office in Fremantle. The customs officials had painted a conspicuous sign on our back bumper that said "Caution, left hand drive"! As I left Margaret's house I gratefully thanked her mother, and Margaret, for my five nights of accommodation. It was great to be back in our own miniature home on wheels again after three months of absence. When I called the American consulate later that morning, I was told that a telegram had arrived from Noble, so I drove to their office and picked it up. The telegram said that he would be arriving in Perth at the airport just after midnight in the very early hours of the coming Saturday, Sept. 23.

That evening (Tuesday) I took Laurie to a movie as we had arranged. Australia was a devoted member of the British Commonwealth and prior to the commencement of the movie the song "God Save the Queen" was played for the audience. I informed Laurie of the telegram from Noble, and she told me that she would like to invite both Noble and me to visit her parents and their farm outside Beverley that weekend, the Saturday and Sunday that Noble arrived. To join us she would also invite her sister Robin, Robin's girlfriend, Jan, and her girlfriend, Maureen, as a companion for Noble, making a total of six persons in our group for the two day, one night outing. We would return to Perth Sunday afternoon, about a two-hour drive. Of course, I was quite agreeable with this plan, and Laurie seemed certain that her parents would approve. Laurie made such decisions confidently on a moment's notice. Three of us would drive in Laurie's little Austin car, and the other three would ride in a car belonging to one of the other girls.

In the meantime, I decided to take a short birdwatching and sightseeing trip in our Jeep for three days (Sept. 20, 21, & 22) through the southwestern forests to the SW tip of the Australian continent at Cape Leeuwin (latitude 34.25°S), a road distance of about 200 miles from Perth. The "Southwest" was much wetter, cooler, and more densely forested than the rest of Western Australia, with considerable endemism in the flora & fauna, particularly the orchids and other wildflowers. My scenic, paved highway would follow a route from Perth to Pinjarra, Harvey, Bunbury, Busselton, Margaret River, Augusta, Cape Leeuwin, Nannup, Pemberton, Manjimup, Bridgetown, Bunbury again, and from there retrace my route back to Perth.

The countryside and scenic attractions along the way included orchards, ornamental red-flowering peach trees, vineyards, green paddocks (pastures) with flocks of sheep and Straw-necked Ibises*, limestone caves, and eucalypt forests of Jarrah, Marri, Karri, and Tuart trees, which like all *Eucalyptus* had evergreen, sclerophyllous (leathery) leaves. Western Gray Kangaroos (one of the three large species of iconic Australian kangaroos, standing 6½ ft. tall and weighing up to 160 lbs.) were conspicuous and widespread in forest clearings and open edges. The coastal heath and sandplain scrub here in southwestern Australia were characteristic of the Mediterranean climate biomes around the world, with "chaparral" type vegetation. Brightening the landscape were yellow-flowered *Banksia* and *Acacia* ("wattle") bushes, and red-flowered coral vines (*Kennedia coccinea*). New Holland Honeyeaters* were the most abundant bird. The rocky seacoast near Margaret River was almost an exact replica in appearance of the

Plate 107: The lovely city of Perth, W. Australia, had a population of 1/2 million people in 1961. The rocky coast with chaparral-like vegetation near Margaret River was remarkably similar in appearance to the coast of central California. It was springtime and the countryside was green from winter rains.

rocky central California coast near Santa Barbara, and Sooty Oystercatchers* along the rocks here were identical ecological counterparts in appearance and behavior of Black Oystercatchers along California coasts.

I interrupted my sightseeing and birding with regular stops to sample the local wines from roadside vineyards, or to relax in a tea shop for a "Devonshire Tea" (with cake, whipped cream, and strawberries). It was an idyllic, relaxing, time.

September was in the springtime, well into the rainy season, and birds were engaged in courtship singing and displaying. White-tailed Black-Cockatoos* and "Twenty-eight" Parrots were among the most conspicuous birds in the forest and along the roadsides. This latter species was the largest and most green-plumaged of the four morphological forms of *Barnardius*, with a prominent red forehead band. I also frequently encountered the endemic Red-capped (King) Parrot*, which used its long, slender hook-tipped bill to extract seeds from capsules of the Marri (Red Gum) tree. Small, wooded ponds provided habitat for Musk Ducks and Purple Swamphens (a widespread species in the Old World). The ten most numerous birds along my route (or at least the most conspicuous) were, in order of abundance, the Welcome Swallow, Australian Raven, Australasian Magpie, White-tailed Black-Cockatoo (subsequently split into two species), Silver-eye (a *Zosterops*), Red-capped Parrot (see above), Gray Fantail (*Rhipidura*), Laughing Kookaburra, Tree Martin, and Red Wattlebird. Seven species of honeyeaters were recorded. Southern Boobooks* (a small owl in the genus *Ninox*) called repeatedly at night with their familiar two-syallabled "boo-book" call, (a vocalization that was originally attributed erroneously to the Tawny Frogmouth*, so that this latter species was popularly called a "mopoke", referring to the sound made by the owl). My 3-day bird list of 36 trip addition species is given in Appendix B, Sept. 20, 21, & 22.

The southwest forests were widely known for their tall Karri trees (*Eucalyptus diversicolor*), and for the wide-buttressed, almost equally tall Red Tingle trees (*Eucalyptus jacksoni*). These trees were ecological counterparts, respectably, of the Redwood and Sequoia trees in California. In both areas, the climate was mild and wet. Karri trees regularly grew up to 250 feet, and rarely as tall as 300 feet. Red Tingle trees were not quite as tall but often developed a very wide spreading base, up to 12-15 ft. diameter at "breast height." They were very long-lived (up to 400 years), and like most eucalypts were relatively fire resistant. The base of these trees sometimes was burned out in bush fires and developed a large hollow, which occasionally went all the way through the bottom of the tree at ground level, as a wide tunnel. Several such giant trees had long been tourist attractions in the southwest (as were Sequoia trees in California). Orchids and other wildflowers added color and attracted tourists to the Karri forests, but avian diversity was low.

A marsupial mammal endemic to the shrubby heath and coastal sandplains of the Southwest was the tiny, enigmatic, mouse-like "Honey Possum" (*Tarsipes rostratus*). It was only 3½" long and 15-20 gm. (0.7 oz.) in weight, and fed almost entirely on the nectar and insects in the blossoms of banksias and eucalypts, where it was an important pollinator. Although largely arboreal and very agile climbers, mouse possums could also run rapidly on the ground, quadripedally. Adaptations for nectar-feeding included an elongated, pointed snout, and a long, brush-tipped tongue. As aids in their arboreal lifestyle, both their long slender tail and some of the digits on all four feet were prehensile, and could be used for grasping and hanging onto flowers or small branches. Mouse Possums were largely nocturnal in their time of activity, and happily they were still relatively common and not yet an endangered mammal (as were so many of the native Australian marsupials). Not being much of a nocturnal biologist, I did not encounter any on this trip. (Note that Australian marsupials are known as "possums", not "opossums" which live only in the New World.)

I arrived back in Perth at 7:00 pm (sunset) on Friday, Sept. 22. Just after midnight, I drove to the airport to pick up Noble, who landed on time and stepped out of customs shortly thereafter with his backpack, as I had done ten days earlier. We embraced warmly and drove to a picnic area on the Swan River estuary to camp for the rest of the night. It was Noble's first night in our Jeep in 15 weeks. We each celebrated with a glass of some red wine I had just purchased on my southwestern trip. In Australian terminology, Noble and I were "cobbers" - good friends.

Plate 108: Characteristic of the flora were tall, magnificent wide-buttressed Red Tingle trees, some with a burned out hollow at their base. Black Swans (the only black swan in the world) frequented lakes, ponds, & marshes. The lighthouse on Cape Leeuwin is located at the SW tip of the continent.

Our next two days were filled with the recently planned visit to "Riverdale", the farm where Laurie grew up, and where her parents still lived and farmed. Noble and I traveled there that very day (Saturday, Sept. 23) with Laurie, her sister Robin, and two girlfriends, Jan & Maureen. Laurie wanted us to meet her parents, and to show us around the farm, which was located 7 miles south of the small country town of Beverley, 80 miles east of Perth. Beverley had a population of perhaps 1,500 residents and was situated in a shire (county) of the same name, at the western edge of the "wheat belt." It was about a two-hour drive to Beverley, via the paved "Brookton highway" and a final short cut on the gravel "Dale Road" into Beverley. Most of the route was through Jarrah forest. Three of us (including myself) rode in Laurie's little "Austin" car, and three rode in another car. Laurie's parents had agreed that the six of us could spend Saturday night with them in the farmhouse.

We arrived in mid-morning. Tall, stately Salmon Gums (*Eucalyptus salmonophloia*) bordered the long entrance road as we neared the farmhouse. These handsome trees were characterized by smooth, pale salmon-colored, bark. The pastures were covered with springtime-blooming, yellow-flowered "dandelions", an exotic perennial herb from South Africa which had inadvertently been introduced into Australia in the early 1900's, and now grew as a rather innocuous, colorful weed. Laurie's parents (Charlie & Mary) greeted Noble and me very warmly. (They were already well acquainted with everyone else.)

The "farm" turned out to be a property of 2,000 acres with a total of 3,000 sheep of 4 different breeds - - Merino, Dorset, Suffolk, and Romney Marsh. In addition, there were 70 Hereford beef cows and several hundred acres each of wheat, oats, and hay (for sale or for local livestock consumption). Thus Riverdale would be referred to as a "ranch" in the USA rather than a "farm." Most of the income was derived from the sale of wool, with additional income from the sale of "fat lambs", stud rams, ewes, beef, and wheat. Charlie was particularly proud of his Romney Marsh stud and every year at the annual, week-long "Royal Show" in Perth he won many more blue ribbons than anyone else. Of course Riverdale possessed, as did all farms, a milk cow, sheep dogs, cats (for catching introduced mice), and chickens ("chooks"). Unwanted pests on the farm included rabbits and foxes that had been deliberately introduced into Australia from Europe many years ago as game animals for hunting!

The average annual rainfall at Riverdale was only 16½", which was marginal for successful farming and income earning. Summers were hot and dry, and winters were mild and wet (without freezing temperatures or snow). Riverdale was situated on the right (E) bank of the very small, northward flowing Avon River, but unfortunately the water was too brackish to be used for livestock consumption. The main farmhouse where Laurie's parents lived was located on the west side of the "Kokeby Road" (which connected Beverley with the tiny community of Kokeby to the south). The entire 2,000-acre farm was almost equally divided on both sides of the road, with the river forming the western boundary. A smaller, newer house ("Merriedale"), was on a small wooded hill on the east side of the road. Charlie employed two permanent farmhands to work for him, Cecil and Bert.

Western Australia in 1961 was about 20-30 years behind the USA in the development of its rural towns and farms. The farmhouse at Riverdale was a typical example. Laurie's parents were the third generation of Weavers on the property. (The sign on the front gate at the road said "J.C.W. Weaver" - Charlie's full name was James Charles Walter Weaver.) The farmhouse had been built about 1907, of brick exterior with a corrugated tin roof. In 1961, there was no external supply of electricity or water to the farm. A diesel engine powered an electrical generator which produced and stored electricity for household use in a bank of twelve, 32 volt DC car-sized batteries in a shed behind the house. The generator was run several times daily, depending on the need for electricity, and had to be running in order to provide enough electrical power for Laurie's mother to iron clothes. There was a small kitchen firewood stove used to boil water for tea ("boiling the billy"), and another firewood stove in the bathroom which heated water for taking a bath. The kitchen oven which was used to roast lamb every day of the week was, until very recently, also heated by firewood, but was now replaced by a propane gas oven which used propane gas stored in a 40 gallon restorable bottle at the back corner of the house, which was regularly replenished by a truck from town. Drinking water came from rainwater on the roof of the house that was collected and stored in a large tank at the side of the house. The telephone was a typical 1920's wall phone, 4-party, with a long mouthpiece sticking out horizontally from the wall. The Riverdale ring was the Morse code for the letter "R", short-long-short (or dot-dash-dot).

Beverley was typical of many small country towns scattered throughout southwestern Australia. It was located on the Avon River, which flowed northward from the Riverdale farm to the town (and eventually westward into the Swan River and the estuary at Perth). In Beverley, there was a single main street bordered on both sides for several blocks with various shops, businesses, hotels, and other enterprises. For a country town it was thriving relatively well in 1961. There was a railroad track which could take people to Perth or (via Northam and Kalgoorlie) all the way several thousand miles east across the vast, arid "Nullarbor Plain" to the "eastern states." Paved highways ran north, east, south, and west from the town. A "Beverley Co-op" provided vegetables, greens, clothing, hardware, and many other items of household necessity. There were two hotels in town (both with a popular bar, as in all Australian hotels), and two banks. A "Chemists" shop provided pharmaceuticals, a butcher shop provided fresh meats, and "Elders Farm Supplies" sold almost everything a farmer needed. On the main street were a town hall, a post office, a "newsagent", two "tearooms" (cafes), a bakery, and a hairdresser. Elsewhere in the town were a police station, a small hospital where for many years Laurie's maternal grandfather, R. P. Hodgson, was the only doctor and surgeon in town (and where Laurie was born), three churches (Catholic, Anglican, and Methodist), and an extensive Sports Ground with facilities for lawn bowling, grass court tennis, cricket, and "Australian rules" football. On the edge of town was a large area for auctioning livestock, and a cemetery. Local farmers always came into the town on weekends to participate in their favorite sport. (Laurie's father was a particularly keen tennis player and later just as keen a lawn bowler.) Beverley was a delightful little town full of pleasant, hard-working people.

All six of us returned to Perth on Sunday afternoon. The next morning (Monday, Sept. 25) Noble and I drove our Jeep camping van to the Jeep agency in the city for some servicing, maintenance, repairs, and outside painting. When the agency owner saw our vehicle, and learned of our adventure around the world, he immediately asked us if we would allow him to exhibit our Jeep for advertising purposes in the annual, one week long "Royal Show" scheduled to begin that coming Friday. In his best bargaining manner, Noble agreed to allow them to exhibit the van in return for not charging us anything for painting the van. The owner readily accepted this idea. Noble and I were now obligated to remain in Perth at least through Thursday, Oct. 5. We would be allowed to sleep in the Jeep at night, both while it was being serviced at the agency and while it was being exhibited on the grounds of the Royal Show. Of course, it would not be available for our transportation or sightseeing anywhere during these next ten days. A large sign placed beside our van on the show grounds read "This Jeep has traveled around the world for 70,000 miles." Our travel in and around the city would now have to be by bus and commuter trains. A side benefit of this arrangement was that it would allow me more time to get better acquainted with Laurie. Life's twists and turns were unpredictable.

I began having lunch with Laurie during her noon hour break from the office of Lang Hancock & Peter Wright, where she worked five days a week as their only secretary. We ate thin-sliced cheese on plain white tasteless bread sandwiches because they were the cheapest thing available at the nearest sandwich shop. We talked about any topic that came to our mind. On Wednesday, Laurie said to me that she had accrued a week's holiday from her work and we could spend this time together next week at Riverdale if I wanted. Of course I said yes. Our Jeep was still going to be unavailable for almost all this time.

Laurie drove me to Riverdale on Saturday, Sept. 30, in her little Austin. We spent most of our week at Riverdale (Sept. 30-Oct. 8), walking over all the paddocks on the entire 2,000 acres, some of which were situated on a little wooded hill and some of which bordered the river, with its riparian vegetation. As we walked and conversed, I kept a list of the birds we encountered. I learned from Laurie that as a young girl she made a collection of 40 different kinds of bird eggs from all the nests she could locate on the property. Many of these she needed help from her father or others to access. Laurie removed all the inside soft parts and saved the eggs in fumigated cardboard boxes filled with cotton. She showed me her collection, and remarkably all the eggs were still intact. They were not labeled but she knew what species they all were at that time. (Her father knew only a few of the birds on the farm when he saw them, and when describing one to me he almost always described it as "looking like a wattlebird"!) On the property were numerous "dams" (small man-made ponds) which not only provided water for the livestock but also attracted a variety of waterfowl and other aquatic birds to the farm, as did the permanent pools in the river. In addition to these sources of water, there were three or four windmills which pumped water from "bores" to surface tanks, and from there to troughs from which livestock could drink. One such "mill" was located at the front of the farmhouse. Habitat variety for birds was provided by small patches of woodland where there were salmon gums, York Gums (*Eucalyptus loxophleba*), wandoo (*Eucalyptus wandoo*), sheoaks (*Casuarina*), *Hakea* bushes, and other woody vegetation.

During my week at Riverdale I documented a total of 50 species of birds, 14 of which were trip additions (Appendix B). My most vivid memories are of the Rainbow Bee-eaters* (a breeding migrant to the farm, which constructed an earthen tunnel in the flat sandy ground for a nesting chamber), Red-capped Robins* (a "flycatcher" in the genus *Petroica*), Banded Lapwings (a noisy *Vanellus* of open paddocks, locally referred to as "plovers"), Whistling Kites (with their pleasant far-carrying whistles, closely resembling Little Eagles in appearance when soaring), White-fronted Chats* (a peculiar, mostly ground dwelling "chat" in the small endemic Australian family Epthianuridae), Yellow-rumped Thornbills (genus *Acanthiza*, a largely ground foraging "warbler", often in small parties and well known by the farmers, who referred to them affectionately as "tomtits"), Varied (Black-capped) Sitellas (filling the nuthatch niche), Brown Songlarks (a plainly-colored, brown, relatively large ground-dwelling sylviid warbler of grassy pastures, in the genus *Cinclorhamphus*, which in its bubbling song and conspicuous courtship flight on fluttering wings was reminiscent of a Bobolink), Brown Honeyeaters (very common along the river), Singing Honeyeaters (one of the most vocal and familiar yard birds), Red Wattlebirds (equally numerous and conspicuous), magpies, Australian Ravens, and Galahs* (a conspicuous, noisy, beautifully colored pink & Gray-colored cockatoo, an Australian icon). Bush Thick-knees (locally referred to as "curlews") were occasionally heard at night, as were Tawny Frogmouths ("mopokes"), which were also sometimes seen after dark sitting on the top of roadside fence posts (like potoos in the neotropics). The abundant Port Lincoln Parrots on the farm were intermediate in their color and size between the all green "twenty-eights" (with a red forehead band) of the wet southwest corner, and the somewhat smaller, yellow-bellied, entirely black-crowned birds inhabiting the hotter, drier Australian interior. Riverdale was located in a wide hybrid zone between these two major morphological forms.

Common Brush-tailed Possums (*Trichosurus vulpecula*), which were arboreal leaf, fruit, and insect-eating marsupials, were sometimes seen at night in trees around the farmhouse, looking remarkably like neotropical kinkajous. As I have previously mentioned, European foxes and rabbits were serious pests everywhere on the farm. Tiger snakes and several other venomous species were common snakes on the farm, and Gould's monitor lizards (*Varanus gouldii*, locally called "goannas") were sometimes encountered, reaching lengths of up to 4-5 feet. These carnivorous reptiles ate a wide variety of vertebrates and invertebrates, and were adept at climbing trees to obtain the eggs or young of cockatoos and other birds.

On our last Sunday morning at the farm (Oct. 8), Laurie and I walked down to the river and sat there on the bank discussing our brief three-weeks together and what the future might hold for us. Noble and I would be leaving in another week. So we decided to get married! Just like that. We walked back to the farmhouse and told Laurie's parents of our decision. Predictably, it was met with great surprise. Laurie's mother was quite happy about the decision. (She and I shared many views.) Her father, understandably, was much less enthusiastic. After all, I was penniless, "bearded and roughly dressed" with no income or any job in sight, and would be taking his beloved daughter to a far-off land. Laurie set Dec. 27 as a date for our wedding, and immediately began making

Plate 109: From Perth, a good paved highway traveled east through the Jarrah forest for 80 miles to the thriving little country town of Beverley. Gray Kangaroos frequented the forest edges and clearings.

Plate 110: The 2,000 acre "Riverdale" farm was just south of Beverley, bordering the Avon River. The farmhouse was built in the early 1900's, and was somewhat behind the times. Laurie loved the little Romney Marsh lambs. It was here that we decided to get married!

plans. Whenever Laurie made up her mind to do something, she followed it through with great commitment and determination. We left the farm that afternoon and returned to Perth.

The next day (Oct. 9), we went to a jewelry shop together in Perth and Laurie picked out and purchased an engagement ring -- with her money since I had none! Noble was as disbelieving as anyone. We still had plans to drive together around northern Australia, 7,000 miles from Perth to Darwin to Cairns to Sydney. I would fly back to Perth from Sydney on Dec. 17, in time for my wedding, and Noble would fly back in time to participate as my best man. Such an event could never have been predicted by anyone, and was beyond all logic. Would it really happen?

For our last Saturday in Perth (Oct. 14) Laurie arranged for 4 of us, Noble, me, Maureen, and her, to all go for a two-hour horseback ride in the "bush" on the northwest side of the city, from a local riding stable familiar to Laurie and Maureen where horses could be rented. At the stable before we started out, we were asked if we were experienced riders. Laurie, Maureen, and Noble all said they were and I made the mistake of saying nothing at all, although I had been on a horse only once in my whole life. Therefore, we were all given "spirited" horses. It was a most memorable experience. Immediately after we started out, all four horses began to run. From the start, I had absolutely no control over my horse at all. It did whatever it wanted, whenever it wanted. All I could do was hang on. Almost instantly it belted away from the other horses and set off through the bush on its own, about as rapidly as it could. It required all my effort and strength not to fall off. As my horse ran through the thick bush, I had to duck under low branches to keep from being swept off. I was certainly not having any fun. I was told by Laurie, afterwards, that Noble, too, experienced his own problems, also with very little control over his horse. According to Laurie, at one point early in his ride he lost his balance and fell over sideways, and would have fallen off his horse to the ground if one foot had not become stuck in the stirrup. He thus had one leg stuck in the stirrup on one side of his horse and the rest of him dangling over the horse on the other side, with his head hanging downward not very far above the ground. Fortunately, the girls managed to stop his horse and help Noble back to an upright position. They then set off to see if they could find me.

Somehow, miraculously, I had managed not to fall off, and the others eventually found me. Thereafter, we all rode together relatively quietly for quite some time, until Laurie determined it was time for us to start back toward the stable. She decided that we could return in couples, since both she and Maureen said they knew the way back, having ridden in this area several times previously. However, after 15 minutes or so Laurie decided she did not know exactly where we were, and that she was lost. Thus, we were on our own to try to find the way back, and our two-hour time period was expiring. I was quite exhausted from all my efforts to avoid falling off. Ten minutes later, we came to a residential hilly area, with paved streets, and Laurie realized where she was, still some distance to go back to the stable, up and down city streets. Sure enough, my horse suddenly decided it wanted to run again, and set off at full gallop down a long, moderately sloping hill. I pulled so hard on the reins that I turned the horse's head almost fully around and we were looking at each other face to face, but it continued to run straight ahead down the hill.

With great horror, I then noticed that at the bottom of the hill, not far ahead, was a cross street with a signal light! Predictably, just before we reached the intersection, the light turned red. My only thought was that I was going to die while riding a horse full speed through a red light, as a big lorry (truck) came through from one side or the other. Miraculously, no vehicle came along. Shortly thereafter, my horse suddenly stopped running and began walking again. My life had been spared one more time. Laurie soon caught up with me and before very long we arrived back at the stables, where the other two were worriedly waiting. I had been in the saddle for just over 2 hours and was completely exhausted, and already beginning to feel sore all over. I vowed never to get on a horse again - - and I didn't.

On Monday, Oct. 16, I said goodbye to Laurie, and Noble & I commenced our 7,000 mile, two month arched route of travel around the north of Australia, from Perth to Broome to Wyndham to Darwin to Cairns to Sydney.

Our route began on the "North West Coastal Highway" which ran north from Perth to Port Hedland. (For a reason I never understood the Australians, in 1961, did not identify any of their roads or highways by a number, rather than a name, which certainly would have made them a whole lot easier to follow for a person unfamiliar with the names, as I was, and furthermore a number would have taken up less room on a road map or road sign.) Our route followed just inland from the Indian Ocean, northward along the west coast to the towns of Geraldton, Carnarvon, and Onslow. It was paved for the first 600 miles, to just beyond Carnarvon, at which point it became graded gravel. The road surface was good at times and very rough at others, with varying degrees of maintenance. Nowhere was there very much traffic. The few people we encountered in the countryside were always cordial and friendly. If I were walking along the roadside looking at birds or picture-taking, any passing motorist would always stop and inquire, in a strong Aussie accent, "Are you right, myte?" I always thanked the driver, and assured him that I was.

As we traveled north from Perth, we first crossed the Murchison River about 75 miles north of Geraldton, then the Gascoyne River at Carnarvon, the Ashburton River 13 miles before Onslow, and finally the Fortescue River, inland at Millstream. These important rivers flowed from the dry interior to the Indian Ocean. In periods of summer drought, they ceased flowing and existed only as permanent pools ("billabongs") in the riverbeds (which were always bordered by River Red Gums, *Eucalyptus camaldulensis*). There were practically no bridges anywhere, and rivers were crossed on concrete floodways (which after one of the few heavy rainfalls during the year would become impassable for several days). Cockatoos and Port Lincoln Parrots nested in the natural tree cavities of the river gums, as did Owlet-nightjars, Kookaburras, Kestrels, Boobook Owls, and other parrots. (An unexplainable fact was that there were no woodpeckers, family Picidae, on the Australian continent, although they occurred on all other non-Antarctic continents.)

On our first morning out of Perth (Oct. 17), near the town of Three Springs, a pair of corellas (all white, medium-sized, cockatoos) flew across the highway in front of us. In 1961, these birds were considered to be a geographically, widely isolated race of the Long-billed Corella, which lived in SE Australia, but later this population in Western Australia was taxonomically elevated to its own species, the Western Corella* (as listed in Clements). Both species possessed very pale lemon yellow under wing linings. I also saw my first Crested Pigeons* and Zebra Finches*, which were common and widespread species of the dry interior. In addition, I documented my first Wedge-tailed Eagle* (which because of its elongated wedge-shaped tail was slightly longer in body length than any of the other 17 species in the widespread genus *Aquila*). Other first time species were the very handsome White-backed Swallow* and the Blue-breasted Fairywren* (which I identified from the almost identical appearing Red-winged and Variegated fairywrens only by its geographic location). We camped at the Murchison River that night, approximately 75 miles N of Geraldton. The river was not flowing and was restricted to pools.

The next day (Oct. 18) we drove north 215 miles all the way to Carnarvon, following the east side of Shark Bay for the last half of this travel, from Hamelin Pool junction. I documented 50-60 species of birds for the day, including 19 trip additions (Appendix B). Some of the birds that were the most pleasing to me were the Australian Hobby* (Little Falcon), Chestnut-rumped Thornbill, Spiny-cheeked Honeyeater*, Pied Butcherbird* (shrike-like in its diet, and very aggressive in its behavior, often chasing larger birds of prey), Emu* (an Australian icon, the continent's largest bird and depicted on the nation's shield, a flightless ratite bird similar in appearance and ecological niche to the Ostrich, though not quite as tall), Brown Falcon (a large, plainly colored, rather lethargic *Falco*, often called the "Brown Hawk"), Galah (always a favorite), Mulga Parrot* (a small green parakeet-like desert parrot in the genus *Psephotus*, with yellow wing coverts and red thighs, frequenting the drier inland woodlands and mulga [*Acacia aneura*] scrub), Regent Parrot* (a long, slender, pale lime-colored parrot with a conspicuous red wing patch, locally called the "Smoker"), Budgerigar* (the familiar "budgie" cage bird, a very small, slender, abundant desert parrot, sometimes occurring in flocks of thousands going to and from a waterhole), Chiming Wedgebill* (in the endemic Australian family Orthonychidae, an uncommon desert bird with a far-carrying,

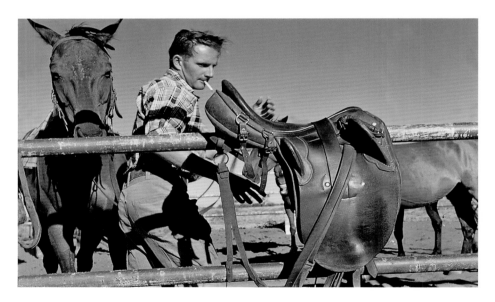

Plate 111: After my engagement to Laurie, Noble and I commenced a two month, 7,000 mile journey around the north of Australia, from Perth to Darwin to Cairns to Sydney. As we started out, we stopped for two days at the 1,000 sq. mi. "Giralia" sheep station south of Onslow (on the W.A. coast).

Plate 112: Windmills provided water for the station (as everywhere in the dry interior of Australia). A few native Aborigines were employed as laborers. The widely distributed, well-known Galah (one of the most colorful cockatoos) was a frequent visitor to inland water tanks.

sweet, ringing series of chiming, bell-like notes descending the scale, one of the most haunting and memorable sounds of the outback), Crimson Chat* (in which males were distinctive red, black, and white in color), and Masked Woodswallow* (a nomadic flock of 400 birds). There was also a wandering flock of 500 Little Crows* (*Corvus bennetti*). In spite of some annoying bush flies, today was one of the more enjoyable days of my Australian travel to date. We camped that night along the Gascoyne River, on the outskirts of Carnarvon, where a Barn Owl watched us from a nearby tree!

Outside our van the next morning (Oct. 19) were four trip birds: Little Corella*, Blue-winged Kookaburra*, Gray-crowned Babbler* (a handsome *Pomatostomus*), and Variegated Fairywren* (*Malurus*). There was also an attractive male White-winged (Blue-and-white) Fairywren. We drove to Giralia Station this day, to acquaint ourselves with a successful outback sheep station. Giralia was located approximately 75 miles south of Onslow, west of our highway just inside the tropics on the southern end of Exmouth Gulf (on the Indian Ocean). Margaret Rankin in Perth had given us the name of a "jackaroo" who worked there (a boy in training to become a stockman). She told us he would show us around on our arrival, and introduce us to those who could explain the operation of the station to us. Noble was particularly excited about our visit. This immense station of semi-moist grasslands and scrub covered an area of almost 1,000 square miles, and supported the astounding total of 44,000 sheep, or approximately 15 acres per sheep. (Further inland, under drier conditions, we were told that it required 25 acres to support one sheep.) Giralia was first established in 1916, when the present homestead was completed. We spent two nights and one full day (Oct. 20) at the station, where I saw my first Bush Thick-knee* (having only heard it at Riverdale).

We found Brian, the jackaroo whose name we had, and he willingly showed us around the horse stable, sheep pens, shearing sheds, machinery repair garage, worker's living quarters (in which aboriginal workers were separated from the others), and the homestead (where the owners lived). We were cordially invited for morning tea and Noble immediately engaged the friendly owner in a pleasant discussion of the financial aspects of running the station. Noble learned that the station produced an average of 700 bales of wool annually (which I calculated was about 63 sheep per bale). Giralia was a very profitable enterprise where everyone involved was hard-working, and managerial decisions were made wisely. Both Noble and I were happy to have had an opportunity to see a functioning sheep station. Wool was still the primary source of income for the country in 1961, but iron ore would soon change that situation. We camped on the station that night (Oct. 20), away from the homestead.

The next morning (Oct. 21) we drove back to the main highway and proceeded northward, soon crossing the Ashburton River en route to the sleepy little coastal town of Onslow, off to the west of the highway at the top end of Exmouth Gulf. Here a few boats exploited the coastal waters and reefs for pearls and for prawns. From Onslow, we proceeded eastward on a secondary road 10 or 15 miles to the Cane River, where we camped for the night. During the day, I added four species to my trip list (Appendix B).

During one and a half hours of birding at our Cane River campsite the next morning (Oct. 22) the most numerous birds I counted were 50 Galahs, 25 Zebra Finches, 20 Tree Martins, 20 White-plumed Honeyeaters, 15 Budgerigars, 12 Diamond Doves, and 10 Painted Firetails* (an estrildid finch). From the Cane River, we continued traveling east for another 12 miles, back to the main coastal highway at Peedamulla Station, where we turned left (NE) and drove 45 miles to the Robe River highway crossing, just before Yaraloola Station. The concrete floodway crossing was dry. We did not cross the river, but turned right off the main highway onto a rather poorly maintained, rough dirt road that wound back and forth along or near the south (left) side of the generally westward flowing Robe River.

We drove only 10 miles along this road until we suddenly came to a lovely, picturesque billabong with an enormous, gnarled, ancient river gum (*E. camaldulensis*) leaning over the pool. The scene was unbelievably magnificent. Nowhere else could there be a billabong with more characteristic features, or which produced in me such an emotional stir as I surveyed it. We immediately decided to have a refreshing swim, to spend the rest of our afternoon here, and to camp for the night. The photograph I took here is one of my all-time favorites, and every time I hear the song "Waltzing Matilda" this picture immediately comes to my mind. No other people or other vehicles were present any of the time we were there. This remote outback billabong on the Robe River in Western Australia is one of my most delightful memories of our three-year adventure.

Birds along the river that first afternoon included Spinifex Pigeons* and Black-tailed Treecreepers*. The very handsome, quail-like pigeons were very striking in their ruddy-buff coloration, black & white throat pattern, bare red area around their eye, and extremely long, pointed plume which stuck straight up from the top of their head. They ran rapidly over the sandy ground and rocks, like a quail, and were a truly fascinating, unusual, terrestrial columbid, an icon of the Australian outback. The term "spinifex" is the name of a widespread desert grass in the genus *Triodia*, which grows in clumps and is characterized by narrow, firm, needle-like "leaves" (which are uncomfortably prickly and should be avoided). The half dozen species of Australian treecreepers (in the genus *Climacteris*) are somewhat larger and more nuthatch-like in their shape and behavior than they are creeper-like. The next morning (Oct. 23) there were 9 species I had not recorded along the river here yesterday: 1 Rufous Night-Heron*, 1 Collared Sparrowhawk, 2 Whistling Kites, 2 Australian Kestrels, 2 Common Bronzewings (a pigeon), 20 Peaceful Doves, 2 Gray (Western) Shrike-Thrushes, 1 Little Woodswallow*, and 1 Brown Honeyeater. These brought my total list of birds seen in the vicinity of this delightful billabong to 35 species, in 4 hours of observing.

We got underway at 9:00 am and continued southeastward on our narrow, untraveled road, gradually gaining in elevation as we ascended slightly into the western Hamersley Range of mountains Eventually our road curved left, to the northeast, and we crossed the dry upper Robe River. Approximately 35 miles later, we came to Millstream Station on the Fortescue River (a much larger stream than the Robe River). Here we crossed the river on a slightly moist paved floodway, above which was a large, deep pool in the river where we stopped an hour for a swim. It was hot in the middle of the afternoon and we were not in a hurry. At this pool, I documented three new trip birds: the Red-browed Pardalote*, the Black Falcon* (*Falco subniger*, a mid-sized, almost entirely dark falcon which often pursued flying birds rapidly just above ground level), and the Australian Bustard* (a very large bustard in the genus *Ardeotis*, which strutted around haughtily with its nose slightly upturned, and drank at the water's edge by "kneeling" on its tarsometatarsi).

As we got underway again, we came to the end of our road at a T-junction with a wider, better maintained gravel road which connected the town of Roebourne on the North West Coastal Highway, to our left, with the asbestos mining town of Wittenoom, to our right. This was an area of undulating hills in the middle of the Pilbara mining region, where manganese, iron ore, and asbestos were all mined. It was a land filled with kangaroos, dingoes, emus, cockatoos, gorges, waterfalls, billabongs, and wildflowers. We turned right (SE) and proceeded toward Wittenoom. Three hours and 75 miles later we stopped to camp for the night.

That night I slept on top of the van because it was too hot inside. All night long I was kept awake by constant twanging of the wires on a nearby sheep fence. The next morning I determined the cause. Sheep fences in the outback were nothing more than three plain straight wires strung horizontally about 12" apart between two fence posts (which somehow deterred sheep from passing). However, kangaroos hopping rapidly across the terrain simply ducked their heads and pushed their way between the top two wires, without slowing down. (Why they did not just jump over the top wire I don't know.) At any rate, their body stretched the top two wires apart as they went through the fence, and afterwards the two wires snapped back to their original position, producing a loud twanging sound as they did so! The landowners, understandably, were not happy about this situation because eventually the fence wires sagged so much that sheep would then just walk through. Our adventure continuously produced new, surprising experiences.

In 1961, Wittenoom was a blue asbestos mining town of several hundred people, on its way out because of recent government regulations and mine closures to protect miners and others from the health problems associated with inhaling mining dust. On Oct. 24, Noble and I took a guided tour through one of the last mines that was still

Plate 113: Colorful foliage and spring wildflowers brightened the magnificent landscape of the Hamersley Range of mountains. Red sandstone cliffs rose above a billabong in the Fortescue River. The 5-6 ft. tall Emu is an Australian icon, pictured on the nation's Coat of Arms.

operating. In a nearby vacant mine shaft there was a colony of Ghost Bats (*Macroderma gigas*, also called False Vampire Bats), a very large, carnivorous, pale-colored bat with a wingspan of two feet. Ghost Bats caught, killed, and ate a wide variety of insects and small vertebrate animals – birds, lizards, frogs, mice, and even other bats, (which were captured in flight). They watched for prey on the ground from a perch in a tree and then flew to the ground, wrapped their wings around the prey, and bit it to death. Ugh!

Just south of Wittenoom, a land owner by the name of Lang Hancock owned an immense property on which he had personally discovered huge, very rich iron ore deposits lying on the surface of the ground. He "pegged" these but did not immediately reveal them to the Western Australian government because at that time the state government claimed all mineral rights on personal property. In 1961, this law was repealed and Lang had recently obtained mining rights to the iron ore on his property, which he had discovered. He began negotiations with the Rio Tinto Mining Company for a royalty on the iron ore, which he was willing to sell to them. (You may remember that Laurie, to whom I was now engaged, worked as the only secretary for the firm of Lang Hancock & Peter Wright in their Perth office.) The Pilbara region of Western Australia was the largest and richest iron ore deposit in the entire world, and before very long iron ore would provide as much income for the country as did wool. Most of this iron ore would eventually be sold to China.

When Noble and I drove through the Hamersley region in October, 1961, the whole countryside was blanketed with gorgeous lavender-pink "mulla mulla" wildflowers. These ephemeral blooming perennials flowered after sufficiently heavy rainfall. We saw euros (mid-sized members of the kangaroo family), rock wallabies, red kangaroos, dingoes (a non-native wild dog first brought to Australia by early aborigines), cockatoos, goannas, skinks, and geckos. Snappy Gum trees (*Eucalyptus lophora*) with smooth white bark were scattered around the landscape, adding picturesque features to the rocky areas and the many red gorges. "Dale's Gorge" was not yet well known, and Noble and I walked down to the bottom for a cool refreshing swim (au naturel) in one of the delightful pools at the bottom, unbothered by any other tourists. The long gorge, with its northward flowing stream, was 125' deep at its upper (southern) end, and 200' deep at its lower (northern) end. The two highest elevations in all of Western Australia (Mt. Bruce and Mt. Meharry) were both just over 4,000' above sea level, and were located not far away in the eastern Hamersleys. Much of this region would one day be set aside as a national park.

I documented a total of 44 species of birds in the vicinity of Dale's Gorge on Oct. 25 & 26, six of which were trip additions (Appendix B). Particularly memorable was the haunting, lovely bell-like song of the Crested Bellbird (*Oreoica gutturalis*, related to whistlers in the family Pachycephalidae). The series of notes were ventriloquist-like in quality and were given at various speeds and with different accentuations, usually ending with a distinct mellow, cow-bell like note. It was a song which once heard could never be forgotten, and reminded me of the vastness and loneliness of the Australian outback. I also encountered my first Spinifex-birds* (plainly colored sylviid warblers in the monotypic genus *Eremiornis*, reminiscent in appearance of an *Acrocephalus* Reed-Warbler), Black Honeyeaters*, and Black-chinned Honeyeaters* (sometimes considered a distinct species, the Golden-backed Honeyeater).

We drove north from Dale's Gorge to Port Hedland on Oct. 26. The remote, very small town of Marble Bar was a short distance to the northeast of our road, and was famous for its claim to have the longest heat wave on record anywhere in the world, when the daytime high temperature was over 100°F for 100 consecutive days, from Oct. 31, 1923 - Apr. 7, 1924! We camped just outside Port Hedland that night, on the northwest coast of Australia on the Indian Ocean, just south of 20°S latitude. The next morning, I scoured the mangroves and beaches for birds, documenting 28 species. These included the mangrove specialists of Bar-shouldered Dove*, Collared (White-collared or Mangrove) Kingfisher, Mangrove Fantail*, Mangrove Robin*, White-breasted Whistler*, Dusky Gerygone* (a warbler), and Australian Yellow White-eye* (Yellow Silver-eye). There were also the following Palearctic shorebirds ("waders" in British terminology): Ruddy Turnstone, Black-bellied Plover, Greater (Large) Sandplover, Whimbrel, Bar-tailed Godwit, Terek Sandpiper, Red-necked Stint, Gray-tailed Tattler, and Common Sandpiper.

Our highway from Port Hedland up the coast to Broome, and all the way beyond to Wyndham, was called the "Great Northern Highway", covering a total distance of 1,000 miles, excluding the short side trip to Derby and back. It was entirely gravel, with both some relatively smooth, well graded sections, and some horribly corrugated bone-rattling sections (particularly between Fitzroy Crossing and Halls Creek where we broke a front wheel hub and spindle and had to stop in Halls Creek for repairs). We started out from Port Hedland on Oct. 27, and arrived in Wyndham 11 days later, on Nov. 7 (for an average rate of travel of only 90 miles per day). For the first day and a half we traveled along "80 Mile Beach" on the Indian Ocean at the western end of the extensive "Great Sandy Desert", which stretched inland for 500 miles. Sheep stations were very widely scattered along this section near the coast, but once we reached Broome they had been replaced entirely by cattle stations (which then continued eastward across the northern part of the continent). I stopped once to photograph a *Tiliqua* skink crossing the road.

Along 80 Mile Beach, the coast was lined for mile after mile with non-breeding "waders" from the Palearctic, in mindboggling numbers, which migrated south many thousand miles to escape the harsh winter in their arctic breeding grounds (as they do also in the New World). On Oct. 28, I estimated 10,000 individuals along a half mile stretch of beach 12 miles southwest of Anna Plains Station. My attempt to identify and count some of these produced the following tally: 2 Ruddy Turnstones, 3 Black-bellied ("Gray") Plovers, 250 Greater Sandplovers ("Large Sand Dotterels"), 150 Caspian Plovers, 1,000 Bar-tailed Godwits, 2 Gray-tailed ("Oriental") Tattlers, 2 Marsh Sandpipers, 1 Common Greenshank, 50 Terek Sandpipers, 100 Curlew Sandpipers, 12 Far Eastern Curlews, 2,500 Great Knots, and 3 Broad-billed Sandpipers*. The vast majority of individuals were not identified to species. There were also two Australian breeding species: four Pied Oystercatchers* and a single Australian Pratincole*. I did not know when these shorebirds had arrived, or for how long they might remain here along the beach. Two passerine birds new for my list were the brilliantly colored Orange Chat* (genus *Ephthianura*) and the Pied Honeyeater*, a highly nomadic black-and-white species of the arid interior of Australia, irregular and unpredictable in its occurrence.

On Oct. 29, we arrived in Broome, a coastal town in northwestern Australia best known for its pearling industry and its "Boab" trees. As we approached the town I added 3 Brolgas* (a breeding, resident crane) to my trip list, along with my first flocks of noisy, conspicuous Red-tailed Black-Cockatoos* (a familiar bird of northern Australia, which also occurred in disjunct populations in the interior and the SW of the continent). Boab trees (*Adansonia gregoni*, also called "bottle trees") were a close relative of the African Baobab tree (in the same genus), and occurred in northwestern Australia from the Kimberley region of northern Western Australia eastward to the Katherine Gorge region of the Northern Territory. They were characterized by a massive swollen trunk, from the top of which branches projected upward and outward in all directions. Trees grew up to 65' in height and lived up to 2,000 years old. The trunk could reach as much as 16 ft. in diameter, and in older trees it would sometimes develop an enormous basal hollow which could be used as a shelter for people, or on several well documented occasions even as a prison for aborigines! The trees were deciduous and lost their leaves in the dry season.

While Noble and I were in Broome (Oct. 29-Nov. 2), I added 12 birds to my trip list (Appendix B). These included the Pheasant Coucal* (more often heard than seen), Great Bowerbird* (of which I will say more later), and 4 species of honeyeaters. I also saw my first White-bellied Sea-Eagle and my first Brahminy Kites in Australia (both species of which I had documented in Asia). On Nov. 2, we left Broome and drove to a campsite 75 miles northeast, 50 miles from Derby.

The next morning (Nov. 3), we drove the remaining 50 miles into Derby, which was situated 25 miles to the north (left) of the Great Northern Highway, on the southern tip of King Sound at a latitude of 17.25°S. Along the way, I documented eight species of parrots & cockatoos in the dry open woodlands, and around water tanks. These were 15 Red-tailed Black-Cockatoos, 25 Galahs, 50 Little Corellas, 5 Rainbow Lorikeets* (*Trichoglossus haematodus*, a widespread, polymorphic species in which the local race is often regarded as a separate species, the

"Red-collared Lorikeet"), 10 Varied Lorikeets*, 2 Red-winged Parrots*, 10 Cockatiels, and 10 Budgerigars. The diversity and number of parrots in Australia was second in the world only to that of the Neotropics. I also saw my first Masked Lapwing* and Red-backed Kingfisher*, plus 10 Gray-crowned Babblers, 25 Black-faced Cuckoo-shrikes (an intra and intercontinental migrant), 10 Pied Butcherbirds, and 30 Black-faced Woodswallows.

Derby was famous for its very high tides of 30-40 feet, second in the world only to those in Nova Scotia. We strolled along the waterfront of this small, impoverished mining and pastoral town, and visited another boab tree that had once been a prison, but there was little of interest for us in Derby. We spent only this one day in the town, relaxing and having a beer in the local hotel. That evening there was a pretty, orange-pink sunset over the bay. We drove back out of town to a wooded campsite 15 miles south of town, not far from the very southern tip of the sound. Mopokes called regularly throughout the night.

I birded for three hours the next morning (Nov. 4) while Noble brought his journal up to date. Among the 42 species of birds that I documented, there was only one addition for my trip list, the Oriental Pratincole*. The most numerous species were 750 Cockatiels, 200 Peaceful Doves, 125 Diamond Doves, 100 Little Corellas, 30 Galahs, 25 Rufous-throated Honeyeaters, and 20 White-winged Trillers.

At 10:00, Noble and I packed up and headed east on the rough, poorly maintained, rapidly deteriorating Great Northern Highway. It was necessary to drive quite slowly, at speeds between 10 and 20 mph. The town of Fitzroy Crossing, on the Fitzroy River, was 140 miles ahead. We stopped for the night 10 miles before reaching the town.

The next morning (Nov. 5) we drove into Fitzroy Crossing, and I counted birds along the Fitzroy River for two and a half hours (from 9:30-12:00), including the lower reaches of Geike Gorge a short distance above the town. Birds on my list were 75 Plumed Whistling-Ducks*, 15 Magpie Geese* (the most primitive of all living waterfowl in the world), 2 Brolgas, 1 Red-kneed Dotterel* (a very handsome plover in the monotypic genus *Erythrogonys*), 1 Black-breasted Kite* (in the monotypic genus *Hamirostra*; an appealing raptor which was endemic to inland areas of Australia), 1 Black Falcon (which pursued birds by flying rapidly at low levels above the ground), 10 Sulphur-crested Cockatoos* (a large, familiar, all white cockatoo of northern & eastern Australia), 1 Dollarbird (my 1st Australian sighting of this widespread Asian "roller"), 5 Rainbow Bee-eaters, 6 species of honeyeaters, and 6 Great Bowerbirds. Yellow-tinted and Rufous-throated honeyeaters were particularly numerous, and I identified my first White-gaped Honeyeaters*.

After stopping for vehicle repairs in Halls Creek on Nov. 6, Noble and I continued our journey toward Wyndham. Our remote, rarely traveled road could hardly have been more badly corrugated (similar to our roads in the Chilean desert) and it was necessary for me to drive at a very slow speed. As we navigated northward in the eastern Kimberleys, we entered sparsely populated, arid, stony spinifex country with dry streambeds and scattered small eucalypts. Birds along this route included 25 Crested Pigeons, 1 Australian Bustard, 20 Black Kites, 30 Red-tailed Black-Cockatoos, 75 Little Corellas, 25 Gray-crowned Babblers, 50 Black-faced Woodswallows, 75 Yellow-throated (Dusky) Miners, and 60 Little/Torresian crows (which I had difficulty separating most of the time). We crossed the dry upper Ord River about 60 miles north of Halls Creek (where it was flowing from west to east), and camped there for the night. This was largely unpopulated, hot, dry, northwestern Australia.

Around our campsite the next morning (Nov. 7) were 100 Peaceful Doves, 500 Varied Lorikeets, 30 Black-faced Woodswallows, 14 Banded Honeyeaters, 10 Little Friarbirds (a large honeyeater), and 5 Long-tailed Finches* (a very attractive estrilded with long, pointed, central tail feathers). We departed at 7:00 and drove slowly north for nine hours to the Dunham River. The countryside varied in elevation between 750 and 1,250 ft., and was featured by dry rocky hills, sparse grasslands, scattered small trees, and dry streambeds bordered by trees and scrub. Birds were varied and numerous, and I documented almost 50 species. These included the following species: 90 Black Kites, 4,000 Varied Lorikeets (wow, in flocks of 10-50 individuals), 40 Red-tailed Black-Cockatoos, 50 Sulphur-crested Cockatoos, 3 Northern Rosellas* (genus *Platycercus*), 35 White-winged Trillers, 50 Gray-crowned Babblers, 120 Black-faced Woodswallows, 7 Varied (White-winged) Sitellas, 15 Black-tailed Treecreepers, 4

Banded Honeyeaters, 15 Rufous-throated Honeyeaters, 10 Pied Honeyeaters (seen at "McPhee's Well", my 2nd record of this nomadic honeyeater of unpredictable occurrence, which was rare this far north), 25 Yellow-tinted Honeyeaters, 30 Yellow-throated (Dusky) Miners, 25 Long-tailed Finches, and 25 Magpie-larks. Rosella species are virtually non-overlapping (allopatric) with each other throughout Australia, and with the genus *Barnardius*.

We camped that night along the Dunham River, at 500' elevation. In the hour before sunset, I added six more trip additions: Australian Koel* (a cuckoo), Blue-faced Honeyeater* (one of the larger, more attractive honeyeaters, with a vivid blue face, black-and-white patterned head, olive green upperparts, and white underparts), Crimson Finch*, Double-barred (Banded) Finch*, Pictorella Munia* (Finch), and Masked Finch*. It had been a very productive day of birdwatching. Wallabies were common around our campsite that evening.

I continued to search for birds along the Dunham River for two hours before breakfast the next morning (Nov. 8), and it was again a very lucrative site. I tracked down 4 new trip birds: Purple-crowned (Lilac-crowned) Fairywren* (sadly, only a female, and the only time I identified this local, uncommon species in which the male had a distinctive, wide lilac-colored band on the side of its crown), Bar-breasted Honeyeater*, Silver-crowned Friarbird*, and Gouldian Finch* (perhaps the most colorful of all the 141 species of estrildid finches, with its combination of lilac, green, red, black, & yellow plumage; it was the least common of the endemic Australian *Erythrura*, and declining). Honeyeaters were overwhelmingly the most numerous family, with 9 species and 124 individuals. The Yellow-tinted Honeyeater was by far the most abundant species, with 50 individuals. A lone aboriginal man came walking past.

After breakfast, we drove into Wyndham, a dying coastal port town at the upper (southern) tip of Cambridge Gulf, near the mouth of the Ord River. The town had very little to offer, and we stopped long enough only to fill up our gas tank from a 50 gallon drum, with a hand pump. We were now at the end of the Great Northern Highway, 2,270 miles from Perth. Wyndham was not a tourist destination. In 1961, the Western Australian government was building a small diversion dam across the lower Ord River to produce a lake which could be used to channel irrigation water, by gravity, to the extensive Ord River floodplain, where it was envisioned the land could be converted to productive cropland of sugarcane, rice, or some other crop. (Unfortunately, for a variety of reasons, this never succeeded and the area was eventually set aside largely as a wetland for wildlife.)

When Noble and I visited this area on Nov. 9, there were open grassy woodlands and large pools of water above the dam construction site, to which many aquatic birds, and others, were attracted. With the help of my telescope, I came up with such birds as cormorants, darters, dotterels, stilts, godwits, cranes, Royal Spoonbills*, ibises, egrets, herons, Green Pygmy-geese*, Magpie Geese, Radjah Shelducks* (a very handsome species), and Oriental Pratincoles. There were also White-bellied Sea-Eagles, Whistling Kites, dozens of Black Kites, a Brown Falcon, Rainbow (Red-collared) Lorikeets, corellas, Galahs, Red-winged Parrots, Budgerigars, Dollarbirds, Blue-winged Kookaburras, Sacred Kingfishers, Rainbow Bee-eaters, Australian Bustards, coucals, and such passerine birds as White-winged Trillers, Black-faced and White-breasted woodswallows, 7 kinds of honeyeaters, 3 kinds of estrildid finches, and a single Great Bowerbird. Obviously, the area in 1961 was already a haven for a great diversity of birds.

Just after noon, Noble and I passed through the small town of Kununurra, and shortly thereafter we departed Western Australia, and crossed the border into the Northern Territory. During my 8 weeks in the state of Western Australia, I added 215 additions to my trip list of birds (Appendix B).

Our road in the Northern Territory took us east from the W.A. border for 120 miles, through open woodland and scrub, to the town of Timber Creek. We camped for the night a few miles beyond. This stretch of road produced two additions to my trip list, the Yellow-billed Spoonbill* and the Pink-eared Duck* (an unusual looking little duck with barred flanks and a spatula-like bill which was widened laterally at the tip). The most numerous roadside birds along this stretch of road were 100 Varied Lorikeets, 75 Black-faced Woodswallows, 50 Rainbow Lorikeets, 50 Black Kites, and 40 Banded Honeyeaters.

Plate 114: Passing motorists were rarely encountered in the Kimberleys. The badly corrugated road near Halls Creek caused us yet another breakdown. A lone Aboriginal man came walking past our campsite on the Dunham River.

The next morning (Nov. 10) we drove 180 miles northeast on a rough gravel road to the bitumen surfaced Stuart Highway, which connected Darwin (to the north) with Alice Springs (to the south). We turned left (north) and several miles later we came to a bridge crossing the Katherine River. Just above the bridge to our right was a lovely big pool of water. It was a hot day and there were no vehicles or pedestrians in sight, so we stripped off our clothes and went for a very refreshing 15-minute swim. We then drove the short distance into the town of Katherine where we found a tearoom for a sandwich and a cup of tea. The pleasant owner of the establishment came up to our table to visit with us. We relayed our swim in the river to him and he was horrified. He informed us, in a very stern manner, that the river was inhabited by aggressive salt water crocodiles, and that under no circumstances should one ever swim there, some persons having lost their lives. These animals were the largest in body weight of all terrestrial predators in the world, individuals regularly weighing as much as 2,000 pounds and rarely even 3,000 pounds, and occasionally attaining lengths of up to 20 feet. They were to be avoided at all times. My thoughts on this matter were that the town should put up warning signs at the highway bridge alerting naive tourists such as ourselves of their presence, and warning persons not to swim there.

Noble and I continued northward from Katherine to Darwin, arriving that afternoon. This tropical coastal town of perhaps 10,000-15,000 people in 1961 was named for Charles Darwin by Cmdr. John Wickham, on board the third voyage of the HMS Beagle in Sept., 1839, when the port was first visited by a British ship. (Wickham had sailed with Darwin on the second expedition of the Beagle from 1831-1836, during which time the ship did not visit the north coast of Australia, and Darwin himself never visited this port.) The town was characterized by a warm tropical climate and heavy monsoonal rains, and it frequently experienced strong winds and extended lightning storms. It had survived many tropical cyclones (and Japanese bombings in WW2). Darwin was situated on a small bluff overlooking the Timor Sea, and in 1961, aborigines made up about 10% of the total population. The town was a gateway to Asia, and was brightened by many red-flowering Poinciana trees and colorful gardens. Surrounding areas of mangroves in the littoral and estuarine lowlands were bordered with casuarinas and Screw Pine (*Pandanus*) trees (monocots with aerial prop roots and a palm-like appearance). Along the coast were sandy beaches, and in the ocean were such popular fish as snapper, jewfish, and mackerel. In coastal rivers were widely prized barramundi fish, and crocodiles! This isolated town on Australia's northern shore did not yet attract very many tourists, and most of the income was provided by mining (zinc, bauxite, gold, and manganese) and military establishments.

We spent four days in Darwin (Nov. 10-13). On the morning of Nov. 14, Noble and I departed Darwin and headed south on the Stuart Highway, southward toward the town of Tennant Creek. Not far south of Darwin, we stopped for 15 minutes at the War Memorial Cemetery on the Adelaide River. Here there were five Great Bowerbirds, and I located a recently built "avenue" type bower. It was an open, 2-sided passageway of sticks between 2 arched walls 18" high and 3' long, with a large array of both natural and man-made items displayed on the ground inside and outside the bower - - bones, shells, stones, glass, coins, bottle caps, flowers, leaves, small fruit, and any other small ornamental or attractive item which a bird could find. The bower was the largest avenue type in Australia. (There is a documented account of a man who lost his wristwatch while walking through the bush, and several days later while searching for it he found it inside the bower of a Great Bowerbird!) The only purpose of the bower was to attract females for mating, which usually took place in the bower. In addition to the bower, males also solicited females by loud vocalizations and elaborate dances. Females were very fickle in their choice of a mate, although males were polygynous and would mate with as many females as possible. Males did not participate in nest building, incubation, or raising the young. I pondered how natural selection could possibly arrive at such a system of breeding behavior. Of the 20 species of bowerbirds in the world, half of them were found in Australia (and the other half was confined to New Guinea).

Noble and I continued on to Katherine, where we camped for the night along the river, wary of any crocodiles! En route, I stopped to identify three Northern Rosellas and three Partridge Pigeons*, a distinctive, terrestrial, quail-like columbid. A Boobook Owl called repeatedly throughout the night (as one frequently did at our campsites). The next morning (Nov. 15), in three hours of pursuing birds in the trees and scrub along the river here, and around the pools, I recorded 44 species of birds. Only one, the White-throated Honeyeater*, was a trip addition. (It was easily confused with the White-naped Honeyeater.) Other birds were a Peregrine Falcon, 200 Rainbow Lorikeets, 100 Varied Lorikeets, 200 friarbirds (the combined total of both Little and Silver-crowned, which I usually found difficult to distinguish), and 7 Blue-faced Honeyeaters.

For 4 days (Nov. 15, 16, 17, and 18) I searched for birds all day in the scenic Katherine Gorge area, 20 miles east of the town. My list for Nov. 18 included two trip additions, the small, rather plainly colored Brush Cuckoo* (genus *Cacomantis*) and the ground-dwelling Sandstone Shrike-Thrush*, with a gorgeous repertoire of beautiful, liquid notes. I was delighted to have found this locally distributed inhabitant of sandstone escarpments and rocky gorges. Honeyeaters (Meliphagidae) were particularly numerous and I listed nine different species on this date. I also recorded five different estrildids, including the Gouldian Finch, which was my second and last record of this rare, stunningly colored, polymorphic finch. In addition to the birdlife, the cliffs and boulders in this region had ancient aboriginal paintings that were of great significance to archaeologists studying the long history of these primitive people on the Australian continent. Prior to the recent colonization by Europeans, the Australian aborigines were perhaps the most primitive people anywhere on earth, with a lifestyle unchanged over many millennia.

The next day (Nov. 19) we continued south on the asphalt Stuart Highway, passing several large cattle stations and "roadhouses" (small establishments where travelers could purchase fuel, food, and sometimes a bed for the night). Just north of Tennant Creek, we came to Banka Banka Station late in the afternoon. It was one of the largest cattle stations in the region. There were flocks of birds going and coming from a windmill, where there were water troughs and spilled pools of water on the ground. I stopped at the homestead and asked the owner if we could camp there for the night, near one of the mills, where I could document the birds. He and his wife warmly gave us permission to do so, and invited us in for a cup of tea. Everywhere Noble and I traveled in Australia we were cordially welcomed. I chose a campsite near one of the mills where there was some surface water. Galahs and Crested Pigeons gathered at the mill for a last drink, and Zebra Finches retired to the nearest scrub in which to roost for the night. Like all columbids, the Crested Pigeon drank by submersing its bill under the water, without raising its head, and pumping water into its crop via "gular pumping" (as did only a few other birds). Some desert mammals also regularly came to drink, and kangaroos, dingoes, and foxes were often observed drinking at water troughs. The artificial sources of water on sheep or cattle stations provided water not only for livestock but also for native wildlife, and were very valuable in maintaining their populations. As the sun went down, there was a vivid orange sunset on the distant horizon. I marveled at the stillness and vastness of the Australian outback, reminding me of our nights in the Sahara.

On Nov. 20, Noble and I drove into Tennant Creek, only long enough to purchase gasoline, a cup of tea, and a "meat pie." We then turned around and drove north back out of town the short distance to the paved Barkly Highway, where we turned right and headed east toward the Queensland border, about 285 miles distant, across bleak cattle country with scattered stations and homesteads. The Gulf of Carpentaria was on our left, 200-250 miles north of us.

Very soon after crossing the Northern Territory-Qeensland border in the remote, far northwest of Queensland, we came to the tiny community of Camooweal, a replica of cowboy towns in the western USA, 40 years ago. There was one main street with a few shops and business enterprises, such as a petrol station, a "Greengrocer", and an ice cream shop from which a young girl emerged with an ice cream cone. Of course, there was the usual two-story, wooden hotel with a balcony, a pub, and a dining room. We arrived there on a very hot afternoon, and I immediately stopped in front of the hotel and went inside for a cold beer at the bar.

There were no lights on in the pub as I stepped inside from the bright sunlight of the street outside, and it was so dark that I had to wait a moment to let my eyes adjust enough to locate the bar. As I paused, I immediately heard, quite clearly, a long series of very rude, impolite slang words and expletives, in a somewhat falsetto sounding voice. I said to myself that someone had been drinking too much, and I took a quick look around the room to see where the sound was coming from, but I saw only the old man behind the bar, which was not the direction from which the words were emanating. I walked up to the bar and ordered a "schooner." (Beer in all Australian pubs was sold in one of three sizes: an 8 oz. "glass", a 10 oz. "middie", and a 12 oz. "schooner.") No sooner had the bartender handed me my beer than I once again heard the long string of foul language. When I looked in the direction from which the words were uttered, I immediately discovered their source, a beautiful, all-white Sulphur-crested Cockatoo, chained to the end of the bar! My conclusion was that this bird had spent its entire life in the bar, and these were the words it most frequently heard, and memorized for repetition! It was a permanent memory for me of Camooweal, in the middle of the remote northern outback of Australia.

It was 120 miles from Camooweal to Mt. Isa, a mining town for the exploitation of lead, zinc, copper, and silver. Roadside birds along our route included Spinifex Pigeons, a Wedge-tailed Eagle, and a Spotted Nightjar* which was heard calling at night (and seen later in our travels). Nightjars (and all other caprimulgids) were uniquely adapted to hot deserts because of their innate ability to keep cool by efficient panting, causing water to evaporate (a cooling process) from their unusually wide throat area, which they moved up and down rapidly, with very little expenditure of energy. This process was called "gular fluttering" by physiologists. (In a laboratory experiment, a Spotted Nightjar survived without food or water for two hours in an environmental chamber, with no ill effects, where the temperature was regulated at 124°F.)

At Mt. Isa, our route across the dry northern interior of Queensland (the "Barkly Highway") became gravel or dirt again, all the way 500 miles to Charter Towers, through the towns of Cloncurry (where it became the "Flinders Highway") and Hughenden. We were still in a region of sparse vegetation, and cattle stations. Black Kites were the most abundant raptor, and Crimson Chats were seen near Mt. Isa. Between Julia Creek and Richmond, the following species were counted: 60 Peaceful Doves, a pair of Pale-headed Rosellas* (the resident *Platycercus*" in NE Australia), a flock of 500 Australian Pratincoles, and 3 Spotted Bowerbirds*. This small geographic area was where the "Cloncurry Parrot" lived (*Barnardius barnardi macgillivrayi*), but I did not encounter this distinct, geographically isolated race of the Mallee Ringneck.

In the vicinity of Hughenden there were 80 Black Kites, 12 Red-tailed Black-Cockatoos, 200 migrating Fork-tailed Swifts (*Apus pacificus*, from the NE Palearctic), 750 migrating Masked Woodswallows (a nomadic, resident species, well known for its mass movements), 75 Yellow-throated Miners (a honeyeater, *Manorina flavigula*, also known as the White-rumped, or Dusky, Miner), and 75 Zebra Finches. On Nov. 26, ten Apostlebirds* were encountered near Torrens Creek (1 of 2 species, along with the White-winged Chough, in the patchily distributed Australian family Corcoracidae). Two unusual looking Channel-billed Cuckoos* (with a very wide bill) were observed near Charters Towers.

On Nov. 27, we arrived in Townsville, on the renowned tropical coral "Great Barrier Reef" of NE Australia. The city streets were lined with bright red-flowering Poinciana trees. In the countryside were many miles of sugarcane fields. We did not visit the reef, but instead drove northward along the coast to experience the tropical rainforests, which were patchily distributed between Townsville, Cairns, and Mossman.

The Australian rainforests had the same definitive features as tropical rainforests elsewhere in the world; high floral and faunal diversity, high endemism, buttressed trees with vines and epiphytes, strangler figs, palms, ferns, orchids, an overwhelming variety of insects, and vertical structured life zones from the forest floor to the canopy. Niches were filled by "ecological counterparts" of rainforest species on other continents. The tropical rainforests along the coast of Queensland were remnants of forests that once covered almost the entire continent 60-70 million years ago, about the time the Australian tectonic plate was separating from the Antarctic tectonic plate at the final breakup of "Gondwanaland."

Plate 115: We departed Western Australia SE of Wyndham, and entered the Northern Territory. We stopped to swim in the Kathleen River, and shortly thereafter learned it was inhabited by 15-20 ft. long man-eating crocodiles!

Plate 116: Rock outcrops occurred locally in arid areas of the Northern Territory. Australian Aborigines at the time Europeans arrived on the continent had perhaps the most primitive of all world cultures. Their ancient rock paintings feature real and mythical creatures.

Plate 117: In northern Australia cattle stations mostly replaced the sheep stations of more southerly regions. Stockmen led a rugged life. Cockatiels, a delightful little parrot often kept in captivity, frequented outback stations for water.

At this ancient time, the components of today's flora and fauna may already have been present. Today's forests may thus have experienced a longer unbroken period of history than tropical rainforests anywhere else, making them, in a sense, the "oldest" tropical rainforests on the planet. Their current remnant, patchy distribution was along the northeastern Queensland coast, including the "Great Dividing Range" of mountains just inland. In these mountains the highest peaks were above 5,000 ft., and the highest rainfall (near the top of Mt. Bellenden Ker) was 335 inches in one year! The forests were characterized by many scenic waterfalls. Because of the oppressive heat I regularly slept on top of our camping van at night, under a mosquito net to protect me from the hordes of mosquitoes.

Characteristic mammals of the rainforest were egg-laying platypuses and echidnas, and such marsupials as tree-kangaroos, swamp wallabies, Red-legged Pademelons (a small macropodid), ringtail possums (an aboreal fruit, leaf, and flower-eater with a prehensile tail coiled into several rings at the tip), Spotted-tailed Quolls (a ferocious cat-sized carnivore), Long-nosed Bandicoots (a rabbit-sized insectivore), the Yellow-footed Antechinus (a mouse-like insectivore & carnivore), and the squirrel-like Striped Possum (an arboreal marsupial that ate wood-boring insects and their larvae). In addition to these "primitive" marsupial mammals, there were placental rodents (murids) and a wide assortment of bats (including very large, fruit-eating "flying foxes").

More than 370 kinds of birds frequented the rainforests or their environs, with 30% being confined exclusively to these forests. The foremost species was the Australian Cassowary, a gigantic forest ratite with a tall, brown, horny casque on top of its head, standing as high as 5½' and weighing up to 130 lbs. (with females a little larger than males). These 3-toed, heavy legged herbivorous birds had a long sabre-like inner claw that was used in defense, and could kill a person if he were kicked in the abdomen with it. The Cassowary's bare head and neck were vivid blue, with a pair of long, bright crimson wattles hanging down both from the lower front of the neck and from the hind sides of the neck. These shy, timid, wary denizens of the forest were best observed as they walked along the outside edge of the forest at daybreak, or across a forest road. Other characteristic rainforest birds included two kinds of megapodes ("mound-building" birds) -- the Orange-footed Scrubfowl and the Brush-turkey. Also noteworthy was the Golden Bowerbird*, the only Australian member of the family which built a "maypole" type bower. Males used pale green lichens and freshly plucked pale orchids to attract females. How vacillating could a female be?

Reptilian inhabitants of the rainforests included crocodiles, amesthistine pythons (very slender, but up to 28' long), carpet pythons (much shorter and more heavy bodied), several varieties of tree snakes, Spotted Tree Monitors (*Varanus timorensis*), and the well-known and much photographed Boyd's Forest Dragon (*Hypsilurus boydii*, an arboreal 1½' long agamid lizard fond of tree trunks, with a conspicuous scaly pointed head crest, ornate head structures, and a long throat pouch, appearing overall much like a New World iguana). The vast variety of rainforest frogs included 30% of all Australian species. Among the myriads of invertebrates were giant golden orb spiders, dozens of vividly colored butterflies, trilling katydids and cicadas, beetles, mosquitoes, ants, termites, and countless more, with new additions constantly being described. Discovering and naming the fauna of rainforests will keep entomologists and others busy for as long as time goes on. By definition, each species inhabited its own niche.

The rainforest flora was equally as challenging and fascinating as was the fauna. The dominant trees were conifers in the Araucariaceae and Podocarpaceae families (trees which the Australians persist in calling "pines", even though they are only distantly related to the Family Pinaceae or the genus *Pinus*). One of the tallest trees in the forest was the N. Queensland Kauri Pine (*Agathis palmerstonii*) which occasionally grew up to 160' tall, with a diameter of 8-10 feet.

In 20 hours of birdwatching in the rainforests over a period of 4 days (Nov. 27-30), I added 42 additions to my trip list (Appendix B). My most memorable ornithological event occurred on the morning of Nov. 29, when I walked off the road into a rainforest south of Mossman. I made my way through the understory, down a hilly slope for 200 yards or so, until I came to a small clearing next to a stream. Here I found a log on which I could sit, and uncharacteristically I chose to sit and wait to see what might birds might come my way. Five minutes later, a medium-sized, relatively slender bird flew to a perch right in front of me at eye level, and perched in clear view, facing me not more than 20 feet away. I gasped in amazement. I had never seen a bird anything like this before, and I could not recall having ever looked at a picture or painting of such a colorful blue, black, buff, and white colored bird, with a conspicuous bright orange-red, moderately heavy, long pointed bill, and with the two narrow central tail feathers greatly elongated, pointed, and all white in color. The bird sat almost perfectly still watching me, seemingly unconcerned by my presence. I thought to myself I must be dreaming, and that I had just discovered a species unknown to science. It was a very thrilling and exciting moment.

After perhaps five minutes or longer the bird finally flew away. I immediately got out my only bird guide (*What Bird Is That?*) from my field bag, and began searching through all the plates, one at a time, starting at the beginning. Cayley's book was not organized in any logical fashion, except that all species illustrated on a particular plate frequented the same HABITAT. Birds of different sizes and taxonomic relationships were all put together on the same plate, which was designated by a Roman numeral. Each species on a plate was given a number, which indicated where you could find its name and description in the chapter identified by the same Roman numeral as the plate. It was a most cumbersome arrangement. However, after less than ten minutes of searching, I found a colored illustration of my unknown bird on Plate IV, between pages 24 and 25, designated only by the number "1" (which peculiarly was placed in the middle of the plate, where it was surrounded by 24 smaller, unrelated birds). When I went to Chapter IV in the text (which was entitled "Birds of the Brushes and Big Scrub"), I located this number, and found out that my unknown bird was called the "White-tailed Kingfisher" (*Tansiptera sylvia*). What a stunning bird it was! (Clements, in his 2007 checklist, calls it the Buff-breasted Paradise-Kingfisher, with the same Latin name as Cayley.) This kingfisher was the only nonpasserine bird on Plate IV. Assigning names to the birds described in my field notes was a challenge, but it was moments such as these that made it worth all the effort.

On Dec. 1, Noble and I went searching for a Cassowary on the Atherton Tableland near Lake Eachum, driving along secondary roads where we had been told we might be lucky enough to see one walking across the road early in the morning. Unfortunately, we didn't. However, we did see many other birds: a half dozen Australian Brush-turkeys*, some Orange-footed Scrubfowls*, Topknot Pigeons*, Superb Fruit-Doves* (also called Purple-crowned Pigeons), Wompoo Pigeons (which I first saw 2 days ago, and one of the most colorful and magnificent of all columbids, a very large fruit dove in the genus *Ptilonopis*), White-headed Pigeons*, Australian King-Parrots* (in which males were colored with a bright red head & body, green wings, and blue rump), Crimson Rosellas*, Rainbow Lorikeets, 3 Gray (or White) Goshawks* (a large accipiter with 2 color morphs, either all Gray or stunningly all white), and more than 50 species of passerines: honeyeaters, whistlers, fairywrens, drongos, trillers, figbirds, flycatchers, fantails, robins, shrike-thrushes, cuckoo-shrikes, bowerbirds, warblers, scrubwrens, woodswallows, silver-eyes, pardalotes, magpie-larks, and Victoria Riflebirds (1st seen on Nov. 28). This latter species was a medium-sized, short-tailed, sexually dimorphic bird-of-paradise with a long, sharply pointed, strongly down-curved bill. It was one of three similar appearing, allopatric, eastern Australian species in the genus *Ptiloris*. Iridescently colored black and green males displayed from a high or low horizontal branch or dead limb, by hanging upside down and vocalizing loudly, shaking and expanding their plumage, and spreading their wide rounded wings. This display could also involve crouching and posturing while standing upright. Females were brown backed and white below with heavy arrow-shaped barring, and crept around on tree trunks and branches looking very much like a treecreeper (which I mistook them for upon my first sighting). There are only four Australian species in the spectacular bird-of-paradise family (Paradisaeidae), which otherwise is confined largely to New Guinea.

We traveled for seven days from Cairns to Brisbane (Dec. 3-9), southward along the wet, populated coast of Queensland, through the coastal cities of Townsville (once again), Mackay, Rockhampton, and Maryborough, a total distance of 1,200 miles. For this week of travel, we picked up two hitchhiking English boys (Gary and

Patrick), who rode on top of our van, at the front. We made very few stops. The 11 trip additions to my bird list during this period included the Ground Cuckoo-shrike*, Black-throated Finch*, White-browed Woodswallow* (one of the more appealing woodswallows), Red Goshawk*, and Superb Fairywren* (Appendix B).

Lamington National Park was located in extreme SE Queensland, on the border of New South Wales in the McPherson Range of mountains, which were part of the Great Dividing Range. The plateau and ravines in the park were covered with a temperate rainforest in which there were many hiking trails and waterfalls, at an average elevation of 3,000 ft. above sea level. The park was a floral and faunal paradise, with affinities extending back in time to the ancient breakup of Gondwanaland. Myrtle Beech trees, *Nothofagus cunninghamii*, and Hoop Pines, *Araucaria cunninghamii*, had affinities with similar species in New Zealand, New Caledonia, New Zealand, and southern South America, all with a Gondwanaland origin.

Noble and I visited the park for two days, on Dec. 10 and 11. I searched for birds during a total of 14 hours, in which time I added 13 species to my trip list (Appendix B). Particularly exciting for me were the Albert's Lyrebird* (one of two species in the endemic Australian family Menuridae) and the Rufous Scrub-bird* (one of two species in the endemic Australian family Atrichornithidae). I pursued the secretive scrub-bird through dense forest understory for 15 minutes, following its loud repetitious song, before I finally got a glimpse that allowed me to identify the elusive songster. Other exciting birds were the stately Wonga Pigeon*, the Noisy Pitta* (equally as challenging to track down as the scrub-bird, also by following its song through the forest understory), the handsome Southern Logrunner* (in the family Orthonychidae, along with the N. Logrunner and the Chowchilla), the gorgeous Regent Bowerbird* (with its bright yellow and black plumage, yellow bill, and yellow eye), the Eastern Spinebill* (a diminutive, distinctly patterned honeyeater with a particularly slender, slightly decurved bill), the Pied Currawong* (in the bellmagpie family, Cracticidae), and the local, uncommon Red-browed Treecreeper*. These two days in Lamington were a perfect ending for my very productive, three weeks of pursuing birds in Queensland, during which time I identified 87 trip birds (Appendix B).

There were approximately 600 miles of paved highways from Lamington National Park (in Queensland) to Sydney (in New South Wales). We followed, in chronological order, the Mount Lindesay Highway (to Tenterfield), the New England Highway (to Armidale and Newcastle), and finally the Pacific Highway (to Sydney). It was a total journey of three days (Dec. 12, 13, & 14). Three birds of particular interest along this stretch (all trip additions) were: (1) Crested Shrike-tit*, in the monotypic genus *Falcunculus* (family Pachycephalidae, with whistlers and allies), (2) White-winged Chough* (1 of 2 species in the endemic family Corcoracidae), and (3) Turquoise Parrot* (genus *Neophema*), a small, slender, local "grass parrot" with a long, pointed tail, green above and yellow below, with a blue face and a red scapular patch.

Sydney was the site where Capt. James Cook, on Apr. 29, 1770, sailed into Botany Bay on the HMS "Endeavor", and with his crew became the first Europeans to set foot on the east coast of Australia (although landings had been made on the west coast much earlier by several Dutch ships, and later one English ship, in the early to late 1600's, and the very first Europeans to document landing anywhere on the Australian continent were apparently the Dutchman Willem Janszoon and his crew on Feb. 26, 1606, on the west side of the Cape York peninsula in northern Queensland). In 1644, the Dutchman Abel Tasman gave the name "New Holland" to the Australian continent. Cook returned to Botany Bay on Jan. 26, 1788 (18 years after his first arrival) with a British fleet loaded with convicts, and established the first European settlement anywhere on the continent, a penal colony! On this date the British flag was raised over the bay, and the event has since been celebrated as "Australia Day."

For Noble and me, Sydney was the end of the road in our never-say-die "Roadrunner" Jeep camping van, which Noble admirably had somehow managed to keep going for almost three full years. It was an amazing accomplishment, and a feat of unmatched perseverance. Kudos to him. My part was to do most of the driving, a none-too-easy task. In spite of innumerable obstacles, we had accomplished what we set out to do. For the last time, we celebrated together with a bottle of wine in our Jeep. It was soon thereafter put on a ship for Los Angeles, where Noble would eventually pick it up, after our arrival there. Since the Jeep had been purchased in his name (even though we both contributed equally to its cost), it was up to him to determine its fate thereafter.

There was still one task left for us to complete together in Australia - my wedding! While Noble remained behind in Sydney for a few days I flew back to Perth, on Dec. 17, ten days before the date on which Laurie and I were going to be married. I felt as if I were dreaming, and would suddenly wake up. But, no, there was Laurie standing in the airport lobby waiting for me as I walked out, with her lovely smile. My heart skipped a beat. She greeted me with a little hug and quick kiss, as were her innate mannerisms. Similar to my first arrival at the airport, I was carrying with me only my backpack, since I had no other luggage of any kind in which to put things. It would be necessary for me to do some shopping in Perth -- with Laurie's financial help! Noble would fly to Perth two days before the wedding, to participate as my best man. We would each need to rent a white shirt, black suit, and bow tie.

Prior to our wedding, Laurie's father, Charlie, invited me to participate on a "duck shoot" with him and five of his Beverley friends, at a small inland shallow lake (Lake Toolibin), in the wheat belt 75 miles SE of Beverley. He would supply me with a double-barreled side by side 12 gauge shotgun, with which I was quite familiar. It was an event I could enthusiastically accept. The lake was margined by dead, dying, and still surviving "paperbark" trees (also known as "tea trees", in the genus *Melaleuca*). The trees had in recent years been slowly dying because of the gradually increasing level of salinity in the lake, caused by greater surface runoff of water from the surrounding saline soils, owing to removal of the natural covering of soil-holding grasses, as the land was plowed and converted to wheat farming. Within a dozen years or so, all the trees around the lake would be dead, and the increased salinity of the lake water would have a strong adverse effect on the aquatic organisms in and around the lake, including the ducks. It was an unfolding ecological disaster (as was the fungus currently destroying the Jarrah forests). The unique Australian flora and fauna were rapidly being eliminated from the Land Down Under. Immediate environmental remedies were needed if they were going to survive for future generations to enjoy.

Not helping the situation was the fact that the legal limit for a duck shooter was 20 ducks a day! Our party of shooters arrived the night before, and we camped on the shore of the lake so as to start shooting the next morning as the sun came up, but legally not before sunrise. One of the members of our party was a very likeable man by the name of Joe Wansbrough, who was the honorary game warden for Beverley Shire (which was not the shire where we were), but Joe was very insistent that all of us follow the legal game laws, which of course everyone did. Since there were a great many ducks on the lake, almost all the members of our party, including myself, shot 15-20 ducks in 3 or 4 hours of shooting that morning, starting legally at sunrise. The most numerous species shot were Pacific Black Ducks, Gray Teals, and Australian Shelducks (which for some peculiar reason unknown to me were locally referred to as "mountain ducks").

Noble arrived in Perth two days before my wedding, which took place as scheduled on Wednesday morning of December 27, 1961, in the St. Mary's Anglican Church in West Perth. More than 150 invited guests were in attendance, almost all of them friends and acquaintances of Laurie's parents from Beverley town and shire. Noble was best man and Laurie's sister, Robin, was one of her two bridesmaids. Of course, her father walked her down the aisle. It was a happy, memorable event for me, and an exciting ending to my 3-year adventure around the world. It was beyond the realm of reality. (My rental trousers were too long and fell baggily around my ankles.)

After the ceremony, all the guests were invited to a reception at the popular, long established, Esplanade Hotel in the city, not far from the Swan River waterfront. (By amazing coincidence, just yesterday at this very same hotel, Laurie was the recording secretary for Hancock & Wright as they signed a contract with the Rio Tinto Mining Company to mine iron ore from Hancock's property in the Pilbara, which would eventually make him the wealthiest man in all of Australia!) At the wedding reception there was lots of champagne for everyone, and there was the traditional toast to the queen by the Master of Ceremonies. After someone whispered in his ear that I was not a subject of the queen, he immediately raised his glass and proposed a second toast "To the President of the United States of America"! (In 1961, this was John F. Kennedy.)

Plate 118: The end of our adventure by Jeep was from the tropical rainforests of northern Queensland south to the Sydney harbor bridge. At the finish, there were a little over 80,000 miles on our odometer, many from the world's worst roads, some of which were unmapped.

Laurie and I spent that night together at a motel on the south side of the river, and the next morning we commenced an 11-day honeymoon through the Southwest, traveling in her parents new 4-door 1961 Ford "Falcon", with the Beverley "number plate" (license plate) of "BE-37." (Noble returned to Sydney and then traveled onward to New Zealand, where Laurie and I would catch up with him on Feb. 1, as passengers on board the "Canberra", from Perth to Los Angeles). In Australia, it was now the middle of summer, and Laurie and I would enjoy our honeymoon in the somewhat cooler coastal regions of the Southwest. We tasted the wines of the Margaret River area, drove along Caves Road, photographed Cape Leeuwin (the most southwestern point on the Australian continent), visited woodcarving shops in Pemberton, drank tea in Walpole on the shore of Nornalup Inlet, and walked among tall karri trees in the Valley of the Giants.

Denmark was a delightful, peaceful little town on the south coast, with friendly people. Here we spent several days with Laurie's Uncle Les and Auntie Jo (a younger sister of her mother) and their four young daughters, sleeping at their house in the back of Les' Land Rover. We all went swimming and sunbathing at William Bay, in frigid water only a penguin could possibly enjoy. Adding color to the areas of coastal heath were yellow-flowered Banksias, Red- flowering Gum trees (*Eucalyptus ficifolia*), and golden-orange "Christmas Trees" (*Nuytsia floribunda*, in the mistletoe family Loranthaceae), so named because of their season of flowering. These trees were unusual in being "hemiparasitic", with their roots spreading out underground and attaching themselves to the roots of nearby trees and shrubs with which they came in contact, and from which the Christmas Tree then obtained its minerals and water.

Birds which I recorded in the vicinity of Denmark on Dec. 30 & 31 included the following species: Emu, Australian Gannet*, Pacific Gull*, Hooded Plover* ("Dotterel", a small black-headed plover with a white collar, red bill, red eye ring, and flesh-colored legs, frequenting sandy and rocky southern Australian seacoasts, with its nearest relative being the Shore Plover in the Chatham Islands), Brown Goshawk, Swamp Harrier, Whistling Kite, White-tailed Black-Cockatoo, Purple-crowned Lorikeet, Port Lincoln (Twenty-eight) Parrot, Red-capped Parrot, Sacred Kingfisher, White-breasted Robin, Scarlet Robin, Golden Whistler, Gray (Western) Shrike-Thrush, Splendid Fairywren, Western Gerygone (White-tailed Warbler), Western Thornbill, Dusky Woodswallow, Silver-eye, Red Wattlebird, New Holland Honeyeater, White-naped Honeyeater, Gray Currawong (which I had first seen 2 weeks ago, on Dec. 20 in the Jarrah forest between Perth & Beverley), Australian Raven, Australasian Magpie, and Red-eared Firetail* (an estrildid finch endemic to the SW forest edges and shrubby clearings). The two most numerous species were Red Wattlebirds and Silver-eyes.

Our honeymoon continued eastward along the coast from Denmark to the scenic, popular tourist town of Albany, which was the site of the very first European settlement in Western Australia, in the year 1826. (This settlement was followed 3 years later by a settlement on the Swan River, at the site of present day Perth, in 1829). Albany was known for its nearby rugged, scenic, rocky coasts, which included a large natural bridge and several tall blowholes (with the right amount of incoming surf). Aquatic birds in the vicinity included the following species: 2 all dark shearwaters flying offshore (probably Flesh-footed), 1 White-headed Petrel (dead on the beach, an unusual summer record), 4 Australian Gannets, 15 Little Pied Cormorants, 5 Australian Pelicans, 1 Pacific Reef-Heron, 1 Caspian Tern, 40 Great Crested Terns, 50 Silver Gulls, 4 Pacific Gulls, 3 Sooty Oystercatchers, 1 Hooded Dotterel, 60 Red-capped Plovers (considered a race of the Snowy Plover by some authors), 2 Black-fronted Dotterels, 3 Pied Stilts, and 75 Red-necked Avocets. In addition, there were two species of visiting Palearctic shorebirds, Curlew Sandpipers and Red-necked Stints. On a freshwater coastal river were four Black Swans and three White-faced Herons.

From Albany we drove north on Jan. 5 to the western edge of Stirling Range National Park, where we slept on a cot next to our car in an open woodland of Wandoo and Powder Bark Wandoo (Eucalyptus accedens) trees. We encountered several roadside Blue-tongued Skinks (*Tiliqua occipitalis*) along the way. A Tawny Frogmouth vocalized at regular intervals throughout the night with its low pigeon-like "oom, oom, oom" notes.

In the vicinity of our campsite the next morning were Purple-crowned Lorikeets, White-tailed Black-Cockatoos, Western Spinebills, Tawny-crowned Honeyeaters, and Red Wattlebirds. Many yellow-flowered Banksias were in bloom, providing nectar for the lorikeets and honeyeaters. The Stirling Range of mountains had two peaks just above 3,500 ft. in elevation, and these mountains were the only place in all of Western Australia that occasionally received light snowfall in the winter. The mountains were best known for their great variety of almost year around wildflowers. Laurie and I only skirted the western edge of the mountains, following secondary roads in several different directions as we proceeded northward toward the town of Gnowangerup. The only new addition to my trip list of birds was the Yellow-plumed Honeyeater*.

Across the southern Australian continent was a broken belt of small eucalypt trees referred to as "mallee", which occurred between the drier mulga vegetation (*Acacia aneura*) to the north and the somewhat wetter eucalypt woodland to the south. Mallee occurred on saline soils where there were hot dry summers and mild wetter winters. The defining characteristic of mallee was the growth form of the small eucalypt trees there, where a clump of slender stems grew from a single underground rootstock. Associated with this habitat were plants and animals found more abundantly here than elsewhere. One such avian species was the much studied Malleefowl (*Leipoa ocellata*), a relatively large megapode that in size and shape reminded me somewhat of a guan. It was cryptically colored in Gray, brown, chestnut, and white.

I very much wanted to see a Malleefowl and I had been told that I should contact Dr. D. L. ("Dom") Serventy, who at that time was living and working for the C.S.I.R.O. (Commonwealth Scientific and Industrial Research Organization) in the town of Gnowangerup, in the drier inland part of southwestern Australia, about 25 miles north of the Stirling Range Nat. Park. He was a 57 years old scientist and ornithologist with an admirable list of achievements, including co-author of *Birds of Western Australia*, past President of the R.A.O.U. (Royal Australian Ornithological Union), and current editor of *The Western Australian Naturalist*, a local periodical journal. Very coincidentally, the name "Gnowangerup" was the aboriginal word meaning "place of the malleefowl." As a person entered this very small town, there was a prominent sign along the roadside with the drawing of a malleefowl! Gnowangerup was situated on the far western edge of the mallee belt.

Laurie and I drove up to Dr. Serventy's house on the afternoon of Jan. 6, and were received very warmly. I explained who I was and that I needed any suggestions he could provide to help me observe a malleefowl. Over a cup of tea, he drew a sketch map of the surrounding area shire roads and marked those where I was most likely to see a malleefowl. He told me that early morning was the best time to look. I thanked him graciously, and we drove a short distance out of town and found an isolated patch of roadside woodland where we camped for the night, sleeping on a blanket on the ground next to our car.

The Malleefowl (family Megapodidae) was a moderately large terrestrial bird, about 2' long and weighing 3-5 pounds. It was a shy, wary, and uncommon bird everywhere but had been extensively studied by scientists and others over many years. Although it roosted in small trees at night, it spent most of the daytime on the ground, where it could run rapidly to escape predators. Malleefowls were omnivorous in their diet, and ate a wide variety of plant and animal matter. However, they were best known for their astounding method of producing young.

Male Malleefowls (which were slightly larger than females) scratched a large depression in sandy soil, averaging about 10' across and 3' deep. They then collected a great quantity of organic matter (sticks, leaves, and the like) with which they filled most of the depression. After sufficient rainfall this "compost" began to rot, a process that released heat. At this time, a female would lay anywhere from 15-30 eggs on top of the compost in an "egg chamber", one at a time every 2 or 3 days. As the egg-laying process began, the male covered the eggs with a huge mound of soil, as wide as 15' in diameter and up to 3' high. The combination of heat from the rotting compost and the hot sand incubated the eggs. Males regularly measured the temperature of the egg chamber with heat sensitive receptors inside their beak, and then removed or added top soil to keep the incubation temperature very constant at about 93°F. It was a time consuming process for the male. The eggs were large and thin-shelled, and required anywhere between 50-100 days to hatch. The young birds hatched one at a time, broke out of their egg, and dug upward through the sandy mound to the ground surface, where they rested motionlessly for about one hour. They were covered with body feathers and almost fully grown wing feathers, so that within 2 hours after hatching they could run rapidly and, remarkably, even flutter off the ground into low lying bushes or trees. Neither parent provided any assistance or care. From the time a "baby" bird hatched it was completely on its own. After only a single day, it could fly strongly. Why did evolution favor such a reptilian-like system of reproduction in a bird?

Happily, Laurie and I were able to briefly observe an adult Malleefowl* the next morning (Jan. 7), as it walked slowly and warily along the edge of a mallee woodlot. Other mallee birds here, which were also trip additions, were the Southern Scrub-Robin*, Chestnut-rumped Heathwren*, Purple-gaped Honeyeater*, and Brush Bronzewing* (a ground-foraging woodland pigeon). These brought my 11-day honeymoon list of trip additions up to 10 species (Dec. 30, 1961 - Jan. 7, 1962, in Appendix B). Viewing the Malleefowl with my new wife was a great way to finish this memorable period. We drove all the way back to Beverley that afternoon.

A favorite summertime retreat for farmers in the Beverley area were the long sandy beaches and bays on the Indian Ocean south of Perth, at the adjoining towns and small communities of Rockingham, Shoalwater Bay, Safety Bay, and Port Kennedy. These beaches were approximately 80 miles from Beverley, less than a 2-hour drive away. Many of the Beverley pastoralists and farmers had a small holiday beach house there, as did some of Laurie's friends and relatives. As a young girl, she stayed there with her family in the Old Rockingham Hotel, which was originally built in the late 1800's, and rebuilt several times since. Laurie and I spent several days in Safety Bay after our honeymoon, prior to our departure for Los Angeles. We visited there with her very close relatives, Dora & Dacre Barrett-Lennard, and their family of four young children, aged between 3 and 10 years, who were holidaying there. Dora was the 18-year younger sister of Laurie's father, and she and her family resided on a large sheep property west of Beverley ("Belhus", which had been in the Barrett-Lennard family for four generations). They were among the Beverley residents who owned a small, simple beach house in Safety Bay. One of their three daughters, Merrie, was 8 years old at the time and volunteered to be my guide to look for a Little Penguin on the small island just offshore, known as "Penguin Island." This island was named for the 50-100 pairs of Little Penguins that nested in the limestone rocks around the edges of the island, and it was easily accessible by a 15 minute ferry boat ride from a pier at Safety Bay. (One could also wade to the island at low tides.)

On January 16 (six days prior to my departure from Western Australia with Laurie), I took the 15-minute ferry boat ride to Penguin Island, with Merrie as my guide. She knew the island from previous visits and promised me she could find a penguin for me. They could be found standing in the back of one of the numerous little crevices or caves in the limestone. Merrie looked diligently, inspecting every possible hidey-hole sometimes by getting on her hands and knees to look inside an entrance. After more then 30 minutes, I was losing some of my hope and beginning to despair of our chances for success, but Merrie never gave up, and sure enough she eventually found one. It was standing quietly about 12 feet back, inside a small cave. The Little Penguin was only 16" tall, and was the smallest of the world's 17 species of penguins (all of which lived only in the southern hemisphere, mostly in Antarctic waters).

The Little Penguin stood motionless facing me, looking directly at me not far away. It looked so cute, harmless, and unaggressive that I decided I would like to pet it, so I got down on my hands and knees and slowly crawled toward it. The penguin watched me intently, but did not move. When I was less than an arm's length away, I gradually reached out my hand to touch it. ZAP! With no warning of any kind, the penguin suddenly lunged with its head toward my outstretched hand, and grasped several of my fingers very securely in its vice-like bite. OUCH! The bite was very painful, and the edges of its mandibles sliced my fingers as would a sharp knife, instantly drawing blood. I jerked my hand back from the grasp of the penguin, and quickly found a clean cloth in my shoulder bag to wrap around my bleeding hand. The moral of this story is that no matter how cute a Little Penguin might look, DON'T PET IT! The penguin never shifted its position in the cave, and continued to look warily directly at me, almost daring me to try again. I didn't. This was the most memorable of my final events in Western Australia, an adventure that had been far beyond my most exciting dreams.

On Tuesday morning, Jan. 22, Laurie, her sister, and I were driven to the passenger pier in Fremantle by her parents in their new white Ford. Everyone was dressed as if we were going to church. Charlie had on a long-sleeved white shirt and a long, narrow, maroon tie, Mary had on wrist length white gloves but was not wearing a hat, Robin was wearing both a wide brimmed blue hat and white gloves, Laurie was wearing a fashionable brown & white wrap-around hat, and holding a pair of white gloves in one hand while carrying a white purse over the same arm, and I was wearing a newly purchased, lightweight, dark blue suit and tie. After all, Laurie and I were going to be boarding the brand new P&O-Orient luxury liner, "Canberra", for its first cruise from England to California, around the southern shore of Australia. As a married man, I was no longer allowed to travel in a "roughly dressed" fashion. Laurie and I boarded the ship and stood alongside other passengers on the side deck, and Laurie waved a tearful goodbye to her family while the ship slowly backed out and away from the pier. The Canberra was a 28,000-ton steamship with twin screws, 700 ft. long with a 30 ft. draft, and rooms for 1,365 passengers. It was the newest, largest, and most modern in the P&O-Orient Line fleet. A new life and adventure were beginning for Laurie and me, very different from anything either of us had ever experienced.

Noble joined us in Auckland (New Zealand), and together the three of us traveled across the Pacific to Hawaii, Vancouver, San Francisco, and finally to Los Angeles, on Feb. 16, 1962, where Noble and I had started out together almost exactly three years and three weeks earlier. Laurie and I temporarily moved in with my parents in Long Beach, and Laurie almost immediately found a job working as a secretary for Hugh Gibbs, a well-known architect. I went to work for Noble for one whole year to help him pay back half his loan, as I had agreed to do when we departed. (Initially my job involved editing his many thousand feet of movie film.)

In his flair for detail, Noble summarized our adventure on a plaque which he gave to me, in appreciation for my contributions. It read, in part: "Roadrunner propelled us to the ends of the earth during our 110,000 mile trek, crossing 6 continents & visiting 83 nations in 1116 days; we broke down 59 times and set two world records: (1) the first vehicle to drive entirely overland to the Panama Canal from the USA, and (2) the only vehicle to negotiate the four continental road extremities on this planet: (a) highest - - 16,000' in Peru, (b) lowest - - minus 1,300' at the Dead Sea, (c) furthest north - - North Cape, Norway, at 71° N. latitude, and (d) furthest south - - Punta Arenas, Chile, at 53° S. latitude." My 3 year trip list of birds totaled almost exactly 3,400 different species, some of them only descriptions in my field notes to which I could not assign a name until many years later. The complete cost for our three years of travel was a little over $3,000 per person, per year. To me, our adventure was best summed up by the popular Australian saying, "It was a good go."

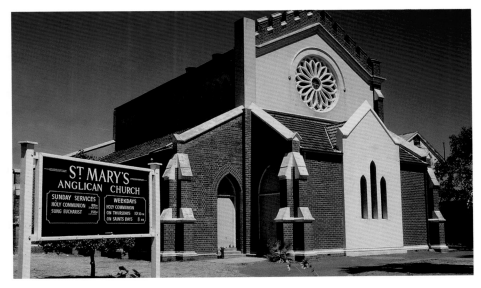

Plate 119: On December 27, 1961, Laurie and I were married in St. Mary's Anglican Church in West Perth. A reception was held afterward in the Esplanade Hotel. On a few occasions during our honeymoon, we camped in Wandoo woodlands.

Plate 120: On the last morning of our honeymoon I was fortunate to locate a Malleefowl just outside a pristine patch of mallee not far from the town of Gnowangerup. I also discovered a Tawny Frogmouth ("mopoke") on a nest.

Plate 121: Our final two weeks in Western Australia were spent on the farm at Riverdale (Dorset sheep are pictured). We played weekend tennis in Beverley, and searched for a Little Penguin with Merrie, Laurie's cousin, on Penguin Island, just offshore in Safety Bay.

Plate 122: On Jan. 22, 1962, we sailed from Perth for Los Angeles on board the P&O-Orient luxury liner, "Canberra". As was the British tradition, Laurie, her parents, and her sister all got dressed up for our departure. We left the land of Waltzing Matilda, and billabongs.

Appendix A: List of all species seen in Central & South America

English Name	1	2	3	4	5	6	7	8	9	10	11
RHEIDAE (2)											
Greater Rhea*									X		
Lesser Rhea*									X		
TINAMIDAE (11)											
Great Tinamou*		X[1]	X[2]								
Little Tinamou*		X[1]	X								
Brown Tinamou*											X
Thicket Tinamou*		X									
Tataupa Tinamou*										X	X
Red-winged Tinamou*									X		
Ornate Tinamou*							X				
Chilean Tinamou*								X			
Andean Tinamou*							X				
Spotted Nothura*									X	X	
Elegant Crested Tinamou*									X		
SPHENISCIDAE (1)											
Humboldt Penguin*							X	X			
PODICIPEDIDAE(6)											
Pied-billed Grebe		X						X	X		
White-tufted Grebe*							X	X	X		
Short-winged (Titicaca) Grebe*							X				
Great Grebe*								X	X		
Eared Grebe	X										
Silvery Grebe*								X	X		
DIOMEDEIDAE (3)											
Wandering Albatross*								X			
Black-browed Albatross*								X	X		
Shy Albatross*								X			
PROCELLARIIDAE (8)											
Antarctic Giant Petrel*								X	X		
Southern Fulmar*								X	X		
Cape Petrel*								X			
Antarctic Prion*								X			

	1	2	3	4	5	6	7	8	9	10	11
Slender-billed Prion*								X			
White-chinned Petrel*								X			
Sooty shearwater							X				
Audubon's Shearwater				X							
HYDROBATIDAE (1)											
Wilson's Storm-Petrel								X			
PELECANOIDIDAE (1)											
Peruvian Diving-Petrel*								X			
PELECANIDAE (2)											
Peruvian Pelican*							X	X			
Brown Pelican			X[2]	X	X	X	X				
SULIDAE (4)											
Peruvian Booby*							X	X			
Masked Booby				X							
Red-footed Booby				X							
Brown Booby	X										X
PHALACROCORACIDAE (6)											
Double-crested Cormorant	X										
Neotropic Cormorant		X	X		X	X	X[3]	X	X		X
Rock Shag*								X	X		
Guanay Cormorant*							X	X			
Imperial Shag*								X	X		
Red-legged Cormorant*							X	X			
ANHINGIDAE (1)											
Anhinga							X[3]				
FREGATIDAE (1)											
Magnificent Frigatebird	X	X	X	X	X	X	X				X
ARDEIDAE (16)											
Whistling Heron*									X		
Capped Heron*							X[3]				
Great Blue Heron	X	X									
Cocoi Heron*						X		X	X		
Great Egret		X	X		X	X	X[3]	X	X		
Reddish Egret	X										
Tricolored Heron	X				X		X				
Little Blue Heron	X	X[1]	X	X	X	X					

434

	1	2	3	4	5	6	7	8	9	10	11
Snowy Egret	X	X	X		X	X		X	X		
Cattle Egret				X	X						
Striated Heron*					X		X				
Green Heron	X	X[1]	X[2]								
Black-crowned Night-Heron							X	X	X		
Fasciated Tiger-Heron*							X[3]				
Rufescent Tiger-Heron*					X						
Stripe-backed Bittern*									X		
CICONIIDAE (3)											
Wood Stork	X	X	X				X[3]		X		
Maguari Stork*									X		
Jabiru*									X		
THRESKIORNITHIDAE (6)											
Andean Ibis*						X					
Black-faced Ibis*								X			
Bare-faced Ibis*					X				X		
White Ibis	X					X					
White-faced Ibis	X								X		
Puna Ibis*							X				
PHOENICOPTERIDAE (1)											
Chilean Flamingo*							X	X	X		
ANHIMIDAE (1)											
Southern Screamer*									X		
ANATIDAE (27)											
Fulvous Whistling-Duck						X			X		
Black-necked Swan*								X	X		
Coscoroba Swan*									X		
Upland Goose*								X	X		
Kelp goose*									X		
Flightless Steamerduck*								X			
White-headed Steamerduck*									X		
Flying Steamerduck*								X	X		
Ringed Teal*									X		
Brazilian Teal*									X		
Chiloe Wigeon*									X		
Speckled Teal*						X	X	X	X		
Spectacled Duck*								X			

	1	2	3	4	5	6	7	8	9	10	11
Crested Duck*							X	X	X		
Yellow-billed Pintail*							X	X	X		
Puna Teal*							X	X			
Silver Teal*									X		
Blue-winged Teal						X					
Cinnamon Teal								X			
Red Shoveler*									X		
Rosy-billed Pochard*									X		
Redhead	X										
Lesser Scaup	X	X									
Bufflehead	X										
Red-breasted Merganser	X										
Andean Duck*							X				
Lake Duck*									X		
CATHARTIDAE (5)											
Black Vulture	X	X[1]	X[2]	X	X	X	X[3]	X	X	X	X
Turkey Vulture	X	X[1]	X[2]	X	X	X	X	X			X
Lesser Yellow-headed Vulture*									X		
Andean Condor*						X	X	X			
King Vulture*				X							
PANDIONIDAE (1)											
Osprey		X	X				X				
ACCIPITRIDAE (21)											
Swallow-tailed Kite*			X	X			X[3]				
White-tailed Kite				X	X				X	X	X
Snail Kite*				X	X	X			X		
Long-winged Harrier*									X		
Cinereous Harrier*						X	X	X	X		
Northern Harrier	X										
Sharp-shinned Hawk	X										
Cooper's Hawk	X										
White Hawk*			X				X[3]				
Common Black-Hawk	X	X									
Great Black-Hawk*		X[1]	X								
Savanna Hawk*						X			X	X	
Harris' Hawk	X		X		X	X		X			
Black-collared Hawk*									X		

	1	2	3	4	5	6	7	8	9	10	11
Black-chested Buzzard-Eagle*							X	X	X		
Roadside Hawk*	X	X[1]	X	X	X	X	X[3]		X		X
Broad-winged Hawk		X									
Gray Hawk	X				X						
"Variable" Hawk*						X	X	X	X		
Red-tailed Hawk	X	X									
Black Hawk-Eagle*			X								X
FALCONIDE (15)											
Black Caracara*							X[3]				
Red-throated Caracara*								X[3]			
Carunculated Caracara*				X	X						
Mountain Caracara*							X	X			
White-throated Caracara*									X		
Crested Caracara	X	X		X	X	X	X				
Southern Caracara*								X	X	X	X
Yellow-headed Caracara*			X	X	X					X	X
Chimango Caracara*								X	X		
Laughing Falcon*		X[1]	X			X			X		
Collared Forest-Falcon*		X[1]									
American Kestrel	X	X[1]		X	X		X	X	X		X
Aplomado Falcon*							X	X			X
Bat Falcon*						X	X[3]				
Peregrine Falcon		X					X	X	X		
CRACIDAE (5)											
Plain Chachalaca		X[1]									
Rufous-vented Chachalaca*				X							
Rufous-bellied Chachalaca*	X										
Crested Guan*		X[1]	X[2]								
Dusky-legged Guan*											X
ODONTOPHORIDAE (2)											
Buffy-crowned Wood-Partridge*		X									
Marbled Wood-Quail*		X									
ARAMIDAE (1)											
Limpkin									X		
RALLIDAE (11)											
Gray-necked Wood-Rail*							X[3]				

	1	2	3	4	5	6	7	8	9	10	11
Giant Wood-Rail*									X		
Blackish Rail*					X						
Plumbeous Rail*							X				
Purple Gallinule					X						
Common Moorhen							X		X		
American Coot	X	X									
White-winged Coot*								X	X		
Slate-colored Coot*							X				
Red-gartered Coot*								X			
Red-fronted Coot*								X			
JACANIDAE (2)											
Northern Jacana*		X	X								
Wattled Jacana*					X	X			X		
HAEMATOPODIDAE (2)											
Magellanic Oystercatcher*									X		
Blackish Oystercatcher*								X			
RECURVIROSTRIDAE (4)											
Black-necked Stilt	X				X						
White-backed Stilt*							X		X		
American Avocet	X										
Andean Avocet*								X			
BURHINIDAE (1)											
Double-striped Thick-knee*				X							
CHARADRIIDAE (11)											
Southern Lapwing*					X			X	X	X	X
Andean Lapwing*						X	X	X			
Black-bellied Plover	X										
Wilson's Plover	X										
Killdeer	X	X[1]					X				
Snowy Plover								X			
Collared Plover*							X[3]		X		
Puna Plover*							X	X			
Two-banded Plover*								X	X		
Rufous-chested Dotterel*								X			
Tawny-throated Dotterel*								X	X		
SCOLOPACIDAE (15)											
South American Snipe*								X	X		

	1	2	3	4	5	6	7	8	9	10	11
Puna Snipe*							X				
Short-billed Dowitcher	X										
Marbled Godwit	X										
Whimbrel	X		X			X		X			
Spotted Sandpiper	X	X	X		X	X	X[3]				
Solitary Sandpiper		X[1]			X						
Greater Yellowlegs	X						X	X	X		
Willet	X							X			
Lesser Yellowlegs		X					X				
Ruddy Turnstone							X	X			
Sanderling							X				
Western Sandpiper	X										
Pectoral Sandpiper							X				
Wilson's Phalarope							X				
THINOCORIDAE (2)											
Gray-breasted Seedsnipe*								X			
Least Seedsnipe*									X		
LARIDAE (13)											
Dolphin Gull*									X		
Belcher's Gull*							X	X			
Gray Gull*							X	X			
Heermann's Gull	X										
Ring-billed Gull	X										
Kelp Gull*							X	X	X		X
Am. Herring Gull	X										
Gray-headed Gull*						X	X				
Brown-hooded Gull*								X	X		
Bonaparte's Gull	X										
Andean Gull*						X	X	X			
Laughing Gull	X		X	X	X	X					
Franklin's Gull						X	X				
STERNIDAE (7)											
Caspian Tern	X		X								
Inca Tern*							X	X			
South American Tern*								X	X		X
Forster's Tern	X										
Royal Tern	X				X				X		X

	1	2	3	4	5	6	7	8	9	10	11
Sandwich Tern			X								X
Elegant Tern	X						X				
RYNCHOPIDAE (1)											
Black Skimmer							X	X			
STERCORARIIDAE (3)											
Chilean Skua*								X			
Pomarine Jaeger				X							
Parasitic Jaeger								X			X
COLUMBIDAE (27)											
Picazuro Pigeon*									X		X
Spot-winged Pigeon*							X				
Band-tailed Pigeon			X								
Pale-vented Pigeon*							X[3]				
Red-billed Pigeon	X										
Plumbeous Pigeon*											X
Short-billed Pigeon*		X[1]	X[2]								
Mourning Dove	X										
Eared Dove*				X	X	X	X	X	X		
White-winged Dove	X										
Pacific Dove*							X	X			
Common Ground-Dove	X	X	X	X	X	X					
Plain-breasted Ground-Dove*		X	X				X				
Ecuadorian Ground-Dove*						X					
Ruddy Ground-Dove*		X[1]	X	X	X						X
Picui Ground-Dove*								X	X	X	
Croaking Ground-Dove*						X	X				
Inca Dove	X	X									
Blue Ground-Dove*			X								
Bare-faced Ground-Dove*							X	X			
Black-winged Ground-Dove*						X	X	X			
Golden-spotted Ground-Dove*								X			
White-tipped Dove		X	X	X	X	X	X				
Gray-headed Dove*		X[1]									
Gray-chested Dove*			X[2]								
Violaceous Quail-Dove*			X[2]								
Ruddy Quail-Dove*		X[1]									
PSITTACIDAE (34)											

	1	2	3	4	5	6	7	8	9	10	11
Blue-and-yellow Macaw*							X[3]				
Military Macaw*	X										
Scarlet Macaw*			X								
Chestnut-fronted Macaw*						X	X[3]				
Blue-winged Macaw*											
Blue-crowned Parakeet*				X							
Green Parakeet*		X									
Scarlet-fronted Parakeet*					X		X				
Dusky-headed Parakeet*							X[3]				
Brown-throated Parakeet*			X								
Orange-fronted Parakeet*	X										
Nanday Parakeet*										X	
Burrowing Parrot*									X		
Maroon-bellied Parakeet*										X	X
Austral Parakeet*									X		
Slender-billed Parakeet*								X			
Monk Parakeet*									X		
Andean Parakeet*							X				
Mountain Parakeet*							X	X			
Green-rumped Parrotlet*				X							
Blue-winged Parrotlet*											X
Spectacled Parrotlet*					X						
Pacific Parrotlet*						X	X				
Gray-cheeked Parakeet*						X					
Orange-chinned Parakeet*		X	X[2]								
Brown-hooded Parrot*		X[1]	X								
Blue-headed Parrot*			X[2]								
Scaly-headed Parrot*										X	
White-crowned Parrot*		X[1]									
White-fronted Parrot*	X	X[1]									
Red-lored Parrot*		X[1]									
Blue-fronted Parrot*									X		
Yellow-crowned Parrot*				X							
Mealy Parrot*		X[1]	X	X							
CUCULIDAE (11)											
Dwarf Cuckoo*						X					
Yellow-billed Cuckoo						X					

	1	2	3	4	5	6	7	8	9	10	11
Dark-billed Cuckoo*					X						
Squirrel Cuckoo*		X	X[2]	X	X		X[3]				X
Smooth-billed Ani			X	X	X	X				X	X
Groove-billed Ani	X	X			X	X	X	X			
Guira Cuckoo*									X	X	X
Striped Cuckoo*			X								X
(Lesser Ground-Cuckoo*)		X									
Greater Roadrunner	X										
Lesser Roadrunner*	X										
STRIGIDAE (9)											
(Guatemalan Screech-Owl*)		X									
(Vermiculated Screech-Owl*)			X								
(Long-tufted Screech-Owl*)											X
Great Horned Owl	X	X					X				
(Northern Pygmy-Owl)	X										
Ferruginous Pygmy-Owl*		X								X	X
Austral Pygmy-Owl*								X	X		
Burrowing Owl								X	X	X	X
Short-eared Owl					X			X	X		
CAPRIMULGIDAE (6)											
Lesser/Common Nighthawk	X	X	X	X	X		X				
Pauraque*	X	X[1]	X	X	X	X	X[3]				X
(Whip-poor-will)	X										
(Dusky Nightjar*)			X								
Band-winged Nightjar*							X				
Little Nightjar*									X		
APODIDAE (12)											
Sooty Swift*											X
Great Dusky Swift*											X
Chestnut-collared Swift*				X			X[3]				
White-collared Swift*		X	X	X	X	X	X				
Band-rumped Swift*			X								
Gray-rumped Swift*											X
Chimney Swift						X					
Vaux's Swift		X[1]									
Short-tailed Swift*						X					
White-throated Swift		X									

442

	1	2	3	4	5	6	7	8	9	10	11
Andean Swift*							X	X			
Lesser Swallow-tailed Swift*		X[1]		X							
TROCHILIDAE (63)											
White-tipped Sicklebill*			X								
Buff-tailed Sicklebill*							X[3]				
Bronzy Hermit*			X								
White-whiskered Hermit*						X					
Green Hermit*					X						
W. Long-tailed Hermit*			X				X[3]				
Scale-throated Hermit*										X	
Planalto Hermit*											X
Sooty-capped Hermit*				X							
Little Hermit*			X								
Wedge-tailed Sabrewing*		X[1]									
Rufous Sabrewing*		X									
Lazuline Sabrewing*				X							
Swallow-tailed Hummingbird*											X
White-necked Jacobin*			X[2]			X					
Black Jacobin*											X
Green Violet-ear*	X	X					X				
Sparkling Violet-ear*					X		X				
White-vented Violet-ear*										X	X
Black-throated Mango*			X[2]		X						X
Plovercrest*										X	X
Golden-crowned Emerald*	X										
Garden Emerald*			X								
Glittering-bellied Emerald*									X	X	X
Short-tailed Emerald*				X							
Broad-billed Hummingbird	X										
Violet-crowned Woodnymph*			X[2]								
Fork-tailed Woodnymph*										X	
Violet-capped Woodnymph*											X
Sapphire-throated Hummingbird*			X[2]								
White-eared Hummingbird	X										
White-throated Hummingbird*											X
White-bellied Hummingbird*							X				

	1	2	3	4	5	6	7	8	9	10	11
Rufous-tailed Hummingbird*		X[1]	X[2]			X					
Amazilia Hummingbird*						X	X				
Cinnamon Hummingbird*		X									
White-bellied Emerald*		X[1]									
Glittering-throated Emerald*				X							
Snowy-bellied Hummingbird*			X[2]								
Berylline Hummingbird*	X										
Blossomcrown*					X						
White-vented Plumeleteer*				X	X						
Bronze-tailed Plumeleteer*			X[2]								
Brazilian Ruby*											X
Chimborazo Hillstar*						X					
Collared Inca*							X				
Giant Hummingbird*							X	X			
Green-backed Firecrown*								X			
Sapphire-vented Puffleg*						X					
Black-tailed Trainbearer*					X	X					
Green-tailed Trainbearer*					X		X				
Purple-backed Thornbill*					X						
Tyrian Metaltail*					X						
Long-tailed Sylph*					X						
Plain-capped Starthroat*		X									
Oasis Hummingbird*								X			
Amethyst Woodstar*											X
Costa's Hummingbird	X										
Purple-collared Woodstar*							X				
White-bellied Woodstar*							X				
Rufous-shafted Woodstar*				X							
Volcano Hummingbird*			X								
Broad-tailed Hummingbird		X									
TROGONIDAE (9)											
Black-headed Trogon*		X[1]									
Baird's Trogon*			X								
Violaceous Trogon*		X[1]	X[2]								
Collared Trogon*		X[1]		X		X					
Elegant Trogon		X[1]									
Black-throated Trogon*			X[2]							X	X

	1	2	3	4	5	6	7	8	9	10	11
Surucua Trogon*											X
Blue-crowned Trogon*										X	X
Slaty-tailed Trogon*		X[1]	X[2]								
ALCEDINIDAE (4)											
Belted Kingfisher	X		X								
Ringed Kingfisher*			X				X[3]		X		X
Amazon Kingfisher*			X				X[3]		X		X
Green Kingfisher*			X				X[3]				
MOMOTIDAE (5)											
Russet-crowned Motmot*	X										
Blue-crowned Motmot*		X[1]	X								
Highland Motmot*							X				
Rufous Motmot*			X[2]								
Turquoise-browed Motmot*		X									
GALBULIDAE (2)											
Rufous-tailed Jacamar*		X[1]	X								
Bluish-fronted Jacamar*							X[3]				
BUCCONIDAE (5)											
White-necked Puffbird*				X							
White-eared Puffbird*											X
White-whiskered Puffbird*			X[2]								
Black-fronted Nunbird*							X[3]				
Swallow-wing*				X			X[3]				
RAMPHASTIDAE (9)											
Chestnut-eared Aracari*										X	
Collared Aracari*		X[1]				X					
Fiery-billed Aracari			X								
Red-breasted Toucan*										X	X
Keel-billed Toucan*		X[1]									
Channel-billed Toucan*				X							
Toco Toucan*										X	X
Red-billed (Cuvier's) Toucan*							X[3]				
Black/Chestnut-mandibled Toucan*		X[2]	X		X						
PICIDAE (38)											
Scaled Piculet*				X							
White-barred Piculet*											X

	1	2	3	4	5	6	7	8	9	10	11
Ochre-collared Piculet*											X
White Woodpecker*										X	
Acorn Woodpecker					X						
Black-cheeked Woodpecker*		X[1]	X[2]								
Golden-naped Woodpecker*			X								
Yellow-tufted Woodpecker*							X[3]				
Yellow-fronted Woodpecker*											X
Red-crowned Woodpecker*			X	X							
Gila Woodpecker	X										
Golden-fronted Woodpecker	X	X[1]									
Yellow-bellied Sapsucker		X[1]									
Striped Woodpecker*								X			
Checkered Woodpecker*									X		
Ladder-backed Woodpecker	X										
Strickland's Woodpecker*	X										
Scarlet-backed Woodpecker*						X					
Smoky-brown Woodpecker*		X[1]		X							
Little Woodpecker*							X[3]				
White-spotted Woodpecker*										X	X
Yellow-browed Woodpecker*											X
Golden-olive Woodpecker*		X[1]									
Black-necked Woodpecker*							X				
Green-barred Woodpecker*									X		X
Northern Flicker	X										
Gilded Flicker	X										
Chilean Flicker*								X	X		
Andean Flicker*							X	X			
Campo Flicker*									X	X	X
Pale-crested Woodpecker*									X		
Blond-crested Woodpecker*										X	X
Helmeted Woodpecker*											X
Lineated Woodpecker*		X[1]	X	X			X[3]				X
Robust Woodpecker*										X	X
Pale-billed Woodpecker*		X[1]									
Crimson-crested Woodpecker*			X[2]				X[3]				
Magellanic Woodpecker*								X			
FURNARIIDAE (78)											

446

	1	2	3	4	5	6	7	8	9	10	11
Common Miner*							X	X	X		
Grayish Miner*								X			
Coastal Miner*							X				
Scale-throated Earthcreeper*								X	X		
Plain-breasted Earthcreeper*							X				
Striated Earthcreeper*							X				
Band-tailed Earthcreeper*									X		
Crag Chilia*								X			
Bar-winged Cinclodes*						X	X	X			
Dark-bellied Cinclodes*								X			
Peruvian Seaside Cinclodes*							X				
Chilean Seaside Cinclodes*								X			
White-winged Cinclodes*							X	X			
Pale-legged Hornero*						X	X				
Rufous Hornero*									X	X	X
Wren-like Rushbird*							X				
Thorn-tailed Rayadito*								X			
Tufted Tit-Spinetail*									X		
Plain-mantled Tit-Spinetail*								X	X		
Andean Tit-Spinetail*						X					
Araucaria Tit-Spinetail*											X
Des Murs' Wiretail*								X			
Itatiaia Thistletail*											X
Chotoy Spinetail*									X		
Rufous-capped Spinetail*										X	X
Gray-bellied Spinetail*										X	X
Sooty-fronted Spinetail*									X		
Azara's Spinetail*							X				
Pale-breasted Spinetail*			X	X	X				X		
Chicli Spinetail*											
Slaty Spinetail*					X						
Cabanis' Spinetail*							X[3]				
Necklaced Spinetail*							X				
Stripe-crowned Spinetail*									X		
Cordilleran Canastero*								X			
Streak-backed Canastero*						X					
Hudson's Canastero*									X		

	1	2	3	4	5	6	7	8	9	10	11
Lesser Canastero*								X			
Dusky-tailed Canastero*								X			
Short-billed Canastero*									X		
Patagonian Canastero*									X		
Common Thornbird*				X							
Little Thornbird*									X		
Greater Thornbird*									X		
Firewood-gatherer*									X		
Lark-like Brushrunner*									X		
White-throated Cacholote*									X		
White-browed Foliage-gleaner*											X
Buff-browed Foliage-gleaner*										X	X
Striped Woodhaunter*			X								
Ochre-breasted Foliage-gleaner*										X	X
Black-capped Foliage-gleaner*											X
Buff-fronted Foliage-gleaner*				X							X
Buff-throated Foliage-gleaner*			X								
Scaly-throated Leaftosser*			X^2								
Sharp-tailed Streamcreeper*											X
Sharp-billed Treehunter*										X	X
Plain Xenops*		X^1	X^2								X
Streaked Xenops*										X	X
White-throated Treerunner*								X			
Plain-brown Woodcreeper*				X							
Thrush-like Woodcreeper*										X	X
Tawny-winged Woodcreeper*		X^1									
Ruddy Woodcreeper*		X^1	X								
Long-tailed Woodcreeper*			X^2								
Olivaceous Woodcreeper*		X^1		X					X		X
Wedge-billed Woodcreeper*			X			X					
Strong-billed Woodcreeper*				X							
White-throated Woodcreeper*										X	X
Planalto Woodcreeper*											X
Lesser Woodcreeper*										X	X
Cocoa Woodcreeper*			X^2								
Ivory-billed Woodcreeper*		X^1									

	1	2	3	4	5	6	7	8	9	10	11
Black-striped Woodcreeper*			X^2								
Narrow-billed Woodcreeper*									X		
Streak-headed Woodcreeper*			X				X				
Scaled Woodcreeper*											X
Red-billed Scythebill*							X^3		X		
THAMNOPHILIDAE (26)											
Great Antshrike*			X				X^3		X		
Barred Antshrike*			X	X			X^3				
Chestnut-backed Antshrike*							X^3				
Rufous-winged Antshrike*											X
Rufous-capped Antshrike*											X
Black-hooded Antshrike*			X								
Western Slaty-Antshrike*			X^2								
Variable Antshrike*									X	X	X
Plain Antvireo*		X^1	X							X	X
Spot-crowned Antvireo*			X^2								
Pacific Antwren*						X					
Checker-throated Antwren*			X^2								
White-flanked Antwren*			X^2								
Rufous-winged Antwren*										X	
Dot-winged Antwren*			X^2				X^3				
Ferruginous Antbird*											X
Scaled Antbird*											X
Streak-capped Antwren*										X	X
Dusky Antbird*		X^3	X								
White-browed Antbird*							X^3				
Warbling Antbird*							X^3				
Spot-winged Antbird*							X^3				
White-lined Antbird*							X^3				
Chestnut-backed Antbird*			X^2			X					
Spotted Antbird*			X^2								
Scale-backed Antbird*							X^3				
FORMICARIIDAE (1)											
Short-tailed Antthrush*											X
CONOPOPHAGIDAE (1)											
Rufous Gnateater*										X	X
RHINOCRYPTIDAE (3)											

	1	2	3	4	5	6	7	8	9	10	11
Black-throated Huet-huet*								X			
Moustached Turca*								X			
(Magellanic Tapaculo*)								X			
COTINGIDAE (8)											
White-tipped Plantcutter*									X		
Rufous-tailed Plantcutter*								X			
Rufous Piha*			X								
Plum-throated Cotinga*							X[3]				
Purple-throated Fruitcrow*			X[2]								
Red-ruffed Fruitcrow*											X
Amazonian Umbrellabird*							X[3]				
Long-wattled Umbrellabird*						X					
PIPRIDAE (9)											
Red-capped Manakin*		X[1]	X[2]								
Blue-crowned Manakin*			X								
Helmeted Manakin*										X	
Blue Manakin*										X	X
Pin-tailed Manakin*											X
Serra Tyrant-Manakin*											X
Wing-barred Piprites*										X	X
Thrush-like Schiffornis*		X[1]									
Greenish Schiffornis*											X
TYRANNIDAE (117)											
Yellow-crowned Tyrannulet*			X								
Yellow-bellied Elaenia*			X		X						
White-crested Elaenia*							X	X			
Lesser Elaenia*					X						X
Sierran Elaenia*							X				
Yellow-bellied Tyrannulet*		X[1]									
Northern Beardless-Tyrannulet		X[1]									
Southern Beardless-Tyran-nulet*			X				X				
White-throated Tyrannulet*					X	X					
Yellow-billed Tit-Tyrant*								X			
Tufted Tit-Tyrant*					X	X	X		X		
River Tyrannulet*							X[3]				
Sooty Tyrannulet*									X		

450

	1	2	3	4	5	6	7	8	9	10	11
White-crested Tyrannulet*									X		
White-bellied Tyrannulet*									X		
Mouse-colored Tyrannulet*					X		X				
Yellow Tyrannulet*										X	X
Sharp-tailed Tyrant*									X		
Southern Antpipit*										X	X
Paltry Tyrannulet*		X	X								
Mottle-cheeked Tyrannulet*										X	X
Bay-ringed Tyrannulet*									X		
Southern Bristle-Tyrant*										X	X
Sepia-capped Flycatcher*										X	X
Slaty-capped Flycatcher*				X							
Ochre-bellied Flycatcher*			X								
Gray-hooded Flycatcher*											X
Ornate Flycatcher*						X					
Many-colored Rush-Tyrant*							X	X	X		
Eared Pygmy-Tyrant*										X	X
Southern Bentbill*			X^2								
Eye-ringed Tody-Tyrant*											X
Hangnest Tody-Tyrant*											X
Ochre-faced Tody-Flycatcher*										X	
Black-backed Tody-Flycatcher*							X^3				
Common Tody-Flycatcher*			X	X	X						
Eye-ringed Flatbill*		X^1	X								
Yellow-olive Flycatcher*		X									X
Yellow-margined Flycatcher*			X^2			X					
Stub-tailed Spadebill*		X^1									
White-throated Spadebill*											X
Golden-crowned Spadebill*			X^2								
Bran-colored Flycatcher*			X			X	X^3				X
Tawny-breasted Flycatcher*						X					
Sulphur-rumped Flycatcher*		X^1	X								
Ruddy-tailed Flycatcher*		X^1	X^2								
Cinnamon Flycatcher*							X				
Euler's Flycatcher*						X				X	
Yellow-bellied Flycatcher		X^1									
Buff-breasted Flycatcher		X									

	1	2	3	4	5	6	7	8	9	10	11
Greater Pewee	X	X									
Tropical Pewee*											X
Tufted Flycatcher	X										
Black Phoebe	X		X	X		X	X[3]				
Say's Phoebe	X										
Vermilion Flycatcher	X			X	X	X	X[3]		X		
Austral Negrito*								X			
Andean Negrito*							X	X			
Spectacled Tyrant*									X		
Cinereous Tyrant*									X		
Blue-billed Black-Tyrant*											X
White-winged Black-Tyrant*							X				
Velvety Black-Tyrant*											X
Rufous-breasted Chat-Tyrant*					X						
D'Orbigny's Chat-Tyrant*							X	X			
White-browed Chat-Tyrant*							X				
Yellow-browed Tyrant*									X		
Pied Water-Tyrant*					X						
Strange-tailed Tyrant*									X		
Fire-eyed Diucon*								X			
Black-crowned Monjita*									X		
White Monjita*									X		
Black-billed Shrike-Tyrant*						X	X	X			
Great Shrike-Tyrant*									X		
Plain-capped Ground-Tyrant*						X					
Taczanowski's Ground-Tyrant*							X				
Cinereous Ground-Tyrant*							X				
Rufous-naped Ground-Tyrant*							X				
Dark-faced Ground-Tyrant*								X			
Shear-tailed Gray Tyrant*											X
Long-tailed Tyrant*							X			X	X
Short-tailed Field-Tyrant*							X				
Cattle Tyrant*									X	X	
Rusty-margined Flycatcher*			X		X						
Social Flycatcher*	X	X[1]	X[2]	X	X	X	X[3]				X
Gray-capped Flycatcher*			X								
Great Kiskadee	X	X[1]		X	X		X[3]		X	X	X

452

	1	2	3	4	5	6	7	8	9	10	11
Baird's Flycatcher*						X	X				
Streaked Flycatcher*			X²								
Boat-billed Flycatcher*		X	X²								X
Variegated Flycatcher*				X							
Snowy-throated Kingbird*							X				
Tropical Kingbird	X	X¹	X²	X	X		X³				
Cassin's Kingbird	X										
Thick-billed Kingbird*	X										
Western Kingbird	X	X									
Gray Kingbird			X		X						
Scissor-tailed Flycatcher	X	X									
Fork-tailed Flycatcher*			X	X	X						
Sirystes*										X	X
Dusky-capped Flycatcher	X		X								
Swainson's Flycatcher*									X		
Panama Flycatcher*			X²								
Sooty-crowned Flycatcher*						X					
Great Crested Flycatcher			X²								
Brown-crested Flycatcher		X¹									
Gray-hooded Attila*											X
Bright-rumped Attila*		X¹									
Masked Tityra*		X¹	X²								
Black-crowned Tityra*											X
Green-backed Becard*									X		X
Chestnut-crowned Becard*											X
Cinnamon Becard*		X¹	X								
White-winged Becard*			X								
Black-and-white Becard*				X							
Rose-throated Becard		X¹									
One-colored Becard*						X					
ALAUDIDAE (1)											
Horned Lark	X										
HIRUNDINIDAE (17)											
Tree Swallow	X										
Violet-green Swallow		X									
Mangrove Swallow*	X	X	X								
White-winged Swallow*											X

	1	2	3	4	5	6	7	8	9	10	11
White-rumped Swallow*									X		X
Chilean Swallow*								X			
Gray-breasted Martin*		X	X			X				X	X
Peruvian Martin*							X				
Brown-bellied Swallow*					X	X	X				
Blue-and-white Swallow*			X	X	X	X	X[3]	X			X
Black-capped Swallow*	X	X									
Andean Swallow*							X				
White-banded Swallow*							X[3]				
N. Rough-winged Swallow	X	X	X								
S. Rough-winged Swallow*			X	X	X		X[3]				X
Barn Swallow	X	X	X	X	X	X	X[3]				
Chestnut-collared Swallow*						X	X				
MOTACILLIDAE (4)											
Yellowish Pipit*							X				
Correndera Pipit*							X		X		
Paramo Pipit*						X	X				
American Pipit	X										
REGULIDAE (2)											
Golden-crowned Kinglet	X										
Ruby-crowned Kinglet	X	X									
PTILOGONATIDAE (2)											
Gray Silky-flycatcher		X									
Phainopepla	X										
CINCLIDAE (1)											
White-capped Dipper*							X				
TROGLODYTIDAE (23)											
Band-backed Wren*		X									
Rufous-naped Wren*		X									
Cactus Wren	X										
Bicolored Wren*					X						
Stripe-backed Wren*				X							
Fasciated Wren*						X					
Rock Wren		X									
Canyon Wren	X										
Inca Wren*							X				
Moustached Wren*							X[3]				

	1	2	3	4	5	6	7	8	9	10	11
Spot-breasted Wren*		X[1]									
Rufous-breasted Wren*			X								
Riverside Wren*			X								
Bay Wren*						X					
Carolina Wren		X[1]									
Plain Wren*		X	X								
Superciliated Wren*							X				
House Wren	X	X	X[2]	X	X	X	X	X	X	X	X
Rufous-browed Wren*		X									
Ochraceous Wren*			X								
Sedge Wren						X			X		
White-bellied Wren*		X[1]									
White-breasted Wood-Wren*		X[1]									
MIMIDAE (9)											
Gray Catbird		X[1]									
Northern Mockingbird	X										
Tropical Mockingbird*	X			X	X						
Chalk-browed Mockingbird*					.				X		X
Patagonian Mockingbird*									X		
White-banded Mockingbird*									X		
Long-tailed Mockingbird*						X	X				
Chilean Mockingbird*								X			
Curve-billed Thrasher	X										
TURDIDAE (23)											
Eastern Bluebird	X	X									
Brown-backed Solitaire	X										
Orange-billed Nightin-gale-Thrush*				X							
Russet Nightingale-Thrush*	X										
Gray-cheeked Thrush		X									
Swainson's Thrush					X						
Hermit Thrush	X										
Wood Thrush		X[1]									
Yellow-legged Thrush*											X
Chiguanco Thrush*							X	X			
Sooty Robin*			X								
Great Thrush*				X	X	X					

455

	1	2	3	4	5	6	7	8	9	10	11
Glossy-black Thrush*					X	X	X				
Rufous-bellied Thrush*									X	X	X
Austral Thrush*								X	X		
Pale-breasted Thrush*				X	X					X	X
Creamy-bellied Thrush*									X		
Cocoa Thrush*				X	X						
Clay-colored Robin*	X	X[1]	X[2]								
Bare-eyed Thrush*				X							
White-throated Thrush*		X[1]	X								
White-necked Thrush*											X
American Robin	X										
POLIOPTILIDAE (5)											
Long-billed Gnatwren*		X[1]									
White-lored Gnatcatcher*	X										
Tropical Gnatcatcher*			X			X	X				
Creamy-bellied Gnatcatcher*											X
Masked Gnatcatcher*									X		
AEGITHALIDAE (1)											
Bushtit	X										
PARIDAE (1)											
Mexican Chickadee	X										
SITTIDAE (2)											
Pygmy Nuthatch	X										
White-breasted Nuthatch	X										
CERTHIIDAE (1)											
Brown Creeper	X	X									
LANIIDAE (1)											
Loggerhead Shrike	X										
CORVIDAE (13)											
Steller's Jay	X	X									
Black-throated Magpie-Jay	X										
White-throated Magpie-Jay*	X	X									
Black-chested Jay*				X							
Green Jay				X							
Brown Jay*		X[1]									
Purplish-backed Jay*	X										
Purplish Jay*										X	

	1	2	3	4	5	6	7	8	9	10	11
Violaceous Jay*							X[3]				
Plush-crested Jay*									X	X	X
Sinaloa Crow*	X										
Chihuahuan Raven	X										
Common Raven	X	X									
VIREONIDAE (12)											
White-eyed Vireo		X[1]									
Yellow-throated Vireo		X	X								
Blue-headed Vireo		X									
Yellow-winged Vireo*			X								
Brown-capped Vireo*					X		X				
Red-eyed Vireo				X	X						X
Yellow-green Vireo		X[1]	X[2]								
Rufous-crowned Greenlet*										X	X
Tawny-crowned Greenlet*		X[1]									
Lesser Greenlet*		X[1]	X[2]								
Green Shrike-Vireo*		X[1]									
Rufous-browed Peppershrike*				X					X		X
FRINGILLIDAE (20)											
Scrub Euphonia*	X										
Purple-throated Euphonia*											X
Thick-billed Euphonia*					X						
Yellow-throated Euphonia*		X[1]									
Spot-crowned Euphonia*			X								
Olive-backed Euphonia*		X[1]	X								
Orange-bellied Euphonia*						X	X[3]				
Chestnut-bellied Euphonia*											X
Blue-naped Chlorophonia*											X
House Finch	X										
Red Crossbill	X										
Pine Siskin	X										
Black-capped Siskin*		X									
Black-headed Siskin*		X									
Andean Siskin*					X						
Hooded Siskin*							X		X		
Lesser Goldfinch	X	X			X	X					
Black-chinned Siskin*								X	X		

	1	2	3	4	5	6	7	8	9	10	11
Black Siskin*							X	X			
Yellow-rumped Siskin*							X				
PEUCEDRAMIDAE (1)											
Olive Warbler	X	X								.	
PARULIDAE (46)											
Tennessee Warbler	.	X	X^2								
Orange-crowned Warbler	X										
Nashville Warbler	X										
Virginia's Warbler	X										
Flame-throated Warbler*			X								
Tropical Parula			X	X					X		X
Yellow Warbler	X	X^1	X^2	X	X						
Chestnut-sided Warbler			X^2								
Magnolia Warbler	.	X^1									
Yellow-rumped Warbler	X	X^1									
Black-throated Gray Warbler	X										
Black-throated Green Warbler		X^1									
Townsend's Warbler	X	X					.				
Hermit Warbler		X									
Blackburnian Warbler					X						
Cerulean Warbler				X							
Black-and-white Warbler		X^1	X^2								
American Redstart		X^1		X	X					.	
Prothonotary Warbler			X^2								
Worm-eating Warbler		X^1									
Ovenbird	X	X^1									
Northern Waterthrush					X						
Louisiana Waterthrush		X^1	X^2								
Kentucky Warbler		X^1									
Mourning Warbler			X		X						
MacGillivray's Warbler	X	X									
Common Yellowthroat	X	X									
Masked Yellowthroat*							X			X	X
Gray-crowned Yellowthroat		X									
Hooded Warbler		X^1									
Wilson's Warbler		X^1	X								
Canada Warbler					X						

	1	2	3	4	5	6	7	8	9	10	11
Red Warbler*	X										
Pink-headed Warbler*		X									
Painted Redstart	X										
Slate-throated Redstart*	X			X			X				
Spectacled Redstart*							X				
Golden-fronted Redstart*					X						
Black-crested Warbler*					X						
Golden-crowned Warbler*		X¹								X	X
Rufous-capped Warbler	X	X									
Golden-browed Warbler*	X										
Three-striped Warbler*				X	X						
White-rimmed Warbler*									X	X	X
Buff-rumped Warbler*			X			X	X				
Yellow-breasted Chat		X¹									
COEREBIDAE (1)											
Bananaquit*			X²	X	X		X³				X
THRAUPIDAE (65)											
Chestnut-vented Conebill*											X
Cinereous Conebill*					X	X	X	X			
Brown Tanager*											X
Cinnamon Tanager*											X
Black-faced Tanager*					X						
Magpie Tanager*							X³			X	X
Sooty-capped Bush-Tanager*			X								
Yellow-throated Bush-Tanager*						X					
Chestnut-headed Tanager*											X
Fulvous-headed Tanager*				X							
Rust-and-yellow Tanager*							X				
Guira Tanager*										X	X
Rosy Thrush-Tanager*			X								
Olive-green Tanager*											X
Gray-headed Tanager*		X¹	X²								
White-throated Shrike-Tanager*			X								
White-shouldered Tanager*			X²								
Tawny-crested Tanager*						X					
Ruby-crowned Tanager*										X	X

	1	2	3	4	5	6	7	8	9	10	11
White-lined Tanager*				X	X						
Black-goggled Tanager*										X	X
Red-crowned Ant-Tanager*										X	X
Red-throated Ant-Tanager*		X¹									
Black-cheeked Ant-Tanager*			X								
Hepatic Tanager	X				X						X
Summer Tanager		X¹	X²	X							
Flame-colored Tanager*			X								
Western Tanager		X									
Masked Crimson Tanager*							X³				
Crimson-backed Tanager*				X	X						
Silver-beaked Tanager*				X			X³				
Brazilian Tanager*											X
Cherrie's Tanager*			X								
Flame-rumped Tanager*			X²		X	X					
Blue-Gray Tanager*	X		X²	X	X	X	X				
Sayaca Tanager*									X	X	X
Golden-chevroned Tanager*											X
Blue-capped Tanager*							X				
Blue-and-yellow Tanager*						X	X				
Yellow-winged Tanager*		X¹									
Palm Tanager*			X²	X	X		X³				X
Lacrimose Mountain-Tanager*					X						
Scarlet-bellied Mt.-Tanager*					X						
Diademed Tanager*											X
Fawn-breasted Tanager*							X				X
Plain-colored Tanager*			X²								
Paradise Tanager*							X³				
Green-headed Tanager*										X	X
Brassy-breasted Tanager*											X
Golden Tanager*					X						
Silver-throated Tanager*			X								
Saffron-crowned Tanager*							X				
Bay-headed Tanager*			X			X					
Burnished-buff Tanager*					X						X
Scrub Tanager*						X					
Blue-necked Tanager*				X	X	X					

	1	2	3	4	5	6	7	8	9	10	11
Golden-hooded Tanager*			X^2								
Beryl-spangled Tanager*							X				
Yellow-bellied Dacnis*							X^3				
Scarlet-thighed Dacnis*			X								
Blue Dacnis*							X^3				X
Green Honeycreeper*		X^1	X^2								
Shining Honeycreeper*			X^2								
Red-legged Honeycreeper*		X^1	X^2	X							
Swallow-Tanager*											X
EMBERIZIDAE (82)											
Red-crested Finch*									X	X	X
Crimson-breasted Finch*						X					
Black-hooded Sierra-Finch*								X			
Peruvian Sierra-Finch*							X				
Gray-hooded Sierra-Finch*								X	X		
Patagonian Sierra-Finch*								X	X		
Mourning Sierra-Finch*							X	X	X		
Plumbeous Sierra-Finch*						X			X		
Band-tailed Sierra-Finch*								X			
Ash-breasted Sierra-Finch*						X	X	X			
Canary-winged Finch*								X			
Yellow-bridled Finch*									X		
White-winged Diuca-Finch*							X				
Common Diuca-Finch*								X			
Cinereous Finch*							X				
Bay-chested Warbling-Finch*											X
Black-&-rufous War-bling-Finch*									X		
Red-rumped Warbling-Finch*											X
Collared Warbling-Finch*							X				
Black-capped Warbling-Finch*									X		
Blue-black Grassquit*	X		X^2	X	X	X	X				X
Gray Seedeater*					X						
Variable Seedeater*			X			X					
White-collared Seedeater		X									
Black-and-white Seedeater*							X				
Yellow-bellied Seedeater*			X	X							

461

	1	2	3	4	5	6	7	8	9	10	11
Drab Seedeater*							X				
Parrot-billed Seedeater*							X				
Ruddy-breasted Seedeater*		X		X	X						
Chestnut-throated Seedeater*						X	X				
Chestnut-bellied Seed-Finch*				X	X		X³				
Thick-billed Seed-Finch*			X								
Blackish-blue Seedeater*											X
Band-tailed Seedeater*						X	X				
Plain-colored Seedeater*						X					
Yellow-faced Grassquit			X								
Black-faced Grassquit*				X	X						
Rusty Flowerpiercer*				X				X			
Black Flowerpiercer*				X							
Black-throated Flowerpiercer*								X			
Puna Yellow-Finch*							X	X			
Saffron Finch*				X		X	X		X		
Grassland Yellow-Finch*					X		X	X	X		
Bright-rumped Yellow-Finch*							X	X			
Greenish Yellow-Finch*							X				
Wedge-tailed Grass-Finch*									X		
Great Pampa-Finch*									X		X
Yellow Cardinal*									X		
Red-crested Cardinal*									X	X	
Sooty-faced Finch*			X								
Large-footed Finch*			X								
White-naped Brush-Finch*			X								
Pale-naped Brush-Finch*					X						
Yellow-breasted Brush-Finch*						X					
Rufous-capped Brush-Finch*	X										
Rusty-bellied Brush-Finch*							X				
Ochre-breasted Brush-Finch*				X	X						
Chestnut-capped Brush-Finch*				X							
Stripe-headed Brush-Finch*				X							
Orange-billed Sparrow*			X			X					
Saffron-billed Sparrow*	·								X	X	
Green-backed Sparrow*		X¹									
Black-striped Sparrow*			X²	X	X						

	1	2	3	4	5	6	7	8	9	10	11
Prevost's Ground-Sparrow*		X									
Green-tailed Towhee	X										
Spotted Towhee	X										
Canyon Towhee	X										
Stripe-headed Sparrow*		X									
Tumbes Sparrow*							X				
Rusty Sparrow*		X									
Striped Sparrow*	X										
Chipping Sparrow	X										
Lark Sparrow	X										
Lark Bunting	X										
Grasshopper Sparrow		X			X						
Grassland Sparrow*					X				X		X
Yellow-browed Sparrow*							X³				
Lincoln's Sparrow	X										
White-crowned Sparrow	X										
Rufous-collared Sparrow*	X	X	X	X	X	X	X	X	X	X	X
Yellow-eyed Junco	X										
Volcano Junco*			X								
CARDINALIDAE (23)											
Streaked Saltator*			X	X	X		X				
Grayish Saltator*		X		X			X³		X		
Buff-throated Saltator*			X		X						
Black-headed Saltator*		X									
Slate-colored Grosbeak*							X³				
Black-throated Grosbeak*											X
Green-winged Saltator*											X
Golden-billed Saltator*							X		X		
Thick-billed Saltator*											X
Black-faced Grosbeak*		X¹									
Northern Cardinal	X										
Pyrrhuloxia	X										
Golden-bellied Grosbeak*						X	X				
Black-thighed Grosbeak*			X								
Black-backed Grosbeak*							X				
Rose-breasted Grosbeak		X									
Ultramarine Grosbeak*					X				X		

	1	2	3	4	5	6	7	8	9	10	11
Blue-black Grosbeak*			X[2]								
Blue Grosbeak		X									
Indigo Bunting	X	X[1]									
Orange-breasted Bunting*	X										
Painted Bunting		X									
Dickcissel		X									
ICTERIDAE (45)											
Saffron-cowled Blackbird*									X		
Red-winged Blackbird		X									
Yellow-hooded Blackbird*					X						
Chestnut-capped Blackbird*									X		
Unicolored Blackbird*									X		
Yellow-winged Blackbird*							X	X	X		
Red-breasted Blackbird*				X	X						
White-browed Blackbird*									X		
Peruvian Meadowlark*						X	X				
Long-tailed Meadowlark*								X	X		
Eastern Meadowlark	X				X						
Melodious Blackbird*		X[1]									
Scrub Blackbird*						X	X				
Brewer's Blackbird	X										
Great-tailed Grackle	X	X	X		X						
Nicaraguan Grackle*		X									
Carib Grackle*				X							
Bay-winged Cowbird*									X		
Shiny Cowbird*				X	X		X	X	X		X
Bronzed Cowbird		X									
Giant Cowbird*					X		X[3]				
Yellow-backed Oriole*	X	X[1]		X	X						
Yellow Oriole*				X							
Yellow-tailed Oriole*						X					
Altamira Oriole	X	X									
Hooded Oriole	X										
Baltimore Oriole	X	X[1]	X	X							
Orchard Oriole		X[1]									
Black-cowled Oriole*	X	X[1]									
Bar-winged Oriole*		X									

	1	2	3	4	5	6	7	8	9	10	11
Yellow-rumped Cacique*			X^2	X		X					
Red-rumped Cacique*										X	X
Scarlet-rumped Cacique*			X								
Golden-winged Cacique*									X		X
Solitary Cacique*									X		X
Russet-backed Oropendola*							X^3				
Chestnut-headed Oropendola*			X^2			X					
Crested Oropendola*					X		X^3				
Montezuma Oropendola*		X^1									
Amazonian Oropendola*							X^3				
Oriole Blackbird*				X							
Yellow-rumped Marshbird*										X	X
Brown-and-yellow Marshbird*									X		
Scarlet-headed Blackbird*									X		
Austral Blackbird*								X	X		
COLUMN TOTALS	178	219	214	120	139	125	257	156	195	85	192

KEY

The numbered columns indicate geographical regions as follows: 1: Mexico; 2: Guatemala to Nicaragua (inclusive); 3: Costa Rica & Panama; 4: Venezuela; 5: Colombia; 6: Ecuador; 7: Peru; 8: Bolivia & Chile; 9: Argentina; 10: Paraguay; 11: Brazil

X: indicates a species was seen in this location; X^1 (in column 2) indicates the species was seen at Tikal, Guatemala, Feb. 16-18; X^2 (in column 3) indicates the species was seen on Barro Colorado Island, Panama, Mar. 9-11; X^3 (in column 7) indicates the species was seen on the upper Madre de Dios River, Peru, May 9-11.

* indicates a species new for my "life list"

( ) indicates the species was only heard and not seen

The classification and English names follow that of Clements, *Checklist of Birds of the World*, 6th edition, 2007

Total species recorded: 1,101 (including 854 "life birds")

APPENDIX B
List Of Species From Africa, Europe, Asia, & Australia

SOUTH ATLANTIC OCEAN

9/3/1959	Cory's Shearwater
	Soft-plumaged Petrel*
9/5/1959	Great-winged Petrel*
9/7/1959	Yellow-nosed Albatross*
	White-headed Petrel*
9/8/1959	Greater Shearwater
	White-bellied Storm-Petrel
9/9/1959	Brown Skua*
9/10/1959	Little Shearwater*
	Antarctic Tern*

AFRICA

SOUTH AFRICA

9/11/1959	Cape Gannet*
	Bank Cormorant*
	Hartlaub's Gull*
	Cape Sparrow*
9/12/1959	Cape Francolin*
	Crowned Lapwing*
	Cape Wagtail*
	Rock Martin*
	Laughing Dove*
	Ring-necked Dove*
	Fiscal Shrike*
	Cape Thrush*
	Red-winged Starling*
	Black-winged Stilt*
	Black-shouldered Kite*
	Hammerkop*
	Black Kite
	Alpine Swift*
	Eurasian Hoopoe*
	African Black Duck*
	Eurasian Kestrel
	Piping Cisticola*
	Familiar Chat*
	Cape Canary*
	Jackal Buzzard*
	Karroo Prinia*
	Victorin's Scrub -Warbler*

Ground Woodpecker*
Cape Siskin*
White-winged Seedeater*
Cape Rock-Thrush*
African Swift*
Orange-breasted Sunbird*
Malachite Sunbird*
Cape Sugarbird*

9/13/1959	Larger Striped Swallow*

Cape Weaver*
Cape Robin-Chat*
Black-headed Heron*
Cape Flycatcher*
Cape White-eye*
Reed Cormorant*
Speckled Mousebird*
Pied Crow*
Red-chested Cuckoo*
African Sand Martin*
Cape Widowbird*
White-necked Raven*
Rufous Rockjumper*
Boubou Shrike*
Bokmakierie*
Little Grebe
Red-knobbed Coot*
Maccoa Duck*
Great Cormorant
Little Egret
Egyptian Goose*
Cape Shoveler*
Yellow-billed Duck*
Water Thick-knee*
Red-eyed Dove*
Malachite Kingfisher*
Lesser Swamp-Warbler*
Tinkling Cisticola*
African Dusky Flycatcher*
Cape Longclaw*
African Pied Starling*

9/15/1959	Cape Teal*		Speckled Pigeon*
	Great Crested Tern*		Karroo Chat*
	White-fronted Plover*		African Pipit*
	Cape Bulbul*		Pin-tailed Whydah*
	Red-headed Cisticola*		African Snipe*
	Red-faced Mousebird*		Pale-winged Starling*
	S. Double-collared Sunbird*		Black-headed Canary*
9/16/1959	Cape Cormorant*		Grey Tit*
	Rufous-chested Sparrowhawk*		Mountain Wheatear*
	Southern Pochard*		Karroo Lark*
	African Marsh-Harrier*		S. Anteater-Chat*
	Grey-winged Francolin*		Namaqua Prinia*
	African Stonechat*		Lark-like Bunting*
	Grey Heron		Cape Lark*
	Red-billed Duck*		Black Crow*
	Greater Flamingo*		Stanley's Bustard*
	Lesser Flamingo*		Namaqua Sandgrouse*
	Blacksmith Plover*		Common Quail*
	Kittlitz's Plover*		Black Bustard*
	Three-banded Plover*		Spike-heeled Lark*
	Great White Pelican*	9/21/1959	Sacred Ibis*
9/19/1959	Namaqua Dove*		Karroo Bustard*
	Yellow Canary*		Black-eared Sparrow-Lark*
	White-backed Mousebird*		Yellow-bellied Eremomela*
	Red-capped Lark*		Tractrac Chat*
	Cape Bunting*		Greenshank
	Karroo Scrub-Robin*		Chat Flycatcher*
	Rufous-vented Warbler*		Ferruginous Lark*
	White-throated Seedeater*		Yellow-rumped Eremomela*
	Red Bishop*		S. Masked-Weaver*
	Pearl-breasted Swallow*		Little Stint*
	White-rumped Swift*		Curlew Sandpiper*
	White-throated Swallow*		S. African Swallow*
	Capped Wheatear*		Little Swift*
	Pale Chanting-Goshawk*	9/22/1959	Black-fronted Bulbul*
	Grey-backed Sparrow-Lark*		Cape Crombec*
	Sicklewing Chat*		Rufous-eared Warbler*
	Pied Avocet*		Pririt Batis*
	S. African Shelduck*		Yellow-tufted Pipit*
	Thick-billed Lark*		Karroo Long-billed Lark*
9/20/1959	Fiscal Flycatcher*	9/23/1959	Sombre Greenbul*
	Fairy Flycatcher*		Forest Canary*
	Layard's Warbler*		Rameron Pigeon*

Brimstone Canary*
Bar-throated Apalis*
Greater Double-collared Sunbird*
Black Sawwing*
African Paradise-Flycatcher*
Darter*
African Goshawk*
(Knysna Turaco*)
Common Waxbill*
Streaky-headed Seedeater*
Fork-tailed Drongo*
Olive Woodpecker*
Lemon Dove*
Plain-backed Pipit*
Banded Martin*
Baillon's Crake*
Wood Sandpiper
Purple Heron*
African Rail*
Great Crested Grebe
Pied Kingfisher
Zitting Cisticola

9/24/1959
African Oystercatcher*
Black Sunbird*
African Fish-Eagle*
Grey Cuckoo-shrike*
Terrestrial Brownbul*
Yellow-throated Wood-Warbler*
Afr. Black-headed Oriole*
Scaly-throated Honeyguide*
Green-backed Camaroptera*
Green Woodhoopoe*

9/25/1959
Chorister Robin-Chat*
Narina Trogon*
Eurasian Buzzard
Spur-winged Goose*
Cape Glossy-Starling*
Fiery-necked Nightjar*

9/26/1959
Southern Black-Tit*
Yellow-fronted Canary*
Southern Tchagra*
Black-backed Puffback*
Yellow-breasted Apalis*

Collared Sunbird*
Emerald-spotted Wood-Dove*
Knysna Woodpecker*
Common Bulbul*
Buff-streaked Bushchat*
Hadada Ibis*
Red-collared Widowbird*
Long-tailed Widowbird*
White Stork*
Helmeted Guineafowl*

9/27/1959
Giant Kingfisher*
Common Sandpiper
African Pied Wagtail*
Brown-hooded Kingfisher*
Red-fronted Tinkerbird*
Village Weaver*
Forest Weaver*
Red-backed Scrub-Robin*
Black-collared Barbet*
Tawny-flanked Prinia*
Lesser Striped-Swallow*
Southern Black-Flycatcher*
Greater Honeyguide*
Chinspot Batis*
Cape Griffon*
Secretary-bird*
Grey Crowned-Crane*

9/28/1959
Yellow-throated Petronia*
Golden-breasted Bunting*
Pallid Harrier*
Comb Duck*

9/29/1959
Gurney's Sugarbird*
Cape Grassbird*
Red-winged Francolin*
Wailing Cisticola*
African Yellow Warbler*
Rock-loving Cisticola*
Lanner Falcon*

9/30/1959
Sentinel Rock-Thrush*
Martial Eagle*
Lammergeier*
Verreaux's Eagle*
Orange-breasted Rockjumper*

	Yellow-breasted Pipit*		Shikra*
	Blue Swallow*		Little Bee-eater*
10/1/1959	Blue Crane*		S. Yellow-billed Hornbill*
10/2/1959	White-browed Coucal*		Crowned Hornbill*
	Rattling Cisticola*		Pale Flycatcher*
	Yellow-throated Longclaw*		Olive Bushshrike*
	Rufous-naped Lark*		Magpie nShrike*
	Fan-tailed Widowbird*		White Helmetshrike*
	Bronze Mannikin*		Red-billed Oxpecker*
10/3/1959	Winding Cisticola*		Burchell's Glossy-Starling*
	Blue-breasted Cordonbleu*		Cinnamon-breasted Bunting*
	Rufous-chested Swallow*	10/6/1959	Green-winged Pytilia*
	White-backed Vulture*		Crested Francolin*
	Bateleur*		Yellow-bellied Bulbul*
	White-breasted Sunbird*		Purple-banded Sunbird*
	Lesser Masked-Weaver*		Dideric Cuckoo*
10/4/1959	Black-bellied Glossy-Starling*		Cardinal Woodpecker*
	Crested Barbet*		Square-tailed Drongo*
	Trumpeter Hornbill*		Pink-throated Twinspot*
	Scarlet-chested Sunbird*		Yellow-rumped Tinkerbird*
	Spectacled Weaver*		Sulphur-breasted Bushshrike*
	Striped Kingfisher*		African Penduline-Tit*
	Flappet Lark*		Jameson's Firefinch*
	Common Scimitar-bill*		Yellow-billed Stork*
	Black Cuckoo-shrike*		African Openbill*
	White-throated Robin*		Square-tailed Nightjar*
	Red-capped Robin-Chat*	10/7/1959	Brubru*
	Black-crowned Tchagra*		African Yellow White-eye*
	Lesser Honeyguide*		African Golden-Weaver*
	Purple-crested Turaco*		House Martin
	Black-bellied Bustard*		Yellow-spotted Nicator*
	Crowned Eagle*		Four-colored Bushshrike*
	Spotted Eagle-Owl*		Goliath Heron*
	Ashy Flycatcher*		Intermediate Egret
	African Hawk-Eagle*		Squacco Heron*
	Lilac-breasted Roller*		Black Heron*
	Lappet-faced Vulture*		Glossy Ibis*
	Tawny Eagle*		African Spoonbill*
	African Firefinch*		African Pygmy-goose*
	Black Goshawk*		Hottentot Teal*
	White-headed Vulture*		White-faced Whistling-Duck*
	Golden-tailed Woodpecker*		White-backed Duck*
10/5/1959	Marabou Stork*		Black Crake*

African Jacana*
Lesser Jacana*
Purple Swamphen*
Marsh Sandpiper*
Broad-billed Roller*
Arrow-marked Babbler*

SWAZILAND

10/8/1959
Wahlberg's Eagle*
Grey Go-away-bird*
S. White-faced Owl*
Bearded Scrub-Robin*
African Broadbill*
Bush Pipit*
Three-streaked Tchagra*
Neergaard's Sunbird*
Black-and-white Mannikin*

MOZAMBIQUE

10/9/1959
African Palm-Swift*
Senegal Lapwing*
Kurrichane Thrush*
Klaas' Cuckoo*

10/10/1959
Lizard Buzzard*
Black Cuckoo*
Small Buttonquail*
Gabar Goshawk*

SOUTH AFRICA
Wire-tailed Swallow*
Brown-headed Parrot*
Swainson's Francolin*
Ostrich*
Rufous-crowned Roller*

10/11/1959
Red-billed Hornbill*
African Grey Hornbill*
Coqui Francolin*
Red-billed Buffalo-Weaver*
Long-tailed Starling*
Wattled Starling*
Southern Ground-Hornbill*
Grey-headed Sparrow*
White-crowned Shrike*
Grey-rumped Swallow*
Mariqua Sunbird*

10/12/1959
Saddle-billed Stork*

Natal Francolin*
White-fronted Bee-eater*
Greater Blue-eared Glossy-Starling*
Red-billed Firefinch*
Fawn-colored Lark*
Hooded Vulture*

10/13/1959
Bearded Woodpecker*
Red-headed Weaver*
Retz's Helmetshrike*
African Green-Pigeon*
Grey-headed Bushshrike*
Verreaux's Eagle-Owl*
Common Cuckoo
Grey-headed Kingfisher*
(Freckled Nightjar*)
(African Scops-Owl*)

10/14/1959
White-headed Black-Chat*
White-headed Lapwing*
White-browed Robin-Chat*
Red-faced Cisticola*
Crested Guineafowl*
African Finfoot*
Brown Snake-Eagle*
Black-throated Wattle-eye*

10/15/1959
Marsh Owl*
Great Rufous Sparrow*
Zebra Waxbill*
Scaly Weaver*
Black-throated Canary*
White-browed Sparrow-Weaver*
Temminck's Courser*

10/16/1959
Orange River Francolin*
Spotted Thick-knee*
White-bellied Bustard*
White-quilled Bustard*
Latakoo Lark*
Cloud Cisticola*
Long-billed Pipit*
African Quailfinch*

10/25/1959
Ruff/Reeve
Pied Barbet*
Rufous-necked Wryneck*
Kalahari Scrub-Robin*

470

	Black-chested Prinia*		Southern Hyliota*
10/26/1959	Greater Kestrel*		Yellow-billed Oxpecker*
10/27/1959	Pied Cuckoo*		Eurasian Hobby*
	Red-headed Finch*		Bradfield's Hornbill*
	Lesser Kestrel*	11/3/1959	Yellow Wagtail
	Chestnut-backed Sparrow-Lark*		Red-crested Bustard*
	Violet-eared Waxbill*		European Roller*
	BECHUANALAND		Eurasian Golden Oriole*
10/27/1959	Social Weaver*		Lesser Grey Shrike*
	European Bee-eater*		Red-billed Francolin*
	Crimson-breasted Gonolek*	11/4/1959	Pearl-spotted Owlet*
	Southern Pied-Babbler*		Senegal Coucal*
	Mariqua Flycatcher*		Bronze-winged Courser*
10/28/1959	Green-backed Camaroptera*	11/5/1959	Rufous-bellied Heron*
	Dusky Lark*		African Reed-Warbler*
	Barred Wren-Warbler*	11/6/1959	Red-backed Shrike*
	Sabota Lark*		Copper Sunbird*
10/29/1959	Meyer's Parrot*		Collared Palm-Thrush*
	Three-banded Courser*		Half-collared Kingfisher*
	Dark Chanting-Goshawk*		Rock Pratincole*
	Groundscraper Thrush*		Garden Warbler
	Black-cheeked Waxbill*		Orange-winged Pytilia*
	Black-breasted Snake-Eagle*		Red-billed Quelea*
	Burchell's Sandgrouse*		Little Sparrowhawk*
10/30/1959	Bank Swallow		Booted Eagle*
	Violet-backed Starling*	11/7/1959	Grosbeak Weaver*
	Abdim's Stork*		NORTHERN RHODESIA
	Swallow-tailed Bee-eater*		Pennant-winged Nightjar*
	Double-banded Sandgrouse*	11/8/1959	Grey Tit-Flycatcher*
	SOUTHERN RHODESIA		Pale-billed Hornbill*
11/1/1959	Willow Warbler		Thrush Nightingale
	Harlequin Quail*		Peter's Twinspot*
	Wattled Lapwing*		Rufous-bellied Tit*
	Holub's Golden-Weaver*		Greencap Eremomela*
	Yellow-fronted Tinkerbird*		SOUTHERN RHODESIA
	Spotted Flycatcher	11/9/1959	White-winged Widowbird*
	Burnt-neck Eremomela*		Mosque Swallow*
	African Harrier-Hawk*	11/10/1959	Horus Swift*
	Croaking Cisticola*	11/11/1959	Miombo Wren-Warbler*
	Eastern Paradise-Whydah*		Cabanis' Bunting*
	Southern Carmine Bee-eater*		Shelly's Francolin*
11/2/1959	African Golden Oriole*	11/14/1959	W. Violet-backed Sunbird*
	Black-eared Seedeater*	11/15/1959	Spotted Creeper*

Miombo Rock-Thrush*
White-breasted Cuckoo-shrike*
African Pygmy-Kingfisher*

11/17/1959 Rosy-throated Longclaw*
Locustfinch*
Buffy Pipit*
Pectoral-patch Cisticola*
Parasitic Weaver*
Wing-snapping Cisticola*
Yellow-shouldered Widowbird*
Montagu's Harrier*
Red-necked Francolin*
African Scrub-Warbler*

11/18/1959 Robert's Prinia*
Variable Sunbird*
Yellow-bellied Waxbill*
Wahlberg's Honeyguide*
Wattled Crane*
Bronze Sunbird*

11/19/1959 Stripe-cheeked Greenbul*
White-starred Robin*
White-tailed Crested-Flycatcher*
Black-headed Apalis*

11/20/1959 Long-crested Eagle*
Red-faced Crombec*
Singing Cisticola*
Western Olive Sunb ird*

11/21/1959 African Wood-Owl*
Whyte's Barbet*
MOZAMBIQUE
Silvery-cheeked Hornbill*
African Barred Owlet*

11/22/1959 Tiny Greenbul*
Dickinson's Kestrel*

11/23/1959 Lesser Blue-eared Glossy-Starling*
Red-winged Prinia*
Moustached Grass-Warbler*

11/24/1959 African Mourning Dove*
African Bush-Warbler*
Blue-cheeked Bee-eater*
Banded Snake-Eagle*

11/25/1959 Collared Pratincole*

NYASALAND

11/28/1959 Tambourine Dove*
African Citril*
White-eared Barbet*
Mountain Wagtail*
Bertram's Weaver*
Fischer's Greenbul*

11/29/1959 Cinnamon Bracken-Warbler*
E. Double-collared Sunbird*
Little Greenbul*
Augur Buzzard*

12/1/1959 Red-capped Crombec*
White-headed Sawwing*
Siffling Cisticola*

12/2/1959 Green-backed Woodpecker*
Marsh Warbler*
Souza's Shrike*

12/3/1959 Blackcap*
Eastern Mountain-Greenbul*
Cameroon Scrub -Warbler*
Mountain Yellow Warbler*
Chapin's Apalis*
Olive-flanked Robin-Chat*
Black-lored Cisticola*
White-chested Alethe*
Churring Cisticola*
Amur Falcon*

12/4/1959 Woodland Kingfisher*
Black-winged Lapwing*
Northern Wheatear
TANGANYIKA

12/5/1959 Stout Cisticola*
Baglafecht Weaver*
Dusky Turtle-Dove*
Black Coucal*
Red-cheeked Cordenbleu*
Fan-tailed Grassbird*
Racket-tailed Roller*
Gray Kestrel*
Ashy Starling*
Slate-colored Boubou*
White-rumped Shrike*
Fischer's Sparrow-Lark*

12/6/1959	D'Arnaud's Barbet*		Livingstone's Turaco*
	African Grey Flycatcher*		Yellow-streaked Bulbul*
	Spot-flanked Barbet*		Uluguru Violet-backed Sunbird*
	Grey Wren-Warbler*	12/19/1959	Red-headed Quelea*
	Purple Grenadier*		African Hobby*
	Mocking Cliff-Chat*		Long-tailed Fiscal*
	African Crested-Flycatcher*	12/20/1959	Eurasian Reed-Warbler
12/7/1959	Hildebrandt's Francolin*		Basra Reed-Warbler*
	Brown-breasted Barbet*		Lesser Seedcracker*
12/8/1959	Greater Sandplover*	12/21/1959	Kretschmer's Longbill*
	Sooty Gull*		Chestnut-fronted Helmetshrike*
	Lesser Crested Tern*	12/24/1959	Eurasian River Warbler*
	Crab Plover*	12/25/1959	Yellow-bellied Hyliota*
	Eurasian Curlew		Mottled Spinetail*
	Lesser Black-backed Gull	12/30/1959	Little Bittern*
	Bar-tailed Godwit		Zanzibar Bishop*
	Palm-nut Vulture*	1/3/1960	Von der Decken's Hornbill*
	Golden Pipit*	1/4/1960	Grasshopper B uzzard*
	Rufous Chatterer*		Rufous-tailed Scrub-Robin*
	Northern Brownbul*	1/5/1960	Waller's Starling*
12/10/1959	Zanzibar Puffback Shrike*		Red-headed Bluebill*
	Yellowbill*		Fischer's Turaco*
12/11/1959	Fasciated Snake-Eagle*		Green-headed Oriole*
12/12/1959	Black-and-white Shrike-flycatcher*		African Tailorbird*
12/13/1959	Bare-faced Go-away-bird*		Pale-breasted Illadopsis*
	Caspian Plover*		Sharpe's Akalat*
12/15/1959	Slender-billed Starling*	1/6/1960	Bar-tailed Trogon*
	Blue-spotted Wood-Dove*		Long-billed Tailorbird*
	Common Nightingale		Northern Carmine Bee-eater*
	Striped Pipit*		White-throated Bee-eater*
	Red-faced Crimson-wing*		KENYA
	Barred Long-tailed Cuckoo*	1/9/1960	Dodson's Bulbul*
	Shelley's Greenbul*		Pangani Longclaw*
	Green Barbet*		Isabelline Wheatear*
	Black-fronted Bushshrike*		African Bare-eyed Thrush*
12/16/1959	Eastern Olivaceous Warbler*		Golden Palm Weaver*
	Spotted Morning-Warbler*		Black-necked Weaver*
	Brown-necked Parrot*		Mouse-colored Sunbird*
	Eurasian Nightjar*		Eastern Chanting-Goshawk*
12/17/1959	Green Tinkerbird*		Chestnut Weaver*
	Livingstone's Flycatcher*		White-bellied Go-away-bird*
12/18/1959	Boehm's Bee-eater*		Golden-breasted Starling*
	Great Reed-Warbler		Fischer's Starling*

	Tiny Cisticola*		Buff-bellied Warbler*
1/10/1960	Scaly Chatterer*		Rosy-patched Shrike*
	Black-throated Barbet*		Red-throated Tit*
	Pale Prinia*		Hildebrandt's Starling*
	Pink-breasted Lark*		Grey-headed Social-Weaver*
	E. Yellow-billed Hornbill*		Chestnut Sparrow*
	Crimson-rumped Waxbill*		Speckle-fronted Weaver*
	White-breasted White-eye*		Blue-capped Cordonbleu*
	Yellow-spotted Petronia*		Straw-tailed Whydah*
	Rufous-tailed Rock-Thrush*		TANGANYIKA
	Pygmy Falcon*	1/18/1960	Hunter's Cisticola*
	White-headed Buffalo-Weaver*		Desert Cisticola*
	White-throated Robin*		White-bellied Canary*
	Black-headed Batis*		Yellow-collared Lovebird*
	Yellow-necked Francolin*		Speke's Weaver*
	Superb Starling*	1/19/1960	Streaky Seedeater*
	Taita Fiscal*		Jackson's Widowbird*
1/11/1960	Violet Woodhoopoe*		Northern Anteater-Chat*
	Red-bellied Parrot*		Yellow-throated Sandgrouse*
	Parrot-billed Sparrow*		Chestnut-bellied Sandgrouse*
	Rueppell's Glossy-Starling*	1/20/1960	Rufous-tailed Weaver*
	Black-bellied Sunbird*		Saker Falcon*
	Purple Indigobird*	1/22/1960	Lesser Spotted Eagle*
	Nubian Woodpecker*		Fischer's Lovebird*
1/12/1960	Abyssinian Scimitar-bill*		Woolly-necked Stork*
	White-bellied Tit*		Greater Painted-snipe*
	Red-and-yellow Barbet*		Great Spotted Cuckoo*
	White-headed Barbet*		Black Bishop*
1/13/1960	Corn Crake*	1/23/1960	Swahili Sparrow*
	Bat-like Spinetail*		White-tailed Lark*
	Black-faced Sandgrouse*		Grey-breasted Spurfowl*
	Black Stork*	1/24/1960	Schalow's Wheatear*
	White-backed Night-Heron*	1/25/1960	Tacazze Sunbird*
1/16/1960	Kenya Rufous Sparrow*		Scaly Francolin*
	Pied Wheatear*		Cinnamon-chested Bee-eater*
1/17/1960	Short-tailed Lark*		Grey-headed Negrofinch*
	Kori Bustard*		Abyssinian Crimson-wing*
	Golden-winged Sunbird*		Thick-billed Seedeater*
	Banded Warbler*	1/28/1960	Vulturine Guineafowl*
	Double-banded Courser*	1/30/1960	Golden-backed Weaver*
	Blue-naped Mousebird*		Cut-throat*
	Red-fronted Barbet*	1/31/1960	Red-fronted Warbler*
	Grey Woodpecker*		Mouse-colored Penduline-Tit*

	Black-capped Social-Weaver*		Yellow-whiskered Greenbul*
	Taveta Golden-Weaver*		Chestnut Wattle-eye*
	Somali Bunting*		Blue-shouldered Robin-Chat*
2/1/1960	Ovampo Sparrowhawk*		Equatorial Akalat*
	White-headed Mousebird*		White-chinned Prinia*
	Tsavo Sunbird*		Banded Prinia*
	Grey-headed Silverbill*		Black-faced Rufous-Warbler*
	Southern Grosbeak-Canary*		Purple-throated Cuckoo-shrike*
2/2/1960	Donaldson-Smith's Nightjar*		Velvet-mantled Drongo*
	Pringle's Puffback*		Grey-green Bushshrike*
	Broad-ringed White-eye*		Lueder's Bushshrike*
2/3/1960	Hartlaub's Turaco*		Dusky Tit*
	Orange Ground-Thrush*		Green-headed Sunbird*
	Rueppell's Robin-Chat*		Green Hylia*
	Kenrick's Starling*		Brown-capped Weaver*
2/5/1960	Hunter's Sunbird*		Black-billed Weaver*
2/8/1960	Moustached Tinkerbird*		Red-headed Malimbe*
	African Hill Babbler*		Great Blue Turaco*
	Brown Woodland-Warbler*		White-headed Woodhoopoe*
2/11/1960	Sombre Nightjar* (coll.)		Buff-spotted Woodpecker*
	Plain Nightjar* (coll.)		Black-headed Gonolek*
	Slender-tailed Nightjar* (coll.)	3/3/1960	Black-and-white-casqued Hornbill*
2/14/1960	Mottled Swift*		N. Double-collared Sunbird*
	Beautiful Sunbird*		Viellot's Weaver*
	Steel-blue Whydah*		African Blue-Flycatcher*
2/15/1960	Pink-backed Pelican*		Black-and-white Mannikin*
	Whiskered Tern*		Jameson's Wattle-eye*
	Garganey*		Olive-green Camaroptera*
2/17/1960	African Emerald Cuckoo*		Uganda Wood-Warbler*
2/18/1960	Red-tufted Sunbird*		Buff-throated Apalis*
	Moorland Chat*		African Shrike-flycatcher*
	KENYA		Slender-billed Greenbul*
2/25/1960	Long-toed Lapwing*		Brown Illadopsis*
	African Silverbill*		Pink-footed Puffback*
2/26/1960	Northern Pied-Babbler*		Blue-headed Bee-eater*
3/1/1960	White-eyed Slaty-Flycatcher*	3/4/1960	Turner's Eremomela*
	Grey-capped Warbler*		Black-bellied Seedcracker*
	Brown-backed Woodpecker*		Grey Apalis*
	Yellow-throated Greenbul*		White-tailed Ant-Thrush*
	Grey-backed Fiscal*		Grey Greenbul*
3/2/1960	Petit's Cuckoo-shrike*		Scaly-breasted Illadopsis*
	Dusky Crested-Flycatcher*		Yellow-shouldered Widowbird*
	Joyful Greenbul*		Grey-chested Illadopsis*

	Tullberg's Woodpecker*		Red-tailed Greenbul*
	Mackinnon's Shrike*		Xavier's Greenbul*
3/5/1960	Least Honeyguide*		Green Crombec*
	Golden-crowned Woodpecker*		Yellow-browed Camaroptera*
	Hairy-breasted Barbet*		Green-headed Sunbird*
	Plain Greenbul*		Sooty Boubou*
	Grey-throated Barbet*		Yellow-spotted Nicator*
	Brown-chested Alethe*		Crested Malimbe*
3/6/1960	Double-toothed Barbet*	3/12/1960	Red-shouldered Cuckoo-shrike*
	Angola Swallow*		Rufous Flycatcher-Thrush*
	UGANDA		Leaf-love*
3/7/1960	N. Brown-throated Weaver*		Chestnut-breasted Negrofinch*
	African Thrush*		W. Black-headed Oriole*
	Red-chested Sunbird*		Black-casqued Hornbill*
	Slender-billed Weaver*		Brown-backed Scrub-Robin*
	Grey Parrot*	3/13/1960	Black-crowned Waxbill*
	Eastern Plantain-eater*		Yellow-throated Tinkerbird*
	Blue-headed Coucal*		Black-billed Turaco*
	Red-headed Lovebird*		Piping Hornbill*
	Sooty Chat*	3/14/1960	Many-colored Bushshrike*
	Orange Weaver*		Blue-throated Roller*
	Ross' Turaco*		Brown-eared Woodpecker*
	Compact Weaver*		Blue-headed Crested-Flycatcher*
	Blue-breasted Bee-eater*		Marsh Tchagra*
	Shoebill*	3/15/1960	Chestnut-throated Apalis*
	Fawn-breasted Waxbill*		White-browed Crombec*
	Swamp Flycatcher*		Ruwenzori Batis
3/8/1960	Black-headed Paradise-Flycatcher*		Abyssinian Ground-Thrush*
	Snowy-crowned Robin-Chat*		Blue-headed Sunbird*
	Brown-throated Wattle-eye*		Regal Sunbird*
	Green-throated Sunbird*		Red-throated Alethe*
	Splendid Glossy-Starling*		Red-faced Woodland-Warbler*
	Yellow-rumped Tinkerbird*		Ruwenzori Apalis*
	Superb Sunbird*		Archer's Robin-Chat*
	Northern Black-Flycatcher*		Ruwenzori Turaco*
	Blue-throated Brown Sunbird*		Strange Weaver*
	Olive-bellied Sunbird*	3/16/1960	Stuhlman's Starling*
	Black-headed Weaver*		Purple-headed Glossy-Starling*
3/10/1960	African Pied Hornbill*		Afep Pigeon*
	Speckle-breasted Woodpecker*		White-breasted Negrofinch*
3/11/1960	Chubb's Cisticola*		Dusky-blue Flycatcher*
	Jameson's Antpecker*		Honeyguide Greenbul*
	White-thighed Hornbill*		Elliot's Woodpecker*

	Black-winged Oriole*	4/8/1960	Egyptian Plover*
	Common Bristlebill*		Rufous-bellied Helmetshrike*
	Dusky Long-tailed Cuckoo*		FRENCH EQUATORIAL AFRICA
	Sooty Flycatcher*	4/9/1960	Preus's Swallow*
	Toro Olive-Greenbul*	4/10/1960	Abyssinian Roller*
	African Cuckoo-Hawk*		Blue-bellied Roller*
3/17/1960	White-collared Oliveback*	4/11/1960	Green-backed Eremomela*
	Masked Apalis*		Black Scimitar-bill*
	Speckled Tinkerbird*		Whistling Cisticola*
	Yellow-spotted Barbet*		White-fronted Black-Chat*
3/19/1960	African Crake*		White-crested Turaco*
	Black-lored Babbler*		Red-faced Pytilia*
	RUANDA-URUNDI		Black-bellied Firefinch*
3/22/1960	Black-winged Bishop*		Purple Glossy-Starling*
	BELGIAN CONGO		Little Weaver*
3/24/1960	Olive Ibis*		Long-tailed Nightjar*
	Mountain Buzzard*		Black-shouldered Nightjar*
	Handsome Francolin*	4/12/1960	Sun Lark*
	Grauer's Scrub-Warbler*		Woodchat Shrike*
	Northern Puffback*		Vinaceous Dove*
	Stripe-breasted Tit*		Black-rumped Waxbill*
	Purple-breasted Sunbird*		Bush Petronia*
	Kandt's Waxbill*		Brown Babbler*
	Dusky Crimson-wing*		Black-faced Firefinch*
3/26/1960	Bate's Swift*		Green Bee-eater*
	Cassin's Flycatcher*		Crested Lark*
	White-throated Blue Swallow*		Black-billed Wood-Dove*
	Chocolate-backed Kingfisher*		Grey Pratincole*
	Icterine Greenbul*		Red-throated Bee-eater*
	Red-billed Dwarf Hornbill*		Bruce's Green-Pigeon*
	Bate's Paradise-Flycatcher*		Grey-headed Batis*
	Square-tailed Sawwing*		Rufous Cisticola*
	Cassin's Spinetail*		Senegal Thick-knee*
3/27/1960	Levaillant's Cuckoo*		CHAD
	Sabine's Spinetail*	4/13/1960	Red-necked Buzzard*
	White-crested Hornbill*		Long-tailed Glossy-Starling*
3/28/1960	Blue Cuckoo-shrike*		Hueglin's Wheatear*
	Orange-cheeked Waxbill*		Red-pate Cisticola*
	Red-crowned Malimbe*		Yellow-billed Shrike*
4/3/1960	Magpie Mannikin*		Piapiac*
	Little Green Sunbird*		Brown-rumped Bunting*
4/7/1960	Black Bee-eater*		Abyssinian Ground-Hornbill*
	Lesser Bristlebill*		Bearded Barbet*

	Fine-spotted Woodpecker*	4/26/1960	Fulvous Chatterer*
	Rose-ringed Parakeet*		White-tailed Wheatear*
	Clapperton's Francolin*		Arabian Bustard*
	Red-necked Falcon*		Barn Owl
4/14/1960	Black Crowned-Crane*		Egyptian Vulture*
	Spur-winged Plover*		Brown-necked Raven*
	Black Scrub-Robin*	4/27/1960	Spotted Sandgrouse*
	Chestnut-bellied Starling*		ALGERIA
4/15/1960	Western Plantain-eater*	4/28/1960	W. Bonelli's Warbler*
	Viellot's Barbet*		Black-eared Wheatear*
	White-rumped Seedeater*	5/1/1960	Crowned Sandgrouse*
	Chestnut-crowned Sparrow-Weaver*		W. Orphean Warbler*
	CAMEROONS		Common Redstart
	Black-headed Lapwing*		House Bunting*
	Ethiopian Swallow*		European Pied Flycatcher
	NIGERIA		Trumpeter Finch*
4/16/1960	Pygmy Sunbird*	5/2/1960	Rock Pigeon*
	Beaudoin's Snake-Eagle*	5/3/1960	Long-legged Buzzard*
	African Collared-Dove*	5/4/1960	Common Chiffchaff
4/17/1960	Kordofan Lark*	5/9/1960	Western Marsh-Harrier*
	Common Gonolek*		Green Sandpiper
	Sennar Penduline-Tit*	5/10/1960	Desert Wheatear*
	Senegal Parrot*		Little Owl*
4/18/1960	Yellow Penduline-Tit*		Cream-colored Courser*
	Bar-breasted Firefinch*		Greater Short-toed Lark*
	Melodious Warbler*		Common Swift
	NIGER		Eurasian Turtle-Dove
4/23/1960	Senegal Batis*	5/11/1960	Calandra Lark*
	Black-crowned Sparrow-Lark*		Thekla Lark*
4/24/1960	Singing Bushlark*		House Sparrow
	Quail-plover*		Black Tern
	Scissor-tailed Kite*		Black Wheatear*
	Rueppell's Griffon*		Corn Bunting*
	Cricket Longtail*		Short-toed Treecreeper*
4/25/1960	White-billed Buffalo-Weaver*		European Stonechat*
	Sudan Golden Sparrow*		European Goldfinch
	Southern Grey Shrike*	5/12/1960	Great Tit
	Dunn's Lark*		Blue Tit
	Desert Lark*		European Greenfinch
	Greater Whitethroat		Winter Wren
	Rusty Lark*		Eurasian Blackbird
	Bar-tailed Lark*	5/14/1960	European Serin*
	Greater Hoopoe-Lark*	5/15/1960	Sardinian Warbler*

	Chaffinch		(Red-necked Nightjar*)
	Eurasian Linnet	6/2/1960	Little Ringed Plover
5/19/1960	Eurasian Griffon*		Black Redstart
	Barbary Partridge*		Eurasian Jay
	Yellow-legged Gull*		Eurasian Crag-Martin*
	Blue Rock-Thrush		White Wagtail
5/20/1960	Wood Warbler		Rock Petronia*
	Dartford Warbler*	6/5/1960	Wood Lark*
	Spanish Sparrow*		Red-billed Chough*
	MOROCCO		Alpine Accentor
	Short-toed Eagle*	6/6/1960	Ortolan Bunting*
5/21/1960	Demoiselle Crane*		Eurasian Skylark
	Black-bellied Sandgrouse*		Tree Pipit
	Spectacled Warbler*		Subalpine Warbler*
	Spotless Starling*	6/9/1960	(European Scops-Owl*)
5/22/1960	Cirl Bunting	6/11/1960	Cetti's Warbler*
	Pallid Swift*	6/20/1960	Azure-winged Magpie
	EUROPE		Green Woodpecker*
	SPAIN		Common Wood-Pigeon
5/28/1960	Audouin's Gull*		Carrion Crow
	Red-rumped Swallow		Common Kingfisher
	Crested Tit*		Dunlin
	Tawny Pipit*		Eurasian Spoonbill*
	Rock Bunting*		Red Kite*
5/29/1960	Grey Wagtail		ANDORRA
	Great Spotted Woodpecker	7/23/1960	Rock Ptarmigan*
	Eursian Jackdaw		Yellow-billed Chough*
	Little Tern		Whinchat
5/30/1960	Eurasian Magpie		Dunnock
	Lesser Short-toed Lark*		Citril Finch*
	Northern Lapwing		Eurasian Sparrowhawk
	Marbled Teal*		Coal Tit
	Gull-billed Tern		Mistle Thrush
	Common Redshank		White-throated Dipper*
	Common Tern		European Robin
5/31/1960	Eurasian Thick-knee*		Goldcrest
	Mallard		Red Crossbill
	Red-legged Partridge*		Yellowhammer
6/1/1960	Eurasian Teal	7/24/1960	Water Pipit*
	Northern Shoveler		Reed Bunting
	Eurasian Coot		FRANCE
	Black-tailed Godwit		Eurasian Penduline-Tit*
	Slender-billed Gull*	7/25/1960	(Water Rail*)

	Black-headed Gull	10/4/1960	Northern Goshawk
	Tawny Owl*		Eurasian Capercaillie*
	Bearded Reedling*		Common Redpoll
	Savi's Warbler*	10/5/1960	Lesser Spotted Woodpecker*
	Moustached Warbler*		Siberian Jay*
	Eurasian Tree Sparrow		Grey-headed Chickadee*
8/10/1960	Marsh Tit		Bohemian Waxwing*
	Eurasian Treecreeper*	10/6/1960	Hoary Redpoll*
	Firecrest*		Parrot Crossbill*
	Eurasian Bullfinch		NORWAY
8/13/1960	Willow Tit		Northern Hawk-Owl*
	White-winged Snowfinch*	10/7/1960	Yellow-billed Loon*
8/18/1960	Long-tailed Tit		Common Eider
9/1/1960	Grey Partridge		Purple Sandpiper
	Stock Dove*		Great Black-backed Gull
	Northern Shrike		Long-tailed Duck
	European Starling		Black-legged Kittiwake
	CZECHOSLOVAKIA		Arctic Tern
9/14/1960	Black Woodpecker*		Razorbill
	Eurasian Nutcracker*		Common Murre
	Eurasian Nuthatch		Black Guillemot
	Song Thrush		Atlantic Puffin
	Ring Ouzel*	10/8/1960	Twite*
	Eurasian Siskin	10/9/1960	Snow Bunting
	Golden Eagle	10/10/1960	White-tailed Eagle*
	Hooded Crow	10/11/1960	White-winged Scoter
	RUSSIA	10/15/1960	Hazel Grouse*
9/18/1960	Common Snipe	10/16/1960	Rock Pipit*
	Rook		SWEDEN
9/26/1960	Mew Gull	10/19/1960	Mute Swan
	Red-throated Pipit		HOLLAND
	Brambling	11/15/1960	Horned Grebe
9/29/1960	Black Grouse*		Common Merganser
	Fieldfare*		Tundra Swan
	Redwing*		Eurasian Golden-Plover
	FINLAND	11/29/1960	Arctic Loon*
9/29/1960	Eurasian Wigeon		Greater Scaup
	Northern Pintail		Smew*
	Tufted Duck		Common Shelduck
	Common Pochard		Little Gull
	Common Goldeneye	12/1/1960	Brant
9/30/1960	Rough-legged Hawk		Eurasian Oystercatcher
	Common Crane*		Common Ringed Plover*

	Spotted Redshank	3/3/1961	Dead Sea Sparrow*
12/12/1960	ITALY		IRAQ
	Eurasian Wryneck	3/10/1961	Thick-billed Lark*
	Meadow Pipit	3/11/1961	White-tailed Lapwing*
	BULGARIA		White-eared Bulbul*
1/18/1961	Eurasian Collared-Dove*		White-throated Kingfisher
	GREECE	3/12/1961	Black Francolin*
1/21/1961	Greater White-fronted Goose		Red-wattled Lapwing*
	Merlin		Bluethroat
1/24/1961	Mediterranean Gull*		Common Babbler*
1/28/1961	Red-crested Pochard*		IRAN
	Greylag Goose	3/17/1961	Great Bustard*
	Jack Snipe*	3/24/1961	Fire-fronted Serin*
	Eurasian Woodcock*		Crimson-winged Finch*
1/29/1961	Sombre Tit*	3/24/1961	Caspian Snowcock*
	Rock Nuthatch*	3/28/1961	Hawfinch
1/30/1961	White-headed Duck*	3/29/1961	Bimaculated Lark*
1/31/1961	Pygmy Cormorant*		Variable Wheatear*
	Dalmatian Pelican*		Siberian Stonechat
	Imperial Eagle*	3/31/1961	Pied Bushchat
	ASIA		AFGHANISTAN
	TURKEY	4/1/1961	Common Myna*
2/1/1961	Syrian Woodpecker*	4/2/1961	Red-breasted Flycatcher*
2/8/1961	Ruddy Shelduck*		Desert Finch*
2/11/1961	White-spectacled Bulbul*		Common Rosefinch*
	SYRIA		Persian Nuthatch*
2/12/1961	Chukar*		Menetries Warbler*
	Finsch's Wheatear*	4/3/1961	Hume's Wheatear*
	JORDAN		Lesser Whitethroat
2/20/1961	Mourning Wheatear*		Rufous-sided Shrike*
	Blackstart*	4/4/1961	See-see Partridge*
	Fan-tailed Raven*		Hooded Wheatear*
2/21/1961	Graceful Prinia*		Plain Leaf-Warbler*
	Sand Partridge*		Citrine Wagtail*
	Tristram's Starling*	4/6/1961	Oriental Skylark*
2/22/1961	Temminck's Lark*	4/7/1961	Hume's Lark*
2/23/1961	Streaked Scrub-Warbler*	4/8/1961	Red-tailed Wheatear*
	Pale Rosefinch*	4/10/1961	Black Drongo*
2/24/1961	Red Sea Warbler*		Plumbeous Redstart
2/26/1961	Greater Spotted Eagle*		White-capped Redstart*
	Little Crake*		Greenish Warbler*
	Temminck's Stint*		Sulphur-bellied Warbler*
	Pin-tailed Sandgrouse*		Eurasian Eagle-Owl*

PAKISTAN

4/11/1961	House Crow*
	White-cheeked Bulbul*
	Red-vented Bulbul*
	Bay-backed Shrike*
	Chestnut-breasted Bunting*
	Oriental White-eye*
	Oriental Honey-buzzard
4/12/1961	White-rumped Vulture*
	Indian Roller*
	Jungle Prinia*
	Long-tailed Shrike
	White-eyed Buzzard*
4/13/1961	Russet Sparrow*
4/14/1961	Oriental Turtle-Dove
	Fork-tailed Swift
	Speckled Piculet*
	Lesser Yellownape*
	Brown-fronted Woodpecker*
	Himalayan Woodpecker*
	Scaly-bellied Woodpecker*
	Grey-hooded Warbler*
	W. Crowned Leaf-Warbler*
	Hume's Warbler*
	Pale-rumped Warbler*
	Pale-footed Bush-Warbler*
	Ashy Drongo*
	Short-billed Minivet*
	Black-winged Cuckoo-shrike*
	Rufous-bellied Niltava*
	Grey-headed Canary-Flycatcher*
	Verditer Flycatcher*
	Ultramarine Flycatcher*
	Rufous-breasted Accentor*
	Blue Whistling-Thrush
	Chestnut Thrush*
	Dark-throated Thrush*
	Grey Bushchat*
	Bar-tailed Treecreeper*
	White-browed Shrike-Babbler*
	Streaked Laughingthrush*
	Variegated Laughingthrush*

	White-cheeked Nuthatch*
	Black-throated Tit*
	Black-crested Tit*
	Black-breasted Tit*
	Green-backed Tit*
	Large-billed Crow
	Gold-billed Magpie*
4/15/1961	Steppe Eagle*
	Slaty-headed Parakeet*
	Black Bulbul*
	Fire-capped Tit*
	Black-and-yellow Grosbeak*
	Red-flanked Bluetail
	Pink-browed Rosefinch*
4/16/1961	Blue-fronted Redstart*
	Blue-capped Redstart*
	Jungle Babbler*
	Indian Robin*
	Blyth's Reed-Warbler*
	Ashy Prinia*
	Purple Sunbird*

INDIA

4/18/1961	Asian Paradise-Flycatcher*
	Yellow-bellied Prinia
	Pale Sand Martin*
	Bank Myna*
	Chestnut-shouldered Petronia*
	White-browed Wagtail*
	Indian Bushlark*
	Ashy-crowned Sparrow-Lark*
	Blue-tailed Bee-eater
	Indian Grey Hornbill*
	River Tern*
	Indian Pond-Heron*
4/19/1961	Rufous Treepie*
	Yellow-eyed Babbler*
	Small Minivet*
	Common Woodshrike*
	Rufous-fronted Prinia*
	Crested Bunting
	Thick-billed Flowerpecker*
	Savannah Nightjar*
	Common Hawk-Cuckoo*

	Alexandrine Parakeet*		Pale-billed Flowerpecker*
4/21/1961	Kashmir Nuthatch*		Grey-faced Woodpecker*
	Brown Dipper		Wedge-tailed Pigeon*
	Tickell's Thrush*		Indian Cuckoo*
	Hume's Whitethroat*		Indian Peafowl*
	Plain Mountain-Finch*		Changeable Hawk-Eagle*
	Himalayan Griffon*	5/1/1961	Blue Magpie
	Snow Pigeon*		Black-headed Jay*
4/24/1961	Olive-backed Pipit*		Spot-breasted Scimitar-Babler*
4/25/1961	Kashmir Flycatcher*		Yellow-crowned Woodpecker*
	White-throated Needletail*		Great Barbet
4/26/1961	Rusty-tailed Flycatcher*		Cheer Pheasant*
	Long-tailed Minivet*	5/2/1961	White-tailed Nuthatch*
4/27/1961	Grey-winged Blackbird*		White-throated Laughing-Thrush*
	Slaty-blue Flycatcher*		Striated Laughingthrush*
	(Koklass Pheasant*)		Chestnut-crowned Laughing thrush*
4/28/1961	Brahminy Starling*		Whiskered Yuhina*
	White-throated Munia*		Chestnut-tailed Minla*
	(Oriental Scops-Owl*)		Rufous Sibia*
	Plum-headed Parakeet*		Himalayan Cuckoo*
	Red-headed Vulture*		(Mountain Scops-Owl*)
	Pallas' Fish-Eagle*		Kalij Pheasant*
	Grey Francolin*	5/3/1961	Grey-crested Tit*
	Black-necked Stork*		White-throated Tit*
4/29/1961	White-browed Scimitar-Babler*		White-collared Blackbird*
	Black-chinned Babbler*		Chestnut-bellied Rock-Thrush*
	B lue-throated Flycatcher*		Dark-sided Flycatcher
	Oriental Magpie-Robin		Rufous-gorgeted Flycatcher*
	Asian Brown Flycatcher		Large Hawk-Cuckoo*
	Common Tailorbird		(Grey Nightjar*)
	Chestnut-headed Tesia*		Black Eagle*
	Jungle Myna*		Himalayan Monal*
	Blue-throated Barbet*	5/5/1961	Striated Prinia*
	Brown Fish-Owl*		Sand Lark*
	Spotted Owlet*		Black-rumped Flameback*
	Large-tailed Nightjar*		Asian Koel
4/30/1961	Grey Treepie*		Slender-billed Vulture*
	Black-lored Tit*		Pheasant-tailed Jacana*
	Blue-capped Rock-Thrush*	5/6/1961	Large Grey Babbler*
	White-throated Fantail*		Booted Warbler*
	Hair-crested Drongo*		Asian Pied Starling*
	Spot-winged Starling*		Red Avadavat*
	E. Crimson Sunbird*		Streak-throated Swallow*

	White-breasted Waterhen		Stork-billed Kingfisher*
	Black-bellied Tern*		Silver-backed Needletail*
5/7/1961	Black-headed Ibis*		Crested Treeswift*
	Yellow-wattled Lapwing*		Jungle Owlet*
	Little Cormorant*		Crested Serpent-Eagle*
	Baya Weaver*	5/21/1961	Velvet-fronted Nuthatch*
5/8/1961	Red Collared-Dove		Orange-headed Thrush*
	Sarus Crane*		Common Iora*
	Great Thick-knee*		Plain Flowerpecker*
	Small Pratincole*	5/22/1961	Stripe-throated Yuhina*
	River Lapwing*		Hoary-throated Barwing*
	Indian Skimmer*		Rufous-winged Fulvetta*
	Painted Stork*		Indian Blue Robin*
	Asian Openbill*		Scaly Thrush*
5/9/1961	Dusky Crag Martin*		Purple Cochoa*
5/12/1961	White-browed Fantail*		Snowy-browed Flycatcher*
	Coppersmith Barbet*		Blyth's Leaf-Warbler*
	Spot-billed Duck		Ashy-throated Warbler*
	Asian Palm-Swift		Golden-spectacled Warbler*
	Yellow-footed Pigeon*		Chestnut-crowned Warbler*
5/13/1961	Clamorous Reed-Warbler*		Black-eared Shrike-Babbler*
	Bronze-winged Jacana*		Yellow-browed Tit*
	Red-naped Ibis*		Gould's Sunbird*
	Brown-capped Woodpecker*		Darjeeling Woodpecker*
	Lesser Adjutant*	5/23/1961	Rufous-bellied Woodpecker*
	Ferruginous Pochard*		Pied Thrush*
	Sirkeer Malkoha*		Long-billed Thrush*
5/15/1961	Chestnut-bellied Nuthatch*		Spotted Forktail*
5/16/1961	Blue-winged Leafbird*		Grey-bellied Tesia*
	Brown-headed Barbet*		Red-billed Leiothrix*
5/18/1961	Black-hooded Oriole*		Pygmy Wren-Babbler*
	NEPAL		Brown Bullfinch*
5/18/1961	Large Woodshrike*		Collared Grosbeak*
	Scarlet Minivet*		Greater Yellownape*
	Large Cuckoo-shrike*		Crimson-breasted Woodpecker*
	White-bellied Drongo*		Golden-throated Barbet*
	Grey-breasted Prinia*		Bronzed Drongo*
	Golden-fronted Leafbird*		Lesser Racket-tailed Drongo*
	Common Hill Myna*		Orange-bellied Leafbird*
	Chestnut-tailed Starling*		Rusty-fronted Barwing*
	Lineated Barbet*	5/24/1961	Green Magpie*
	Himalayan Flameback*		White-crested Laughing thrush*
	Chestnut-headed Bee-eater*		Lesser Necklaced Laughing thrush*

	Puff-throated Babbler*		Dark-rumped Rosefinch*
	White-bellied Yuhina*		White-browed Rosefinch*
	Ashy Bulbul*		Red-headed Bullfinch*
	Black-crested Bulbul*		Buff-barred Warbler*
	Black-naped Monarch		Large-billed Leaf-Warbler*
	Pale-chinned Blue-Flycatcher*		Grey-sided Bush-Warbler*
	Bar-winged Flycatcher-shrike*		Verditer Flycatcher*
	Hooded Pitta*		Plain-backed Thrush*
	Greater Flameback		Little Forktail*
	Dollarbird		Golden Bush-Robin*
	Asian Drongo-Cuckoo*		Rusty-flanked Treecreeper*
	Crested Goshawk*		Rufous-vented Yuhina*
	Spotted Dove		Black-faced Laughingthrush*
	Emerald Dove		Spotted Laughingthrush*
	Red Junglefowl*		Fulvous Parrotbill*
	Red-headed Trogon*		Streak-breasted Scimitar-Babbler*
5/25/1961	Striped Tit-Babbler*		Rufous-vented Tit*
	Crow-billed Drongo*	6/4-5/1961	Asian Barred Owlet*
	Streak-throated Woodpecker*		(Brownish-flanked Bush-Warbler*)
	Fulvous-breasted Woodpecker*		White-browed Shortwing*
	INDIA		Black-headed Shrike-Babbler*
5/26/1961	Bengal Weaver*	6/7/1961	Besra*
	Oriental Pipit		Black-throated Sunbird*
	Brahminy Kite		Hill Prinia*
	Cinnamon Bittern		Little Pied Flycatcher
5/28/1961	Cotton Pygmy-goose*		Brown-throated Treecreeper*
	Ferruginous Flycatcher*		Striated Bulbul*
	Small Niltava*		Nepal Fulvetta*
	Large Niltava*		Cutia*
	Ashy Woodswallow*		Silver-eared Mesia*
	Collared Owlet*		Rufous-necked Laughing thrush*
5/29/1961	Blue-winged Minla*	6/8/1961	Thick-billed Pigeon*
	Red-tailed Minla*		Oriental Pied-Hornbill*
	Rufous-capped Babbler*		Blue-bearded Bee-eater*
	Yellow-bellied Fantail*		Green-billed Malkoha*
	Fire-breasted Flowerpecker		Blue-eared Barbet*
5/30/1961	White-browed Fulvetta*		Rufous Woodpecker*
	Black-throated Parrotbill*		Yellow-vented Flowerpecker*
	Aberrant Bush-Warbler*		Yellow-bellied Warbler*
5/31/1961	Ashy Wood-Pigeon*		Greater Racket-tailed Drongo*
	(Hill Partridge*)		Ruby-cheeked Sunbird*
6/1-3/1961	Fire-tailed Sunbird*		White-rumped Shama*
	Dark-breasted Rosefinch*		White-throated Bulbul*

	Asian Fairy-bluebird*		Nutmeg Mannikin
	Abbot's Babbler*		White-rumped Munia*
6/9/1961	Greater Adjutant*	6/29/1961	Grey Junglefowl*
	Great Hornbill*		Greater Coucal
	Bengal Bushlark*		CEYLON
	Chestnut Munia	7/7/1961	Black-headed Cuckoo-shrike*
6/15/1961	Purple-rumped Sunbird*		Ceylon Hanging-Parrot*
6/24/1961	Brown-cheeked Fulvetta*	7/8/1961	White-throated Flowerpecker*
	Dark-fronted Babbler*		Yellow-fronted Barbet*
	Yellow-browed Bulbul*	7/10/1961	White-bellied Sea-Eagle*
	Tickell's Blue-Flycatcher*		Grey-headed Fish-Eagle*
	Black-and-rufous Flycatcher*		Orange-breasted Pigeon*
	Pacific Swallow		Indian Cormorant*
	Crimson-backed Sunbird*		Lesser Whistling-Duck*
	Blue-faced Malkoha*	7/11/1961	Ceylon Grey Hornbill*
	Indian Swiftlet*	7/12/1961	Jerdon's Nightjar*
6/25/1961	Indian Scimitar-Babbler*		Barred Buttonquail*
	Yellow-billed Babbler*	7/14/1961	Ceylon Blue Magpie*
	White-browed Bulbul*		Orange-billed Babbler*
	Malabar Whistling-Thrush*		Ceylon Whistling-Thrush*
	Nilgiri Flycatcher*		Dull-blue Flycatcher*
	Black-throated Munia*		Ceylon Bush-Warbler*
	Common Flameback*		Ceylon White-eye*
	Malabar Parakeet*		Yellow-eared Bulbul*
	Vernal Hanging-Parrot*		Ceylon Junglefowl*
	Malabar Trogon*	7/16/1961	Spot-winged Thrush*
	(Indian Nightjar*)		Ceylon Myna*
6/26/1961	Rufous-breasted Laughing thrush*		Layard's Parakeet*
	White-bellied Shortwing*	7/17/1961	White-faced Starling*
	Spot-breasted Fantail*		**N. INDIAN OCEAN**
6/27/1961	Rufous Babbler*	7/19/1961	Wedge-tailed Shearwater
	Tawny-bellied Babbler*		**SE. ASIA**
	Malabar Grey Hornbill*		MALAYA
	Brown-backed Needletail*	7/23/1961	Silver-rumped Needletail*
	Red Spurfowl*	7/24/1961	Plaintive Cuckoo*
6/28/1961	Grey-headed Bulbul*		Blue-throated Bee-eater
	Long-billed Sunbird*		Grey-rumped Treeswift*
	Little Spiderhunter*		Black-naped Oriole
	Heart-spotted Woodpecker*		Golden-bellied Gerygone*
	White-bellied Woodpecker		Scarlet-backed Flowerpecker*
	Crimson-fronted Barbet*		Asian Glossy Starling
	Pompadour Green-Pigeon		Plain-throated Sunbird*
	Painted Bush-Quail*		(Sunda Scops-Owl*)

7/25/1961	Straw-headed Bulbul*		Grey-and-buff Woodpecker*
	Yellow-vented Bulbul		Green Iora*
	White-headed Munia*		Red-eyed Bulbul*
	Rufous-bellied Swallow*		Cream-vented Bulbul*
	Spectacled Bulbul*		Yellow-bellied Bulbul*
	Stripe-throated Bulbul*		Finch's Bulbul*
	Black-headed Bulbul*		Ferruginous Babbler*
	Fluffy-backed Tit-Babbler*		Black-capped Babbler*
	Rufescent Prinia*		Sooty-capped Babbler*
	Little Cuckoo-Dove*		Maroon-breasted Philentoma*
	Wreathed Hornbill*		Crimson-breasted Flowerpecker*
	Fire-tufted Barbet*		Grey-breasted Spiderhunter*
	Maroon Woodpecker*	7/28/1961	Blyth's Hawk-Eagle*
	Grey-chinned Minivet*		(Great Argus*)
	Streaked Bulbul*		Black-bellied Malkoha*
	Streaked Wren-Babbler*		Chestnut-bellied Malkoha*
	Chestnut-capped Laughing thrush*		Bushy-crested Hornbill*
	Long-tailed Sibia*		Rhinocerous Hornbill*
	Mountain Fulvetta*		Gold-whiskered Barbet*
	Mountain Tailorbird		Yellow-crowned Barbet*
	Rufous-browed Flycatcher*		Crimson-winged Woodpecker*
	Streaked Spiderhunter*		Green Broadbill*
	Black-and-crimson Oriole*		Black-and-red Broadbill*
7/26/1961	Black-thighed Falconet*		Black-and-yellow Broadbill*
	Pink-necked Pigeon*		Dark-throated Oriole*
	Zebra Dove*		Great Iora*
	Blue-crowned Hanging-Parrot*		Grey-bellied Bulbul*
	Ruddy Kingfisher*		Grey-cheeked Bulbul*
	Laced Woodpecker*		Chestnut-naped Forktail*
	Olive-winged Bulbul*		Short-tailed Babbler*
	White-chested Babbler*		Scaly-crowned Babbler*
	Ashy Tailorbird*		Grey-throated Babbler*
	Mangrove Whistler*		Grey-headed Babbler*
7/27/1961	Slaty-breasted Rail*		Brown Fulvetta*
	Mangrove Pitta*		Yellow-breasted Flowerpecker*
	Rufous-tailed Tailorbird*		Temminck's Sunbird*
	Mangrove Blue-Flycatcher*		Spectacled Spiderhunter*
	Copper-throated Sunbird*		Everett's White-eye*
	Crestless Fireback*	7/29/1961	Red-billed Malkoha*
	Chestnut-breasted Malkoha*		Raffle's Malkoha*
	Blue-eared Kingfisher*		Black-nest Swiftlet*
	Red-crowned Barbet*		Scarlet-rumped Trogon*
	Brown Barbet*		White-crowned Hornbill*

	Checker-throated Woodpecker*		Rufous-chested Flycatcher*
	Buff-necked Woodpecker*		Grey-chested Jungle-Flycatcher*
	Silver-breasted Broadbill*	8/10/1961	Red-throated Barbet*
	Lesser Green Leafbird*	8/13/1961	Collared Kingfisher
	Greater Green Leafbird*		German's Swiftlet*
	Scaly-breasted Bulbul*		**THAILAND**
	Horsefield's Babbler*	8/21/1961	Buff-breasted Babbler*
	Moustached Babbler*		Red-billed Scimitar-Babbler*
	Striped Wren-Babbler*		Rufous-fronted Babbler*
	Eyebrowed Wren-Babbler*		Chestnut-fronted Shrike-Babbler*
	Spotted Fantail*		Striated Yuhina*
	White-tailed Flycatcher*		Grey-cheeked Fulvetta*
	Yellow-eared Spiderhunter*		Rufous-backed Sibia*
	Long-billed Spiderhunter*		Sooty-headed Bulbul
7/30/1961	Great Slaty Woodpecker*		Flavescent Bulbul*
	Black-throated Babbler*		Pale Blue Flycatcher*
	Chestnut-rumped Babbler*		Hill Blue Flycatcher*
	Plain Sunbird*		White-tailed Leaf-Warbler*
	Black Magpie*		Maroon Oriole*
8/1/1961	Barred Cuckoo-Dove*		Stripe-breasted Woodpecker*
	Golden Babbler*		Bay Woodpecker*
	Yellow-breasted Warbler*	8/22/1961	Collared Falconet*
8/2/1961	Mountain Imperial-Pigeon*	8/24/1961	Javan Pond-Heron*
	Black-browed Barbet*		White-vented Myna*
	Slaty-backed Forktail*		**CAMBODIA**
	Mountain Warbler*	8/30/1961	Rufous-winged Buzzard*
	Sultan Tit*		Plain-backed Sparrow*
	Blue Nuthatch*	9/1-2/1961	Racket-tailed Treepie*
8/5-6/1961	Green Imperial-Pigeon*		Chestnut-capped Babbler*
	Little Green-Pigeon*		Streak-eared Bulbul*
	Long-tailed Parakeet*		Banded Broadbill*
	Blue-rumped Parrot*		Dusky Broadbill*
	Violet Cuckoo*		Banded Bay Cuckoo*
	Malaysian Nightjar*		Red-breasted Parakeet*
	Buff-rumped Woodpecker*		Black Baza*
	Ochraceous Bulbul*	9/6/1961	Moustached Barbet*
	Hairy-backed Bulbul*		Green-eared Barbet*
	Chestnut-winged Babbler*		Pale-legged Leaf-Warbler*
	Red-throated Sunbird*		Grey-eyed Bulbul*
	Purple-naped Sunbird*		Streaked Wren-Babbler*
	Purple-throated Sunbird*		Yellow-breasted Magpie*
8/8/1961	Rufous Piculet*	9/8/1961	Indochinese Bushlark*
	Rufous-crowned Babbler*		

AUSTRALIA

<u>WESTERN AUSTRALIA</u>

9/16-18/61
Dusky Moorhen*
Australasian Grebe*
Hoary-headed Grebe*
Little Black Cormorant*
Pied Cormorant*
Little Pied Cormorant*
Australian Pelican*
Silver Gull*
Red-capped Plover*
Black Swan*
Musk Duck*
Australian Kestrel*
Port Lincoln Parrot*
Sacred Kingfisher*
Welcome Swallow*
Tree Martin*
Grey Fantail*
Willie-wagtail*
Scarlet Robin*
Grey-breasted Robin*
Golden Whistler*
Rufous Whistler*
Grey Shrike-Thrush*
Magpie-lark*
Black-faced Cuckoo-shrike*
Western Gerygone*
Weebill*
Western Thornbill*
Inland Thornbill*
Yellow-rumped Thornbill*
Splendid Fairywren*
White-winged Fairywren*
Rufous Treecreeper*
Mistletoebird*
Striated Pardalote*
Silver-eye*
White-naped Honeyeater*
Western Spinebill*
Tawny-crowned Honeyeater*
Brown Honeyeater*

Singing Honeyeater*
New Holland Honeyeater*
White-cheeked Honeyeater*
Little Wattlebird*
Red Wattlebird*
Australasian Pipit*
Australian Raven*
Grey Butcherbird*
Australasian Magpie*

9/20/1961
Common Bronzewing*
Pied Stilt*
Straw-necked Ibis*
White-faced Heron*
Pacific Heron*
Pacific Reef Heron
Grey Teal*
Pacific Black Duck*
Australian Shoveler*
Whistling Kite*
Brown Falcon*
White-tailed Black-Cockatoo*
Elegant Parrot*
Red-capped Parrot*
Western Rosella*
Tawny Frogmouth*
Fan-tailed Cuckoo*
Shining Bronze-Cuckoo*
White-breasted Robin*
White-browed Scrubwren*
Black-faced Woodswallow*

9/21/1961
Fairy Tern*
Sooty Oystercatcher*
Banded Lapwing*
Swamp Harrier*
Collared Sparrowhawk*
Little Eagle*
Southern Emuwren*
Dusky Woodswallow*
Varied Sitella*
Southern Boobook*

9/22/1961
Brown Goshawk*
Purple-crowned Lorikeet*
White-browed Babbler*

	Red-winged Fairywren*		Redthroat*
	Spotted Pardalote*		Masked Woodswallow*
9/24/1961	Brown Songlark*		White-fronted Honeyeater*
9/30/1961	Red-necked Avocet*		Little Crow*
10/5/1961	Australian Shelduck*	10/19/1961	Little Corella*
	Rainbow Bee-eater*		Blue-winged Kookaburra*
	Horsefield's Bronze-Cuckoo*		Grey-crowned Babbler*
	Red-capped Robin*		Variegated Fairywren*
	White-winged Triller*		Torresian Crow*
	White-fronted Chat*	10/20/1961	Bush Thick-knee*
	Chestnut-rumped Thornbill*	10/21/1961	Spotted Harrier*
	Brown-headed Honeyeater*		Cockatiel*
	Yellow-throated Miner*		White-breasted Woodswallow
10/6/1961	Galah*		Australian Bushlark*
	Black-fronted Dotterel*	10/22/1961	Painted Finch*
	Restless Flycatcher*		Spinifex Pigeon*
	Rufous Songlark*		Black-tailed Treecreeper*
10/7/1961	Australian Kite*	10/23/1961	Rufous Night-Heron*
10/11/1961	Stubble Quail*		Little Woodswallow*
	Australian Reed-Warbler*		Australian Bustard*
10/17/1961	White-eyed Duck*		Black Falcon*
	Crested Pigeon*		Red-browed Pardalote*
	Sharp-tailed Sandpiper*	10/25/1961	Hooded Robin*
	Western Corella*		Black-chinned Honeyeater*
	Blue-breasted Fairywren*		Black Honeyeater*
	White-backed Swallow*		Western Bowerbird*
	Zebra Finch*	10/26/1961	Crested Bellbird*
	Wedge-tailed Eagle*		Spinifex-bird*
10/18/1961	Diamond Dove*	10/27/1961	Bar-shouldered Dove*
	Black-tailed Native-hen*		Terek Sandpiper
	Maned Duck*		Red-necked Stint
	Australian Hobby*		Grey-tailed Tattler
	Southern Whiteface*		Mangrove Fantail*
	White-plumed Honeyeater*		Mangrove Robin*
	Spiny-cheeked Honeyeater*		White-breasted Whistler*
	Pied Butcherbird*		Dusky Gerygone*
	Emu*		Australian Yellow White-eye*
	Little Buttonquail*	10/28/1961	Pied Oystercatcher*
	Regent Parrot*		Far Eastern Curlew
	Mulga Parrot*		Great Knot
	Budgerigar*		Broad-billed Sandpiper*
	Chiming Wedgebill*		Australian Pratincole*
	Crimson Chat*		Fairy Martin*

	Orange Chat*		Gouldian Finch*
	Pied Honeyeater*	11/9/1961	Royal Spoonbill*
10/29/1961	Brolga*		Radjah Shelduck*
	Red-tailed Black-Cockatoo*		Green Pygmy-goose*
	Little Friarbird*		NORTHERN TERRITORY
10/30/1961	White-throated Gerygone*		Yellow-billed Spoonbill*
	Rufous-throated Honeyeater*		Pink-eared Duck*
	Yellow-tinted Honeyeater*	11/12/1961	Pacific Baza*
10/31/1961	Olive-backed Oriole*		Rose-crowned Fruit-Dove*
	Great Bowerbird*		Northern Fantail*
11/2/1961	Little Curlew*		Broad-billed Flycatcher*
	Pheasant Coucal*		Grey Whistler*
	Pallid Cuckoo*		Large-billed Gerygone*
	Red-headed Myzomela*		Green-backed Geygone*
11/3/1961	Masked Lapwing*		Cicadabird*
	Rainbow Lorikeet*		Dusky Myzomela*
	Varied Lorikeet*		Rufous-banded Honeyeater*
	Red-winged Parrot*		Green Oriole*
	Red-backed Kingfisher*		Green Figbird*
	Jacky-winter*		Spangled Drongo*
11/4/1961	Oriental Pratincole*		Black Butcherbird*
11/5/1961	Red-kneed Dotterel*		Chestnut-breasted Munia*
	Plumed Whistling-Duck*	11/13/1961	Pacific Golden-Plover*
	Magpie Goose*		Beach Thick-knee*
	Black-breasted Kite*		Torresian Imperial-Pigeon*
	Sulphur-crested Cockatoo*		Black-eared Cuckoo*
	White-browed Robin*		Forest Kingfisher*
	Papuan Cuckoo-shrike*		Lemon-bellied Flycatcher*
	Red-backed Fairywren*		Shining Flycatcher*
	Banded Honeyeater*		Golden-headed Cisticola
	White-gaped Honeyeater*		Mangrove Gerygone*
11/6/1961	Grey-fronted Honeyeater*		Varied Triller*
11/7/1961	Long-tailed Finch*	11/14/1961	Partridge Pigeon*
	Northern Rosella*	11/15/1961	White-throated Honeyeater*
	Australian Koel*	11/18/1961	Brush Cuckoo*
	Blue-faced Honeyeater*		Sandstone Shrike-Thrush*
	Crimson Finch*	11/20/1961	Spotted Nightjar*
	Double-barred Finch*		QUEENSLAND
	Pictorella Munia*	11/25/1961	Pale-headed Rosella*
	Masked Finch*		Spotted Bowerbird*
11/8/1961	Purple-crowned Fairywren*	11/26/1961	Laughing Kookaburra*
	Bar-breasted Honeyeater*		Satin Flycatcher*
	Silver-crowned Friarbird*		Apostlebird*

	Noisy Friarbird*		(Barking Owl*)
	Channel-billed Cuckoo*	12/1/1961	Orange-footed Scrubfowl*
11/27/1961	Rufous Shrike-Thrush*		Australian Brush-turkey*
	Yellow Honeyeater*		Superb Fruit-Dove*
	Comb-crested Jacana*		Graceful Honeyeater*
	Tawny Grassbird		Topknot Pigeon*
	Macleay's Honeyeater*		White-headed Pigeon*
	Brown-backed Honeyeater*		Australian King-Parrot*
	Metallic Starling*		Crimson Rosella*
	Red-browed Firetail*		Grey-headed Robin*
	Plum-headed Finch*		Bower's Shrike-Thrush*
11/28/1961	Scaly-breasted Lorikeet*		Lewin's Honeyeater*
	Little Kingfisher*		Bridled Honeyeater*
	Australian Swiftlet*		Scarlet Myzomela*
	Grey-headed Whistler*		Grey Goshawk*
	Spectacled Monarch*	12/3/1961	Brown Quail*
	Large-billed Scrubwen*		Ground Cuckoo-shrike*
	Lovely Fairywren*		Black-throated Finch*
	Brown Gerygone*	12/4/1961	(White-throated Nightjar*)
	Yellow-eyed Cuckoo-shrike*		Australian Figbird*
	Victoria's Riflebird*		Yellow-faced Honeyeater*
	Yellow-spotted Honeyeater*		White-browed Woodswallow*
	Blue-faced Finch*	12/5/1961	White-bellied Cuckoo-shrike*
11/29/1961	Red-chested Buttonquail*		Noisy Miner*
	Wompoo Fruit-Dove*	12/6/1961	Red Goshawk*
	Double-eyed Fig-Parrot*	12/9/1961	Superb Fairywren*
	Buff-breasted Paradise-Kingfisher*	12/10/1961	Wonga Pigeon*
	Yellow-breasted Boatbill*		Noisy Pitta*
	Pied Monarch*		Southern Logrunner*
	Black-faced Monarch*		Olive-tailed Thrush*
	Pale-yellow Robin*		Pied Currawong*
	Eastern Whipbird*		Satin Bowerbird*
	Leaden Flycatcher*		Regent Bowerbird*
	Fairy Gerygone*		Paradise Riflebird*
	Green Catbird*		Eastern Spinebill*
11/30/1961	Little Lorikeet*	12/11/1961	Albert's Lyrebird*
	Yellow Robin*		Rufous Scrub-bird*
	Northern Logrunner*		Striated Thornbill*
	Yellow-throated Scrubwren*		Red-browed Treecreeper*
	Fernwren*		NEW SOUTH WALES
	Mountain Thornbill*	12/12/1961	Red-rumped Parrot*
	Tooth-billed Catbird*		Eastern Rosella*
	White-throated Treecreeper*		Crested Shrike-tit*

	Speckled Warbler*	2/3/1962	Sooty Tern
	White-winged Chough*		Brown Noddy
	Bell Miner*		White Tern
	Brown Treecreeper*		**N. PACIFIC OCEAN**
	Diamond Firetail*	2/11/1962	Laysan Albatross
12/13/1961	Turquoise Parrot*		Black-footed Albatross
	Yellow-tufted Honeyeater*		Northern Fulmar
12/14/1961	Yellow Thornbill*		Western Gull
	White-eared Honeyeater*		Tufted Puffin*
	WESTERN AUSTRALIA		**NORTH AMERICA**
12/20/1961	Grey Currawong*		BRITISH COLUMBIA
12/30/1961	Australian Gannet*	2/12/1962	Canada Goose
12/31/1961	Pacific Gull*		Snow Goose
	Hooded Plover*		Western Grebe
	Red-eared Firetail*		Pelagic Cormorant
1/6/1962	Yellow-plumed Honeyeater*		Black Scoter
1/7/1962	Malleefowl*		Glaucous-winged Gull
	Southern Scrub-Robin*		Black-capped Chickadee
	Chestnut-rumped Heathwren*		Black-billed Magpie
	Purple-gaped Honeyeater*		
	Brush Bronzewing*		
1/16/1962	Little Penguin*		
1/23/1962	White-faced Storm-Petrel*		
	Short-tailed Shearwater		
	VICTORIA		
1/25/1962	Fluttering Shearwater*		Total species recorded: 2305 (inclu. 2,078 "life
	Common Diving-Petrel*		birds", denoted with an asterisk)
	Fairy Prion*		Names follow Clements, 2007
1/26/1962	Superb Lyrebird*		Parentheses denote birds heard but not seen
	Pilotbird*		
1/27/1962	Black-faced Cormorant*		
	S. PACIFIC OCEAN		
1/30/1962	Gould's Petrel*		
	Red-tailed Tropicbird		
1/31/1962	Buller's Shearwater		
	NEW ZEALAND		
2/1/1962	White-fronted Tern*		
	Double-banded Plover*		
	Grey Gerygone*		
	Tui*		
	S. PACIFIC OCEAN		
2/2/1962	Black-winged Petrel*		
	White-necked Petrel*		

Appendix C: Three Year Summary of Countries Visited and No. of Trip List Birds

Country	Entry Date	Days	No. of species added to Trip List	
			Total	First Time
C. AMERICA				
Mexico	01-27-59	14	178	34
Guatemala	02-10-59	20	149	98
El Salvador	02-20-59	1	4	2
Honduras	02-21-59	1	0	0
Nicaragua	02-21-59	3	9	6
Costa Rica	02-24-59	10	83	80
Panama	03-05-59	13	48	40
S. AMERICA				
Venezuela	03-21-59	6	60	51
Colombia	03-26-59	12	58	50
Ecuador	04-07-59	10	62	58
Peru	04-17-59	31	147	138
Bolivia	05-18-59	8	5	4
Chile	05-26-59*	40	72	68
Argentina	06-30-59*	41	106	105
Uruguay	08-01-59	1	0	0
Paraguay	08-09-59	6	50	50
Brazil	08-16-59	17	70	70
S. Atlantic Oc	09-02-59	10	10	8
AFRICA				
S. Africa	09-11-59*	44	398	381
Swaziland	10-08-59	1	9	9
Mozambique	10-09-59*	8	20	20
Bechuanaland	10-27-59	4	21	20
S. Rhodesia	10-31-59*	21	76	72
N. Rhodesia	11-07-59	2	7	6
Nyasaland	11-27-59	7	29	28
Tanganyika	12-04-59*	64	149	143
Zanzibar	12-09-59	1	0	0
Kenya	01-07-60*	23	128	128
Uganda	03-07-60	13	87	87

Country	Entry Date	Days	Total	First Time
Belg. Congo	03-20-60	18	30	30
R-Urundi	03-21-60	1	1	1
Fr. Eq. Africa	04-09-60	4	30	30
Chad	04-13-60	3	21	21
Cameroons	04-15-60	1	2	2
Nigeria	04-16-60	6	10	10
Niger	04-23-60	5	23	21
Algeria	04-28-60	23	44	26
Morocco	05-20-60	8	7	6
EUROPE				
Spain	05-28-60*	54	48	21
Portugal	07-16-60	2	0	0
Andorra	07-23-60	2	15	5
France	07-24-60*	41	19	10
Monaco	07-28-60	1	0	0
Switzerland	08-15-60*	10	0	0
Luxembourg	09-07-60	1	0	0
W. Germany	09-07-60*	6	0	0
Czechoslova-	09-10-60	5	8	3
Poland	09-15-60	2	0	0
Russia	09-17-60	12	8	3
Finland	09-29-60	8	16	8
Norway	10-06-60	13	18	6
Sweden	10-19-60	15	1	0
Denmark	11-4-60	12	0	0
Holland	11-15-60*	8	13	3
Belgium	11-18-60*	6	0	0
England	11-20-60	7	0	0
Lichtenstein	12-10-60	1	0	0
Italy	12-12-60*	18	2	0
San Marino	12-17-60	1	0	0
Austria	12-31-60	10	0	0
Yugoslavia	01-16-61	2	0	0
Bulgaria	01-18-61	1	1	1
Greece	01-19-61	12	13	10
ASIA				
Turkey	02-01-61	11	3	3
Syria	02-12-61*	5	2	2
Lebanon	02-14-61	3	0	0

Country	Entry Dare	Days	Total	First Time
Jordan	02-19-61	21	15	15
Iraq	03-10-61	6	8	6
Iran	03-16-61	16	9	6
Afghanistan	03-31-61	11	22	20
Pakistan	04-11-61	6	62	55
India	04-17-61*	70	250	224
Nepal	05-18-61	8	74	69
Ceylon	07-03-61	15	24	24
N. Indian Oc.	07-18-61	4	1	0
Malaya	07-22-61*	25	144	137
Thailand	08-14-61*	13	18	17
Laos	08-26-61	3	0	0
Cambodia	08-30-61	10	17	17
Vietnam	09-09-61	3	0	0
AUSTRALIA				
W. Australia0	09-14-61*	95	230	222
N. Territory	11-09-61	12	32	31
Queensland	11-21-61	21	87	86
New S. Wales	12-12-61*	7	12	12
Victoria	01-25-62	4	6	6
S. Pacific Oc.	01-30-62*	6	8	3
New Zealand	02-01-62	1	4	4
N. Pacific Oc.	02-05-62	7	5	1
N. AMERICA				
Br. Columbia	02-12-62	1	8	0

Common names of all species can be found in Appendices A and B (names follow Clements, 2007)
An asterisk after the first date of entry indicates there were more than one entry into the country
Names of countries are those that were in use at the time of my travel, and the boundaries at that time.
Total species encountered and named so far: 3,406: (including 2,932 "life birds")

About the Author

Dean was born on August 28, 1933, in Topeka, Kansas, to school teacher parents. He had a twin sister, and later a 15 months younger brother. His family struggled financially through the great depression, and moved to several different towns and locations in eastern Kansas.

In January, 1947, he was 13 years old and living at 1309 Kentucky Street in Lawrence, Kansas. As he was walking one morning on his way up to a small high school which he attended on the University of Kansas campus, he happened to notice a little bird foraging at eye level on the trunk of a large elm tree in his back yard. He paused briefly and studied it carefully, thinking that he had not previously seen such a bird. That afternoon after school he went to the science classroom where there was a wooden bar on which were hanging 15-20 large canvas pages depicting in color almost all of the North American birds. Starting at the beginning he examined each page, one by one, until he finally came to a page where he found the bird he had seen on his elm trunk that morning. It was a Red-breasted Nuthatch.

Dean was elated, and a thought suddenly came to his mind. Since he had never seen such a bird before, he asked himself the question "How many different kinds of birds could he see if he started looking for them?" So he began looking, and what followed was a lifetime passion of searching for and assigning a name to as many species as he could find. His pursuit was a personal challenge, and provided him with a sense of accomplishment and adventure. He was competing only with himself, and he had no magic number of how many he wanted to see, or even how many there were to see. Somewhat surprisingly, he was unaware as he first started out with his list that anybody else did this sort of thing. He soon discovered otherwise when he became a member of the Topeka Audubon Society. His father bought him a used pair of German army surplus 6X30 binoculars for $5.00. Dean had no field guide to assist him, only "The Birds of America" with copies of Audubon's original paintings, published by the Macmillan Company in 1946, a very large book his grandmother gave to him.

In the spring of 1948 when Dean was 14 years old and a sophomore at University High School he took up a competitive sport by running the mile race on the school track team. At the end of the season he had done well enough to qualify for competition in the Kansas state championships (with other schools of the same size throughout the state). Dean was not only the youngest runner competing in the mile but also the only one wearing tennis shoes because his father could not afford to buy him track shoes. He finished overall in tenth place in the final standings (which won his team one point toward the team championship, the only point his team one).

Dean's family moved to Decatur, Illinois, that summer (1948), and he graduated from high school there in the spring of 1950, at age 16.

He entered the University of Michigan as a sophomore in the fall of 1951, and enrolled in the School of Forestry with a combined major in both forestry and wildlife management. It was his first year in a four year program as a "midshipman" in the regular Navy ROTC. Dean completed a B.S. degree in the spring of 1954, and received the "Oreon E. Scott Award for Distinguished Scholarship in Science" (a brand new, 3,194 page "Merriam Webster Dictionary of the English Language") as the highest academically ranked graduating student in the new School of Natural Resources (which replaced the earlier School of Forestry).

For his extra year at Michigan (1954-55) which was needed to complete NROTC requirements, Dean enrolled in several graduate level classes and worked as an assistant curator in both the Bird and Mammal Divisions of the University of Michigan Museum of Zoology (UMMZ), where he learned to prepare and catalog specimens for scientific study. He took an ornithology class from the well-known ornithologist, Josselyn Van Tyne.

Dean was commissioned an Ensign in the regular U.S. Navy in Aug. 1955, and began a required three year period of active duty. Fortunately, this was the brief period of time between the Korean War and the Vietnam War. He served on board two different attack aircraft carriers in the Pacific Fleet, the USS Oriskany (CVA 34) and the USS Kearsarge (CVA 33), on both of which he worked in the ship's Combat Information Center (CIC). His primary responsibilities were as the CIC watch officer and as one of seven air controllers on board. In 1958, Dean was one of two air controllers (out of 49 on active duty in the Pacific Fleet) to be awarded a certificate as an "Outstanding Air Controller."

At the end of his three years of active duty Dean formed an unlikely partnership with Noble Trenham, one of the night fighter pilots on board the Kearsarge who was also finishing his Navy obligations at this time, and who told him that he had once saved his life. At Noble's invitation the pair then formed a comradeship to travel around the world together. Noble's original intention was to fly, but Dean convinced him to travel instead by land. They departed in a 4WD Jeep camping van from Los Angeles on Jan. 25, 1959. This book is the story of their 3-year adventure together. Noble was in charge of mechanical repairs and all financial transactions, and Dean did most of the driving.

Near the end of their travels Dean met Laurie Weaver in Perth, Western Australia, on Sept. 16, 1961. She was a 24 year old office secretary who had grown up on a sheep ranch in the dry wheat belt country near the little town of Beverley, 80 miles east (inland) from Perth. Three weeks after meeting each other they got engaged, parted company for two months, and then got married in Perth on December 27, 1961! Noble was best man. They all three traveled together via ship to Los Angeles, arriving there on Feb. 16, 1962.

After a year in Long Beach, California, Laurie and Dean returned to her parent's ranch for 16 months, in 1963-64. In Oct. 1963, Laurie gave birth to their first daughter, Kerryn, in the very same hospital where she herself was born 26 years earlier (and where her maternal grandfather for many years was the town's only doctor and surgeon)!

Dean returned to the University of Michigan with his family in the fall of 1964 to begin graduate study for a Ph.D in zoology, under the direction of Harrison B. ("Bud") Tordoff. Two years later, by extremely good fortune he was able to return to Perth with his family for two years of zoological research, working through the University of Western Australia, from 1966-68. He assisted in physiological research for a Michigan professor, William R. Dawson, and at the same time gathered data for his doctoral thesis, "Geographic Variation and Evolution in the Australian Ringneck Parrots (*Barnardius*)." In Aug. 1966, a second daughter, Donna, was born, in Perth. (There were now three Australian born females in his family.)

Dean completed his Ph.D. in the summer of 1970, at which time Dean and Laurie added a son, Scott, to their family. In the fall of that year Dean began a 25 year teaching career in the Biology Department at Stephen F. Austin State University, situated among the pine trees of eastern Texas in the small town of Nacogdoches (which advertised itself as the "oldest town in Texas").

Dean taught a dozen different undergraduate and Master level courses and topics, creating the first undergraduate course in "Evolution" and such graduate level courses as "Tropical Ecology" and "Birds of the World." The stories of his adventure around the world were always popular diversions to his classroom lectures. During the summertime Dean often took students on 4-5 week camping trips for course credit, on many occasions to Mexico and one time each to Costa Rica and Ecuador. Dean served as President of the Texas Ornithological Society from 1977-79. His last graduate student, Cliff Shackelford, took every course offered by Dean (totalling 36 credit hours)! Cliff completed his Masters degree in 1994 and went to work for the Texas Parks & Wildlife Dept. as an ornithologist and non-game biologist, commencing a very successful career in which he published a book on Texas hummingbirds and frequently gave lectures and radio talk programs to the general public.

From 1984-2011 many of Dean's worldwide pursuits for birds were with Craig Rudolph, a Ph.D. research biologist working for the U.S. Forest Service at their research laboratory in Nacogdoches. Craig and Dean were very compatible companions. Their 3-5 week trips usually involved renting a car at the airport upon their arrival and then driving to birding destinations, stopping whenever they liked. Occasionally they sought assistance from a local person to do the driving or in finding a particular bird. Dean also regularly took independent birding trips.

Beginning in 1995 when Dean retired from teaching he set two birdwatching challenges for himself: (1) to see at least one species in every bird family and (2) to see all species in the worldwide family Corvidae. For taxonomic names and the classification of species and families he chose to follow the I.O.C world bird list published online (but science was not static and all bird lists were constantly changing, and were out of date almost as soon as they were published.) Dean stopped trying to keep up with taxonomic changes after July, 2011.

Dean fulfilled his first goal in May, 2011, when with the aid of a local guide in Uganda he saw (and quickly photographed as it ran across the road) a Red-chested Flufftail, in the family Sarothruridae. It was his last family.

He finally completed his second goal in May, 2015, when he observed several Iranian (Pleske's) Ground-Jays in a desert 150 miles east of Isfahan (Iran), having been driven to this site by a young environmental science lecturer, Hadi Radnezhad, at the Islamic Azad University in Isfahan. It was his last corvid (following the July, 2011, I.OC. list). Dean had been issued a two-week tourist visa to visit Iran as a guest lecturer at the university, in spite of the strained relations between Iran and the U.S.A. at this time. Dean could now hang up his binoculars, with a world list of almost exactly 6,500 species, a very modest number by modern techniques (to which he never adapted). He now continued the arduous task of writing this book, which he had commenced several years earlier.

Dean at his office desk in the late 1970's

A LETTER
as Dean's teaching career neared an end

4-15-94

Hello.

    I was watching some peregrine falcons near the top of Seattle's tallest building this morning and I thought about you, and later I decided to write. Well first, I moved to Seattle soon after my graduation in December. If you have any problems remembering me, I took your bird and evolution class last summer. Anyhow, Texas' climate was too hot for me and I thought this area of the world would be a nice place to reside, and so far it has been. I have seen so many species of birds around here, mostly waterfowl. I still really want to see an albatross, or better yet many of them. I will get around to it one day.

    Of my 17.5 years of school and of all the classes I have taken in that time I found your ornithology class to be in the top three, maybe the best, but I wouldn't want to think back and analyze every class I have taken. One anthropology class, "Magic, Ritual, and Religion", that I took at Southwest Texas State University would go blow for blow with the bird class but not too many others come to mind when I think of sheer quality. As a teacher, you were superb. Very objective and you communicated near perfectly, qualities scarce among teachers. If I ever get into a teaching profession I will certainly think of your approach and utilize aspects of it. Your stories were classic, especially the one where that pilot crash landed his plane and some natives paddled out to him and you had to relay the reports to the Navy honchos. Of all my teachers and mentors, you certainly rate in the top few percent.

Always a student,
Matthew Orange